HANDBOOK OF TRANSPORTATION ENGINEERING

ABOUT THE EDITOR

MYER KUTZ is president of Myer Kutz Associates, Inc., a publishing and information services consulting firm. He is the editor of numerous books, including *Biomedical Engineering and Design Handbook*, Second Edition.

HANDBOOK OF TRANSPORTATION ENGINEERING

Volume I: Systems and Operations

Myer Kutz   Editor

Second Edition

New York Chicago San Francisco Lisbon London Madrid
Mexico City Milan New Delhi San Juan Seoul
Singapore Sydney Toronto

The McGraw-Hill Companies

Library of Congress Cataloging-in-Publication Data

Handbook of transportation engineering / [edited by] Myer Kutz—2nd ed.
   p.   cm.
   Includes bibliographical references and index.
   ISBN 978-0-07-161492-4 (v. 1 : hardback)—ISBN 978-0-07-161477-1 (v. 2 : hardback)
   1. Transportation engineering—Handbooks, manuals, etc.   2. Traffic engineering—
Handbooks, manuals, etc.
   I. Kutz, Myer.

TA1151.H34   2011
629.04—dc22   2010050750

McGraw-Hill books are available at special quantity discounts to use as premiums and sales promotions, or for use in corporate training programs. To contact a representative please e-mail us at bulksales@mcgraw-hill.com.

Handbook of Transportation Engineering, Volume I: Systems and Operations

1 2 3 4 5 6 7 8 9 0   DOC/DOC   1 9 8 7 6 5 4 3 2 1

ISBN 978- 0-07-161492-4
MHID 0-07-161492-3

The pages within this book were printed on acid-free paper.

Sponsoring Editor	**Proofreader**
Larry S. Hager	Surendra Nath Shivam, Glyph International
Acquisitions Coordinator	**Production Supervisor**
Michael Mulcahy	Richard C. Ruzycka
Editorial Supervisor	**Composition**
David E. Fogarty	Glyph International
Project Manager	**Art Director, Cover**
Vastavikta Sharma, Glyph International	Jeff Weeks
Copy Editor	
Erica Orloff	

For Jayden, now well on his way on his life's journey

CONTENTS

Chapter 16. Airline Management and Operations   *Saad Laraqui*   **16.1**

Chapter 17. The Marine Transportation System   *James J. Corbett*   **17.1**

Chapter 18. Global Logistics and the Maritime Transport System
John Mangan   **18.1**

Chapter 19. Freight Transportation Planning   *Kathleen Hancock*   **19.1**

Chapter 20. Management of Transportation Organizations   *George L. Whaley*   **20.1**

CONTRIBUTORS

Panagiotis Ch. Anastasopoulos   *Purdue University, School of Civil Engineering, West Lafayette, Indiana* (CHAP. 4)

Michael S. Bronzini   *Department of Civil, Environmental and Infrastructure Engineering, George Mason University, Fairfax, Virginia* (CHAP. 1)

Mark L. Burton   *Department of Economics, Marshall University, Huntington, West Virginia* (CHAP. 15)

Arun Chatterjee   *Department of Civil and Environmental Engineering, The University of Tennessee, Knoxville, Tennessee* (CHAP. 7)

John Collura   *University of Massachusetts at Amherst, Amherst, Massachusetts* (CHAP. 12)

James J. Corbett   *Department of Marine Studies, University of Delaware, Newark, Delaware* (CHAP. 17)

Bart Egeter   *Bart Egeter Advies, Rotterdam, The Netherlands* (CHAPS. 2 AND 3)

Konstadinos G. Goulias   *University of California, Santa Barbara, California* (CHAP. 10)

Kathleen Hancock   *Civil and Environmental Engineering Department, University of Massachusetts at Amherst, Amherst, Massachusetts* (CHAP. 19)

Ben Immers   *TRAIL Research School for Transport, Infrastructure and Logistics, The Netherlands; Department of Mechanical Engineering, Centre for Industrial Management, Traffic & Logistics, Katholieke Universiteit Leuven, Belgium* (CHAPS. 2 AND 3)

Kara Kockelman   *Department of Civil, Architectural and Environmental Engineering, The University of Texas at Austin, Austin, Texas* (CHAP. 8)

Saad Laraqui   *Business Administration Department, Embry-Riddle Aeronautical University, Daytona Beach, Florida* (CHAP. 16)

D. John Mangan   *School of Marine Science and Technology, Newcastle University, United Kingdom* (CHAP. 18)

Anthony M. Pagano   *Department of Managerial Studies, University of Illinois at Chicago, Chicago, Illinois* (CHAP. 11)

Zhong-Ren Peng   *Department of Urban and Regional Planning, University of Florida, Gainesville, Florida; School of Transportation Engineering, Tongji University, Shanghai, China* (CHAP. 9)

Sarah Perch   *Department of Urban and Regional Planning, University of Florida, Gainesville, Florida* (CHAP. 9)

Elena Shenk Prassas   *Polytechnic Institute of NYU, Brooklyn, New York* (CHAP. 5)

Amelia Regan   *Computer Science and Civil and Environmental Engineering, University of California, Irvine, California* (CHAP. 14)

Jon Ross   *MPSA Partners, Chicago, Illinois* (CHAP. 13)

Maaike Snelder   *Netherlands Organization for Applied Scientific Research, TNO Mobility and Logistics, The Netherlands* (CHAP. 3)

Gary S. Spring   *Department of Civil Engineering, Merrimack College, North Andover, Massachusetts* (CHAP. 6)

Chris Tampere   *Department of Mechanical Engineering, Centre for Industrial Management, Traffic & Logistics, Katholieke Universiteit Leuven, Belgium* (CHAP. 3)

Andrew P. Tarko   *Purdue University, School of Civil Engineering, West Lafayette, Indiana* (CHAP. 4)

Rob van Nes *Delft University of Technology, Faculty of Civil Engineering and Geosciences, Transport & Planning, The Netherlands* (CHAP. 2)

Mohan M. Venigalla *Civil, Environmental, and Infrastructure Engineering Department, George Mason University, Fairfax, Virginia* (CHAP. 7)

Carrie Ward *Capital District Transportation Authority, Albany, New York* (CHAP. 12)

George L. Whaley *San Jose State University, San Jose, California* (CHAP. 20)

P. Buxton Williams *MPSA Partners, Oak Park, Illinois* (CHAP. 13)

Wesley W. Wilson *Department of Economics, University of Oregon, Eugene, Oregon, Upper Great Plains Transportation Institute* (CHAP. 15)

Fei Yang *Department of Urban and Regional Planning, University of Florida, Gainesville, Florida* (CHAP. 9)

Bin (Brenda) Zhou *Department of Engineering, Central Connecticut State University, New Britain, Connecticut* (CHAP. 8)

PREFACE

Volume I of the Second Edition of the *Handbook of Transportation Engineering* focuses on systems and operations. It is divided into two parts:

Part I: Networks and Systems, which contains 9 chapters

Part II: Operations and Economics, which consists of 11 chapters

Of the 20 chapters in Volume I, 4 are entirely new to the handbook, 8 are updated from the first edition, and 8 are unchanged. The purpose of these additions and updates is to expand the scope of the parts of the volume and provide greater depth in individual chapters.

The four new chapters in Volume I are

Chapter 3: Reliability of Travel Times and Robustness of Transport Networks

Chapter 8: Transportation and Land Use

Chapter 9: Sustainable Transportation

Chapter 18: Global Logistics and the Maritime Transportation System

Among the eight: chapters that contributors have updated are

Chapter 2: Transport Network Planning: Methodology and Theoretical Notions

Chapter 4: Transportation Systems Modeling and Evaluation

Chapter 5: Software Systems and Simulation for Transportation Applications

Chapter 6: Applications of GIS in Transportation

Chapter 10: Transportation Planning and Modeling

Chapter 12: Innovative Information Technology Applications in Public Transportation

Chapter 15: The Economics of Railroad Operations: Resurgence of a Declining Industry

Chapter 20: Management of Transportation Organizations

Nearly all chapters in this volume have been contributed by academics, the sole exception Chapter 13: Parking Management. Nearly all contributors are located in universities in the United States, except for a group from the Netherlands and a single contributor from the United Kingdom. I would like to express my heartfelt thanks to all the contributors for having taken the opportunity to work on this book. Their lives are terribly busy, and it is wonderful that they found the time to write thoughtful and complex chapters. I developed this handbook because I believed it could have a meaningful impact on the way many engineers and other transportation professionals approach their daily work, and I am gratified that the contributors thought enough of the idea that they were willing to participate in the project. I should add that most of the contributors to the first edition were willing to update their chapters, and it's interesting that even though I've not met most of them face to face, we have a warm relationship and are on a first-name basis. They responded quickly to queries during copyediting and proofreading. It was a pleasure to work with them—we've worked together on and off for nearly a decade. The quality of their work is apparent. Thanks also to my editors at McGraw-Hill for their faith in the project from the outset and to the personnel at Glyph International who guided the manuscript through production. And a special note of thanks to my wife, Arlene, whose constant support keeps me going.

MYER KUTZ
Delmar, New York

VISION STATEMENT

The first edition of the *Handbook of Transportation Engineering* was published in the fall of 2003. It was a substantial reference work, with 38 chapters addressing major areas of interest to transportation engineers and other professionals working in any phase of this civil engineering subdiscipline, including civil engineers, city and regional planners, public administrators, economists, social scientists, and urban geographers. The handbook was divided into five parts: transportation networks and systems, with five chapters covering topics where theory and practical application often intersect; traffic, streets, and highways, with 10 chapters dealing with issues involved in moving automobiles, buses, and trucks efficiently; safety, noise, and air quality, with five chapters; nonautomobile modes of transportation, with eight chapters covering movement by foot, bicycle, rail, and air; and a section on transportation operations and economic issues, with 10 chapters covering issues of particular interest to planners and administrators. Despite the breadth of issues addressed in detail in the handbook, coverage was not as broad as I would have liked mainly because not all the assigned chapters could be delivered in time to meet the publication schedule, as is often the case with large contributed works (unless the editor keeps waiting for remaining chapters to stagger in while chapters already received threaten to become out of date). So, even as the first edition was being published, I looked forward to a second edition, when I could secure more chapters to fill in gaps in the coverage and allow contributors to add greater depth to chapters that had already been published.

The overall plan for the Second Edition of the *Handbook of Transportation Engineering* was to update or retain as-are most of the chapters that were in the first edition and add half as many new chapters, including chapters with topics that were assigned for the first edition but were not delivered, plus chapters with entirely new topics. Specifically, I was looking for new chapters on the relation of transportation to land use, sustainability, and climate change; hazardous materials transportation and emergency response; and engineering design aspects of pavement, bridges, tunnels, ships, pipelines, and people movers. Because of the size of the Second Edition, I recommended splitting it into two volumes, with 20 chapters in Volume I and 35 chapters in Volume II. The split is uneven but natural: The first volume covers systems and operations, and the second volume covers applications and technologies.

The two volumes have been arranged as follows:

Volume I: Transportation Systems and Operations

　Part I: Networks and Systems

　Part II: Operations and Economics

Volume II: Transportation Engineering Applications and Technologies

　Part III: Automobile Transportation: Traffic, Streets, and Highways

　Part IV: Non-Automobile Transportation

　Part V: Safety, Noise, and Air Quality

Overall, two-thirds of the 55 chapters in the Second Edition are new or updated—17 chapters cover topics not included in the First Edition and are entirely new, and 20 chapters have been updated. The Preface to each volume provides details about the parts of the handbook and individual chapters.

The intended audience for this handbook, as noted earlier, includes transportation engineers, civil engineers, city and regional planners, public administrators, economists, social scientists, and urban geographers, as well as upper-level students.

To meet the needs of this broad, audience, I have designed a practical reference for anyone working directly, in close proximity to, or tangentially to the discipline of transportation engineering and who is seeking to answer a question, solve a problem, reduce a cost, or improve the operation of a system or facility. The two volumes of this handbook are not research monographs. My purpose is much more practice-oriented: It is to show readers which options may be available in particular situations and which options they might choose to solve problems at hand. I want this handbook to serve as a source of practical advice to readers. I would like this handbook to be the first information resource a practitioner or researcher reaches for when faced with a new problem or opportunity—a place to turn to before consulting other print sources or even, as so many professionals and students do reflexively these days, going online to Google or Wikipedia. So the handbook volumes have to be more than references or collections of background readings. In each chapter, readers should feel that they are in the hands of an experienced and knowledgeable teacher or consultant who is providing sensible advice that can lead to beneficial action and results.

MYER KUTZ
Delmar, New York

HANDBOOK OF
TRANSPORTATION
ENGINEERING

P · A · R · T · I

NETWORKS AND SYSTEMS

CHAPTER 1
NATIONAL TRANSPORTATION NETWORKS AND INTERMODAL SYSTEMS*

Michael S. Bronzini
Department of Civil, Environmental and Infrastructure Engineering
George Mason University
Fairfax, Virginia

1.1 INTRODUCTION

Transportation systems of regional and national extent are composed of networks of interconnected facilities and services. It follows that nearly all transportation projects must be analyzed with due consideration for their position within a modal or intermodal network, and for their impacts on network performance. That is, the network context of a transportation project is usually very important. Thus, it is appropriate to begin a volume on transportation engineering with a chapter on national transportation networks.

The subject of national transportation networks may be approached from at least two different perspectives. One approach, common to most introductory transportation textbooks, describes the physical elements of the various transport modes and their classification into functional subsystems. A second approach focuses on the availability of national transportation network databases and their use for engineering planning and operations studies. The latter approach is emphasized in this chapter, with the aim of providing the reader with some guidance on obtaining and using such networks. In describing these network databases, however, some high-level descriptions of the physical networks are also provided.

The modal networks considered are highway, rail, waterway, and pipeline and their intermodal connections. Airports and airline service networks are deliberately excluded, as air transport is markedly different in character from the surface transportation modes. Likewise, urban highway networks and bus and rail public transportation networks are not covered, since the emphasis is on national and state-level applications. For reasons of space and focus, only transportation networks in the United States are included, although the general concepts presented apply to any national or regional transportation network.

The chapter begins with a general consideration of the characteristics and properties of national transportation networks and the corresponding network databases. The modal networks are then described, followed by a section on multimodal networks and intermodal connections. The concluding section discusses national and local applications of network databases for practical planning studies.

---

*Reprinted from the First Edition.

1.2   NATIONAL TRANSPORTATION NETWORK DATABASES

1.2.1   The U.S. Transportation Network

Table 1.1 indicates the broad extent of the U.S. surface transportation system. The national highway network (FHWA 2009) includes over 4 million miles of public roads, and total lane-miles are more than double that, at 8.2 million miles. The vast majority of the total highway mileage, 77.5 percent, is owned and operated by units of local government. States own 19.3 percent and the federal government owns only 3 percent. The interstate highway system, consisting of 47,011 miles, accounts for only 1.2 percent of total miles but carries 24 percent of annual vehicle-miles of travel. Another important subsystem is the National Highway System (NHS), a congressionally designated system that includes the interstate highways and 117,084 miles of additional arterial roadways. The NHS includes about 4 percent of roadway miles and 7 percent of lane miles but carries over 44 percent of total vehicle-miles of travel. Highways are by far the dominant mode of passenger travel in the United States, and trucks operating on the vast highway system carry 29 percent of domestic freight ton-miles (BTS 2010b).

The class I railroad network in the United States presently consists of 94,082 miles. This mileage has been decreasing over the past 50 years; in 1960 the class I railroads owned 207,334 miles of track (BTS 2010a). Railroad mergers, rail line abandonment, and sales to short-line operators account for the decrease. While this mileage is limited, the rail mode continues to provide vital transportation services to the U.S. economy. For example, railroads carry 39 percent of domestic freight ton-miles, which exceeds total truck ton-miles, and Amtrak provides passenger service over 21,178 miles of track (BTS 2010a).

The other modes of transportation listed in Table 1.1 are probably less familiar to the average citizen. The inland waterway system includes 25,320 miles of navigable channels. Of this total, about 12,600 miles are commercially significant shallow-draft waterways (BTS 2010a), consisting primarily of the Mississippi River and its principal tributaries (notably the Ohio River system and the Gulf Intracoastal Waterway). To this could be added thousands of miles of coastal deep-draft shipping routes serving domestic intercoastal shipping (e.g., routes such as New York to Miami) and providing access to U.S. harbors by international marine shipping. Nearly totally hidden from view is the vast network of oil and gas pipelines. In fact, at 1.5 million miles, gas pipelines are second in extent only to the highway network. The water and oil pipeline modes each carry about 12 and 20 percent respectively of domestic freight ton-miles (BTS 2010b).

1.2.2   National Transportation Network Model Purposes and Uses

Motivating the development of national transportation network databases has been the need to consider broad national and regional policies and strategies, and projects for meeting critical needs for mobility and economic development. Assessing the benefits of such projects often requires considering their role within the national transportation infrastructure. For example, consider the new

TABLE 1.1   U.S. Transportation Network

Transportation mode	Statute miles in the U.S. (2008)
Highways	4,042,778
Class I rail	94,082
Inland waterways	25,320
Crude petroleum pipeline	50,214
Petroleum products pipeline	84,914
Natural gas pipeline	1,530,012

[a]Trunk lines only.
Source: BTS 2010a.

highway bridge crossing the Potomac River on I-95, recently opened near Washington, DC. When this project was nearing a critical funding decision, the question arose as to how much of the traffic using the existing bridge and other regional crossings was interstate truck traffic versus local traffic. Local modeling based on historical truck counts simply could not provide the requisite information. Answering this question (BTS 1998) required a regional or national network model of broad enough scope to capture a diverse set of commercial truck trips (BTS 1997).

Other examples of national network modeling are numerous. An early use of national rail networks was for analyzing the impacts of railroad mergers. The initial proposal to impose a diesel fuel tax on domestic inland waterway transportation was analyzed, in part, with a waterway system network model (Bronzini, Hawn, and Sharp 1978). Subsequent to the energy crisis of the mid-1970s, USDOT used national rail, water, highway, and pipeline networks to examine potential bottlenecks in the movement of energy products (USDOT/USDOE 1980). The potential impacts of spent fuel shipments from nuclear power plants to the proposed waste repository in Nevada have been estimated with the aid of rail and highway network models (Bronzini, Middendorf, and Stammer 1987). Most recently, the Federal Highway Administration (FHWA) has developed the Freight Analysis Framework (FAF), which is a network-based tool for examining freight flows on the national transportation system. Information on the FAF may be found at http://www.ops.fhwa.dot.gov/freight/. Examples of state and local uses of network models are covered at the end of this chapter.

What these examples have in common is that the demand for using specific segments of the transportation system arises from a set of geographically dispersed travelers or shippers. Likewise, the impacts of improving or not improving critical pieces of the network are felt by that same set of diverse network users. Building network models for these types of applications used to be a daunting prospect, due to the lack of available network data. As will be seen later, much of this impediment has been overcome.

1.2.3 Characteristics of Large-Scale Transportation Networks

A network model of the transportation system has two basic analytic requirements: (1) it must be topologically faithful to the actual network; and (2) it must allow network flows along connected paths. A network model that included every mile of every mode would obviously be very unwieldy. Constructing the initial database would be very time-consuming, the quality of the data would likely be compromised, and maintaining and updating the model would be equally difficult. Hence, no such undertaking has yet been attempted, at least not for a model that fulfills both analytic requirements. Topographic databases, as used for mapmaking, do not satisfy the second requirement and hence are not entirely useful for computer-based transportation analyses.

Since the entire system cannot be directly represented in the network model, some judgment must be exercised in determining the model's level of detail. This is referred to as the granularity of the model, which is a relative property. A particular network model can only be characterized as coarser or finer than some other model of the same network, i.e., there is no accepted "granularity scale." Figure 1.1 displays two possible models of a simple highway intersection. In panel (a) the intersection is represented as four links, one for each leg of the intersection, meeting at one node. In panel (b) each direction of travel and each movement through the intersection is represented as a separate link. (In fact, many different types of detailed intersection network coding have been proposed.) The level of granularity adopted will depend upon whether the outcome of the analysis is affected by the details of the within-intersection traffic flows and upon the capabilities of the analytical software to be used in conjunction with the network database.

Related to network granularity is the granularity of the spatial units that contain the socioeconomic activity that generates transportation demands. It is customary to divide the analysis area into zones or regions and to connect these regions with the transportation network model so as to allow analysis of the flows between the zones. For example, in a statewide model the spatial units could be counties and cities. Obviously, the zones and the network must have complementary degrees of granularity.

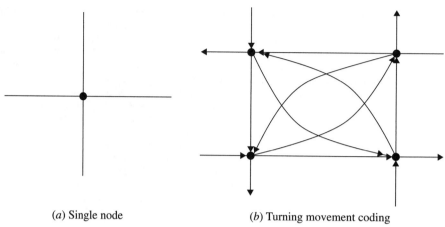

(*a*) Single node   (*b*) Turning movement coding

FIGURE 1.1   Representation of intersections in network models.

1.2.4   Typical Network Data Elements

Transportation networks inherently have a node and link structure, where the links represent linear features providing for movement, such as highways and rail lines, and the nodes represent intersections. Thus, the principal data content of a node is its name or number and location. Links usually have characteristics such as length, directionality, number of travel lanes, and functional class. Flow capacity, or some characteristics enabling ready estimation of the capacity, are also included. Of course, the whole assemblage of nodes and links will also be identified with a particular mode.

Another representational decision to be made is whether the network links will be straight lines or will have "shape points" depicting their true geography. Early network models were called "stick networks," which is topologically accurate but lacking in topographic accuracy. For many types of analyses this is of no concern; a software system that deals only with link-node incidences, paths, and network flows will yield the same answer whether or not the links have accurate shapes. For producing recognizable network maps and for certain types of proximity analysis, however, topographically accurate representations are needed (see Figure 1.2). Hence, most large-scale network models currently utilize shape points. This comes at a price, in that much more data storage is required, and plots or screen renderings are slowed. Fortunately, advances in computing power and geographic information systems (GIS) software have minimized these drawbacks to a large extent.

The idea of link capacity was mentioned above. In some networks this is stated directly for each link, in units such as vehicles per hour or tons per day. In others the functional class of a link points to an attribute table that has default capacity values. In the case of an oil pipeline, for example, the

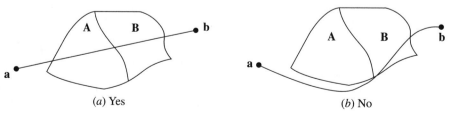

(*a*) Yes   (*b*) No

FIGURE 1.2   Does link *ab* enter region *A*?

diameter of the pipe could be used to estimate flow capacity for various fluid properties. Nodes seldom are modeled as capacity-constrained, but in principal can be (and have been) treated in the same way as links.

1.3 EXAMPLES OF NATIONAL MODAL NETWORKS

The principal source of national transportation network data in the public domain is the National Transportation Atlas Database (NTAD), developed and distributed by the USDOT Bureau of Transportation Statistics (BTS). Information on the NTAD may be obtained at http://www.bts.gov/gis/. As stated there: "NTAD presents a set of geographic databases of transportation facilities. These databases include geospatial information for transportation modal networks and intermodal terminals, and related attribute information."

Figure 1.3 is a plot of a portion of the U.S. transportation system (excluding pipelines), centered on the state of Ohio, drawn from the NTAD. As could be seen by comparing this figure with state-level highway and rail maps, the NTAD does not contain data for the entire system. In particular, facilities that largely serve local traffic are not represented. Nonetheless, the facilities included carry the great bulk of intercity traffic, hence the networks have proven valuable for conducting national and regional planning studies.

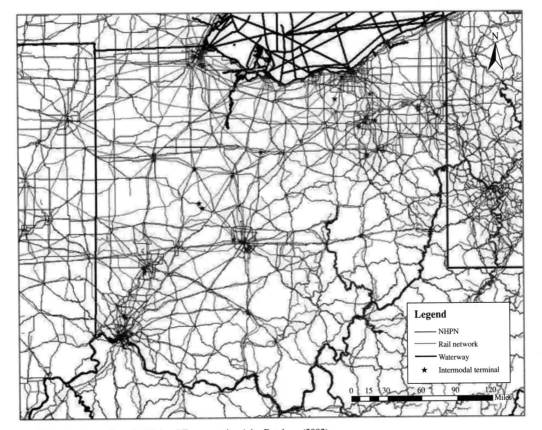

FIGURE 1.3 Extract from the National Transportation Atlas Database (2002).

1.3.1   Highway Networks

For the highway mode, the NTAD includes the National Highway Planning Network (NHPN), shown in Figure 1.4, which is a comprehensive network database of the nation's major highway system. Data for the NHPN are provided and maintained by the Federal Highway Administration (FHWA). The NHPN consists of over 400,000 miles of the nation's highways, including those classified as rural arterials, urban principal arterials, and all NHS routes. Functional classes below arterial vary on a state-by-state basis. The data set covers the 48 contiguous states plus the District of Columbia and Puerto Rico. The nominal scale of the data set is 1:100,000 with a maximal positional error of ±80 m. The NHPN is also used to keep a map-based record of the NHS and the Strategic Highway Corridor Network (STRAHNET), which is a subnetwork defined for military transportation purposes.

Highway nodes are labeled with an identification number and located by geographic coordinates, FIPS code, and other location identifiers. Links are designated by the nodes located at each end, a scheme common to all of the databases discussed in this section, and also have identifiers such as a link name or code, sign route, and street name. Other link attributes include length, direction of flow permitted, functional class, median type, surface type, access control, toll features, and any special subnetworks (such as the NHS) to which the link belongs. Each link also has a shape point file.

The NHPN originated at Oak Ridge National Laboratory (ORNL), which has gone on to develop further and maintain its own version of a national highway network database, the Oak Ridge National Highway Network. This is nearly identical in structure and content to the NHPN. For details see http://www-cta.ornl.gov/transnet/Highways.html. Like the NHPN, this database is in the public domain.

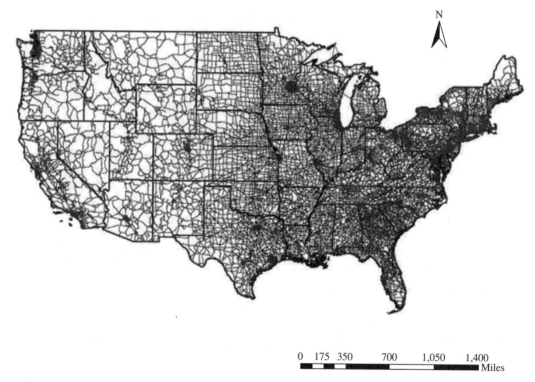

| 0 | 175 | 350 | 700 | 1,050 | 1,400 |

Miles

FIGURE 1.4   National Highway Planning Network (2002).

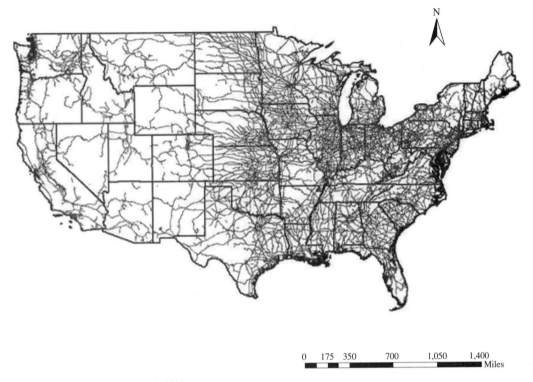

N

0   175   350   700   1,050   1,400
                                              Miles

FIGURE 1.5   National rail network (2002).

1.3.2   Rail Networks

The Federal Railroad Administration (FRA) has developed and maintains a national rail network database. The BTS compiled and formatted the rail network data for release as part of NTAD. The rail network (Figure 1.5) is a comprehensive data set of the nation's railway system at the 1:2,000,000 scale. The data set covers the 48 contiguous states plus the District of Columbia. Nodes and links are identified and located in the usual fashion. Link attributes include the names of all owning railroads and all other railroads that have trackage rights, number of main tracks, track class, type of signal system, traffic density class for the most recent year of record, type of passenger rail operations (e.g., Amtrak), and national defense status. FRA also has developed a 1:100,000 scale network, which is now part of NTAD.

As in the case of highways, ORNL also maintains and makes available its own version of the national railroad network database. This network is an extension of the Federal Railroad Administration's national rail network. In addition to the network attributes listed above, the ORNL rail network includes information on the location and ownership (including ancestry) of all rail routes that have been active since 1993, which allows the construction of routable networks for any year since then. The geographic accuracy of this network is generally 100 m on active lines.

1.3.3   Waterway Network

The National Waterway Network is a comprehensive network database of the nation's navigable waterways. The data set covers the 48 contiguous states plus the District of Columbia, Puerto Rico, ocean routes for coastwise shipping, and links between domestic and international ocean routes and inland harbors. The majority of the information was taken from geographic sources at a scale of 1:100,000, with larger scales used in harbor/bay/port areas and smaller scales used in open waters. Figure 1.3 shows segments of the National Waterway Network database in and around the state of Ohio.

Links in the waterway network represent actual shipping lanes or serve as representative paths in open water where no defined shipping lanes exist. Nodes may represent physical entities such as river confluences, ports/facilities, and intermodal terminals, or may be inserted for analytical purposes. Approximately 224 ports defined and used by the U.S. Army Corps of Engineers (USACE) are geo-coded in the node database.

The National Waterway Network was created on behalf of the Bureau of Transportation Statistics, the USACE, the U.S. Census Bureau, and the U.S. Coast Guard by Vanderbilt University and Oak Ridge National Laboratory. Additional agencies with input into network development include Volpe National Transportation Systems Center, Maritime Administration, Military Traffic Management Command, Tennessee Valley Authority, U.S. Environmental Protection Agency, and the Federal Railroad Administration. In addition to its general uses, the network is used by the USACE to route waterway movements and compute waterborne commerce ton-miles for its *Waterborne Commerce of the United States* publication series.

1.3.4 Pipeline Networks

Pipeline network data are available from PennWell MAPSearch, an information provider to the oil, gas, electric, and related industries. Information is published as paper map and CDROM products, or licensed in either GIS or CAD formats. The oil and gas database provides pipeline logistical information, including diameter, owner/operator, direction of flow, storage terminals, gas processing facilities, refineries, truck loading/unloading, compressor/pump stations, marketing hubs and other facilities related to crude oil, LPG/NGL, natural gas, refined products, petrochemicals/olefins, and other petroleum-related commodities transported by pipeline. Further information is available at http://www.mapsearch.com/home.cfm.

The USDOT Office of Pipeline Safety (OPS) has underway a joint government-industry effort called the National Pipeline Mapping System. However, at this juncture it appears that the OPS project will not provide a public domain pipeline database, at least not in the near future.

1.4 MULTIMODAL NETWORKS AND INTERMODAL CONNECTORS

There are many applications of national network models that require consideration of traffic that uses more than one mode of transportation for travel between origin and destination areas. In most cases the exact routes and transfer locations of the individual movements are unknown, and hence a multimodal network model must be used to estimate these results. A good example is the processing system used to estimate ton-miles of traffic by commodity and mode for the national commodity flow surveys (CFS) conducted by the USDOT and the U.S. Census Bureau. The procedures used are described by Bronzini et al. (1996). The CFS collected information from shippers about specific intercity freight shipments, including the commodity, origin, destination, shipment size in tons, and the mode or modes of transportation used. Shipment distance by mode was not collected, so a multimodal network model was used to find routes through the U.S. freight transportation network, thereby allowing estimation of mileage by mode for each shipment in the survey. To allow for multimodal routings, the separate modal networks were connected at appropriate locations using intermodal transfer links.

Establishing analytically correct intermodal transfer links for a multimodal network is not a simple undertaking. To a first approximation, one could use GIS software to find nodes of different modes that are within some threshold distance of each other, and simply establish mode-to-mode connectors at all such locations. This, however, ignores the investment cost and special-purpose nature of intermodal transfer facilities, and tends to overestimate the number of intermodal connectors.

To assist with these types of applications, the NTAD includes a file called the Intermodal Terminal Facilities data set. The Oak Ridge National Laboratory developed the intermodal terminal facility data from which this database was derived. This database contains geographic data for trailer-on-flatcar (TOFC) and container-on-flatcar (COFC) highway-rail and rail-water transfer facilities in the United States. Attribute data specify the intermodal connections at each facility; i.e., the modes involved in the intermodal transfer, the AAR reporting marks of the railroad serving the facility, the type of cargo, and the direction of the transfer. These latter two attributes are extremely important. Even though two modes may have an intermodal connection at a given point, it does not follow that all commodities carried by the two modes can interchange there. Typically, each such connector handles only one commodity or type of commodity. For example, a coal terminal will not usually handle grain or petroleum products. Further, the transfer facility may serve flows only in one direction. A waterside coal transfer terminal, for example, may allow dumping from rail cars to barges but may not provide facilities for lifting coal from barges into rail cars. These examples illustrate why a simple proximity analysis method is unlikely to yield correct identification of intermodal connector links.

Attribute data for the Intermodal Terminal Facilities data set were extracted from the Intermodal Association of North America (IANA) 1997 Rail Intermodal Terminal Directory, the Official Railway Guide, the TTX Company Intermodal Directory, the Internet home pages of several railroads, the U.S. Army Corps of Engineers Port Series Reports, Containerization International Yearbook, the 1996 Directory of the American Association of Port Authorities (AAPA), and various transportation news sources, both in print and on the Internet. Attribute data reflect conditions at TOFC/COFC facilities during 1995–96 and are subject to frequent change. The database does not include TOFC/COFC and marine container facilities known to have been closed before or during 1996. However, because of the frequent turnover of this type of facility, some of the terminals included in the database may now be dormant or permanently closed.

The locations of TOFC/COFC facilities were determined using available facility address information and MapExpert, a commercial nationwide digital map database and software package, and recording the longitude/latitude of the approximate center of the facility. Facility locations are not bound to any current or previous highway, railway, or waterway network models. This is an advantage in that the facility locations in the database will be unaffected by changes in the other networks. Figure 1.3 shows some of the intermodal terminals that are included in the NTAD.

Further work for the CFS has validated the use of modal and multimodal networks for national and regional commodity flow studies. A recent paper by Qureshi, Hwang, and Chin (2002) documents the advantages.

1.5 NETWORK MODEL APPLICATIONS

Section 1.2.2 briefly described use of transportation network models for national-level studies, an area of activity that dates back more than 20 years. Recent transportation studies carried out by states and Metropolitan Planning Organizations (MPOs), however, demonstrate that this type of analytical work is now within the reach of engineers and planners at those levels.

The prototypical use of network modeling at the state level is for statewide transportation planning. Horowitz and Farmer (1999) provide a good summary of the state-of-the-practice. Statewide passenger travel models tend to follow the urban transportation planning paradigm, using features such as separate trip generation and trip distribution models, and assignment of traffic to a statewide highway network. Michigan has one of the most well-developed statewide passenger models (KJS Associates, Inc. 1996). Statewide freight models also tend to follow this paradigm, with a focus on truck traffic on highways. Indiana (Black 1997) and Wisconsin (Huang and Smith 1999; Sorratini 2000) have mature statewide freight models, and Massachusetts (Krishnan and Hancock 1998) recently has done similar work.

Sivakumar and Bhat (2002) developed a model of interregional commodity flows in Texas. The model estimates the fraction of a commodity consumed at a destination that originates from each production zone for that commodity. The model includes the origin-destination distances by rail and truck, which were determined using the U.S. highway and rail networks that are included in TransCAD.

Work by List et al. (2002) to estimate truck trips for the New York City region is representative of freight network analysis activity at the MPO level. The model predicts link use by trucks based on a multiple-path traffic assignment to a regional highway network composed of 405 zones, 26,564 nodes, and 38,016 links. The model produced an excellent match between predicted and observed link truck volumes ($R^2 > 95\%$).

Switching back to the national level, Hwang et al. (2001) produced a risk assessment of moving certain classes of hazardous materials by rail and truck. They used national rail and highway network routing models to determine shipping routes and population densities along the routes for toxic-by-inhalation chemicals, liquid petroleum gas, gasoline, and explosives. Their work is fairly representative of network-based risk assessment methods. They assessed the routing results as follows: "Although the modeled routes might not represent actual routes precisely, they adequately represented the variations in accident probability, population density, and climate that characterize the commodity flow corridors for each hazardous material of interest." A similar statement could be made about most transportation network analysis results.

ACKNOWLEDGMENT

The figures in this chapter were prepared by Mr. Harshit Thaker.

REFERENCES

Black, W. R. 1997. *Transport Flows in the State of Indiana: Commodity Database Development and Traffic Assignment, Phase 2.* Transportation Research Center, Indiana University, Bloomington, IN, July.

Bronzini, M. S., S. Chin, C. Liu, D. P. Middendorf, and B. E. Peterson. 1996. *Methodology for Estimating Freight Shipment Distances for the 1993 Commodity Flow Survey.* Bureau of Transportation Statistics, U.S. Department of Transportation.

Bronzini, M. S., A. F. Hawn, and F. M. Sharp. 1978. "Impacts of Inland Waterway User Charges." *Transportation Research Record* 669:35–42.

Bronzini, M. S., D. P. Middendorf, and R. E. Stammer, Jr. 1987. "Analysis of the Transportation Elements of Alternative Logistics Concepts for Disposal of Spent Nuclear Fuel." *Journal of the Transportation Research Forum* 28(1):221–29.

Bureau of Transportation Statistics (BTS). 1997. *Truck Movements in America: Shipments From, To, Within, and Through States.* BTS/97-TS/1, Bureau of Transportation Statistics, U.S. Department of Transportation, Washington, DC, May.

———. 1998. *Truck Shipments Across the Woodrow Wilson Bridge: Value and Tonnage in 1993.* BTS/98-TS/3, Bureau of Transportation Statistics, U.S. Department of Transportation, Washington, DC, April.

———. 2010a. *National Transportation Statistics.* Bureau of Transportation Statistics, U.S. Department of Transportation, Washington, DC.

———. 2010b. *Pocket Guide to Transportation 2010.* Bureau of Transportation Statistics, U.S. Department of Transportation, Washington, DC.

Federal Highway Administration (FHWA). 2009. *Highway Statistics 2008.* Federal Highway Administration, U.S. Department of Transportation, Washington, DC.

Horowitz, A. J., and D. D. Farmer. 1999. "Statewide Travel Forecasting Practice: A Critical Review." *Transportation Research Record* 1685:13–20.

Huang, W., and R. L. Smith, Jr. 1999. "Using Commodity Flow Survey Data to Develop a Truck Travel-Demand Model for Wisconsin." *Transportation Research Record* 1685:1–6.

Hwang, S. T., D. F. Brown, J. K. O'Steen, A. J. Policastro, and W. E. Dunn. 2001. "Risk Assessment for National Transportation of Selected Hazardous Materials." *Transportation Research Record* 1763:114–24.

KJS Associates, Inc. 1996. *Statewide Travel Demand Model Update and Calibration: Phase II*. Michigan Department of Transportation, Lansing, MI, April.

Krishnan, V., and K. Hancock. 1998. "Highway Freight Flow Assignment in Massachusetts Using Geographic Information Systems." 77th Annual Meeting, Transportation Research Board, Washington, DC, January.

List, G. F., L. A. Konieczny, C. L. Durnford, and V. Papayanoulis. 2002. "Best-Practice Truck-Flow Estimation Model for the New York City Region." *Transportation Research Record* 1790:97–103.

Qureshi, M. A., H. Hwang, and S. Chin. 2002. "Comparison of Distance Estimates for Commodity Flow Survey; Great Circle Distances versus Network-Based Distances." *Transportation Research Record* 1804:212–16.

Sivakumar, A., and C. Bhat. 2002. "Fractional Split-Distribution Model for Statewide Commodity Flow Analysis." *Transportation Research Record* 1790:80–88.

Sorratini, J. A. 2000. "Estimating Statewide Truck Trips Using Commodity Flows and Input-Output Coefficients." *Journal of Transportation and Statistics* 3(1):53–67.

USDOT/USDOE (1980). *National Energy Transportation Study*. U.S. Department of Transportation and U.S. Department of Energy, Washington, DC, July.

CHAPTER 2
TRANSPORT NETWORK PLANNING: METHODOLOGY AND THEORETICAL NOTIONS

Ben Immers

TRAIL Research School for Transport
Infrastructure and Logistics, The Netherlands

Department of Mechanical Engineering
Centre for Industrial Management
Traffic & Logistics, Katholieke Universiteit Leuven, Belgium

Bart Egeter

Bart Egeter Advies, Rotterdam
The Netherlands

Rob van Nes

Delft University of Technology
Faculty of Civil Engineering and Geosciences
Transport & Planning, The Netherlands

2.1 INTRODUCTION

Mobility is undergoing constant change, in terms of both volume and spatial patterns. The traffic infrastructure has to respond to this continual process of change. Where bottlenecks emerge, improvements can be made from a whole palette of measures, varying from traffic management and pricing to the expansion of capacity in stretches of road and junctions.

This kind of bottleneck-oriented approach has offered some degree of solace for some time but occasionally the need arises to completely review and rethink the whole structure of the network: Does the existing structure come to terms with changing mobility patterns? Are structural modifications necessary, like reconsidering the categorizing of roads and the associated road design, expanding the robustness of the network, disentangling traffic flows, or changing the connective structure of urban areas? In other words, there is a need to *redesign* the network. The problem of network design that emerges then is a very complex one that requires a consideration of the (vested) interests of various parties.

In the Netherlands, a methodology has recently been developed for the integral design of the transport networks of different modalities (Egeter et al. 2002). The focus lies on networks at a regional scale. Parties that have worked with this methodology cite the following key features:

1. As a basis of the analysis, separate from the current infrastructure, an "ideal network" is designed.

2. Design occurs together with the stakeholders on the basis of clear, practicable steps.

By creating an ideal network separate from the current network, a very clear insight is gained into the structure of the network because it is not obscured by the existing situation that has emerged historically, and therefore not always ideally. By then confronting this ideal situation with the existing situation, weaknesses in the structure come to light. A second function of the ideal network is that of a long-term horizon within which short-term measures have to fit.

By reducing the theoretically highly complex design problem to a number of successive design steps or decisions, this methodology provides insight and is applicable in practical situations. What is important in this respect is that for each step there is *commitment* from the stakeholders before the next step is taken. It is, then, most effective when the methodology is used in a workshop-type situation whereby these parties themselves participate in the design process.

The result of the methodology is that stakeholders gain a clear picture of the crucial dilemmas and decisions. The methodology prevents thinking in terms of end solutions. Instead, the functions of the different parts of the network can be analyzed in terms of whether they actually fulfill the functions for which they were designed or for which they are now ascribed. The function of a particular part of the network is thereby the leading factor for form and technique. Analysis may result per situation in a whole palette of possible recommendations, varying from no action, through to traffic management, function adjustment coupled with modification of the road design, and disentangling or expanding existing connections, or through the construction of new junctions or new connections. This can be phased in, for instance, by first applying traffic management and then in the longer term building new junctions or connections.

2.2 A FUNCTIONAL CLASSIFICATION OF TRANSPORT SYSTEMS

The approach described in this chapter is based on a classification of transport systems (Egeter et al. 2002). This classification (see Table 2.1) is used to emphasize that what matters is the *quality* that is offered, not the modes and technologies used. It distinguishes five levels of scale (represented by

TABLE 2.1 Functional Classification of Transport Systems by Scale Level and Organization Type

Scale level (trip length)[*]	Individual, private transport	Collective, transport service supplied	Design speed	Accessibility, distance between access nodes
< 1 km neighborhood	1-0 e.g., walking	—	0–10 km/h	—
1–10 km district, medium-sized village, (part of) a town	1-1 e.g., bicycle, in-line skates, car on the local road network	C-1 e.g., local bus/tramway/ scheduled taxi service	10–30 km/h	0.2–1 km
10–50 km agglomeration, area, region	1-2 e.g., moped/scooter, urban car, car on regional road network (highway, expressway)	C-2 e.g., subway, light rail, commuter train service	30–80 km/h	2–5 km
50–300 km county, state	1-3 e.g., car on highway/ freeway network	C-3 e.g., long-distance train and bus services	80–200 km/h	10–30 km
> 300 km state, interstate	—	C-4 e.g., high-speed train, airplane, Greyhound bus	> 200 km/h	60–150 km

[*]1 km = 0.62 mile; 1 mile = 1.61 km

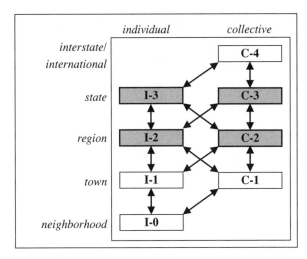

FIGURE 2.1   Scale levels the design method focuses on.

their trip length), and two different types of organization (individual or collective transport). Roughly speaking, the term *individual systems* refers to road networks, and the term *collective systems* refers to public transportation networks.

The design method focuses on the national (state) and regional level (I-3, I-2, and C-3, C-2), but is not limited to this level. Figure 2.1 shows the different subsystems, as well as the connections (the arrows in Figure 2.1) between different scale levels and between the individual and collective systems. The focus of this chapter is highlighted in gray.

2.3   KEY CHARACTERISTICS OF THE DESIGN METHOD

Designing successful transportation networks requires more than the application of the functional classification. In order to assist stakeholders in the design process, a step-by-step design process was set up. It is not a blueprint that tells stakeholders exactly what to do, but merely a framework within which they make decisions. The stakeholders get to make the designs, but the method brings structure to the design process, by indicating which decisions need to be made at what point in the process.

It is based on a number of important characteristics, which are listed in random order in Figure 2.2.

2.3.1   First Structure, then Elements

First, a perspective on the complete *structure* of the network must be developed, such as which cities must be connected by the network, which scale levels are distinguished, etc. Only then can a decision be made about the *elements* (road sections, junctions, and routes/alignment). In practice, problems are usually solved at the element level: bottleneck by bottleneck. This kind of bottleneck-oriented approach has offered some degree of solace for some time, but occasionally the need arises to completely review and rethink the whole structure of the network: Does the existing structure come to terms with changing mobility patterns? Are structural modifications necessary, like a reconsideration of the categorizing of roads and the associated road design, expanding the robustness of the network, disentangling traffic flows, or changing the connective structure of urban areas?

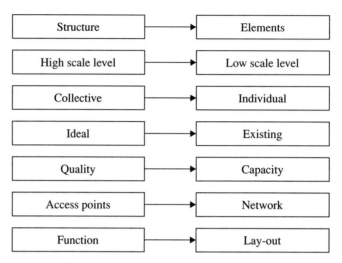

FIGURE 2.2   Main characteristics of the design methodology (in random order).

2.3.2 First the Higher Scale Level, then the Lower Scale Level

Networks for every scale level are designed independently, following a top-down approach: from the *higher* to the *lower scale level*, with a feedback loop bottom-up. Each network is designed to optimally meet its functional requirements. In order to achieve coherence between networks of different scale levels, access points of higher scale level are automatically included in the lower scale level.

2.3.3 First the Collective Networks, then the Individual Networks

Access to collective transport systems is much more cumbersome than access to individual transport; therefore the situation of the access points of the *collective* system (public transport stops) requires more careful consideration than the situation of access points of the *individual* networks (e.g., highway and freeway entry points). This is because in the case of collective transport, unlike individual transport, access and egress by lower-level transport either requires physical transfers from and to other modes, or takes place on foot. Therefore, important public transport nodes are preferably situated within a short distance of main origin and destination points. When integrating collective and individual networks (per scale level), the collective transport system gets priority in the design, for instance when it comes to the situation of intermodal transfer points.

2.3.4 First Ideal, then Existing

First, an *ideal* network is designed, ignoring the existing network. Subsequently, this ideal structure is confronted with the existing situation. The actions that need to be taken to change the existing situation into the ideal situation can then be prioritized. This way, improvements in the existing networks will be coherent; the ideal structure serves as a long-term perspective.

2.3.5 First Quality, then Capacity

The desired level-of-service, or *quality*, of the connections in the network needs to be defined clearly. Quality concerns characteristics such as speed, reliability, and comfort, but also pricing policies and traffic management strategies that are applied to the network. An acceptable volume-capacity ratio (*capacity*) is a prerequisite, but capacity should be considered separately from the desired quality. In practice, capacity is more often than not the primary aspect, which means that quality aspects get less attention.

2.3.6 First Access Points, then the Network

A transport network serves to connect access points. Therefore, it is logical to define first which *access points* should be connected, and then design the connections between these points (*the network*). In practice, it is often done the other way round. A well-known example in Europe is the discussion about which cities should get high-speed train stations on the line from Amsterdam to Paris. Whether or not the train was going to stop in The Hague became dependent on the choice for one route or the other, while the choice of "whether or not to stop in The Hague" should have been made before a route was chosen.

2.3.7 First Function, then Layout and Technique

Before the *layout* of the various components of the networks (access points, links, and junctions) is defined, it must be clear what the *function* of this component is. By gearing the layout to the functional requirements, it is more likely that this road will be used in accordance with the objectives set for this road. As a consequence, changing the function of a road (e.g., from national to regional) can lead to changing the layout (e.g., from highway to regional main road). The same principles apply to collective networks. For example, the choice between bus and rail should depend on the function; in some cases both techniques can meet the requirements.

2.4 DILEMMAS ENCOUNTERED IN DESIGNING A TRANSPORTATION NETWORK

A transportation system is made up of links, nodes, and a number of other design variables. Designing a transportation system is then a matter of assigning values to each variable. This sounds simple, but in practice, because of the different objectives set (by transport authorities, services providers, and users), there will always be conflicting variables, resulting in so-called *design dilemmas*. The design method distinguishes four major design dilemmas:

1. The number of systems: differentiation versus cost reduction
2. Access point density: quality of a connection versus accessibility
3. Access structure: accessibility versus differentiation in use
4. Network density: quality of a connection versus cost reduction

These dilemmas are implicitly processed in the functional categorization used in transport systems.

2.4.1 Dilemma 1: The Number of Systems

Several subsystems make up the total transportation system (see Table 2.1). The reason for having several subsystems is that this makes it easier to fulfill the different functions a system may have. The more subsystems there are, the better their functions can be geared toward the needs of the traveler. Offering more subsystems increases the user benefit. On the other hand, reducing the number of subsystems reduces the investor costs, which means the capacity offered can be used more efficiently. A practical example of this dilemma is the question of whether short- and long-distance travel should be combined on the same ring road: This means a high-quality road for short-distance travel, but disturbance of the long-distance traffic flow caused by the short distance between access points. In general, more subsystems can be offered in more urbanized areas, where the transport demand is higher.

2.4.2 Dilemma 2: Access Point Density

For any given subsystem, there is the question whether there should be few or many access points. The more access points, the better its accessibility. This means that a smaller part of the trip needs to be made on the lower scale level (and therefore slower) networks. On the other hand, the quality of connections (how fast, and how reliable from one access point to another) provided by the subsystem is higher when there are few access points. This dilemma plays a major role in the design of public transport networks, but it is also becoming more and more important in road networks. In many countries, long-distance traffic often encounters congestion near urbanized areas caused by regional, or even local traffic, entering and exiting the freeway and frequently causing disturbances in doing so. In general, higher scale level networks have fewer access points–this has to do with the fact that access points are usually found near cities, and fewer cities will be connected to the higher-order networks.

2.4.3 Dilemma 3: Access Structure

Apart from defining the ideal structure of the connections between towns, there is the question of where to situate the access points: one access point in the middle (as is usual for train stations), or one or more at the edges of the built-up area (as is usual for through roads). The first option maximizes the accessibility of the system, but this often leads to 'misuse' of the system by traffic that could use a lower-order network. Also, it may affect livability in the area. Finally, it undermines the intended differentiation in systems. Although this dilemma plays a role in individual as well as in collective systems, the outcome of the question is different for each type:

- In the collective systems, the access point is preferably situated in the center of the urban area. This is because changing from one collective system to another always involves a physical transfer (from one vehicle to the next). Transfers should be kept at a minimum, which means that it is desirable to concentrate access points of all collective subsystems in one location.

- In contrast, a transfer from one individual system to the next is almost seamless: passengers do not change vehicles. With livability issues in mind, access points are usually planned outside built-up areas. This also helps in fighting the undesired use of through roads (and sometimes congestion) by short distance traffic.

2.4.4 Dilemma 4: Network Density

Once it has been established which cities need to be connected, it still has to be decided whether these cities should be connected by direct links or by way of another city. More links means higher-quality connections because there are fewer detours. In public transport, however, limiting the number of links makes higher frequencies possible. Obviously, more links mean higher costs, not only in infrastructure investments, but also in the effects on the environment.

What network density will be acceptable depends chiefly on two factors:

1. The amount of traffic: High volumes justify the need for extra infrastructure.
2. The difference of quality between two subsystems: A greater difference (in design speed) between scale levels means that a greater detour is acceptable when using the higher order system.

2.5   *FEASIBILITY OF DESIGN*

In practice, there will almost always have to be a trade-off between the ideal network design and the realistic network design. The difference between both networks is mainly related to the resources that are available to lay the new infrastructure. The term feasibility of design has to be interpreted, however, in relation to the gradual development of a network and the wish to have a long-term view. On the basis of such a view of the ideal structure of the network, the various investment steps can be better substantiated and the network coherence better guaranteed. The absence of a long-term view results in an incoherent bottleneck approach that poses questions. The risk is then considerable that all kinds of short-term utilization measures will form the basis of a long-term infrastructure policy.

2.6   *THE DESIGN PROCESS*

2.6.1   Rules of Thumb

Designing means that certain choices are made with regard to each dilemma. To help the designer, the design method includes a number of "rules of thumb." Certain values to variables are proposed (different for each scale level), and the designer is free to use or discard these values. Per scale level we have defined what the "optimal" values are for:

- The number and size of the cities the network is meant to connect
- The expected travel distance over the network
- The desired distance between access points
- The desired distance between (center of) built-up area and access points
- The acceptable detour factor (the distance traveled over the network divided by the distance as the crow flies)

These variables determine, to a large extent, what the design is going to look like. Moreover, the design sessions held so far have shown that these variables are strongly interconnected within a scale level. Inconsistent combinations of values for these variables lead to inefficient networks. The optimal values (derived from the design speed for each scale level) depend on local circumstances.

2.6.2   The Design Method Step-by-Step

Applying the design method results in designs for the collective and individual networks for each scale level distinguished, and the interchange points where the networks are connected. Every network at every scale level is designed independently, thereby ensuring that each network is optimally geared toward its function. Possibly, in a later stage of the design process, some of the connections from different scale levels will be combined on one route, or even on one road or railway line. In that case, however, it is a conscious choice, a trade-off between the advantages and disadvantages of combining functions on that particular connection. Because the situation of the access points for the collective systems is much more important than for the individual systems, the collective network for a scale level is always designed before the individual network.

Step 1: Distinguish Urbanization Levels (urban/rural).   The edges of urban areas provide good locations for intermodal transfer points, so the border between urban and rural area must be indicated on the map for later use.

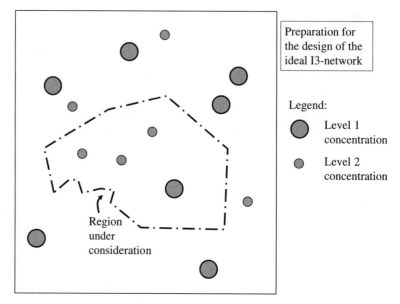

FIGURE 2.3   Definition of hierarchy of nodes.

Step 2: Define the Hierarchy of Cities and Towns.   In this step, the rule of thumb for the number and size of the nodes (cities and towns) the network is meant to connect is used to define which towns should be accessible via the network, and in what order of importance. In doing so (for the scale level under consideration), first-, second-, and third-level nodes are selected and indicated on the map. Large cities are split up into several smaller units. As an example, the results of steps 1 and 2 are illustrated in Figure 2.3.

Step 3: Design Desired Connections.   The desired connections (heart-to-heart) are drawn on the map, as depicted in Figure 2.4, according to the following rules:

- First connect first-level nodes.
- Add connections to second-level nodes.
- Include third-level nodes when they are close to an already included connection. When adjusting a connection to include a third level node it should be checked that this does not result in unacceptable detours in the network.

Step 4: Design the Ideal Network.   This is the most difficult and intuitive stage in the design method. The existing situation must be ignored. The desired connections must be translated into an efficient network with the right density. The access points must be put in the right place. Step-by-step this stage involves, for the individual network:

a. For the super-regional scale levels: drawing circles around first- and second-level concentrations, to indicate the desired distance between built-up area and through roads

b. Identifying main flow directions past first-level concentrations ("at which side of town should the road pass")

c. Defining the optimal routes past concentrations ("accessibility structures")

d. Connecting the selected concentrations

e. Checking to see whether the network density is right and detours in the network are acceptable; if not, add (or remove) connections

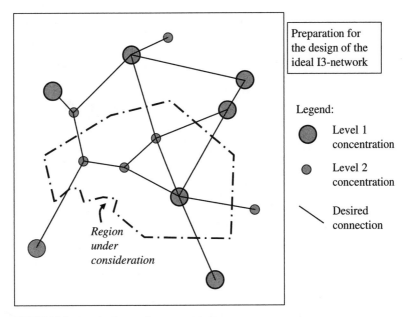

FIGURE 2.4   Drawing heart-to-heart connections.

The result of substeps *a-c* is illustrated in Figure 2.5 (based on a design of an I3 or national road network for a province in The Netherlands).

Substeps *d* and *e* result in an ideal I3-network as depicted in Figure 2.6. It must be noted that many other designs are possible; the network in Figure 2.6, however, is the one that resulted for this region. This I3 (national) network formed the basis for the regional network that was subsequently designed.

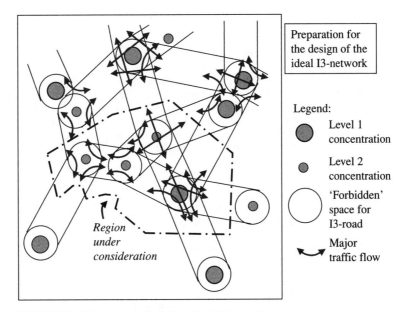

FIGURE 2.5   Substeps *a-c* in the design of the ideal network.

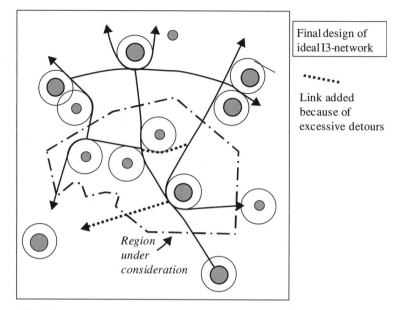

FIGURE 2.6 Final design of ideal I3-network.

The process is less complicated for the collective network, because the stops should be as much in the center of the built-up area as possible.

Step 5: Assess Current Network. The ideal network will differ from the existing network in these aspects:

- The connections that have been included
- The major traffic flows (which have implications for the layout of the interchanges of roads)

Step 5 has been included to assess how much of the existing network meets the requirements set by the method. This is done by looking at the existing connections that would most likely serve as a connection in the ideal network. The information gathered here can be used in a later stage, when it must be decided which part of the ideal network is given up in order to create a feasible network or to establish which parts of the network should be adapted first.

Design requirements to look at include:

- The distance between access points (too small?)
- The design speeds (too high? too low?)
- The requirements with respect to a logical layout of the network (do the through lanes at interchanges cater for the major flows?)

This step results in a map with connections on, over, or under the desired level-of-service and illogical points in the network.

Step 6: Design Realistic Network. We now have an ideal network and an assessment of where the existing network falls short of the ideal network. It must now be decided what is an acceptable amount of new infrastructure. Also, the individual and collective networks must be connected to each other. Likewise, the networks of the different scale levels must be connected. This means:

- Selecting routes: following the ideal or existing network
- Choosing main flow directions (so illogical points will be avoided)
- Selecting access points for collective and individual networks of all scale levels and for connecting collective and individual networks

Depending on the time horizon chosen, a "realistic" network can be selected that is closer to either the existing or the ideal network. In our case, two variants have been elaborated in this manner. Policymakers were quite pleased with the design that stays closer to the ideal network. It gave them many new ideas for their long-term plans. Interestingly, when the effects of these two designs were evaluated, it was found (with the help of an integrated land use and transport model) that the second (more "ideal") design performed better in many respects (e.g., it was more sustainable).

2.7 THEORETICAL BACKGROUND

This section discusses some theoretical issues related to the network design methodology. Topics that are discussed are the network design problem, hierarchical network structures, and some special issues.

2.7.1 Network Design Problem

A network consists of access nodes, nodes, and links connecting these nodes. In the case of transit, networks lines are included as well. The network design problem in it simplest form is to find a set of links that has an optimal performance given a specific objective. Basically, there are two kinds of network design problems:

1. Designing a new network, for instance a new higher-level network or a transit network
2. Improving an existing network, for instance increasing capacities or adding new roads

In this chapter, the focus is on designing a new network.

The network design problem is known to be very complicated for three reasons. First, there is the combinatorial nature of the problem. Given a set of access nodes the number of possible link networks connecting all access nodes increases more than exponentially with the number of access nodes. Therefore, there are no efficient methods available for solving large-scale network design problems.

Second, the perspective on the design objectives might be very different. The key conflict is that between the network user (i.e., the traveler) and the investor or network builder. The traveler prefers direct connections between all origins and destinations, whereas the investor favors a minimal network in space (see Figure 2.7).

Third, there is a strong relationship between the demand for transport networks and transport networks themselves. Changes in transport networks lead to changes in travel behavior, and changes in travel behavior set requirements for the transport network. As such, the network design problem can be seen as a Stackelberg game in which one decision maker, i.e., the network designer, has full knowledge of the decisions of the second decision maker, the traveler, and uses this information to achieve his or her own objectives (see Figure 2.8).

Methods to reconcile these opposing perspectives could be:

- Formulating an objective that combines the interests of both parties involved. Typical examples of such design objectives are maximizing social welfare and minimizing total costs.
- Focusing on the perspective of one of the parties, usually the traveler, while using the opposing perspective as a constraint, e.g., minimizing travel time given a fixed budget.

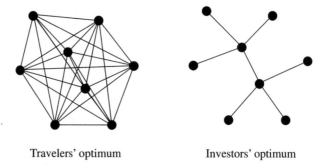

Travelers' optimum   Investors' optimum

FIGURE 2.7 Illustration of the difference in optimal network structures between the traveler's and the investor's point of view.

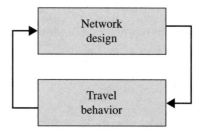

FIGURE 2. 8 Network design problem as a Stackelberg game.

- Choosing a specific objective, in this case usually the investor's perspective, but at the same time taking into account the behavior of the other party involved, i.e., the traveler. An example of this approach is a transit operator maximizing profit while considering the fact that inadequate quality will reduce patronage and thus revenues.

These three complicating factors (combinatorial nature, conflicting perspectives, and relationship between transport network and transport demand) explain the huge amount of literature on transport network design. Most of the scientific research deals with mathematical models that can be used to solve the network design model. For transport planners, however, design methodologies such as those presented in this chapter are more suitable.

2.7.2 Hierarchical Network Structures

Hierarchy as a Natural Phenomenon. It can easily be demonstrated that hierarchy is a common phenomenon in transport networks. Let us assume a perfectly square grid network in which all origins and destinations are located at the crossings, all links being equal in length and travel time. The demand pattern is uniformly distributed; that is, at every origin the same number of trips start in all directions having the same trip length, leading to the same number of arrivals at all destinations coming from all directions. Since it is a grid network, the traveler may choose between a number of routes having the same length and travel time. In this hypothetical situation, no hierarchies in demand or supply are assumed, and at first sight no hierarchy in network usage results.

However, if small deviations to these assumptions occur, a process is started that leads at least to a hierarchical use of the network. Examples of such small changes are:

• Travelers might prefer specific routes, even though all routes are equal in time and length from an objective point of view. Such a preference might be because of habit, because of the traveler's own perception of the routes or perception regarding the crossings, or be based on information provided by other travelers.

• Link characteristics might differ slightly leading to objective differences in route characteristics.

• Travelers might prefer to travel together, bringing in the stochastic element of travelers passing by and having an overlap with one of the possible routes.

• Some origins and destinations might be more attractive than others.

All of these deviations have the same effect regardless of the size of the change: Namely, some routes will become more attractive than others. This effect is mainly caused by the demand side of the transport system. The higher usage of some routes, however, also influences the supply side of the transport system. In the long run, the most intensively used routes will receive better facilities and become more attractive, while the less used routes will be neglected. The supply side of the transport system thus strengthens the hierarchy started by the demand side. In fact, the process described here is an example from economics based on increasing returns (see e.g., Waldrop 1992; Arthur Ermoliev, and Kaniovski 1987), which is a fundamental characteristic in all kind of evolutionary processes, be they in economics or in biology. The final result in this case is a hierarchical network structure consisting of two link types, or put in other words, a higher-level network is superimposed on the original lower-level network. A similar line of reasoning can be found in the work of Levinson and Yerra (2006) and Zhang and Levinson (2004).

Hierarchy in settlements stimulates hierarchical network structures. Furthermore, the introduction of faster modes speeds up the processes leading to hierarchical networks. Similarly, hierarchical transport networks lead to concentration of flows, and if these flows are large enough they allow for more efficient transport leading to lower travel costs per unit traveled (economies of scale), and reduce negative impact on the environment, which also stimulates the development of hierarchical network structures. Hierarchical networks are thus a natural phenomenon resulting from the interaction between demand and supply that, because of technological developments and modern decision processes focusing on environmental impact, are becoming more common in transport networks (see Figure 2.9).

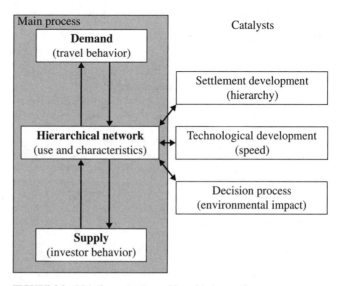

FIGURE 2.9   Main factors leading to hierarchical networks.

Development of Hierarchical Network Structures.   The main process, that is the interaction between demand and supply, might have self-organizing characteristics. Many networks, however, have been developed over a long period of time and are, therefore, influenced by many factors. Hierarchy in spatial structure has always been such a factor. The importance of technology has substantially increased in the last two centuries. Rail networks were developed early in the nineteenth century and were a true accelerator for hierarchical network development in transport networks and spatial structures. The introduction of high-speed trains today will have a similar effect. The introduction of the private car at the beginning of the twentieth century led to more ambiguous developments. Private cars improved space accessibility and thus had a reverse effect with respect to spatial structure. At the same time, however, the private car allowed substantially higher speeds given the quality of the infrastructure, and can thus be seen as an accelerator for hierarchical road network development. In the second half of the twentieth century, a strong focus on planning processes, especially with regard to environmental impact, and the concept of bundling of transport and thus of infrastructure became dominant issues. Hierarchical networks can therefore be seen as a result of a continuous interaction process between demand and supply, which has a strong correlation with spatial development, and which is influenced over time by other developments, such as technological advances and decision processes.

Hierarchical Network Levels.   A hierarchical network structure is a multilevel network in which the higher-level network is characterized by a coarse network, limited accessibility, and high speeds, and is especially suited for long-distance trips. The lower-level networks are intricate, have high accessibility, and low speeds, making them suitable for short-distance trips and for accessing higher-level networks. It can be shown that the hierarchy in transport network levels is linked with the hierarchy in settlements (Van Nes 2002). Each network level then offers connections between cities of a specific rank and offers access to cities and networks of a higher rank. Figure 2.10 shows this concept as proposed by Schönharting and Pischner (1983) for the German road network guidelines (see also FGSV 2008). Table 2.2 presents a classification for road networks as proposed by Van Nes (2002). Please note that currently no higher speeds are possible for the two highest network levels. These network levels will therefore need more attention with respect to directness and traffic quality (i.e., reliability). For transit networks, however, high-speeds trains really make it possible to provide higher network levels.

Plausibility Scale-Factor 3.   A logical criterion for a higher-level network is that the lower-level network is not considered as an alternative for a trip using the higher-level network. Another way of formulating this criterion is the elimination of shortcuts, which is a criterion that is primarily based

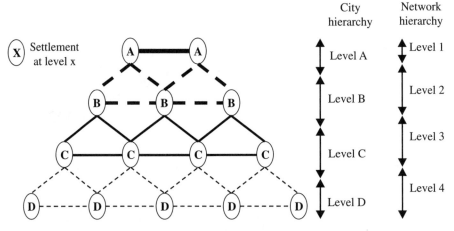

FIGURE 2.10   Road network structure according to Schönharting and Pischner (1983).

TABLE 2.2 Classification for Road Networks

Network level	Spatial level	Distance [km]	Road spacing [km]	Speed [km/h]
International	Metropolis	> 300	300	*
National/Interstate	Agglomeration	100–300	100	*
Interregional/Freeway	City	30–100	30	100–120
Regional highway	Town	10–30	10	60–70
Local highway	Village	3–10	3	35–40

*Theoretically, these network levels should have higher speeds, however, these are not yet technically feasible
Source: Van Nes (2002).

on travelers' perspectives. A possible approach in this case is to look at the maximum detour for a shortcut in a single grid. This detour determines the necessary difference in travel speed between the network levels. The most realistic scale factor for road spacing can be found by calculating this travel speed ratio for a set of scale factors, and selecting the scale factor resulting in the lowest travel speed ratio. The choice for the lowest value is based on the intuitive notion that the lower the travel speed ratio, the easier it will be to develop a higher-level network.

In this approach, the only assumption that is necessary is that the trip length is equal to or longer than the road spacing of the higher-level network. Within a grid network the trip having the maximum detour using the higher-level network can be defined as the trip between two nodes that are located at the middle of two opposing sides of the grid of the higher-level network (see Figure 2.11*a*). In the case that the scale factor for road spacing sf is uneven, this trip is located between two nodes as close to the middle as possible (Figure 2.11*b*). The trip distance using the lower-level network is always equal to the road spacing of the higher-level network.

If the scale factor for the road spacing sf is even, the trip distance using the higher-level network is twice as large, which implies that the travel speed for the higher-level network should be at least twice as high in order to have a shorter travel time using the higher-level network. This implies that in this case no choice for the most realistic scale factor can be made. In case sf is uneven, the trip distance for the higher-level network becomes $(2 \cdot sf - 1)/sf$ as large. In order to have a shorter travel time using the higher-level network, the travel speed should increase accordingly. It can easily be shown that the smallest increase of travel speed is found if the scale factor for road spacing sf equals 3: the speed of the higher level network then is 1.67 times the speed of the lower level network. As sf increases the necessary increase in travel speed converges to a factor 2. In both cases the maximum travel speed ratio is 2. Apparently, it is not necessary to have larger travel speed ratios to avoid shortcuts.

This analysis clearly shows that the existence of a scale-factor 3 for the road spacing of hierarchical road networks can be explained using a simple and plausible mechanism based only on network characteristics. The corresponding scale-factor for network speed is 1.67 and should not be larger than 2.

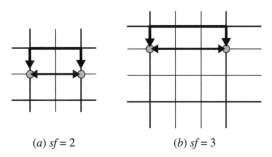

(a) sf = 2   (b) sf = 3

FIGURE 2.11 Maximum detours in a hierarchical grid network.

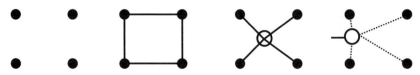

FIGURE 2.12   Application of Steiner nodes.

2.7.3   Special Issues

Steiner Nodes.   When building a network, planners usually consider only the nodes that have to be connected, i.e., cities or agglomerations. However, it might be an interesting option to introduce extra nodes that make it possible to reduce network length and thus investment costs. The impact of these so-called Steiner nodes is illustrated in the Figure 2.12.

On the left-hand side of the figure, we have four nodes that have to be connected. Using only these four nodes a grid network might be a proper solution. Introducing an additional node in or near the center, however, reduces the network length significantly (about minus 30 percent) while travel times are reduced in some cases and are increased in other. The net effect on investment costs and travel times depend strongly on the demand pattern and the location of the additional node. Finally, it is possible to introduce an additional node where a specific road type ends and which is connected to the surrounding cities by links of lower-level networks.

Integrating Functions.   The notion of hierarchical transport networks is primarily functional. In urbanized areas, however, there is a strong tendency to integrate functionally different network levels within a single physical network. In urban areas the distance between access nodes for freeways, i.e. on and off ramps, is clearly shorter than in more rural areas. Integrating network levels might be attractive since they reduce the necessary investments. There is, however, an important pitfall for the quality of the transport network on the long run (Bovy 2001). Medium- and short-distance trips that theoretically would be served by lower-level networks, experience a higher quality because of the higher accessibility and higher speed of the higher-level network. This higher quality influences all kinds of traveler choices such as location, destination, mode, and route. The net result will be a relatively large increase of these medium- and short-distance trips using the freeway network, in quantity as well as in trip length. The resulting congestion reduces the transport quality for the long distance trips for which the freeway network was originally designed. In some cases, the impact on location choice of individuals and companies might even limit the possibilities to increase the capacity in order to restore the required quality for long distance trips. This unwanted impact of integrating functions requires special attention when planning higher level networks in urbanized areas.

2.8   APPLICATION OF THE DESIGN METHOD

To illustrate the design method described, two concrete cases have been elaborated: the design of the national network of Hungary and the federal network of the state of Florida. The following sections present the results of the various design steps.

2.8.1   Hungary: Design of the National Road Network

This section is based on the situation in the year 2002; see also Monigl (2002) and Buckwalter (2001).

Step 1: Hierarchy of Nodes.   The first step in the design process is the decision of how many (and which) nodes (cities) will have to be incorporated in the national network. Two approaches for this can be considered:

1. An approach based on distance classes (a quality approach)

2. An approach based on size of the various nodes (a user approach)

The national network is meant to be used for trips ranging from 50 to 300 km. Accommodating these trips adequately requires an optimal network density and access point density, and this should match the density of nodes.

For example, if we apply the same density of nodes as used in the Netherlands then we need to select approximately 40 nodes (30 in the Netherlands as the size of the country is smaller). The consequence of this assumption is that we have to select nodes with a number of inhabitants of approximately 30,000.

In the second approach, we assume that including a node in the national network is determined by the number of inhabitants. The number should be higher than 50,000. For the Hungarian situation, this would result in the inclusion of 21 nodes.

Of course the situation in Hungary differs from the situation in the Netherlands. For example:

- The size of the country (area) is greater than the Netherlands.

- The population density is lower than in the Netherlands.

- The distribution of the population is quite unbalanced in Hungary, as the largest node (Budapest) has 1.8 million inhabitants and the second-largest node (Debrecen) has only 210,000.

It was decided to make a network design starting with 26 nodes (see Figure 2.13): The minimum number of inhabitants per node is 40,000. The adopted approach is more or less a combination of the quality approach and user approach. The consequence of this is that the national network will not be used as intensively as in more densely populated countries like the Netherlands.

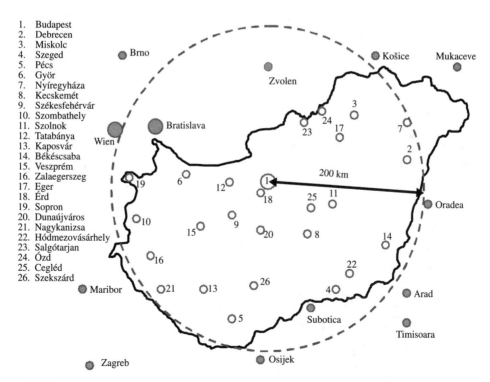

1. Budapest
2. Debrecen
3. Miskolc
4. Szeged
5. Pécs
6. Györ
7. Nyíregyháza
8. Kecskemét
9. Székesfehérvár
10. Szombathely
11. Szolnok
12. Tatabánya
13. Kaposvár
14. Békéscsaba
15. Veszprém
16. Zalaegerszeg
17. Eger
18. Érd
19. Sopron
20. Dunaújváros
21. Nagykanizsa
22. Hódmezővásárhely
23. Salgótarjan
24. Ózd
25. Cegléd
26. Szekszárd

FIGURE 2.13 Nodes and ranking number.

Step 2: International Connections. These connections have to be dealt with before designing the national network

Budapest is connected with the following large cities abroad (city and direction):

- Bratislava/Brno–direction Slovakia, Czech Republic, and Poland
- Vienna–direction Austria, Germany
- Maribor–direction Slovenia
- Zagreb–direction Croatia/Slovenia/Italy
- Subotica–direction Serbia
- Arad/Timisoara–direction Romania
- Oradea–direction Romania
- Mukačeve/L'viv–direction Ukraine

Based on the criterion for accessibility, we also assume an international connection to the center of Slovakia and Poland (direction Zvolen/Krakow). The result of this step is shown in Figure 2.14.

Step 3: Design of the Ideal National Network. The main structure of the national network is based on the international corridors and the 10 largest cities (central nodes) in Hungary (minimum of 80,000 inhabitants):

1. Budapest
2. Debrecen

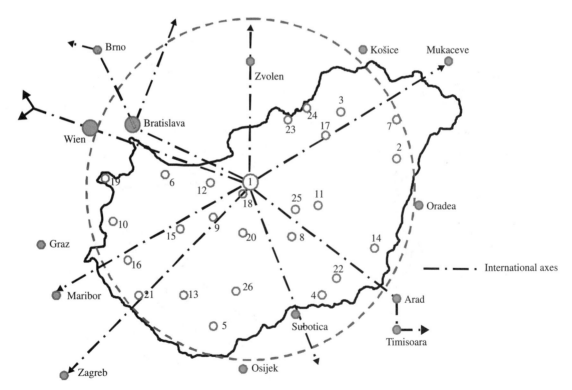

FIGURE 2.14 International axes.

3. Miskolc

4. Szeged

5. Pécs

6. Györ

7. Nyíregyháza

8. Kecskemét

9. Székesfehérvár

10. Szombathely

In addition to these nodes, Lake Balaton is also indicated as a rather important leisure attraction node (especially during summer time). We further assume that all national connections are situated on national territory (no bypasses via neighboring countries).

Because of the dominant position of Budapest in the list of central nodes, the minimum spanning tree (connecting all cities) has a radial structure, starting in Budapest. We can add some remarks:

- Connection with Pécs (5): directly from Budapest [along river Danube via Dunaújváros (20) and Szekszárd (26) or via Székesfehérvár (9) and Kaposvár (13)]

- Only between Budapest and Oradea (Romania) does a shortcut does seem obvious; for all other connections the minimum spanning tree will do

Now we add the nodes 11 to 26, extending the national network with lower-order roads:

11. Szolnok

12. Tatabánya

13. Kaposvár

14. Békéscsaba

15. Veszprém

16. Zalaegerszeg

17. Eger

18. Érd

19. Sopron

20. Dunaújváros

21. Nagykanizsa

22. Hódmezővásárhely

23. Salgótarjan

24. Ózd

25. Cegléd

26. Szekszárd

Some remarks:

- Some of these nodes are already accessible via the main network structure.

- Additional links to nodes not yet connected are necessary in the western part [Sopron (19), Zalaegerszeg (16), and Nagykanizsa (21)], and, if Pécs (5) is connected via Székesfehérvár (9), we need to connect one node south of Budapest [Dunaújvaros (20)]. This further supports the realization of a direct connection with Pécs (5).

- In the northern part, an additional link connects Budapest to Salgótarjan (23), Ózd (24) (via Vác). In the eastern part of the country, the international connection to Oradea (Romania) goes via Cegléd (25) and Szolnok (11). Békéscsaba (14) is connected to Kecskemét (8).

- In addition we need to establish a few shortcuts:

  - South of Budapest: Györ (6)–Székesfehérvár (9)–Dunaújvaros (20)–Kecskemét (8)–Békéscsaba (14) or Szombathely (10)–Veszprém (15)–Dunaújvaros (20)–Kecskemét (8)–Békéscsaba (14)

  - Along the south-east border: Pécs (5)–Szeged (4)–Hódmezővásárhely (22)–Békéscsaba (14)–Debrecen (2)

  - Eger (17)–Debrecen (2)

The final result of this design step (the ideal national network) is presented in Figure 2.15.

Step 4: Analysis of (Comparison with) Existing Road Network (Including Roads under Construction); Situation 2002.   Currently, four motorway corridors are under construction (all starting in Budapest):

- Direction West: Györ (6)–Vienna and Györ (6)–Bratislava (Slovakia)
- Direction North-East: Eger (17)–Miskolc (3)–Nyíregyháza (7)–Debrecen (2)
- Direction South-East: Kecsemét (8)–Szeged (4)–Serbia
- Direction South-West: Székesfehérvár (9)–Lake Balaton–Zagreb (Croatia)

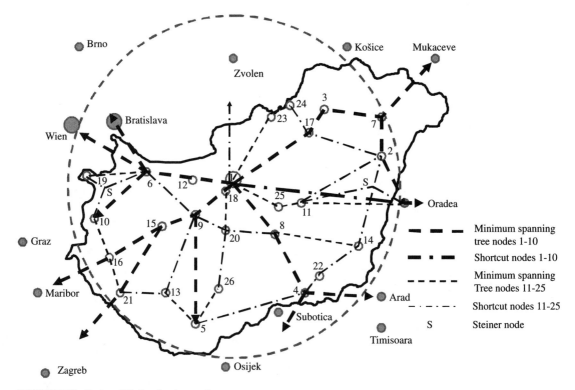

FIGURE 2.15 Design of ideal national network.

This network still needs a few extensions:

- Direction North-East: extension in the direction Nyíregyháza (7)–Debrecen (2) (possible via a Steiner node, a node added solely for improving network efficiency)
- Direction South-East: extension to Szeged (4) and Subotica (Serbia)
- Direction South-West: missing link along South bank of Lake Balaton to Croatian border via Nagykanizsa (21) and to Slovenian border via Zalaegerszeg (16)

After completion of this network the following regions are not yet connected:

- East corridor: Cegléd (25)–Szolnok (11)–Békéscsaba (14) and possible extension to Arad/ Timisoara (Romania)
- South corridor: direction Pécs (5) [directly via Dunaújvaros and Szekszárd (26), or via Lake Balaton]; this is important because Pécs (5) is the fifth-largest city (170,000 inhabitants)
- The Western part: Győr (6)–Sopron and Győr (6)–Szombathely (10) and Zalaerszeg (16) (via Steiner node)
- The Northern part: direction center of Slovakia and the nodes Salgótarjan (23) and Ózd (24)

Possible shortcuts:

- Szombathely (10)–Zalaerszeg (16)–Kaposvár (13)–Pécs (5)
- Along the South-East border: Pécs (5)–Szeged (4)–Hódmezővásárhely (22)– Békéscsaba (14)
- Szolnok (11)–Debrecen (2)–Oradea (Romania) (via Steiner node)
- Kecskemét (8)–Szolnok (11)
- Ózd (24)–Eger (17)

Transit traffic Budapest:

- South Ring road (connection between the motorways heading for Vienna, Lake Balaton and Szeged (4)): this part of the ring road already exists
- East Ring road (connection with motorway heading for Eger (17) and further and the Northern route toward Slovakia): this part doesn't exist yet
- The availability of the two above mentioned ring roads, excludes (diminishes the necessity for) the construction of a ring road heading to the South Győr (6)–Székesfehérvár (9)–Dunaújvaros (20)–Kecskemét (8), although a complete ring road would significantly increase the robustness of the network in the Budapest region

A realistic network design for 10 largest nodes is shown in Figure 2.16.

National Roads.   The question presents itself whether all suggested extensions to the existing motorway network and motorways under construction are of the same type. It does make sense to investigate whether some extensions/connections with low volumes (less than 1200 veh./h/direction) could be constructed according to a lower standard. For example, an extension could have the following characteristics:

- Dual carriageway (2*1) with possibilities to overtake
- Speed limit 90 to 110 km/h
- Matching horizontal and vertical curve radius
- Limited number of intersections (access points), preferably grade separated

Should the occasion arise, this type of road can combine the national and regional function. Figure 2.17 shows a further extension of the national network with lower-order roads.

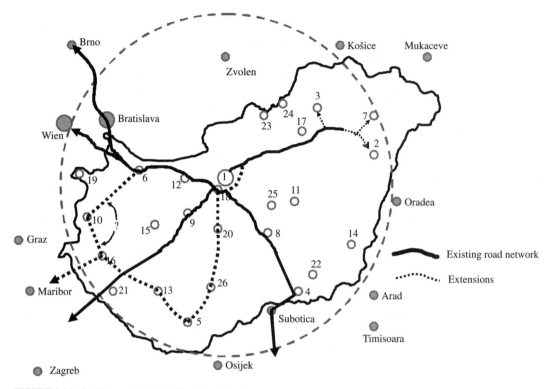

FIGURE 2.16 Realistic network design for 10 largest nodes.

Step 5: Check on Opening-up Function. Besides the function of connecting economic centers, the transport system also needs to open up areas (provide access to as many travelers as possible within an acceptable distance or timeframe). In this step, additional access points are selected that contribute to this function. Possible criteria that can be used to find additional stops are:

- Ninety-five percent of the population should live within a specific distance (25 km for national network) of an access point; areas that are not yet served will get a connection if the area represents a least a specific number of inhabitants or departures/arrivals.

- Cities with a specific rank (e.g., county capital) lying outside the influence area of the network (e.g., 25 km or 30 min of travel from already existing access point) could get a stop.

Looking at the map and the network, there are a few a few areas (places) with poor access to the national road network, e.g., Jászberény, Esztergom, and the North-West part of Lake Balaton. Some of these cities could be selected as a national node. Consequently they should be given access to the network. This would result in a further extension of the national network.

Comparison with the Actual Situation. The network design presented here is based on the situation in 2002. Since then, over 7 years passed by and during this period the Hungarian

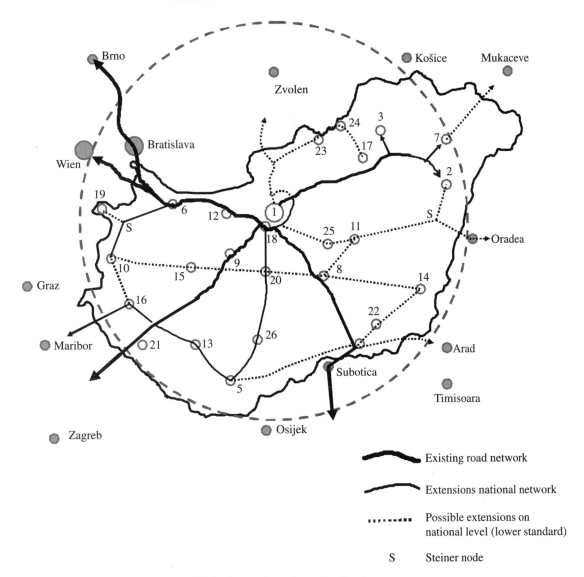

FIGURE 2.17   Realistic network design including lower-order roads at national level.

national network also further evolved. Therefore, it is interesting to see (check) to what extent the further evolution of the national network matches the network design as presented in this chapter.

Figure 2.18 shows the actual situation regarding the further completion of the Hungarian national road network. The similarities between the actual network and the presented network design (Figure 2.17) are striking, although in some cases connections do not completely overlap.

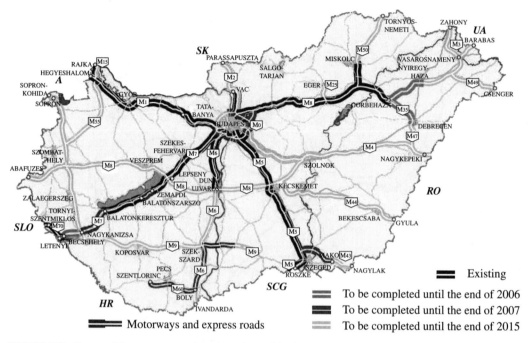

FIGURE 2.18 Current and future motorway and express road network in Hungary (*Source:* Ministry of Economy and Transport, 2007).

2.8.2 Network Design for Florida

Step 1: Node Hierarchy. A summary of the municipalities with 100,000 inhabitants or more reveals that there are nearly 20 municipalities for level I3.

Municipality	Rank	Count
Jacksonville	1	807,815
Miami	2	413,201
Tampa	3	340,882
St. Petersburg	4	245,314
Orlando	5	230,519
Hialeah	6	210,542
Fort Lauderdale	7	183,126
Tallahassee	8	171,922
Cape Coral	9	156,835
Port St. Lucie	10	154,353
Pembroke Pines	11	145,661
Hollywood	12	141,740
Coral Springs	13	125,783
Gainesville	14	114,916
Miami Gardens	15	109,346
Miramar	16	108,484
Clearwater	17	105,774
Pompano Beach	18	101,943
Palm Bay	19	100,786

Source: U.S. Census 2008

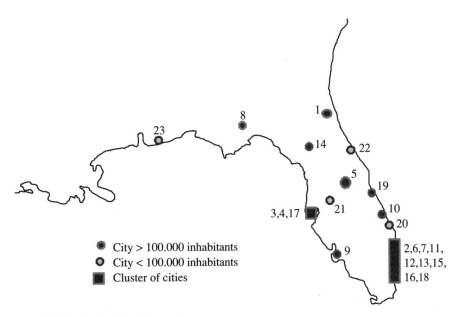

FIGURE 2.19   Nodes with ranking number.

Analysis of the map also reveals the presence of urban clusters. This is particularly true of Miami/ Fort Lauderdale (2, 6, 7, 11, 12, 13, 15, 16, and 18) and the Tampa/Saint Petersburg/Clearwater cluster (3, 4, and 17). The number of nodes is considerably reduced by this and so nodes with less than 100,000 inhabitants and which do not lie in clusters have also been examined. On this basis, the following secondary nodes have been determined for the network:

- 20 West Palm Beach (99,000)
- 21 Lakeland (93,000)
- 22 Daytona Beach (64,000)
- 23 Pensacola (53,000)

With these nodes included a fair coverage of the area is achieved at the same time. It is also possible to include other criteria in selecting nodes, like employment or large tourist attractions (for example, in Orlando). Figure 2.19 shows the selected nodes with ranking number.

Step 2: Interstate Connections.   These connections are focused on the largest (primary) nodes [Jacksonville (1), Miami (2), Tampa (3), Orlando (5), and Tallahassee (8)] and the largest urban areas around Florida [New Orleans, Atlanta, and East Coast (toward Savannah)]. A point of attention in the final design is the detour from New Orleans to southern Florida: It runs in this network via Jacksonville. The results of this step are shown in Figure 2.20.

Step 3: Ideal Network
Step 3a: Connections.   This step investigates how the other selected nodes can be incorporated in the ideal by modifying the interstate connections or introducing new connections. Examples of possible network adaptations are Palm Bay (19) and Port Saint Lucie (10), Coral Springs (9), Lakeland (21), Tallahassee (8), and Pensacola (23). Extra connections are necessary to link up Gainesville (14). For the time being, the preference is for the Gainesville-Jacksonville connection only. For Daytona Beach (22), possibly the connection with Orlando will suffice. The result is shown in Figure 2.21.

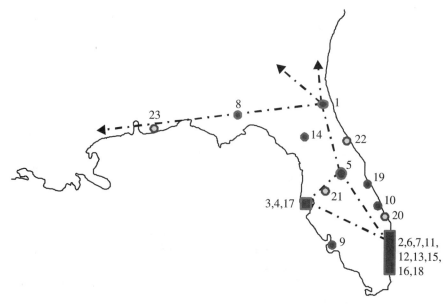

FIGURE 2.20 Interstate connections.

Step 3b: Accessibility of Nodes. Analysis of the connections indicates quite clearly what the desired location of the I3 network at the nodes is:

- Jacksonville (1): westerly
- Orlando (5): northerly from Orlando splitting in a westerly branch (toward Tampa) and easterly branch (Miami); a point of attention is probably the accessibility of the tourist area south-west of Orlando in relation to Miami

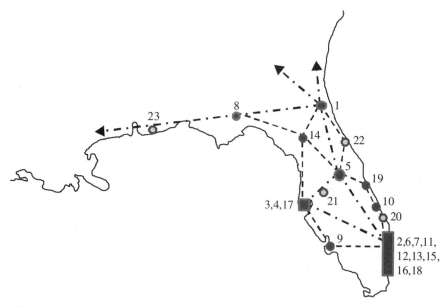

FIGURE 2.21 Ideal network: connections.

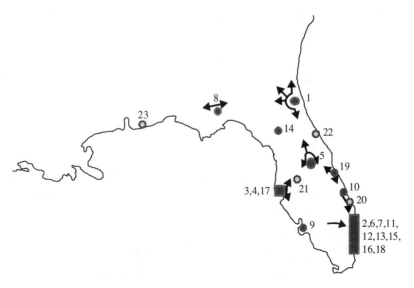

FIGURE 2.22   Accessibility structure of the various nodes.

- Cluster Tampa/Saint Petersburg/Clearwater (3): easterly
- Palm Bay (19): westerly
- Miami (2): no through connection necessary, so only municipal access roads from the west and north, possibly westerly ring-road

Figure 2.22 presents the accessibility structure of the various nodes.

Step 3c: Ideal Network.   In the ideal network (see Figure 2.23) all nodes are included in the network, with the location around the nodes having no influence on the network. Discussion points/

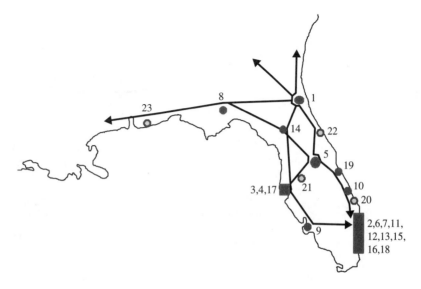

FIGURE 2.23   Ideal network.

alternatives in respect of the ideal network are specifically the connections around Gainesville (14) and the Daytona Beach (22) and Jacksonville (1) connection.

Step 4: Confrontation of Existing Network.   In this step, the relationship with the existing network is central and the acceptability of incorporating secondary nodes in this network is examined. In comparison with the existing network, there are three main points for discussion:

- Position of Gainesville (14)
- Location of the I3 network at Orlando (5)
- Connection along the east coast

A direct connection from Orlando (5) to Atlanta via Gainesville (14) with a link from the Tampa cluster (3) reduces a number of detours, especially for the New Orleans-Tampa cluster and Tampa cluster-Atlanta connections, while no really new detours are created. In addition, this connection makes it possible to go from Jacksonville (1) via Daytona Beach (22) to Orlando (6), though the proposed Gainesville-Jacksonville connection does not have the required quality. This would need to be upgraded.

In Orlando (5) the I3 network lies relatively centrally. With the new structure northward, the branching structure proposed previously northward from Orlando is less suitable. Keeping the existing structure would be suitable here if the number of on-/off-ramps in the neighborhood of Orlando could be restricted.

The Orlando (5)-Miami (2) connection can go via Port Saint Lucie (10). This does mean, however, that the Jacksonville (1)-Miami (2) connection has a clear detour and that Palm Bay (19) cannot be incorporated in the network. The alternative is that the Jacksonville (1)-Miami (2) route runs completely along the coast with two special connections (municipal access roads) in the direction of Orlando from Daytona Beach (22) and Palm Bay (19). For the latter connection, the road to Cape Canaveral can be used.

Conclusion.   The existing I3 network works well. The interstate 75 Orlando-Atlanta, in particular, has a number of interesting advantages. Weak spots are the Gainesville-Jacksonville connection and the central location of the I3 network in Orlando.

The final result (realistic network design) is shown in Figure 2.24.

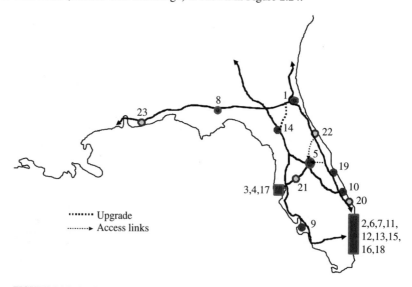

FIGURE 2.24   Realistic network design.

2.9 EVALUATION NETWORK DESIGN

Finally, this section provides additional information regarding the assessment of various effects of a new network design, as well as the complexity of the policy process with which we are dealing.

2.9.1 Costs-benefits

In the design process, the main emphasis lies on defining a network that connects and/or links up selected nodes (hierarchy) with a certain quality. It is self-evident that the accessibility of these nodes is thereby guaranteed. Of equal importance, is the quality of the designed network as well as a verification of other sustainability criteria,[*] like:

- Environment and livability
- Traffic safety
- Accessibility
- Economy
- Costs

In principle, such verification includes a cost-benefit ratio. Because the costs and benefits are spread over a longer period of time, the costs and benefits will have to be reduced to a net cash value with due regard to a discount rate and a time horizon. This chapter does not intend to go into the technical details of a cost-benefit analysis [please read volume I, part II (Operations and Economics) of this handbook for a comprehensive review]. But we will describe in brief what effects will have to be taken into account and how these effects can be calculated.

The cost-benefit analysis can be performed for a separate (newly designed) network but it is also possible to compare alternative network designs with each other.

Load on the Network.   To carry out a cost-benefit analysis, one needs to know the traffic load on the links and intersections from which the network is built up. For this purpose an Origin-Destination Matrix (O-D matrix) will have to be assigned to the network using a static or dynamic equilibrium assignment. The dynamic model is preferred because use of this model allows for a more correct calculation of the congestion losses in the network. As the effects are also calculated for a future situation, it is recommended highly to use O-D matrices for the longer term. If this approach becomes too complicated, an average growth percentage of the traffic load on the network links will suffice.

For a more extensive explanation of traffic assignment models, see volume I, Chapter 10 (Transportation Planning and Modeling) and volume II, part III (Automobile Transportation—Traffic, Streets, and Highways) of this handbook.

2.9.2 Calculating Effects

Environment and Livability.   Effects that come under this header are:

- Emissions of harmful substances (harmful to humans, fauna, and flora)
- Noise nuisance
- Fragmenting the landscape

Emissions.   Traffic emits a number of harmful substances like carbon monoxide (CO), carbon dioxide (CO_2), volatile organic compounds (VOC), nitrogen oxides (NO_x), particulate matter (PM_{10}, $PM_{2.5}$), toxic metals, e.g., lead (Pb), etc. The size of these emissions depends on various factors like fuel usage, type of fuel, speed, driving cycle, gradient, etc. If we want to take account of all of these factors, this implies a highly complex method of calculation. This level of detail is undesirable at this stage in the design process and largely not feasible given the absence of required data. A global approach consists of a calculation of the emissions based on the kilometers covered per road type.

The fuel (energy) consumption per kilometer per road type differs from region to region because of differences in numerous elements, for example, the composition of the vehicle fleet. Emission factors per road type are often published by national authorities. Some background information can be found in Wikipedia, under "fuel economy in automobiles." (http://en.wikipedia.org/wiki/Fuel_economy_in_automobiles).

Noise nuisance. To quantify noise nuisance, the following data is required:

- Traffic volumes per stretch of road
- Composition of the traffic
- Speed of the traffic
- Distance of road axis from building facades
- Building density (number of premises/residents along the side of the road)

This data helps determining the equivalent noise level, which can be compared with the norm [e.g., 55 dB (A) threshold value]. Subsequently, the effect can be quantified by multiplying transgression of the norm per affected person by the cost [e.g., $21 per dB (A)). [Source: ECMT 1998]. For more information on the external damage cost of noise emitted from motor vehicles, see Delucchi and Hsu (2004).

If one does not have access to the building density data, the number of vehicle kilometers (distinguishing between cars and trucks) or passengers and ton kilometers (road and rail) can be used as a starting point.

Fragmenting the landscape. The laying of new roads or widening existing roads can significantly impact the landscape. For example:

- Slicing through the landscape can result in valuable landscapes and/or ecosystems being disturbed.
- Dividing up areas can mean they are subsequently too small to function as a habitat for a species.
- Finally, roads can create a barrier that hinders crossing, for both humans and animals.

This effect is difficult to quantify generically; each case will largely have to be examined on its own merits. One way of calculating the effect is to determine which remediation measures (investments) are desirable to retain the original situation.

Traffic Safety. The safety of a journey in a network depends on many factors. To ascertain the effects of the proposed network structure on safety, the following procedure can be applied:

- The transfer kilometers* on the various types of road are multiplied by risk factors that indicate the possibility of an accident with (fatal) injury as a function of the distance covered. Per road type a distinction is made between the numbers of dead and injured.
- Depending on the type of road, the risk factors as presented in Figure 2.25 can be used.

Accordingly (SWOV 2007), the costs of a fatal accident are estimated at $2.8 million, an accident with severe injury at $0.350 million and an accident with slight injury at $0.03 million.

Network Accessibility. Network accessibility is related to the distance or travel time between origin and destination and a point of access in the network (system). A possible requirement for network accessibility could be that a certain percentage of the population lives within a radius of 10 km or a journey time of 15 min from an access point to the freeway network. By using a network accessibility requirement, this may result in additional access points having to be incorporated into the network.

In calculating the effects in terms of accessibility, a distinction can be made in terms of the components: travel time, robustness, and average journey speed.

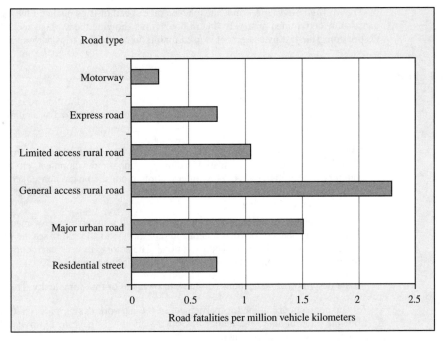

FIGURE 2.25 Fatality risk (number of road fatalities per million vehicle kilometers) by road type (situation 1998) (*Source:* SWOV 2006).

The *travel times* are calculated, preferably on the basis of the resistance of the loaded road network, so it is necessary to assign an Origin-Destination matrix to the network. Use of a dynamic assignment model enables realistic travel times to be calculated.

The laying of new roads will tend to have a positive effect on the total travel time. The net cash value of the gain in travel time can be determined by the gain in hours (calculated on an annual basis) multiplied by the valuation of one hour's travel time.

The valuation of travel time differs per country (region), and it is, moreover, essential to make a distinction in the valuation between the travel time losses that result from normal congestion (peak traffic) and travel time losses that result from incidental congestion (e.g., accidents). In this latter case, the travel time valuation will be higher.

To determine the *robustness* of the network, there are various possibilities that are open, for example:

• Calculating the effects of fluctuations in demand and/or supply (e.g., as a result of the occurrence of an incident). By temporarily restricting the capacity of one or more links in a network, the effects of these fluctuations on the quality of the traffic flow can be calculated.

• A second factor that has a significant impact on the robustness of a network concerns the traffic load of the "vulnerable road sections" in the network. As a result of the "spillback" effect, the tail of a traffic jam can seriously obstruct the traffic flow in other parts of the network.

The robustness of a network can be investigated by varying both demand and supply. The network's capacity to accommodate varying load patterns can be determined on the basis of journey times (total and relative) and intersection loads (load on various parts such as slip roads, acceleration lanes, etc.). The robustness of transport networks is further elaborated in Volume I, Part I, Chapter 3 of this handbook.

The *average travel speed* indicates the average speed of trips made on the various networks and can also be determined relatively (in km/hour), thus showing the average speed between origin and destination. This last measurement is used mainly in comparing trips across various distances.

2.9.3 Process

Proposals to modify the network are often sensitive issues given the numerous effects that will occur. Actually, we are confronted with a situation that can be characterized as a complex policy problem, exhibiting two dimensions:

- A knowledge-intrinsic complexity derived from the relationship between the various intrinsic themes–traffic and transport, spatial planning, economic development, environment, nature, and landscape–that are interwoven. Any intervention in one theme has consequences for the other. But all too often a swings-and-roundabouts mechanism occurs; as a solution is found for one theme, this causes a problem for another.
- A process-related complexity that is expressed in the involvement of many public and private actors, at various levels of government and/or scales, who all want to see their vested interests covered in the problem definitions and solutions. The actors are very interdependent in the problem-solving and consequently they form changing coalitions to this end.

In the design process, account will have to be taken of this complexity. This implies that:

- All stakeholders are invited to participate in the network design process. This means that, apart from the "traditional" road authorities other parties will have to be invited like environmental agencies, chambers of commerce, etc.
- The effects of the proposed measures (network alternatives) are fed back to the stakeholders. By ensuring that the feedback of effects is communicated in a direct (fast) and clear mode with everyone, all stakeholders are stimulated to participate in a structured discussion on potential solutions.

The following requirements are essential for the success of such an approach:

- All stakeholders participate on the basis of equality.
- Everyone has access to the same information.
- The issue/problem is not made more complicated than necessary.

ACKNOWLEDGMENTS

We would like to thank Isabel Wilmink, Arno Hendriks, and Janos Monigl for their inspiring contribution in writing this chapter.

REFERENCES

Arthur, W. B., Y. M. Ermoliev, Y. M. Kaniovski. 1987. "Path Dependent Processes and the Emergence of Macrostructure." *European Journal of Operational Research* 30:294–303.

Bovy, P. H. L. 2001. "Traffic Flooding the Low Countries; How the Dutch Cope with Motorway Congestion." *Transportation Reviews* 21(1):89–116.

Buckwalter, D. W. 2001. "Complex Topology in the Highway Network of Hungary, 1990 and 1998." *Journal of Transport Geography* 9:125–135.

Delucchi, M., and S. Hsu. 2004. "The External Damage Cost of Noise Emitted from Motor Vehicles." University of California, Davis. UCD-ITS-RR-96-3 (14) rev. 1.

ECMT. 1998. Efficient Transport for Europe, Policies for Internalisation of External Costs, European Conference of Ministers of Transport, OECD Publications Service, Paris.

Egeter, B., I. R. Wilmink, J. M. Schrijver, A. H. Hendriks, M. J. Martens, L. H. Immers, and H. J. M. Puylaert. 2002 (in Dutch). IRVS: Ontwerpmethodiek voor een integraal regionaal vervoersysteem (IRVS: design methodology for an integrated regional transport system). TNO Inro, report nr. 02 7N 200 71621. Delft, October 2002.

FGSV (Forschungsgesellschaft für Strassen- and Verkehrswesen). 2008 (in German). Richtlinien für integrierte Netzgestaltung (German Road Research Organisation, Guidelines for Integrated Network Design), FGSV-nr. 121, FGSV Verlag GmbH, Köln.

Levinson, D., and B. Yerra. 2006. "Self Organization of Surface Transportation Networks." *Transportation Science* 40(2):179–188.

Ministry of Economy and Transport. 2007. "Transport Infrastructure Development in Hungary." Ministry of Economy and Transport, Republic of Hungary.

Monigl, J. 2002. "Highway Network Planning in Hungary Including Tolled Motorways." TRANSMAN Consulting, Budapest.

Schönharting, J., and Th. Pischner. 1983 (in German). Untersuchungen zur Ableitung van Strassenkategorieabhängigen Soll-Qualitäten aus Erreichbarkeitsanalysen. (Research into Road Categorization based on Target Qualities Derived from Accessibility Analyses). *Forschung Strassenbau und Strassenverkehrstechnik,* Heft 399.

SWOV Institute for Road Safety Research. 2006. "Advancing Sustainable Safety: National Road Safety Outlook for 2005–2020." SWOV, Leidschendam, The Netherlands.

SWOV Institute for Road Safety Research. 2007. "SWOV Fact Sheet: Road Crash Costs." SWOV, Leidschendam, The Netherlands.

Van Nes, R. 2002. "Design of Multimodal Transport Networks, A Hierarchical Approach." *TRAIL Thesis Series* T2002/5, DUP Science, Delft.

Waldrop, M. Mitchel. 1992. *Complexity: The Emerging Science at the Edge of Order and Chaos.* New York: Simon & Schuster.

Zhang, L., and D. Levinson. 2004. "A Model of the Rise and Fall of Roads. Engineering Systems Symposium at the Massachusetts Institute of Technology, 2004." Internet http://en.wikipedia.org/wiki/Fuel_economy_in_automobiles.

CHAPTER 3
RELIABILITY OF TRAVEL TIMES AND ROBUSTNESS OF TRANSPORT NETWORKS

Ben Immers

Department of Mechanical Engineering
Centre for Industrial Management, Traffic & Logistics
Katholieke Universiteit Leuv en, Belgium

TRAIL Research School for Transport
Infrastructure and Logistics, The Netherlands

Bart Egeter

Bart Egeter Advies, Rotterdam, The Netherlands

Maaike Snelder

Netherlands Organization for Applied Scientific Research
TNO Mobility and Logistics, The Netherlands

Chris Tampere

Department of Mechanical Engineering
Centre for Industrial Management, Traffic & Logistics
Katholieke Universiteit Leuven, Belgium

3.1 INTRODUCTION

The ever increasing mobility of the population in many countries means the arterial road network is subject to heavy usage; in many places, it is almost fully used during rush hours. This not only causes many daily recurring traffic jams, but also leads to increased vulnerability of the road network. A small accident or heavy rainfall may be all it takes to trigger huge delays affecting large parts of the network in a short period of time. The result is that travel times are increasingly unpredictable in heavily congested networks.

Road users place a high value on the predictability of their travel times. Studies indicate that road users consider predictability to be even more important than solving the congestion problem. The development of a robust road network meets this need for predictability. A robust network is much less susceptible to disruptions and thus increases the predictability of travel times.

In this chapter, we elaborate the relationship between road network robustness and travel time reliability. In Section 3.2 (definitions, relationships, and examples), we provide definitions for the various terms related to network robustness and travel-time reliability. Furthermore, we analyze the relationships between the various factors that, in the end, influence travel-time reliability. By means of an example, we also show how road networks are becoming more and more vulnerable. In Section 3.3, we demonstrate how vulnerable parts (roads and intersections) of a road network might be identified by using selection criteria and modeling techniques. In a subsequent section (Section 3.4), we describe how the robustness of a road network might be improved, i.e., how a road network can be made less vulnerable. Based on the general principles outlined, a robust arterial structure is drawn up for the road network in the Rotterdam—The Hague metropolitan area (Section 3.5.) Finally, in Section 3.6, the major conclusions of this chapter are presented.

In this chapter, the focus is on road infrastructure. This does not negate the fact that other measures such as, for instance, transportation management, mobility management, incident management, traffic management and road maintenance, form an integral part of a robust road network. Also the availability of other modalities (e.g., public transport and bicycles for individuals, and railways and inland shipping for goods) play an important role—whether or not in combination with the roads.

We illustrate our methods by using examples from The Netherlands and Belgium, but obviously these methods could easily be applied to other urban areas throughout the world.

3.2 DEFINITIONS, RELATIONSHIPS, AND EXAMPLES

3.2.1 Robustness and Reliability Explained

The terms robustness and reliability are often used in similar contexts. This is because there is a strong relation between both concepts. However, they do not have the same meaning. In the following text, we define the terms reliability and robustness and elaborate on the relation and difference between both concepts.

The most generally accepted definition of reliability is given by (Billington and Allan 1992) and (Wakabayashi and Iida 1992):

> Reliability is the probability of a road network performing at its proposed service level adequately for the period of time intended under the operating conditions encountered.

Some authors (e.g., Chen, Yang, and Tang 2002) distinguish between different types of reliability. *Connectivity reliability* is only concerned with the probability that network nodes remain connected. *Capacity reliability* assumes a fixed demand and considers the probability that the capacity of one or more network arcs is insufficient to handle this demand. Finally, *travel-time reliability* considers the probability that a trip can be made within a specified time interval.

The reliability of travel time is typically expressed using one of the following indicators:

- *Variation coefficient*: The variation coefficient indicates how great the variation is, in percentages, compared with the (arithmetic) average travel time (standard deviation/average).
- *Buffer time index*: This index calculates how much extra time is required to arrive on time with a probability of 95 percent.
- *Misery index*: The misery index focuses on the duration of the delays in the worst situations. This delay is calculated by subtracting the average travel time from the slowest 10, 15, or 20 percent of the travel times.

In comparison with the research devoted to reliability, the study of robustness and vulnerability has received less attention. The terms robustness and vulnerability have a strong relation, but they are actually each other's opposites. Vulnerability describes the weakness of a network and robustness describes the strength of a network.

We define robustness as the extent to which a network is able to maintain the function that it was originally designed for and this under all kinds of circumstances.

This definition has two components which are explained below:

- *Function:* The most general function of a road network is to enable trips from origins to destinations. To achieve an adequate road network design it has to be known for which kind of trip a network link or network node fulfills a function. Trips can be categorized by length, by type of transport (passenger, freight transport) and by trip purpose. The level at which a network has to function is usually defined by government organizations.

- *All kinds of circumstances:* In practice demand and supply are not constant. There may be variations in demand and supply due to irregular, unexpected and exceptional events like natural disasters (e.g., earthquakes, hurricanes, floods, landslides), extreme weather, incidents, road works, social events (e.g., football matches, big fairs), malicious attacks, and signal failures. Besides these irregular causes of variation there are also regular and expected variations, like fluctuations in times of day, days of the week, and seasons of the year.

To enhance the robustness of a network we should carefully consider the following five aspects:

1. *Prevention:* The best way to maintain the function of a traffic system is by preventing the occurrence of disturbances. However, in this case prevention refers to preventing the congestion that occurs in case of disruptions, instead of preventing the occurrence of disruptions themselves. An example of a preventive measure is heating the roads. This avoids snowfall or freezing rain disturbances that cause capacity reductions and thus congestion.

2. *Redundancy:* The robustness of a system can be increased by including spare capacity in the network. This spare capacity is often referred to as redundancy.

3. *Compartmentalization or partitioning:* The degree to which congestion is limited to the link involved or a small part of the network. If a network is subdivided in a number of more or less independent parts, congestion is prevented from spreading, like an oil spill, over the entire network.

4. *Resilience:* Resilience refers to the capability of the network to recover over and over again and as quickly as possible from temporal disturbances.

5. *Flexibility:* The robustness of the transport system can partly be measured by the degree to which the system is able to fulfill other functions, in addition to those for which the system was originally designed. Or, in other words, the ability to adjust is a characteristic that enables a system to grow with the new demands that are made on it.

Figure 3.1 shows the relation between network characteristics and robustness and between robustness and reliable travel times. Disturbances such as accidents, special weather conditions, road works, public events, and seasonal variations, lead to variations in demand and supply. The effect of these variations depends on the behavior of drivers and network managers and on the robustness of the network. For instance, in a robust network, deviations from the regular demand-and-supply pattern will result in less variation in travel time as compared to a network with a lower robustness level. Finally, the experienced reliability of travel time is influenced by the characteristics of the travel time itself and by aspects related to the traveler.

The variations in travel time can be expressed by a probability density function of travel time. Stability is the degree to which the travel time changes as the flow level increases and/or the capacity drops. Ideally, the change in travel time should be limited; after all, a sudden "collapse" of speeds or "explosion" of travel times should be avoided if at all possible. The second component of travel-time reliability is a more subjective component. It explains how a traveler experiences the disturbances. If the traveler is informed about delays and if route alternatives are available, then the travel times are less unreliable in the perception of the traveler then they would be in a more insecure situation. Nicholson et al. (2003) state that if the user is well informed (i.e., information is provided well in advance, and route guidance is available once the trip has commenced), the range of available options is greater and the consequences of degradation are reduced.

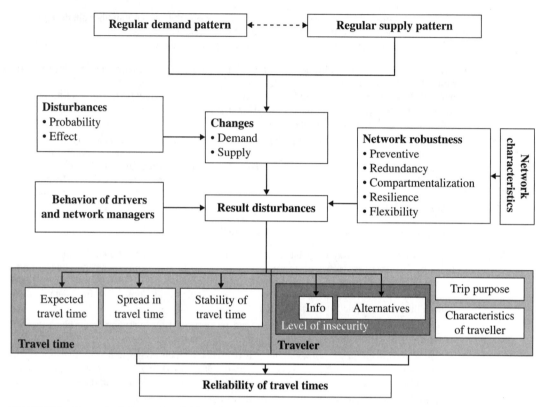

FIGURE 3.1 Factors that influence the reliability of travel times. (Based on Immers and Bleukx 2008.)

In Figure 3.1 we have indicated that, apart from the level of security experienced by the traveler, other aspects such as trip purpose and personal characteristics of the traveler are also important issues.

Several actions can be undertaken to improve (the perception of) the reliability of travel times:

- The spare capacity is determined by the difference between the regular supply pattern and the regular demand pattern. Measures that influence the regular demand-and-supply pattern, therefore, influence the spare capacity.

- Prevention of disturbances is a very effective measure to keep the travel times reliable. However, some disturbances are very difficult to prevent, such as bad weather conditions. Placing trees at a sufficient distance from the motorway so that they do not fall on the road during storms is an example of a preventive measure that can be taken.

- Professionalization is another measure. Drivers and network managers can be trained to react in a more professional way when disturbances occur. This will influence the effect that disturbances have.

- Network management is needed in order to maintain network performance at an acceptable level during disturbances. The ability to maintain network performance in adverse conditions is a measure of the level of robustness of the network.

- Graceful degradation measures should be taken that ensure that travel times degrade gracefully when disturbances occur.

- Providing information is imperative. If information is provided to travelers about the duration of congestion, the reason for congestion, as well as alternatives, travel times will be perceived as less unreliable.

3.2.2 The Vulnerability of Road Networks

In this section, the vulnerability of the existing road network is illustrated by an example of an accident that occurred around 7.00 AM on September 11, 2007 at the A12 freeway into The Hague (the Netherlands), at the off-ramp of Voorburg, a suburb of The Hague. This location is marked with a star in Figures 3.2 and 3.3a. Of course, similar effects could occur in other urban areas around the world. Figure 3.2 also shows the names of the roads and cities that are used throughout this section.

Figure 3.3a to f are based on traffic counts. They show how queues build up and dissolve after an incident has occurred. From Figure 3.3a, it can be seen that at 7.15 AM the congestion spills back over the A12–A4 intersection. In the period thereafter, the congestion spills back over a large part of the A12 and on the A13. At 8.00 AM the A13 is completely blocked, and the traffic on the A12 has come to a complete standstill up to Gouda (the junction between the A12 and the A20). At 8.45 AM the traffic at the A13 also comes to a complete standstill. Of course, this situation is enhanced by the fact that the accident happened just before the start of the regular peak period. In The Netherlands, the peak period lasts on average until about 9.00 AM, but in this case, the network as shown in the picture remains completely congested until about 9.50 AM. Figure 3.3e shows that congestion starts to dissolve on the A12 at the head of queue and the queue moves backwards. A short while later, the congestion on the A13 also starts to dissolve and, as can be seen in Figure 3.3f, at 11.00 AM the situation is almost back to normal. This implies that the effects of the incidents could still be felt 4 h after the incident had occurred.

In Figure 3.4a and b, the travel time that is needed to travel over the complete A13 and the A12 between The Hague and Gouda is shown. The dark-gray line is the travel time on September 11th.

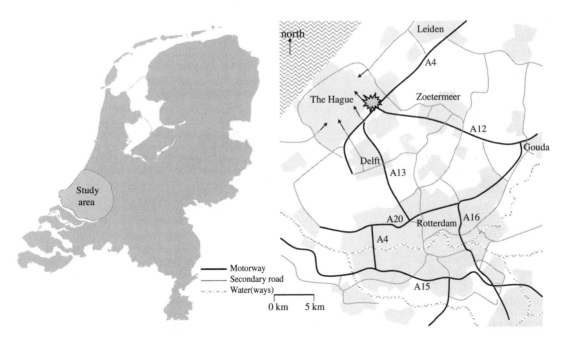

FIGURE 3.2 Map of the study area with the names of the roads.

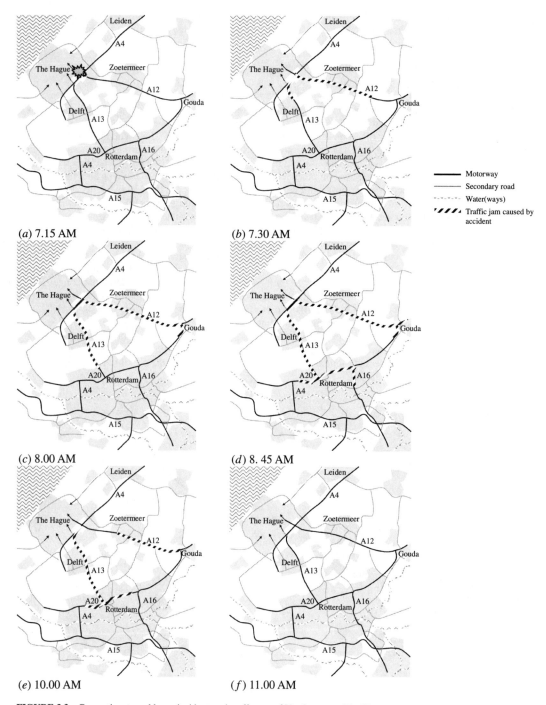

FIGURE 3.3   Congestion caused by an incident on the off-ramp of Voorburg near The Hague.

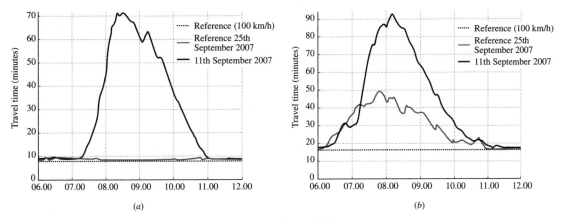

FIGURE 3.4   Travel time on the A13 (*a*) and the A12 (*b*) on September 11, 2007.

The light-gray line is the travel time on a regular Tuesday (without incidents) in September 2007, and the black dotted line is the free-flow travel time at a speed of 100 km/h. These figures show that on September 11th, individual travelers experienced a delay of up to 1 h on the A13 and of up to 40 min on the A12.

In total the incident on the off-ramp causes about 10,000 extra vehicle loss hours on the A13 and A12. This number multiplied with an average value of time of $22.30/h results in an economic damage of about $223,000. The real economic damage of this incident on the off-ramp is considerably higher because the extra travel time attributed to taking longer alternative routes is not included in the vehicle loss hours. Furthermore, delays on the local roads, the costs of emergency services, repair costs, medical costs, and environmental costs are not included either.

This example demonstrates that the existing road network is vulnerable because a small accident on an off-ramp might cause many hours of congestion. Because levels of congestion keep growing, it can be expected that the road network, especially in major urban areas, will become more and more vulnerable to unforeseen disturbances. Furthermore it will be more difficult to recover from unforeseen disturbances, as spare capacity in the network is reduced because of more spatially concentrated flows around cities. This is because of concentrated spatial development, new land use policies (bundling), and market-led agglomeration forces combined with widening of congestion periods because of pricing measures and a gradual reorganization of household and working schedules.

Snelder et al. (2008) demonstrated, by multiplying the expected number of incidents with duration of more than 1 h with the effects of these incidents, that the expected costs of vulnerability in The Netherlands might grow from about $1.5 billion in 2008 up to $6 billion in 2030 if no measures are taken.

3.3   THE ROBUSTNESS OF A NETWORK ON THE MICRO LEVEL: IDENTIFYING VULNERABLE NETWORK ELEMENTS

3.3.1   Motivation

A vulnerable network element is defined as a part of the network that is responsible for a sharp decrease in traffic operation performance caused by capacity restrictions caused by an incident or circumstances of unusually high traffic demand.

The identification of vulnerable road sections is useful for operators for several reasons. First, it shows where capacity increase, buffers, or parallel alternative routes in the network are most needed. Second, it identifies sections that are candidates for small improvements to control or infrastructure, like increased accessibility for incident response teams, improved monitoring for rapid incident verification, etc. Third, it shows where incident-related traffic management actions are most needed, as well as what kind of action should be taken.

In this section, we use the region west of Brussels in Belgium, covering about 40 by 20 km, as an example (seen in Figure 3.6). The motorway E40 runs right through it, connecting the cities of Brussels and Ghent. The corridor is characterized by serious congestion problems, especially at the entrances to Brussels.

Identifying the vulnerability of network links is a topic that has received extensive research attention. Berdica (2001, 2002) investigated the effect of a restricted set of input variables, such as load variations and disruptions, on the performance of a network. Cassir and Bell (2000) used an approach based on game theory. Tamminga et al. (2005) initially used a static traffic assignment to estimate the effects of an incident and the probability of it occurring. Next they used a sequence of static and dynamic simulations to arrive at a ranking of vulnerable network links. The approach employed here is similar to theirs, however, we extend the number of factors taken into account and we use dynamic traffic assignment right from the start.

The obvious way to solve the problem of finding vulnerable network links is to simulate an incident on each and every network link. The next step then would be to compute the impact of this incident, for example in terms of lost travel time. Because of the amount of computation involved, static assignment is usually applied. But this may cause problems. As is well known, congestion and, more specifically, spillback are not very well represented in static models.

For this reason, we attack the problem in a different way. Before simulating incidents on certain network links we first try to derive a list of possible vulnerable candidate links using one basic simulation of the traffic on the network, not involving incidents. This basic simulation is carried out running a dynamic traffic assignment model. Any state-of-the-art dynamic assignment model would do; we applied the INDY model (Bliemer 2005, 2007; Yperman 2007) for this purpose.

3.3.2   Methodology

The proposed methodology proceeds along a number of steps:

1. Draw up a long list of vulnerable network elements.
2. Reduce the long list to a short list.
3. Simulate incidents on the links in the short list.

Step 1: Draw up a long list of vulnerable network elements.   The aim of this step is to arrive at a first rough selection of possibly vulnerable links. The primary idea is not to miss any candidate links, but this may entail the inclusion of some false alarms. A network link is deemed to be vulnerable if an incident on it carries large consequences in terms of lost travel time and if there is a high probability of an incident occurring on this link. In this first step, the focus is on the consequences; incident probability is considered in the next step.

Analysis of the stages of an incident   There are four stages in the progress of an incident as is shown in the figure below. These stages can also be recognized in Figure 3.5.

PHASE 1: COMPLETE BLOCKAGE, TAILBACK ONLY IN THE LINK AFFLICTED BY THE INCIDENT.   For the sake of simplicity, we first consider the situation where no blocking back occurs to the upstream links (Figure 3.5a top left). The line at slope I from the origin in Figure 3.5b represents the inflow into the link. The line at the same slope shifted along the t-axis shows the outflow from the link. An incident reduces outflow to a value of zero. When the incident location is cleared after a period of $\Delta t1$, outflow resumes at a rate of C, the initial capacity of the link, until (after a period T) all

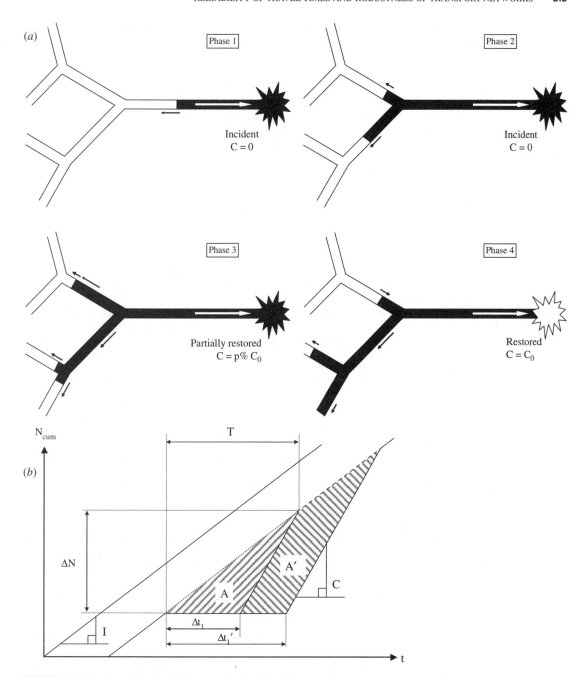

FIGURE 3.5   (*a*) Four stages of an incident and (*b*) calculation of its impact.

cumulated traffic has left the link. At that point, the initial outflow at a rate I is restored again. The hatched area A indicates the vehicle-hours lost by this incident:

$$A = \frac{\Delta t_1^2 I}{2(1 - I/C)}$$

The formula shows the impact of an incident to be proportional to the square of incident duration, highlighting the importance of adequate incident management. This is illustrated in Figure 3.5*b* by the increase of the area A to A' caused by an increased duration of the incident to $\Delta t_1'$.

Links having a high value for $I/(1 - I/C)$ are potentially vulnerable links. We call this factor the *incident impact factor* for the link.

PHASE 2: BLOCKING BACK OF TAILBACK TO UPSTREAM LINKS.   When the tailback reaches the upstream links, it is as if these links also are hit by an incident with complete blockage (Figure 3.5*a* top right). With jam density k_j, length of road section L, the number of lanes equal to n and free speed V_f, the maximum time period T_b to blocking back can be shown to be:

$$T_b = L/I(nk_j - I/V_f)$$

The amount of vehicle-hours lost by the incident is equal to the vehicle-hours lost on the link where the incident occurred, augmented by the vehicle-hours lost on the upstream links struck by the blocking back phenomenon. We have a problem if these upstream links, in their turn, are characterized by a high impact factor.

PHASE 3: PARTIAL RECOVERY OF THE LINK AFFLICTED BY THE INCIDENT.   Usually some time after the incident took place, the road is partially cleared, and part of the original capacity is recovered. Flow in the afflicted link gradually resumes, and after some time, traffic starts moving in the upstream links also. However, the recovered capacity will not be distributed proportional to demand on these upstream links, but proportional to the capacity or number of lanes of these upstream links (Bar-Gera and Ahn 2010). It, therefore, often happens that flow recovery is relatively better in the links that need it least because of the relatively low traffic demand on these links. And at the same time, the tailback grows fastest in those directions that lead to the greatest loss in vehicle-hours, because the high demand links may block other flows that did not want to pass along the incident location at all. This unfavorable state of affairs gets worse at lower capacities of the link struck by the incident.

This means that special attention should be paid to those links having a high incident impact factor *combined with a relatively low capacity*. Such links, for instance one-lane ramps of an interchange, might have a disproportionately high impact on the number of lost vehicle-hours, especially if they are located near busy links. For this reason, this type of link is retained in the long list of vulnerable links. In Step 2 of the methodology, it is decided, on the basis of topology, if these links will also be retained in the short list.

PHASE 4: COMPLETE RECOVERY OF INITIAL CAPACITY.   Finally, full capacity is restored at the location of the incident. As in the previous stage the recovery will be propagated to the upstream links. And here too, we find that the tailbacks in the busiest links are the last ones to dissolve. This last phase, however, has no effect on the drawing up of the long list of vulnerable links.

Selection criteria for drawing up a long list of vulnerable road sections.   The analysis of the course of a typical incident in the preceding section results in a number of selection criteria that are used to draw up a long list of vulnerable road sections. A key role is played by the incident impact factor $\frac{I}{1 - I/C}$. Road sections with a high incident impact factor will certainly be included in this list. The list will be extended by those links that may have a somewhat smaller value for the incident impact factor, but where blocking back to the upstream links occurs comparatively fast and those links that slow down the recovery of an incident because of low capacity. The selection criteria are:

c_1:   the incident impact factor

c_2:   the inverse of the blocking back time period

c_3:   the incident impact factor, restricted to links with $C \leq 2500$ pcu/h

The values for these criteria are computed for each link. The output from a basic run of a dynamic assignment model is used for this purpose. The links with the highest scores on each of the criteria are retained for inclusion in the long list. The number of links to include is arbitrary, depending on circumstances.

Application to the study area.   A simulation of the morning peak period generated by the dynamic assignment model INDY was used to select a top 100 of network links for each of the selection criteria (see Figure 3.6, the darker the color, the higher in the ranking).

There naturally is a certain overlap in the links selected, which can be seen from Figure 3.6*d*. Because of this we finally obtain a long list of 209 distinct and potentially vulnerable network links for the study area.

Step 2: Reduce the Long List to a Short List.   The aim of this step is to reduce the number of network links in the long list to a manageable short list. Now aspects are considered that were not taken into account while drawing up the long list: the probability of an incident occurring on the link, the location of the link with respect to other (busy) links, and the possibility of avoiding the incident location by changing routes. Whereas drawing up the long list is a fully automated process, reduction to a short list needs some expert judgment from the part of the user. We now outline the appraisals to be carried out in this step.

Analysis of risks

PROBABILITY OF AN INCIDENT OCCURRING ON A LINK.   The severity of an incident is determined by the probability of an incident occurring multiplied by the consequences of the incident, measured for example in lost vehicle-hours. Many aspects could be taken into consideration, such as the type of road, the surrounding area (urban, rural), prevailing traffic conditions, etc. For sake of simplicity, we used the number of vehicles present on a link as a proxy to the incident probability. Because of relatively safe conditions on motorways, known from traffic accident statistics, we reduced the incident probability on motorways as compared to other types of road. We combine the probability

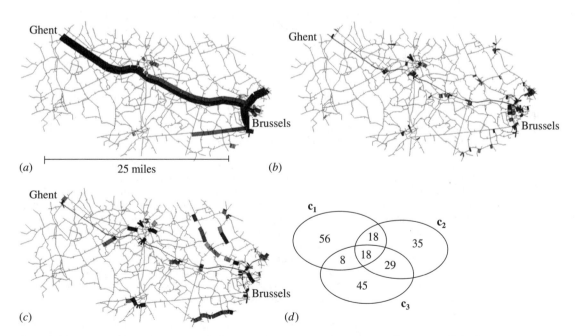

FIGURE 3.6   Links selected, (*a*) by the c_1 criterion; (*b*) by the c_2 criterion; (*c*) by the c_3 criterion; (*d*) overlap in links selected by the three criteria.

of an incident with each of the previously mentioned selection criteria. This combination is found as the intersection of the set of links with high incident probability with the set of links scoring high on criteria c_1, c_2, and c_3, respectively.

HIGH RISK ON THE LINK ITSELF. The links scoring high on criterion c_1 and also having high accident risk are retained in the short list. From the case study in the test area it appeared that one often finds a stretch of links meeting both criteria. For such a stretch of links it suffices to simulate an incident on the most upstream (best case) and most downstream link (worst case).

HIGH RISK BECAUSE OF BLOCKING BACK. Here we first select all links meeting criterion c_2 and having high incident probability. From this selection, we only keep those links in the short list that have one or more busy links in their upstream tree that score heavily on criterion c_1.

HIGH RISK BECAUSE OF BLOCKING BACK FOLLOWED BY A LOW RECOVERY RATE. A network may comprise links that create an extra risk by a combination of limited capacity, their location with respect to other links and the route flows traversing them. In those conditions, an incident may cause disproportionate hindrance in the final stages of an incident when the capacity is partially recovered. Here we select those links for the short list that have a high score for criterion c_3, a high-incident risk and, as in the preceding section, contain busy links in their upstream tree. The low recovery rate should affect precisely those busy links. This is the case if there is an upstream merge between the busy link and the link where the incident occurred and (a) if the I/C ratio of the busy link is higher than that of the other links in the merge, or (b) if the busy link competes with many other links in the merge.

HIGH RISK OF GRIDLOCK. Gridlock occurs when a blocking back tailback follows a closed circuit and finally hits the source where the original tailback started. If the restriction imposed by the end of the tailback is stronger than the restriction that caused the tailback in the first place, a negative spiral may arise bringing all the traffic in the circuit to a complete standstill.

A discussion of the gridlock phenomenon is outside the scope of this chapter. The methodology contains a number of heuristic rules to help recognize parts of the network where gridlock, or some weak form of gridlock, is likely to occur. Naturally these links in the network are added to the short list.

REDUCING RISK BY AVAILABILITY OF ALTERNATIVE ROUTES. In all of the foregoing, we have assumed that travellers will persist in their once-chosen route. Of course a number of travellers, depending on their familiarity with the local situation, will deviate from this route. From this viewpoint, reducing the links contained in the long list means assessing:

- Whether there are routes that avoid the incident location and the area affected by the incident

- Whether these routes are accessible from the original routes (because many drivers will be en route when the incident occurs)

- Whether the remaining capacity of these alternative routes suffices to handle the additional traffic without creating too many lost vehicle-hours themselves

Currently, no method is available to automate these assessments. It is left to the user of the methodology to do these assessments on the basis of expertise and common sense.

Application to the study area. The proposed methodology entails carefully examining all network links using the guidelines described above. This will lead to the elimination of quite a number of network links from the long list. Obviously, many isolated links satisfying criterion c_1 and c_2, but which are located far from any busy traffic artery, are removed from the long list. On the other hand, the application of the methodology to the study area also unearthed some potentially vulnerable links in the network that were not so obvious at first sight. An example is the network link joining the E40 to the southbound part of the Brussels Ring Road. Here an incident would have great consequences during all stages of the incident; the third stage (partly recovered capacity) especially would seriously affect traffic on the motorway trying to reach the center of Brussels. Some locations prone to gridlock were also identified in the network.

The possibility of reducing the number of potentially vulnerable links because of the availability of alternative routes was also examined. In particular, links lying in dense parts of the network usually have alternatives available.

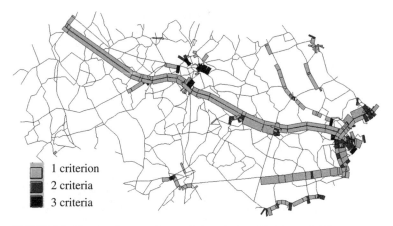

FIGURE 3.7   The short list created by combining criteria and incident probability.

A good approximation of the short list could be obtained by graphically depicting the links satisfying one or more of the criteria c_1, c_2, and c_3 in combination with the probability p of an incident occurring on the link (Figure 3.7). Figure 3.7 shows the links satisfying only one, two, or all three criteria. In addition, the figure indicates if the incident probability is low (darker) or high (lighter). Although some care must be taken when interpreting this figure, the diagram has the merit that it highlights nonobvious vulnerable links in the network.

Step 3: Simulate Incidents on the Links in the Short List.
Worst- and best-case analysis.   In principle, we could do a simulation run for every network link in the short list. As we indicated before, it often happens that a sequence of links is found in the short list. In these cases, it often suffices to simulate an incident in the most downstream link of the sequence.

For the incident simulations in our case study, we chose an incident implying complete closure of the link for 30 min, followed by a stage lasting for 60 min, during which capacity at the incident location is partially restored. Obviously many other incident scenarios could be envisaged.

We used the dynamic model INDY to do the simulations. This model, unfortunately, does not allow for drivers to change their route during the journey if they meet with an unexpected incident. Lacking a better model, however, we approached the problem by simulating a "worst-case" and "best-case" scenario. The worst case implies all drivers persisting in their once-chosen route, despite the incident. This corresponds to using the routes from dynamic user equilibrium without incidents. These routes are then the only ones that are considered on the network where capacity has been reduced at the incident location. The "best case" means computing a user equilibrium directly on the network with the capacity reduction. This last option is not a very realistic one. For one, most drivers are hardly aware of alternatives to their usual route. Second, an equilibrium approach in the calculation means that drivers in the model start changing their route even before the incident has happened and, third, most path- and route-generation modules in existing packages would probably not be capable of generating the seemingly illogical routes chosen in case of an incident.

Application to the study area.   Figure 3.8a shows some locations taken from the short list for the study area where an incident was simulated. Figure 3.8c gives an impression of the consequences of one such "worst case" incident for location number 1, on the approach of the E40 motorway to Brussels. The "normal" situation without an incident is given for comparison in Figure 3.8b.

In Figure 3.9, we see the lost vehicle-hours (worst case) incurred at all the incident locations of 3.8a, split up by road type. Total vehicle-hours on the whole network without an incident come to about 50,000. The graph shows that an incident at location 1 adds about 20,000 extra vehicle-hours to this figure.

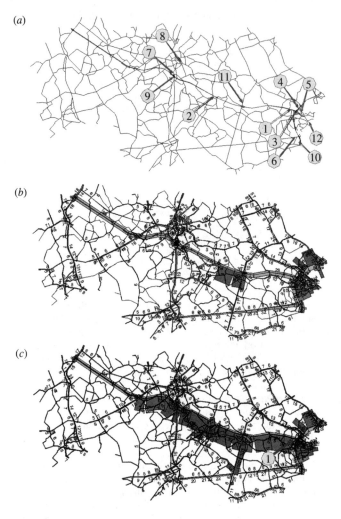

FIGURE 3.8 (*a*) Some vulnerable network links in the short list; (*b*) regular congestion in the study area without incidents (bandwidth = density, colour = speed: the darker, the slower); (*c*) congestion caused by an incident at location 1.

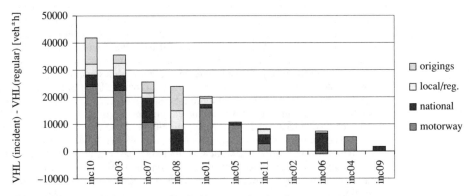

FIGURE 3.9 Lost vehicle-hours caused by incidents at various locations.

For some locations we also carried out a best-case analysis. Some of the incident locations then lose their prominent position in Figure 3.9. We already noticed that the assumption that all drivers deviate from their once-chosen route in the case of an incident (some even before the incident has happened!) is an overly optimistic and unrealistic one. But if, even in this overly optimistic case, the number of vehicle-hours remains very high, this is a clear indication of a very vulnerable network link.

This section presented a methodology for identifying vulnerable road sections without imposing the need for repetitive (Monte Carlo) simulation. The methodology considers the flows from a basic dynamic traffic assignment (which, in practice, is often already available from other projects in the studied region). Based on these flows and the capacity, three criteria are defined that estimate the impact in terms of lost vehicle hours in the case that an incident would occur on the considered road section. These criteria reveal not only the most obvious vulnerable points (including the most severe bottlenecks and most heavily loaded sections), but also some more unexpected points that deserve special attention. Once these are identified, only a limited number of simulations with incidents on the most critical points are needed to select the most critical sections in the short list of candidate locations. This analysis can help to improve incident management programs, identify candidate sections for improvement projects (infrastructure, traffic control, network structure), and to refine incident-related traffic management.

3.4 THE ROBUSTNESS OF A NETWORK ON THE MACRO LEVEL: SOME GENERAL PRINCIPLES FOR A ROBUST ROAD NETWORK

In this section, some general principles for the design of a robust road network are presented. These principles are derived from Egeter et al. (2007) and Schrijver et al. (2008).

3.4.1 A Healthy Balance between Demand and Supply

A robust road network must be able to stand up to a good deal of adverse conditions. Part of the road capacity can fail for several reasons, e.g., road accidents or bad weather conditions. A robust network must be able to compensate for this to some degree. A healthy balance between demand and supply should, therefore, be the point of departure in the normal situation. The occurrence of a limited amount of congestion (localized and temporary) is acceptable under these circumstances: The largest peaks in demand are thus spread out over a somewhat longer period of time so that the expensive infrastructure needs not to be sized according to the "peak within the peak."

But what is a "healthy balance"? The principle we have used is that the direct and indirect costs of the creation of extra capacity on the road must be balanced with the benefits for consumers and producers in that less time is lost through incidents [see Meeuwissen (2002) and Immers et al. (2005)]. The optimization of accessibility is not the only important issue in this consideration—the consequences for quality of life and space utilization are also important.

At the same time, we want to prevent this extra capacity from generating increased mobility, thus causing the extra space on the road to "evaporate" within a few years. This mechanism is at odds with the principles of a robust road network, because the network also needs to continue to function well in the long term. The core of the solution is a careful selection of the design speed of the infrastructure, thus limiting the growth of travel distances. Most freeway traffic in urbanized areas remains within the area and travels relatively short distances. For this short distance traffic however, a steady traffic flow is more important than a high design speed. For long-distance traffic, the requirements are different: This kind of traffic should be enabled to bypass urban centers without significant loss of speed.

3.4.2 Integrating Flexibility and Options into the Network

A robust road network offers multiple ways of getting from A to B. This results in a leveling off of peaks, because road users on a certain route have the option of taking another route during congestion, be it one of somewhat lower quality, or one that is a little longer. These alternative routes are also allowed to use other modes of transport—public transport, for instance. We do not consider different forms of transportation to be in competition with each other, but see them as an increase in options for the transport consumer.

In addition to freedom of choice, alternative routes through the road network also offer fallback options in case a road segment or intersection becomes (entirely or partially) unavailable, whether for foreseen or unforeseen reasons. Moreover, extra flexibility that is only used during abnormal situations can be built into the network at strategic locations. Examples of this include removable barriers that can be opened or roads/links that are normally not used, but also extra capacity that is only utilized in abnormal situations, e.g., road shoulders used as peak lanes. Traffic flow over the alternative route can be handled well by providing information to road users and through dynamic traffic management.

3.4.3 Designs for Both Flow and Buffering

Since we anticipate a certain amount of congestion in a robust road network, we must not only consider the flow function, but also—far more than has been done so far—integrate the *buffering function* of road segments into the design of roads, intersections, and junctions. Every road segment can accommodate a certain number of vehicles: the "natural" buffer capacity. Wherever the natural buffer capacity is insufficient, we can intentionally offer extra buffer capacity. These buffers prevent congestion spillback to other routes; they also regulate the traffic, so that certain roads, like through roads or the urban road network, for instance, remain as congestion-free as possible (see Figure 3.10).

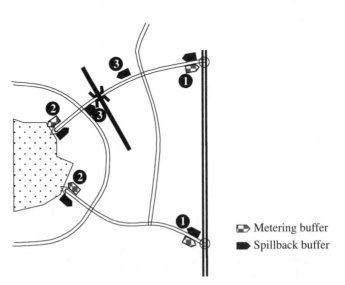

FIGURE 3.10 Example of buffers in the metropolitan network.

As indicated in Figure 3.10, it can be useful to apply buffers at the following locations:

1. Prevent congestion on the long-distance network by creating traffic metering buffers at access ramps, but also prevent congestion on the lower-level network from spilling back on the freeway by creating spillback buffers at exit ramps.

2. Prevent congestion on the urban road network by creating traffic metering buffers at the urban area "access points" (and spillback buffers at the "exits," if needed); this includes keeping urban ring roads as congestion-free as possible.

3. Add spillback buffers upstream from bottlenecks within the metropolitan network, if needed.

3.4.4 Form Follows Function

In general, freeway networks were originally designed for long-distance traffic: high-speed design, wide lanes, weak curves, and spacious intersections. However, we notice that freeways in metropolitan areas, to a considerable extent, are being used for short trips, simply because the freeway is the fastest and often the only route available. Thus, the two traffic types, using the same routes, tend to get in each other's way. The form is thus not tailored to the function anymore, leading to inefficient usage of traffic space.

In a robust road network, form follows function: Whenever advisable, we make a physical distinction between long-distance and short-distance road networks, so that we can tailor the network characteristics (mesh size and design speed) to their function. This makes the total network more efficient (e.g., in terms of space utilization); especially in urban areas with heavy traffic, and ensures that the different functions are not in conflict with each other.

3.4.5 Balance in the Network

If the traffic flows and the capacity are distributed over the network in a balanced way, it is easier to cope with disturbances. One route can serve as a backup option for the other. Figure 3.11 shows an example of a balanced and unbalanced distribution of lanes over alternative routes. In the situation at the top, there are two routes to travel from north to south. The first has four lanes and the other two. If an incident occurs on the route with four lanes, there are only two lanes on the other route left as a backup option minus the capacity that is already used. In the situation in the middle, there are two routes with three lanes. This situation is a bit more balanced. The most balanced situation is the situation in which three routes each have two lanes. If something happens on one route there are four lanes left as a backup option minus the capacity that is already used.

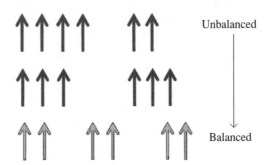

FIGURE 3.11 (Un)balanced distribution of capacity (lanes) over routes.

3.5   AN EXAMPLE: THE ROTTERDAM—THE HAGUE URBAN AREA

Based on the general principles outlined above, a robust arterial structure is drawn up for the road network in the Rotterdam—The Hague urban area. This should not be seen as a "plan" in which all design choices have been worked out in detail, but as an illustration of how a vision on the development of a robust road network could evolve.

3.5.1   Functional Analysis of the Road Network

"Form follows function"; therefore, the first step in this development should be a "functional analysis": an analysis of the current and future use of the network in the Rotterdam—The Hague urban area. The area, which is around 26 km in diameter, has a population of about 1.5 million. The network was shown in Figure 3.2. The figure shows that the main structure of the network consists of a system of freeways that also has to accommodate through traffic. There is only a limited regional network, mainly designed to guide regional traffic to the freeways as quick as possible.

Table 3.1 gives the distribution of the total traffic over three functional categories:

1. Metropolitan traffic, remaining within the Rotterdam—The Hague urban area
2. Incoming and outgoing traffic, exiting or accessing this area
3. Through traffic, with origin and destination outside of the area, but passing through the area

Metropolitan Traffic.   A large part of the traffic on the main routes in the Rotterdam—The Hague urban area remains within this area ("metropolitan traffic"). This traffic uses many different types of roads, each having its own speed regime, capacity, type of intersections, and access forms. The differences between all these road types are not (any longer) functional: The road user (and transporter) simply chooses the fastest route at random. The example below (Figure 3.12) gives possible routes between the northeastern part of Rotterdam (Rotterdam–Alexander) and the city of Leiden. It is easy to see that choosing the freeway leads to a huge detour, whereas alternative routes using the regional network (if existing at all) are often time-consuming and not suited for heavy metropolitan use. Most of the regional network was originally designed for a local or rural function, and was not sufficiently adapted to the urbanization process that took place in the past decades.

So in most cases, the freeway network acts as the main structure for metropolitan traffic. There are only a limited number of freeway routes available for metropolitan trips through the region (A12, A13, A15, A20), with few or no alternatives. This makes the network structure very vulnerable. Because the mesh of freeways has remained relatively rudimentary in comparison with the ever expanding and in density decreasing urban area, it is understandable that it has become very tempting for consumers to take "shortcuts."

Through Traffic.   There are two important through traffic flows in the Rotterdam—The Hague urban area. First, the north-south route runs from Amsterdam to Belgium, using the A4–A13–A20–A16

TABLE 3.1   Traffic in the Rotterdam—The Hague Urban Area[*]

Metropolitan traffic	44%
Through traffic	11%
Incoming and outgoing traffic	45%

[*]As percent of the total number of kilometers driven in the area.

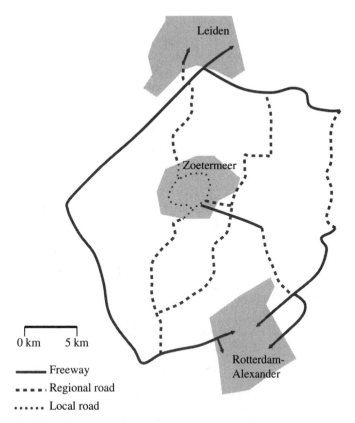

FIGURE 3.12 Routes between Rotterdam-Alexander and Leiden.

route. Second, the traffic flow to and from Rotterdam harbor in the west runs partly along the east-west axis to and from Germany, mainly using the A15 freeway. Other harbor-bound flows divert to the north (Amsterdam) and south (Belgium). Both flows of through traffic use the same roads as the metropolitan traffic, which leads to relatively low speeds, unreliable travel times, and chaotic traffic. This is especially the case on the routes passing the central urban areas of Rotterdam and The Hague, where the freeways are even used by strictly local traffic.

In case of incidents or calamities, only a few alternative routes are available, including the A44 (backup for the A4 north of The Hague) and the eastbound A20–A12 route (backup for the A15). The situation is expected to improve by around 2020, when two new freeway links are meant to be completed: the A4 section between The Hague and Rotterdam (as an alternative to the A13), and a new connection between the A13 and the A16, bypassing the A20 north of Rotterdam. Even then, no real alternatives are available for several crucial road segments—for instance, the A4 passing The Hague between Ypenburg (A13) and the Leidschendam (N14) intersection, or the A15 to the west of the Benelux tunnel (A4 intersection).

The most vulnerable links in the existing freeway network are shown in Figure 3.13.

Incoming and Outgoing Traffic. Incoming and outgoing traffic makes up a large part of the total traffic in the area, which should be no surprise if we consider the heavy population of the area, being bordered to the west by the North Sea shore. Because of the size of the area, a lot of incoming and

FIGURE 3.13 Vulnerable links.

outgoing traffic is traveling considerable distances through the metropolitan area. For example, outgoing traffic from The Hague to the south has to pass the freeways through the Rotterdam urban area first, and there encounters the same problems as through traffic. So incoming and outgoing traffic can be largely considered as through traffic, and the same analysis applies.

3.5.2 Building Blocks for a New Network Structure

The functional analysis may be best summarized as follows: The problem is not so much the congestion of freeways, but much more the fact that the metropolitan traffic almost exclusively depends upon a few freeway routes, and what's more, must share these routes with through traffic.

Therefore, we must look for the core solution in the *adaptation of the structure* of the whole road network in the Rotterdam—The Hague urban area, and in a way more compatible with the nature of mobility in the area. The main components of this new structure are:

- A separate, robust, and balanced network of "metropolitan arterial roads" (80 km/h), specifically targeted at metropolitan traffic that remains within the Rotterdam—The Hague area. This "metropolitan arterial network" consists of regional (through) traffic roads that are a hybrid between freeways and the underlying road network.

- A limited number of congestion-free freeways specifically targeted at through traffic (120 km/h). The through traffic would be physically separated from the metropolitan traffic (it would be "unbraided," so to speak). There would be only a limited number of locations where traffic can access or exit the through routes.

No Large-Scale Asphalting.   This new structure does not require large-scale asphalting: the overall amount of extra asphalt required for the *Robust Road Network Vision* is roughly the same as for the current transport policy plan, but the total traffic space is divided differently, i.e., a larger part of it is reserved for metropolitan arterial roads than is currently the case. Some existing freeways may even be "metropolitanized": converted into metropolitan arterial roads. Since this type of road is more space-efficient (narrower lanes, simpler intersections and junctions), the same amount of asphalt covers more lane kilometers.

Creating the new structure can best be characterized as a renovation operation, with minor additional new development. The following types of measures are needed:

- The "metropolitanization" of freeways in areas where these are mainly being used by metropolitan traffic
- The "unbraiding" of through routes, so that through traffic and metropolitan traffic no longer hinder each other
- The upgrading where necessary of underlying roads, so that they offer sufficient safety and capacity to be part of the metropolitan arterial network
- The restructuring of intersections and junctions so that they fit within the new structure
- The realization of several crucial additional connections, mainly within the metropolitan arterial structure
- Adding extra buffer space at smart locations to prevent congestion spillback and to control the inflow of traffic
- Creating extra facilities at specific locations for abnormal situations, e.g., removable barriers

Example Using the Rotterdam—The Hague Section.   We will now illustrate this approach using the Rotterdam—The Hague section as an example. Theoretically, a lane can handle 2,200 cars per hour. In order to be able to handle the traffic demand expected in 2020, an average of seven lanes will be needed per direction. This figure includes the limited extra growth in demand attributed to improved traffic performance; however, this growth will be slowed down as a result of the introduction of a pricing policy. These seven lanes can be divided into five lanes for metropolitan traffic, and two lanes for through traffic. For metropolitan traffic, as we have seen, a steady traffic flow is more important than a high-speed design. Through traffic however should be enabled to bypass urban centers without significant loss of speed.

But we need to build extra capacity into a robust network. We believe this can be done by assuming a "robust capacity" of 1,900 cars per lane per hour in our calculations—a figure at which the additional investment costs for more lanes are roughly in balance with the reduction of time loss through incidents (Immers and Bleukx 2008). This leads to a requirement for one extra lane in the Rotterdam—The Hague section. Figure 3.14 shows the total capacity needed in 2020, with a distinction made between metropolitan traffic, through traffic and a "robust spare capacity."

According to current policy plans, a total capacity of six lanes has been planned for 2020 for this section (see Figure 3.15): the current A13 (3 lanes) and N471 (one lane), plus the new A4 section between The Hague and Rotterdam which is yet to be built (two lanes). This means that, based on the theoretical capacity, we are short of one lane in the current policy plans, and based on the "robust capacity" we are even two lanes short. Moreover, the distribution of the lanes over road types does not match the traffic demand: five freeway lanes are being provided, whereas only two such lanes are needed for through traffic—it would suffice to size the others as metropolitan arterial roads.

When we redesign these road sections according to the principles of Robust Network Design discussed earlier, we offer a robust capacity of six lanes for metropolitan traffic that is equally

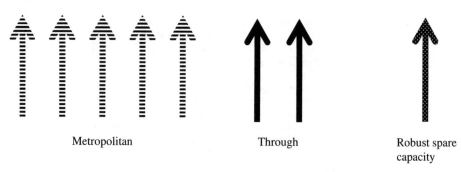

Metropolitan   Through   Robust spare
                                       capacity

FIGURE 3.14   Rotterdam—The Hague: number of lanes needed in 2020.

A4   A13   N471

FIGURE 3.15   Rotterdam—The Hague: lane provision in 2020 (current policy plan).

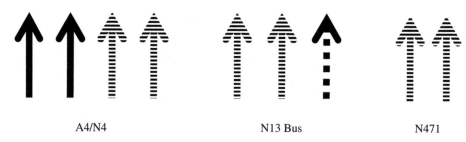

A4/N4   N13 Bus   N471

FIGURE 3.16   Rotterdam—The Hague: lane provision in 2020 (*Robust Road Network Vision*).

distributed across the three 80 km/h routes, the N4, the N13, and the N471.* Because of the equal distribution of the capacity, a reasonable processing of traffic is still possible, should an incident on one of the three routes lead to stoppages there. This means that the N471 needs to be expanded to two lanes (in each direction), but that the A13 can have one lane less in each direction, while the remaining two lanes could also be a little narrower. The space gained could be used for the realization of a bus transit lane beside the N13, leading to a more robust public transport network in this corridor (see Figure 3.16).

In addition to these six lanes (in total) for metropolitan traffic, we still need two lanes per direction for through traffic. These could be situated along either route, but rather not along the central

---

*Using N4 and N13 instead of A4 and A13 symbolizes the function shift of these roads from through traffic to metropolitan traffic: In the Dutch road numbering system A stands for freeway, N for regional road.

A13 route (which is the current through route), because this requires more turning off of through traffic (and consequently, complicated intersections) when approaching the urban areas of Rotterdam and The Hague. We have chosen to locate the lanes for through traffic next to the A4, since that complies best with current policy. It means that the A4/N4 would get a total of four lanes in each direction—two for metropolitan traffic (80 km/h) and two for through traffic (120 km/h). That is two more than the number projected in the current plans.

3.5.3 Description of a Robust Network Structure

Based upon these general principles and building blocks for a robust network structure, we can now design a new robust network structure for the Rotterdam—The Hague metropolitan region. In doing so, we make use of the theoretical framework, discussed in Volume I, Chapter 2 (Transport Network Planning: Methodology and Theoretical Notions).

Metropolitan Arterial Network.   To meet the needs of the large amount of traffic remaining inside the Rotterdam—The Hague metropolitan region, we propose to set up a separate metropolitan arterial network. This has a finer mesh than the current freeway network: the mesh size is around 5 km. For the most part, the new network consists of existing links: either existing (or already planned) freeway links, or existing regional roads. The difference is made by entirely redesigning those links: rather than the current mish-mash of road types and capacities, ranging from multi-lane freeways to single-lane regional roads, all routes in the metropolitan arterial network will have an 80 km/h speed limit and comparable capacities (usually 2×2 lanes). Because of the lower speed limit, road sections and junctions can be designed in a less space-consuming way, which is a big advantage in this heavily built-up area. Wherever it makes sense, this network can be combined with public transport lanes.

Metropolitan arterial road intersections can be a mix between (turbo) roundabout(s), fly-over(s) and/or traffic lights, depending on the situation. However, it is undesirable to have roundabouts in the main traffic flow at locations where local roads connect to this metropolitan arterial network. It is preferable to use simple fly-over solutions, or, if need be, traffic lights with a "green wave." The chance of congestion on the metropolitan arterial network will be lower than it is now because of the design-level inclusion of extra network capacity. If a traffic jam occurs, the intentional inclusion of buffer areas in the road design ensures that little or no spillback to other routes will take place any more. Figure 3.17 shows the proposed structure of the metropolitan arterial network.

Through Traffic Routes.   Two primary backbone routes have been planned for through traffic: a north-south one and an east-west one. Theoretically (see also Volume I, Chapter 2), through routes should *bypass* urban centers by some distance. In this region however, this is practically impossible. So we keep the existing routes through the area as they are, but in order to optimize traffic performance on these through routes, they will be physically separated from the metropolitan traffic when passing through the central urban areas of Rotterdam and The Hague. The access to the through route will be very limited—roughly one access ramp in every 10 km—whereas currently there is one every 2 km or even less in urban areas. Through routes consisting of two lanes in each direction are likely to be sufficient for the amount of through traffic. All through routes have a design speed of 120 km/h and no congestion in the normal situation. This is also achieved by offering buffer areas at the through route access and exit ramps.

In addition to these primary through routes, several secondary arterial routes have been planned that offer additional access and exit routes for traffic to and from the urban area. These routes will also function as a backup for through traffic. If further backup is needed, it can be offered through the metropolitan network, which will be supported by traffic management when used for through traffic.

FIGURE 3.17   Metropolitan arterial network.

North-South Through Route.   In principle, two solutions are conceivable for the primary north-south route (see Figure 3.18*a* and *b*):

1. Primary north-south route using all of the A4 (including sections that are currently still missing)
2. Primary north-south route using the A16 that will be joined to the A4 northeast of The Hague, which requires constructing a new connection from the A16-A20 intersection (Terbregseplein) to the A4 The Hague–Leiden

Figure 3.18*a* and *b* show the possible alternative routes using dotted and dashed lines. The extended A16 option offers benefits from a network structure viewpoint: the through route avoids the urban area, and what's more, there is space for a fully adequate alternative route that does not interfere with the primary route (by way of A44-N14-A4). However, creating a new A16-A4 link through an area that is already fully occupied by urban and recreational functions is expected to be a huge and expensive operation that would take a long time to decide upon. Thus the variant that fits better with current policy was chosen after all: a through route using all of the A4.

However, this implies a suboptimal solution for the secondary (alternative) route. The northbound secondary access route is the A44/N44, which is situated north of The Hague, to the west of the A4;

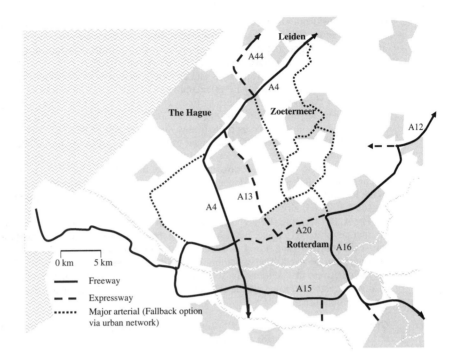

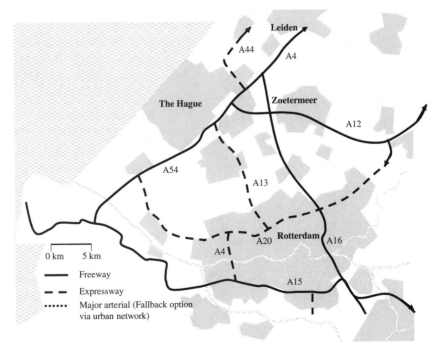

FIGURE 3.18   (*a*) Principal options for through routes, (*b*) principal options for through routes.

to the south of Rotterdam, the A16 (situated to the east of the A4) serves as the southbound secondary access route. These two secondary routes are satisfactory for incoming and outgoing traffic to and from The Hague and Rotterdam respectively. But in order to establish a backup for the entire north-south through route, we have to join these two main access roads together using several metropolitan arterial network routes. These backup routes need to be made suitable for processing through traffic in a reasonable way in case of calamities.

East-west through route. The A15 freeway (south of Rotterdam) acts as the primary east-west route, connecting the harbor with the hinterland. To serve also the northbound and southbound harbor traffic, the harbor access route (the westernmost part of the A15) has to be connected to the north-south through route. The southbound traffic simply makes use of the intersection with the A4 (to the southwest of Rotterdam), or with the A16 in the alternative solution discussed above. For northbound traffic, a new river crossing has to be made to the west of Rotterdam, in order to make a connection with the A4 The Hague—Amsterdam, bypassing the existing congested river crossings in the Rotterdam urban area. The optimal location of this new river crossing depends on the solution chosen for the primary north-south-route, see Figure 3.18.

The A20-A12 route (through the northern part of Rotterdam, eastbound to Utrecht) serves as an alternative for the A15 route; for destinations in the central and northeastern part of The Netherlands, this is the primary route. For the connection to the harbor, either of the above-mentioned new river crossings can be used. The westernmost crossing however, can be considered as the most robust solution, because this shortens the section of the harbor access road with no backup, and enables an additional eastbound route via the edge of The Hague (A4-A12) as an alternative of the A20 through the Rotterdam urban area.

Overview. The map (Figure 3.19) shows the details of the robust network; both metropolitan arterial roads and through roads are included.

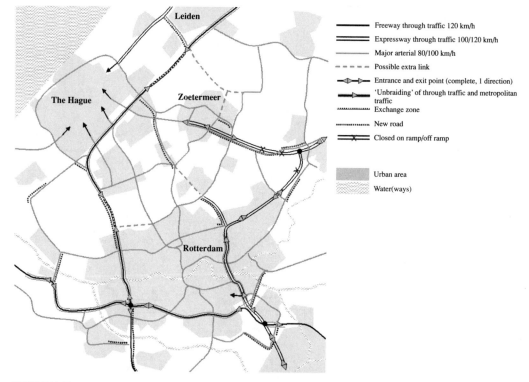

FIGURE 3.19 *Robust Road Network Vision:* overview map.

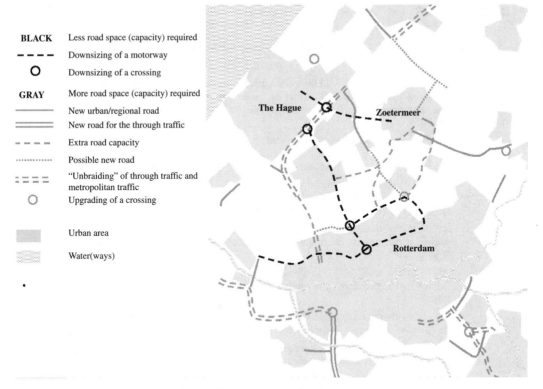

BLACK	Less road space (capacity) required
– – – –	Downsizing of a motorway
O	Downsizing of a crossing
GRAY	More road space (capacity) required
———	New urban/regional road
═══	New road for the through traffic
– – – –	Extra road capacity
··············	Possible new road
═ ═ ═	"Unbraiding" of through traffic and metropolitan traffic
O	Upgrading of a crossing
▓	Urban area
▒	Water(ways)
•	

FIGURE 3.20   Comparison *Robust Road Network Vision* to current policy plans.

Comparison to Current Policy Plans.   Figure 3.20 gives insight into the ways in which the *Robust Road Network Vision* suggests the way investments in the road network are planned, according to current policy, should be changed. Sometimes, the robust road network calls for more traffic space than the current plans (striped and dotted lines in grey). In other cases however, it actually uses less traffic space (indicated by striped lines in black). In these cases, it's not just about road segments (less and/or narrower lanes), but also about the restructuring of intersections (sharper curves, simpler design).

Combining both effects, the total amount of traffic space used in the *Robust Road Network Vision* turns out to be roughly the same as in current policy plans—only the distribution of traffic space is different, and apparently more efficient.

3.5.4 Impact Assessment

To enable a comparison between the robust road network and the current policy plans for 2020, an impact assessment has been made of both situations using the dynamic assignment model INDY (Bliemer 2005, 2007; Yperman 2007), a computer model that includes both pricing policies and all currently planned projects. The results of the model calculations show that the robust road network not only reduces the average travel time in the regular rush hour by 2.3 percent but also reduces the distance traveled by 0.8 percent through the elimination of detours.

However, the robust network is especially better equipped for handling incidents—which is what it was designed for: additional travel time due to incidents is almost 30 percent lower. The modeling also shows that a large-scale application of buffers can reduce total network travel times by up to 12 percent.

3.6 CONCLUSIONS

Based on this chapter's analyses, network elaborations, and model calculations, the following conclusions can be drawn:

- As further extension of road network capacity lags behind the expected growth pace of mobility, density of traffic flows will further increase. Consequently network vulnerability will increase. This increasing network vulnerability will generate a serious (negative) impact on travel-time reliability.
- Travel-time reliability can be improved by increasing the robustness of the network (apart from other measures, e.g., demand-oriented measures). Network robustness is defined as the extent to which a network is able to maintain the function that it was originally designed for and this under all kinds of circumstances.
- The robustness of a network can be assessed on the micro level (individual links and intersections) and on the macro level (network structures).
- On the micro level, network robustness can be improved by tracking the most vulnerable road sections (intersections) accompanied by measures aiming at further reducing the vulnerability of these sections.
- On the macro level, network robustness can be assessed on the basis of five (network-related) characteristics: prevention, redundancy, compartmentalization/partitioning, resilience, and flexibility.
- The above characteristics are further elaborated into network design principles that allow us to improve the robustness of the network.
- Model calculations show that applying the introduced network design principles generates a significant improvement (increase) of travel-time reliability, this despite the limited extension of network capacity.

ACKNOWLEDGMENTS

We would like to thank James Stada, Rob van Nes, Jeroen Schrijver, and Tariq van Rooijen for their inspiring contribution in writing this chapter.

REFERENCES

Bar-Gera, H., and S. Ahn. 2010. "Empirical Macroscopic Evaluation of Freeway Merge-Ratios." *Transportation Research Part C: Emerging Technologies,* Vol. 18(4), pp. 457–470.

Berdica, K. 2001. "Vulnerability—A Model-Based Case Study of the Road network in the City of Stockholm." In Papers Presented at the 1st International Symposium on Transportation Network Reliability (INSTR), Kyoto, Japan.

Berdica, K. 2002. An Introduction to Road Vulnerability: What Has Been Done, Is Done and Should Be Done. *Transport Policy* 117–127.

Billington, R. and R. Allan. 1992. *Reliability Evaluation of Engineering Systems: Concepts & Techniques.* New York: Plenum Press.

Bliemer, M. C. J., 2005. "INDY 2.0 Model Specifications." Delft University of Technology working report, The Netherlands, 2005.

Bliemer, M. J. C. 2007. "Dynamic Queuing and Spillback in an Analytical Multiclass Dynamic Network Loading Model." *Transportation Research Record: Journal of the Transportation Research Board, No. 2029,* Transportation Research Board of the National Academies, Washington D.C.

Cassir, C., and M. G. H. Bell. 2000. "The N + M Person Game Approach to Network Reliability." In Bell, M. G. H. and C. Cassir. Eds., *Reliability of Transport Networks* (pp. 91–102). Baldock, Herthordshire, England: Research Studies Press Ltd.

Chen, A. H., Lo Yang, W. H. Tang. 2002. Capacity Reliability of a Road Network; An Assessment Methodology and Numerical Results. *Transportation Research Part B*. (36):225–252.

Egeter, B., T. Vonk, G. Tertoolen, T. Van Rooijen, and L. H. Immers. 2007. (In Dutch) ARKO: Architectuur. Wegenknooppunten (ARKO, Architecture for Road Network Intersection Design), TNO Mobiliteit en Logistiek, Delft, report nr. 034.65169.

Immers, L. H., J. E. Stada, and I. Yperman. 2004. Robustness and Resilience of Transportation Networks; Problem Survey and Examples. Presented at NECTAR Cluster Meeting on Reliability of Networks, Amsterdam.

Immers, L. H., J. Stada, I. Yperman, and A. Bleukx. 2005. "Towards Robust Road Network Structures." *Slovak Journal of Civil Engineering*. (XII):2004/4, 10–17.

Immers, L. H., and A Bleukx. 2008. "On the Robustness of Transportation Network Structures." In: *Network Strategies in Europe: Developing the Future for Transport and ICT*. P. Nijkamp and M Giaoutzi (Eds.). Ashgate, 37–50.

Meeuwissen, A. 2002. Exchange Principle of Congestion Cost and Road Infrastructure Cost, TNO Internal report, Delft.

Nicholson, A. J., Schmöker, J. D., Bell, M. G. H. and Y. Iida, 2003. "Assessing Transport Reliability: Malevolence and User Knowledge." In: Bell, M. G. H.; Iida, Y. (Eds.), The Network Reliability of Transport. 1-22. Oxford, UK: Elsevier Science.

Schrijver, J. M., B. Egeter, L. H. Immers, and M. Snelder. 2008. (In Dutch) Visie Robuust Wegennet (Vision on a Robust Road Network). TNO Mobiliteit en Logistiek, Delft, TNO-report nr. 034.75210.

Snelder, M., J. Schrijver, R. Landman, J. Mak, and M. Minderhoud. 2008. (In Dutch) De kwetsbaarheid van Randstedelijke vervoernetwerken uit verkeerskundig perspectief, (The Vulnerability of Road Networks in Randstad, The Netherlands, looked at in a Transport Related Perspective). TNO-report 2008-D-R0882/A, Delft.

Tamminga, G. F., J. C. Matton, R. Poorterman, and J. Zee. 2005. (In Dutch) De Robuustheidscanner. Robuustheid van Netwerken: Een Modelmatige Verkenning. (The Robustness Scanner: Robustness of Networks: A Model Based Exploration). Report I&M-99366053-GT/mk Grontmij, The Netherlands.

Yperman, I. 2007. "The Link Transmission Model for Dynamic Network Loading." PhD-thesis, University of Leuven, Faculty of Applied Sciences, Department of Traffic and Infrastructure, Leuven, U.D.C. 656.021.

Wakabayashi, H., and Y. Iida. 1992. "Upper and Lower Bounds of Terminal Reliability of Road Networks: An Efficient Method with Boolean Algebra." *Journal of Natural Disaster Science*. (14):29–44.

CHAPTER 4
TRANSPORTATION SYSTEMS MODELING AND EVALUATION

Andrew P. Tarko
Purdue University, School of Civil Engineering
West Lafayette, Indiana

Panagiotis Ch. Anastasopoulos
Purdue University, School of Civil Engineering
West Lafayette, Indiana

4.1 INTRODUCTION

Transportation includes its infrastructure, administration, vehicles, and users and can be viewed from its various aspects, which include engineering, economics, and societal issues. A *transportation system* can be scoped narrowly as a single driver/vehicle with its second-by-second interactions with the road and other vehicles. A system also can be defined broadly as a regional transportation infrastructure with its year-by-year interactions with the regional economy, the community of transportation users and owners, and its control components, such as transportation administration and legislature. These two extremes show the range of transportation systems, with a variety of possible intermediate scenarios.

Transportation models are a formal description of the relationships between transportation system components and their operations. Knowledge of these relationships allows for estimating or predicting unknown quantities (outputs) from quantities that are known (inputs). As our knowledge of possible transportation relationships is limited, transportation models are subsequently imperfect and selective. Awareness of the models' limitations facilitates using the models according to the need, the required accuracy, and the budget.

Evaluation has two distinct meanings: *calculate approximately* and *form an opinion about*. Indeed, both the meanings are reflected by two basic steps of transportation systems evaluation:

1. *Quantify* by applying a model
2. *Qualify* by applying evaluation criteria

The first step requires a valid model while the second step uses preferences of decision-makers and transportation users. Modeling, in most cases, is a required part of transportation systems evaluation.

A transportation model is a simplification of transportation reality. It focuses only on what is essential at the level of detail appropriate for its application. If one wants to improve traffic at a specific location by redesigning signals, then optimal signal settings is the solution. This component has negligible economic effect on the regional economy and therefore should not be considered in the model. The situation changes if one wants to program transportation improvements in the region that must compete with large-scale highway projects for funding. Then, the economic impact of the decision is important and detail signal settings are not considered; instead, the overall effect of the typical control is represented in the analysis. These two cases require two distinct models that differ in scope and detail. A specific job requires a specific model. Understanding the basics of modeling in transportation engineering is helpful in selecting an adequate model, using it properly, and interpreting the results correctly.

This chapter aims to help: (1) decide whether a model is needed, (2) select an adequate model, and (3) use it effectively. The reader will find neither endorsements nor a complete overview of the existing modeling software packages, and specific references are mentioned for illustration of the points raised in the presentation without any intention to either compliment or criticize.

Although this chapter has been written with all the areas of transportation engineering in mind, examples are taken from surface transportation, which is the authors' area of expertise. The authors believe that this focus does not constrain the generality of the chapter.

4.2 TYPES OF TRANSPORTATION MODELS AND MODELING PARADIGMS

Classification of transportation models is challenging because there is a wide variety of transportation models and a considerable number of ways the models can be categorized. We will try to classify the models in a way that helps transportation engineers select models adequate to the job.

Transportation models are applied to individual highway facilities, groups of facilities, and to entire transportation systems at the city, state, and national levels. Transportation models are also applied to time horizons, ranging from the present to 20 and more years ahead. Depending on the use and scope, transportation models can focus on long-term prediction of demand for various transportation modes with adequately incorporated economic impacts; can focus on routing and scheduling using choice models with properly represented connectivity between various network components; or can focus on a faithful representation of traffic flows at various transportation facilities. From this perspective, transportation models can be divided into the following categories:

1. *Transportation demand models* (econometric analyses, demand generation, trips distribution, etc.)
2. *Transportation network models* (interaction between flows, modal split, traffic assignment, etc.)
3. *Traffic flow models* (traffic control, interaction between vehicles, delays, traffic perception, safety performance, etc.)
4. *Transportation and traffic impact models* (fuel consumption, air-pollution generation, noise generation, etc.)

4.2.1 Transportation Demand Models

Transportation demand models forecast the number of trips at origins, their split between destinations, and the way of travel (modes and routes) within a transportation network. The most widely used approach to model transportation demand is a sequential four-step process (see Mannering, Washburn, and Kilareski 2009) that involves the following components: trip generation, choice of mode of transportation, choice of destination with the use of the gravity model (Meyer and Miller 2001), and choice of route (also referred as traffic assignment). More advanced models combine some of these steps to better reflect the travelers' decision making.

Transportation demand depends strongly on the regional economy as the majority of trips are work- or business-related. On the other hand, economic growth in the area can be stimulated by the business-generating ability of transportation infrastructure or hindered by excessive transportation costs. *Econometric models* attempt to grasp these complex impacts through simultaneous statistical equations that represent the relationships between transportation, regional economy, land use, regional policies (laws, zoning, pricing, subsidies, etc.), and travel preferences of the residents (an example can be found in Johnston et al. 2001). These relationships are developed from the historical data for the region and then applied to predict future transportation demands (see for example Washington,

Karlaftis, and Mannering 2003). These models are typically highly aggregated and give total annual or daily numbers of trips between zones by various transportation modes. Although commercial packages provide a necessary modeling suite, models are developed to properly reflect the local conditions, the objective of the analysis, and the available data. The set of equations are developed on a case-by-case basis.

Econometric models are the "high end" of demand modeling while *extrapolation methods,* including the trends of the growth rate method, are the "low end." According to the growth rate method, future volumes are predicted by multiplying the current volumes with the growth factors (growth method for highways is mentioned in Robertson 2000). Separate growth factors are developed for different regions, transportation modes, and types of transportation facilities (road category, airport category, etc.). This method is not able to reflect unusual changes either in the economy or in the land use.

The third category of demand models is *traffic generation models* (for example, see Meyer and Miller 2001). Traffic generation models require knowledge of the future land use and household characteristics because they link the number of trip ends in a traffic zone with particular zonal characteristics, such as the number of workplaces and the number of households. These models include the land use and the economic impact through the zonal characteristics required as an input.

Once the generated traffic is known, the traveler mode and destination can be determined with a mode-destination choice model (typically logit; see Washington, Karlaftis, and Mannering 2003) and the use of the decision-making theories on utility maximization (Simon 1959). Input required for this task involves travelers' mode-destination preferences, traveler-specific information (e.g., income, occupation, etc.), and mode-destination specifics (e.g., travel time of mode, destination's commercial floor space, etc.). With the demand and mode-destination choice information in hand, the final step is traffic assignment in which mathematical programming is typically utilized (see Patriksson 1994). In a more traditional approach, the destination split can also be modeled with the mentioned gravity model (Meyer and Miller 2001).

4.2.2 Transportation Network Models

Econometric models can and should consider all major transportation modes (air, rail, highway, pipes, and water). On the other hand, the network models have typically been developed to deal with certain transportation modes. The interactions between the transportation modes presently are attracting the interest of researchers who are trying to incorporate transportation intermodalism into regional planning. The primary focus of highway network modeling is the connectivity of the transportation infrastructure and on the travelers' path choices across the network and of the transportation modes (for example, highways with public transit routes). Network models require travel demand to be known as a table of one-way flows between all possible pairs of network nodes or traffic zones. Modeling frameworks for travel choices include deterministic user equilibrium, stochastic user equilibrium, system optimum performance, and mixed flows (Sheffi 1985). Some models consider static equilibrium applied to hourly or daily network operations, while other models assign travelers to paths dynamically in short time intervals (Peeta and Ziliaskopoulos 2001). The latter, which are considered to be the state-of-the-art, use simulation models combined with real-time traffic and origin-destination information to predict the effects of various management strategies, thus allowing more effective management and providing better traffic information than is currently possible, In other words, instead of reacting to existing conditions, this feature enables dynamic control and management systems to anticipate problems before they occur. To that end, there are two software systems for online traffic prediction, DynaMIT-R and DYNASMART-X, and two for offline traffic operations planning, DynaMIT-P and DYNASMART-P (for DynaMIT see Ben-Akiva et al. 1997; for DYNASMART see Mahmassani et al. 1995). All four have been co-developed and tested by the Federal Highway Administration (FHWA). The online systems are field tested at traffic management centers for online traffic management, whereas the offline systems are used in several region-wide traffic management projects.

Another large class of network models includes logistic models that optimize transportation of commodities. These models are focused on routing commodities across a multimodal transportation network to minimize transportation costs and meet time constraints. Logistic models are of particular interest to private transportation companies and to large manufacturers that use supply-chain analysis to reduce transportation and storage costs. A good introduction to the logistics modeling and practice can be found in Fredendall and Hill (2000) and Greene (2007).

4.2.3 Traffic Flow Models

With a few exceptions, network models represent transportation facilities in a relatively simplified manner through the analytical relationships between traffic demand and travel cost components; for example, the traditional BPR function (Bureau of Public Roads 1964). These simplifications do not allow for designing facilities and may raise concerns about the accuracy of the simplified demand-cost relationships. To allow design and traffic management applications through detail analyses of traffic operations at these facilities, a group of models is available that we call here "traffic models." They incorporate the impact of facility geometric characteristics and traffic control on speed, delay, travel time, queue, etc. One of the best-known depositories of traffic models for design and traffic analysis is the *Highway Capacity Manual* (Transportation Research Board 2000). These models are transportation facility-specific and require knowledge of the traffic demand, geometry, and control characteristics of the facility.

A typical microscopic, time-step, and behavior-based simulation traffic model is VISSIM (brief for Verkehr In Städten–SIMulationsmodell, which in German means "simulation model for urban traffic"). The model was developed in 1992 by PTV AG to analyze the full range of functionally classified roadways (including roundabouts, railroad crossings, and toll plazas) and transit operations. It is capable of modeling traffic with various traffic control measures in a three-dimensional environment and to assist in comparing different alternates in designing roundabouts, at-grade intersections, and high-type traffic interchanges. VISSIM goes beyond a single road facility, which places it also in the category of network models. It can model integrated roadway networks found in a typical corridor as well as various modes consisting of general-purpose traffic, buses, high-occupancy vehicles, rail (heavy or light), trucks, pedestrians, bicyclists, etc. (Fellendorf 1994). VISSIM also can model variable message signs, ramp metering, incident diversion, transit signal priority, and dynamic lane control signs, among others (for examples, see Gomes, May, and Horowitz 2004; Park and Schneeberger 2003).

4.2.4 Transportation and Traffic Impact Models

Impact models estimate or predict the impact of a transportation system or its components on the affected population, economy, and environment. The impact measures (costs, noise, pollution, and safety) are estimated based on the transportation and traffic characteristics, including transportation demand, traffic flow, speed, traffic density, travel time, delay, etc. The following models are examples of widely known and used impact models:

- Models for evaluating and assessing the safety impacts of geometric design decisions, such as the interactive highway safety design model IHSDM (FHWA 2003)
- Models for economic efficiency analysis of transportation alternatives/investments, such as MicroBenCost (McFarland, Memmott, and Chui 1993b), FHWA's STEAM, IMPACTS, SPASM and HERS (Cambridge Systematics 2000; FHWA 2002), the highway development management model HDM (University of Birmingham 2005), and California's Cal-B/C model (Booz Allen & Hamilton Inc. 1999)
- Models of air dispersion, such as the hybrid roadway model HYROAD (System Application International 2002), the atmospheric dispersion modeling system ADMS-3 (Carruthers et al. 1994), and the California line source dispersion model CALINE4 (Benson 1988)
- Models of vehicle emission, such as MOBILE (Jack Faucett Associates 1994)
- The mobile source emission factor model EPA MOBILE6 (USEPA 2002)
- Emission models based on vehicle operating modes, such as the mobile emissions assessment system for urban and regional evaluation MEASURE (Guenslar, Washington, and Bachman 1998)
- Microscopic emission models, such as the transportation analysis and simulation system TRANSIMS (Smith et al. 1995), the traffic simulation and dynamic assignment model INTEGRATION (USEPA 1998; Rouphail et al. 2001), and FHWA's TRAF-NETSIM (Rathi and Santiago 1990)

- Fuel-based emission models, such as the SYNCHRO traffic model (Rouphail et al. 2001)
- Greenhouse gas emission models, such as USEPA's score-based model for estimating greenhouse gas emission amounts (USEPA 2006)
- Models for noise analysis and cost-effectiveness evaluation of alternative noise barrier designs, such as the traffic noise mode TNM (Menge et al. 1998)
- Models of aircraft noise generation and propagation, such as the INM (Volpe Center 2002), the NOISEMAP computer model (Moulton 1990), and the Federal Aviation Administration's Airport Noise Compatibility Planning Toolkit (FAA 2006)
- Models to address the functional features of wetland ecosystems, such as the wetland functional analysis WET II (Adamus et al. 1991), the hydro-geomorphic classification method HGM (Evink 2002), and the habitat evaluation procedures software HEP (Schamberger and Krohn 1982)
- Models that assess the water quality impacts of transportation activities (see Sinha and Labi 2007)

A close relative of traffic impact is traffic safety modeling. Safety models are typically econometric models and are particularly designed to depict the traffic safety impact of critical traffic, geometry, and socioeconomic characteristics. There are two broad categories of safety models: those that look into crashes (e.g., crash frequencies), and those that investigate injury severities. Typically, crash frequency safety models use count data modeling techniques, such as Poisson and Negative Binomial (see Venugopal and Tarko 2000) and their derivatives (see Anastasopoulos and Mannering 2009). However, recently, Anastasopoulos, Tarko, and Mannering (2008) investigated crash rates instead of frequencies, using a tobit model. Injury severity models typically utilize discrete choice modeling techniques, such as logit/probit models (see Savolainen and Mannering 2007). The first group of models (crash frequencies and rates) requires volume and geometry information as input, whereas the second (injury severities) requires driver, vehicle, and crash-specific information.

4.2.5 Models Integration

Figure 4.1 depicts the traditional modeling paradigm used in transportation engineering. Modeling is carried out in three separate modeling phases: demand modeling, network modeling, and traffic flow modeling. Each of the modeling phases provides input to performance estimation and to impact

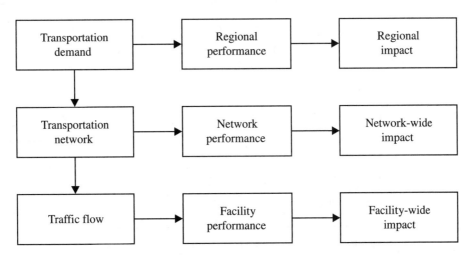

FIGURE 4.1 Traditional transportation modeling paradigm.

models. The outcome of the demand modeling phase is used to model transportation and traffic assignment in the transportation networks, while the outcome of the network modeling provides necessary input to the traffic flow facility modeling of transportation facilities. Each phase can be performed by a different transportation engineering unit and the results can be transferred from one unit to another.

The traditional modeling paradigm was proven to provide a manageable modeling framework using data that are available and providing computational demands that are reasonable, which was accomplished in each phase through adequate definition and representation of the transportation systems. Demand models represent the transportation and economy of the studied region with synthetic characteristics of subregions. Network models must be more detailed and present all of the important components of the transportation infrastructure; however, they represent only part of the regional transportation infrastructure. Although they require more data, the task is distributed among multiple planning organizations. The geometric design and operational analysis of a transportation facility requires the most detailed information, but such an analysis is performed only for selected facilities. In the traditional modeling paradigm, the outputs from higher models are used and are inputs to lower models, with no mutual interaction.

Traditional transportation modeling has three phases, with a gradually growing level of detail and a narrowing geographical scope. The phases are executed sequentially and the outcome from the previous phase feeds the next phase. This modeling approach cannot properly treat many Intelligent Transportation Systems (ITS) where traffic operations are important component of the ITS representation, while many strategic decisions about the system are made at the system level in real-time. These decisions may even influence demand for traveling. A need for modeling such systems requires a new modeling paradigm where demand, network, and traffic are modeled simultaneously and dynamically in an integrated single phase. This paradigm is presented in Figure 4.2. The full interaction between different levels is obtained by adding feedback between the layers or developing models that truly integrate all the components into one model. It should be noted that the type of interaction between the demand and the network layers in traditional transportation management is different than in the ITS-based dynamic management. In the traditional framework, the demand layer is used to predict the aggregate average demand based on high-level policy and planning decisions while the changes in disaggregate demand in ITS-based management are short-term and are caused by ITS operations.

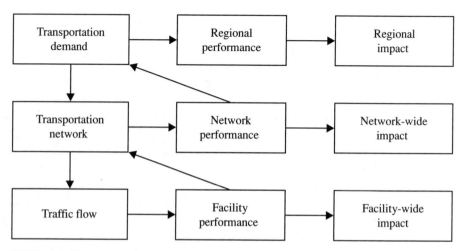

FIGURE 4.2 New transportation modeling paradigm.

4.3 EXAMPLE STUDIES AND MODELS

The number of available models and their computer versions is impressive. Most of the existing computer software for transportation modeling and evaluation include multiple model types integrated in packages, almost all of which include performance models to produce results useful for evaluation and decision making. Some of these packages integrate models and various levels of detail within the new modeling paradigm. This section presents selected models and studies to illustrate the model types and their applications.

4.3.1 Maine Statewide Travel Demand Model

The Maine DOT funded development of a modeling procedure to predict future travel demand in Maine (U.S. Department of Transportation 1998). Future travel demand was predicted based on future population and other socioeconomic data for the traffic zones of the region. A gravity model was used to split the trip ends in a zone among all the zones. The traffic zones were defined inside and outside the study area to enable modeling internal trips with both ends inside the study area and external trips with one or both ends outside the study area. The model used current and assumed future characteristics of the transportation infrastructure. The model was calibrated for current data and demand predicted by changing the model inputs that represented the future. This model is an example of a traditional approach to long-term modeling of transportation demand where knowledge of the future land use is mandatory. The Maine study is a good example of the use of GIS technologies to code data and visualize results.

4.3.2 California Statewide Models

The California Statewide Model was developed at Caltrans using the TP+/VIPER software, GIS, and travel demand modeling (see Jones 1999). Caltrans used the model to provide information on statewide and regional trends in transportation demand, supply, and operations. Its key features involve a nested logit mode choice model to forecast intercity air, rail, and bus travel and trip distribution procedures to forecast typical urban trips and long distance travel.

In that context, the San Francisco Bay Area Metropolitan Transportation Commission (MTC) conducted project evaluation, cost estimating and travel demand forecasting for the 2005 Regional Transportation Plan, utilizing the MTC model (TP+). This model combines all of the base and future road and transit network alternatives into a master network database integrated with project listing databases. Auxiliary lanes, ramp metering, ferries, ITS, and bus rapid transit (BRT) can be evaluated, and detailed performance and cost-benefit measures can be calculated at a system-wide level and for individual corridors.

Also, the City of Pleasanton uses a detailed TransCAD model to evaluate land use changes and innovative traffic management policies, such as ramp and arterial metering. School trips are distributed based on specific school district boundaries, peak hour spreading considers congestion between each origin and destination, and most importantly, the TransCAD model is linked directly to the City's Synchro/SimTraffic network. Furthermore, for the NCHRP 25-21 Air Quality Effects program, a new transportation modeling process was developed for forecasting the long and short-term effects of transportation capacity improvements on mobile source emissions for air quality analysis. This process involved the development of custom software to implement and enhance the Portland activity-based modeling system and apply it to the existing databases.

Finally, a travel demand model was developed for the City of Mountain View with a user interface fully implemented in the existing software system (Cube). The Mountain View model uses a detailed local land use database and traffic generation rates based on ITE and combines the local information with person-trip and mode choice inputs from the Santa Clara County and regional models.

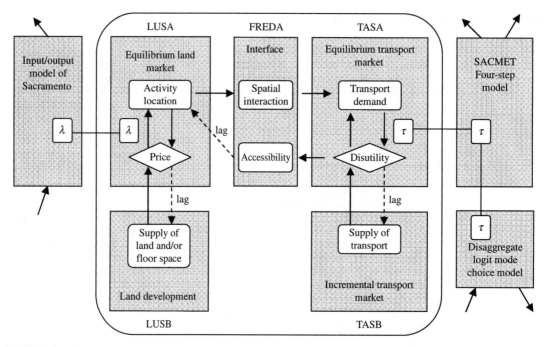

FIGURE 4.3 MEPLAN and other components of Sacramento model with interactions (based on Johnson et al. 2001).

4.3.3 Sacramento Area Travel Demand Model

The Sacramento area was modeled with econometric equations in the MEPLAN modeling frame-work (Johnston et al. 2001). The model includes the interaction between two parallel markets—the land market and the transportation market, as illustrated in Figure 4.3. In the land market, price and cost affect decisions about the location of various activities. In the transportation market, the costs of travel affect both mode and route selection. The interaction between the two markets was accomplished by including transportation costs in the attractiveness of zones for producing activities and including the producing activities in the traffic demand factors. The generated trips were split by mode choice and assigned to routes based on stochastic user equilibrium with capacity restraint. The quasi-dynamic modeling was executed in time steps with a time lag in the interaction between the two markets. This approach is an example of advanced modeling of the two-way impact between land use and transportation where both the land use and the transportation demand are predicted.

4.3.4 ITS Evaluation for the Washington, DC, Metropolitan Area

In this study, the traffic flows for current and future years without ITS were predicted with traditional four-step demand and network modeling (Schintler and Farooque 2001). Several combinations of two sets of scenarios were examined for a selected portion of the arterial network in Washington, DC:

1. Present traffic and future traffic
2. Presence and absence of the *SmarTraveler* system

The traffic was simulated using a mesoscopic simulation package INTEGRATION for integrated arterials and freeways (Van Aerde et al., 2005). INTEGRATION models represent individual vehicles on freeways and on urban streets with traffic signals and ITS, while preserving the macroscopic properties of traffic flows. The program uses dynamic queuing-based traffic assignment, driver diversion, and rerouting during congested conditions. Travelers may behave according to their current experience or in response to information received via the Internet, from Variable Message Signs, or from Highway Advisory Radio Stations. INTEGRATION models use Advanced Transportation Management Systems with real-time surveillance devices and adaptive traffic signal systems. This study is a good example of integrating the traditional planning modeling with computer simulation where the three model types were integrated in a dynamic fashion to allow short-term demand adjustments to traffic conditions, network modeling, and rather sophisticated traffic modeling at the facility level.

4.3.5 Network Analysis: Dynamic Traffic Management Systems

In 1998, two traffic estimation and prediction system prototypes, DYNASMART (see Hawas et al. 1997) and DynaMIT (see Ben-Akiva et al. 1998), were developed by the University of Texas at Austin and the Massachusetts Institute of Technology (MIT), respectively.

DYNASMART is a real-time computer system for traffic estimation and prediction that supports both transportation management systems and advanced traveler information system (ATIS). It continuously interacts with multiple sources of real-time information, such as loop detectors, roadside sensors, and vehicle probes and integrates this information with its own model-based representation of the network traffic condition. DYNASMART provides reliable estimates of the traffic conditions of the network, predictions of network flow patterns and travel times in response to various contemplated traffic control measures and information dissemination strategies, and routing information to guide trip-makers in their travel. The DYNAMSMART modeling features include, among others, a simulation-based dynamic traffic assignment system with micro-simulation of individual user decisions, as well as a mesoscopic traffic flow simulation approach. The system fully integrates transportation management systems and advanced traveler information systems (ATIS) under different operational scenarios and deployment levels and provides guidance information and control actions that are robust under various operational conditions. Applications of the system include, among others, off-line design and on-line generation and evaluation of operational strategies (including real-time signal control strategies, coordination schemes along arterials, and path-based coordination schemes), provision of network status information to private and public advanced traveler information system service providers, and determination of optimal congestion pricing schemes that vary with location, time, and prevailing network state.

DynaMIT is another real-time computer system for traffic estimation, prediction, and generation of traveler information and route guidance. It supports the operation of traffic management systems and advanced traveler information systems (ATIS) at traffic management centers (TMC). The system can provide real-time estimation of network conditions, rolling-horizon predictions of network conditions in response to traffic control measures and information dissemination strategies, and traffic information and route guidance for roadway users. Its features include demand simulation using a micro-simulator (the latter generates individual travelers and simulates their choices regarding whether or not to travel, departure time, mode, and route in response to advanced traveler information), simulation of different vehicle types and driver behaviors, and distinction between informed and uninformed drivers. DynaMIT's applications include generation of unbiased and consistent information to drivers, efficient operation of Variable Message Signs (VMS), off-line evaluation of real-time incident management strategies, and real-time incident management and control.

4.3.6 Highway Capacity Manual

The Highway Capacity Manual is a leading reference for practicing highway transportation engineers who seek guidance in capacity and traffic quality modeling and evaluation for planning, design, and traffic operations applications (Transportation Research Board 2000). The Highway Capacity Manual focuses on traffic operations at various facilities, including freeway segments and ramp intersections; multilane and two-lane rural roads; urban streets; signalized and unsignalized intersections; and pedestrian, bike, and public transit facilities. The Highway Capacity Manual includes performance models that convert traffic measures of effectiveness into levels of service that have become the standard procedure to qualify traffic performance and is an example of traffic models combined with performance models. The models are analytical and are implemented dependably on a computer (McTrans 2010). The models neither predict traffic demand nor traffic diversion between routes. Traffic volume is among the required inputs and must be predicted with demand and network models.

4.3.7 Interactive Highway Safety Design Model

The Interactive Highway Safety Design Model (IHSDM; FHWA 2003) is road safety evaluation software being developed for highway planners and designers (visit http://www.fhwa.dot.gov/ihsdm/). It already includes two-lane rural highways and is ultimately intended to include multilane rural highways. IHSDM will consist of several components, including crash prediction and traffic analysis modules.

The crash prediction module will predict the expected number of crashes along the road segment given the basic geometric characteristics of the segment and the traffic intensity (Harwood et al. 2000). The crash prediction models take the form of equations and are sometimes called safety performance functions.

The objective of the traffic analysis module is to provide practitioners with a tool to investigate the operational effects of alternative road designs, marking schemes, traffic controls, and vehicle size and performance characteristics for current and projected future traffic flow (Wade et al. 2000). TWOPAS, a traffic simulation model for two-lane rural highways, forms the basis for this module. The TWOPAS model simulates traffic operations on two-lane rural highways by reviewing the location, speed, and acceleration of each individual vehicle on the simulated road every second and advancing those vehicles along the road in a realistic manner. The model takes into account the effects of road geometry, driver characteristics and preferences, vehicle size and performance characteristics, and the presence of oncoming and same direction vehicles that are in sight at any given time.

4.3.8 Micro-Simulation for Regional Planning: TRANSIMS

The Los Alamos National Laboratory is the leader in the development of a model called the Transportation Analysis and Simulation System or TRANSIMS (Los Alamos National Laboratory 2002). TRANSIMS consists of integrated micro-simulation forecasting models designed to give transportation planners information on traffic impacts, congestion, and pollution in a metropolitan region. TRANSIMS models travelers' activities and the trips they take to carry out their activities and then builds a model of household and activity demand. Trips are planned to satisfy the individual's activity patterns. TRANSIMS simulates the movement of individuals across the transportation network, including their use of vehicles such as cars or buses, on a second-by-second basis. The model tries to forecast how changes in transportation policy or infrastructure might affect those activities and trips. The interactions of individual vehicles produce traffic dynamics from which the overall performance measures are estimated.

TRANSIMS is a courageous attempt to introduce detail-level modeling to regional planning. It is a departure from the widely accepted modeling framework presented in Figures 4.1 and 4.2 as the model represents traffic in the entire region at the level of individual travelers. To cope with the huge amount of data and computational demand, vehicles move in a simplified manner by jumping between road "cells." This modeling attempt may answer the question of whether data aggregation and representation simplicity in planning can be replaced with detail models that are traditionally applied to study traffic operations.

4.3.9 Benefit/Cost Analysis: MicroBENCOST

MicroBENCOST constitutes a comprehensive and convenient framework for conducting highway user benefit-cost analysis. It is designed to analyze different types of highway improvement projects in a corridor. Benefits are calculated for existing and induced traffic, as well as for diverted traffic in the presence of an alternative route or when a bypass project is evaluated. The program allocates traffic between the highway being evaluated and the alternative routes, if present, equalizing marginal user costs on alternative routes for hour-by-hour traffic volumes (McFarland et al. 1993a).

Vehicle delays and operating costs may be estimated for up to nine passenger vehicle types and up to nine truck types. User time savings vary with the class of highway and the type of improvement being considered. The program incorporates speed and volume-capacity-ratio relationships for rural highways based on the *Highway Capacity Manual*, interchange and intersection delay relationships, railroad grade crossing delays, and incident and work zone delays based on simple deterministic queuing concepts. It evaluates several types of highway projects, such as capacity enhancement, bypass construction, intersection or interchange improvement, pavement rehabilitation, bridge improvement, highway safety improvement, and railroad grade crossing improvement.

MicroBENCOST has been utilized in many studies. For example, Daniels and Stockton (2000) utilized MicroBENCOST to evaluate the cost-effectiveness of high-occupancy vehicle (HOV) lanes. The study used 10 years of HOV lane data in Texas to estimate benefit-cost ratios for all HOV lanes in the state. Using actual cost data, benefit estimates were compared with HCM calculated benefits from the program. HOV lanes were assessed as both a stand-alone project and as an alternative to two additional general-purpose freeway lanes. The results showed that the potential benefit of either alternative varies significantly by corridor. A major finding of the study is that the role and effectiveness of the HOV lane varied significantly by the type of the lane (i.e., concurrent, reversible, contraflow) as the total corridor traffic increased.

4.3.10 FHWA Traffic Noise Model

The Federal Highway Administration (FHWA) developed the FHWA Traffic Noise Model (TNM), a state-of-the-art computer program for predicting noise levels in the vicinity of highways. The program uses advances in acoustics and computer technology to improve the accuracy and ease of modeling highway traffic noise, including the design of efficient cost-effective highway noise barriers (Menge et al. 1998).

TNM predicts traffic noise levels using new acoustical algorithms and newly measured emission levels for five standard vehicle types: cars, medium trucks, heavy trucks, buses, and motorcycles. Its flexible database includes more than 6,000 individual vehicle pass-by events, measured at 40 sites across the country.

Sound levels for locations with and without noise barriers can also be calculated, and analysis of noise from constant-flow and interrupted flow traffic can be conducted. Also, the program determines the effects on noise levels of different pavement types, graded roadways, and rows of buildings, among others.

TNM provides a one-decibel increase in accuracy over FHWA's previous prediction model. This is important, because a one-decibel improvement in traffic noise prediction accuracy can reduce the barrier height needed to control noise by two feet, generating significant savings in noise barrier program costs.

4.4 *WHAT ARE THE USES OF TRANSPORTATION MODELS?*

Modeling is a necessary component of transportation engineering if future traffic conditions are being analyzed or existing conditions are too expensive to observe. These circumstances are present in the majority of transportation studies. A practical approach to transportation studies is to solve transportation problems by first solving their models and then implementing the solutions to the real world. Modeling is applied in all the areas of transportation engineering: planning, design, and operations. Although planning is traditionally the most "model-demanding" area, intelligent technologies and techniques applied to transportation have increased the demand for modeling in the design and traffic operations areas.

4.4.1 Planning

Modeling for planning includes prediction of future travel demands, identification of potential performance problems under the future travel demands, and proposing general solutions for the anticipated transportation problems. Modeling packages for long-range planning may include all the types of models presented and discussed previously: demand prediction, network modeling, traffic flows, and performance modeling; although traffic flows are modeled with much more simplification. From this perspective, long-range planning requires the most comprehensive modeling. Questions frequently addressed with modeling for planning applications are listed below.

1. *Future travel demands:* Travel demand is typically defined as the volume of travelers willing to travel between two locations. A quadratic matrix of one-way traffic flows represents the travel demand between multiple locations. Demand for travel strongly depends on land use and on economic conditions.

2. *Impact of regulations and policies on future travel demands:* Knowledge of this relationship is particularly useful if proper regulatory and economic measures are sought to keep travel demand at the desired level. The desired level of demand is determined by congestion management or environmental concerns.

3. *Identification of components of the existing infrastructure that will need improvements:* Performance of an existing system under future demands is frequently studied to identify the system "bottlenecks" (i.e., the infrastructure components that will cause excessive hazard, congestion, or environmental concerns). This analysis leads to the next one, whereby engineering solutions of the anticipated transportation-related problems are sought.

4. *Identification of projects needed to maintain acceptable performance of the system:* Once the future transportation issues and weak components of the infrastructure are identified, adequate alternative solutions can be proposed and evaluated. Transportation models can be implemented for various scenarios that represent various solutions, and consecutive evaluation of the scenarios is aimed to select the best one.

5. *Identification of improvements is needed to make the existing infrastructure more resilient to damage and more efficient in emergency situations:* Modeling of emergency conditions that follow natural or human-inflicted disasters (earthquake, hurricane, nuclear blast, biological/chemical attack) help identify the critical components of a transportation system. A component of transportation infrastructure is critical if its failure brings severe deterioration of the system performance. Such components, once identified, can be structurally strengthened, better protected through enhanced security means, or supplemented with backup facilities. Scenarios of operations management can be prepared if a critical component is damaged.

4.4.2 Design

A well-designed transportation facility is economical in construction and maintenance, accommodates traffic demand during its lifetime, and does not expose its users to excessive danger. To meet these design criteria, a designer must be able to link design decisions about facility geometry and traffic control with facility performance (delays, speeds, safety, and costs). Although many design decisions related to safety and aesthetics are governed by design policies and guidelines, decisions that determine capacity (e.g., the number of lanes on a freeway) require traffic modeling. Among the many models used in design, two design tools have already been mentioned: the Highway Capacity Manual and the Interactive Highway Safety Design Model.

Evaluation of alternatives is a common approach to geometric design. More complex systems, such as arterial streets or networks with coordinated signals, may require quite advanced optimization tools for proper setting of signal parameters. Traffic volumes are assumed to be insensitive to design decisions and are provided by planning models.

4.4.3 Operations

Studies of existing systems may include direct observations of their performance. Although this is the most desirable method of evaluating existing systems, the costs incurred and the measurement difficulties often prompt for modeling. Transportation analysts try to replicate the existing system through its model and outcome from the model is used to evaluate the system's performance, detect operational problems, and identify possible solutions. Although this approach brings to mind planning studies, there is a considerable difference—only the present traffic conditions are evaluated. Such studies are particularly justified if the considered improvements are low-cost and are implemented quickly. Signal retiming and reorganizing traffic circulation in the area are examples of such improvements.

Intelligent Transportation Systems require a new class of model applications. Advanced Travel Information Systems and Advanced Traffic Management Systems use fast models run in real-time to support realtime decisions about changes in schedules, signal retiming, and traveler rerouting in response to planned events and unplanned incidents. Real systems are simulated with a short time lag using real data. Short-term traffic prediction is desired to foresee the near future. Only then can optimal traffic control and management strategies be applied in a timely manner.

4.5 SELECTING A MODEL

Most of the available modeling tools are packages of models described in Section 4.2. In this section, the term *model* stands for a model package. The variety of transportation studies and the multiplicity of models make selection of a model difficult, and this section therefore is written with inexperienced transportation analysts in mind who seek general guidance for selecting a proper model.

A model is appropriate if it can do the job. Transportation analysts tend to seek a universal model which can do most of the jobs, but it is doubtful that such a model exists. A model is a simplification of reality and is developed as a result of a preassumed class of applications that determine the simplifications that are acceptable and those that are not. If the model is used for an application outside of the assumed class, then seemingly reasonable results may be in fact inaccurate or even useless. Selection of a model therefore must consider the job and, more precisely, the model's applicability for the job. For example, traditional long-term planning models estimate the travel times along links between nodes using the simplified BPR function. These travel-time estimates are sufficient to model network flows but are useless if one wants to analyze, for instance, the performance of urban arterial streets during special events.

A modeling project is specified sufficiently for model selection if the following is known:

1. Studied transportation system
2. Purpose of the modeling
3. Required output scope and format
4. Required accuracy of results (model validation and calibration)
5. Resources available for the modeling

4.5.1 Output Scope and Format

Transportation models are used to produce specific results. If a model does not produce what is needed or the results are insufficient to calculate the needed outcome, then such a model cannot be used.

4.5.2 Model Validity

Another criterion often overlooked and closely related to the validity of the model is the accuracy of the results, which is tested by inspecting the model error. The model is run a sufficient number of times using input collected in the field, and the obtained results are then compared with the measured values. The model is valid if the discrepancies between the calculated and observed results are acceptable. The validity of many available models is not sufficiently documented, part of which can be explained by the fact that validation of transportation models is not easy. Validation requires field-collected inputs and field-collected values corresponding to the model outputs. Not only are the high costs of data collection a problem, but sometimes the measurability of the quantities used in the validated model is as well. For example, a considerable number of micro-simulation models require the percentages of drivers of certain personality types (aggressive, regular, or passive), inter-vehicle time gaps that drivers use when crossing a road, or desired speeds. Models that use a large number of input variables are difficult to validate because validation requires collection of all these variables simultaneously, which is often nearly impossible.

Prediction models present a new category of validation issues. The only possible validation method is retrospection, where past observations are predicted using older observations. The weakness of this approach is that even good replication of the past data does not rule out failure in predicting the future.

Another way of evaluating model validity is to analyze the model's fundamentals and its structure. Sufficient description of the model and explanation of its theoretical basis increase confidence in the model's validity, which is particularly true for computer simulation if the component models are well documented in the literature and fit together in a plausible manner. Proprietary computer codes unavailable to transportation analysts make indirect evaluation of these models difficult if not impossible. Preferable models are those well documented by developers and validated independently by research units.

4.5.3 Model Calibration

Even a valid model requires calibration if the local conditions differ from the conditions for which the model was developed and validated. Frequently, a simulation model is developed from component models and the default values of model parameters (input values that represent certain conditions and remain unchanged as long as the model is used for these conditions) are assumed using common sense or published research. Then, the model is offered to clients with a presumption that users are responsible for model calibration to local conditions.

There are two ways of setting model parameters. If a parameter has a meaning and can be measured in the field, then a measured value should replace the default value. If a parameter is not observable because it has no meaning or is difficult to measure, then it may serve as a "tuning knob." A calibration exercise is often understood as applying measured parameter values and changing the remaining parameters until the model returns results of satisfactory accuracy. Although the formula seems simple, its application is not. The calibration difficulty caused by the required large amount of data was explained previously in the validation issues discussion.

The need for calibration of simulation models has been investigated at the University of Hawaii (Wang and Prevedouros 1998). Three micro-simulation models: INTETRATION, CORSIM, and WATSim were applied to a small congested network of two highway intersections and several segments. The simulated volumes and speeds were compared to the measured values, and the authors concluded that the results were reasonable after a large number of parameters were modified to replicate the real traffic conditions. In no case did the default parameters offer satisfactory results. Further, the authors concluded that the parameter calibration could be tedious and time consuming and the effort for calibration multiplies with the size of the transportation system and the number of model parameters.

Models, particularly micro-simulation, may have a large number of parameters. Replicating real values in a small number of cases is always possible if the number of "tuning knobs" is sufficient. The problem is that a model calibrated for a small number of data points may not perform well for new cases. This issue is similar to a regression model calibrated for a small number of data points but with a large number of parameters. If the number of data points equals the number of parameters, then it is always possible to find parameter values that allow a model to replicate the data exactly. Unfortunately, the predictive ability of the model is nonexistent.

Model validation and calibration require special attention, particularly when the models are complex. Having mentioned difficulties with validation and calibration of complex models, simple models sufficiently validated and with a limited number of parameters should be preferred. Specific methods for planning model validation and calibration can be found in Barton-Aschman Associates and Cambridge Systematics (1997), while calibration issues for microsimulation traffic models are discussed in Hollander and Liu (2008).

Numerous transportation models have been developed for particular analysis. For example, researchers from the University of Texas in Austin proposed a simulation model developed to evaluate the effect of introducing new large aircraft in 2006 (Chiu and Walton 2002). The method simulates the operating characteristics of a new large aircraft and its market share for various scenarios to address prediction uncertainty. The arrival passenger flows are modeled as a queuing network system comprised of a series of passenger processing facilities. The model was validated with survey data and statistics for international airports.

4.5.4 Scope of Input

A model is practical if it requires input that is feasible to obtain. Again, this condition should be considered in connection with a specific job for which a model is needed. In the planning phase, only major transportation characteristics are known or considered, and it is unusual, if not impossible, to consider the details of intersection geometry or the detail settings of a signal controller 20 years in advance. Not only is the information uncertain, but planners are not interested in such detail decisions either. Models that require unavailable input lead to the use of default values, and the correct situation when default values may be used is in itself an important issue. The answer depends on the variability of the input and on the model sensitivity. If the input varies significantly from one case to another and the results are strongly dependent on the input, then using a default value is questionable. A model that requires such an unknown input should not be used.

An excessively detailed model may impose difficulties when using default inputs, but an overly simplified model, on the other hand, may defeat the purpose. For example, if a model is used to evaluate the introduction of a median on an arterial street, the model must incorporate the effect of the median on the capacity of the crossing and turning streams.

4.5.5 Modeling Costs

The costs of transportation analysis include gathering and formatting data, running a model, and documenting and reporting results. Data collection and formatting are frequently the primary component of the total costs, particularly when the model is complex and requires calibration. The costs can be significantly reduced if the data required by the model are routinely collected by a transportation agency and stored in well maintained databases.

The costs can be further reduced if the amount of data required by the model is limited. The input required by models varies dramatically and depends on the level of modeling detail. Micro-simulation requires the largest amount of data compared to other types of models. For example, the TRANSIMS micro-simulation modeling tool was applied to the Portland area (Los Alamos National Laboratory, 2002). Highway network representation included 100,000 nodes, 125,000d links, and 250,000 activity locations. In the introduction to the report presenting intermediate results after nearly 2 years of effort, the authors characterized collating the existing data (no special data were collected for this project) as a tremendous chore and a daunting task. The report disclaims that the purpose of the Portland study was to calibrate the model with the Portland historical data and that the stated objective was to allow the developers of the model to understand transportation planning methods and the model components. The Portland lesson indicates that the hurdle of using micro-simulation for planning is the large effort required for data preparation and model calibration. Another issue that must be resolved is the massive computations needed. TRANSIMS was proposed for the modeling of the Switzerland transportation system, and in the interim report, researchers mentioned the large computational load that required clusters of PCs and Internet connection (Raney et al. 2002).

4.5.6   Other Considerations

Client preferences must be considered when selecting a model and a modeling approach. For example, transportation administration may have its own "standard" models. For years, the *Highway Capacity Manual* was preferred by most state and local highway agencies. Then, simulation models were approved by several states to be used concurrently or instead of the *Highway Capacity Manual*. In the private sector, modeling practice is more diversified and depends on the preferences of individual clients.

Modeling tools with a good visual presentation of the results, including micro-simulation, are attractive as they facilitate communication between transportation analysts, decision-makers, and the public. A danger is that realistically displayed vehicles moving in a pseudo-realistic transportation system, although not guaranteeing realistic results, may be convincing evidence of model validity to the lay person. Another dangerous myth is that the more detailed the representation of a transportation system is, the more accurate the results are. These two myths have continued to proliferate since micro-simulation was introduced and modeling became an industry. Quality control, typically expensive and difficult in transportation modeling, may be easily lost in these transactions. The burden of quality assurance in modeling usually lies with clients that are public agencies.

4.6   TRANSPORTATION SYSTEMS EVALUATION

4.6.1   Steps in Evaluation

Transportation projects affect communities and are usually funded by public agencies. Evaluation of large projects that considerably impact whole communities requires public involvement. This specific requirement of a transportation project determines the evaluation method as the benefits and costs are estimated from the societal point of view. The primary objective of the evaluation effort is to select the best alternative solution, and typically, no-build is one of the alternatives. Inclusion of the no-build scenario allows feasibility checking by comparing proposed solutions with the no-build scenario. The best alternative is selected from feasible solutions.

Step One.   The first step of any evaluation method is developing project alternatives. There are no universal guidelines for development of alternatives. This part depends strongly on the transportation problem, the existing infrastructure, and the public and decision-makers' preferences.

For example, the Central Area Loop study finished in 2001 (OKI 2001) considered four alternatives for the Cincinnati area:

1. Do nothing
2. Transportation System Management including an improvement of the existing shuttle service and a package of other improvements such as transit signal priority, traffic signal optimization, real-time information maps at transit centers, and dedicated bus lanes
3. Streetcar alternative utilizing a circular route connecting CBDs, their attractions, and business centers
4. Personal Rapid Transit—a novel automated transit system based on small cars

Another example is the Southeast Corridor project (T-Rex) in Denver, Colorado (Goetz et al. 2007) where growing traffic volumes caused severe congestion. Past studies repeatedly recommended mass transit to improve travel conditions. A large number of alternatives were considered during the Southeast Corridor Major Investment Study (MIS) conducted between 1995 and 1997. The selected project alternative included highway widening, safety improvements, and light rail transit components. The Southeast Corridor project, also known as the Transportation Expansion Project or T-REX, represents a true multimodal undertaking and exemplifies the innovation that is taking place today as transportation agencies meet the challenge of limited resources and growing infrastructure needs. By the fall of 2006, the T-REX project concluded—nearly 2 years earlier than scheduled and a lot less expensive than expected.

Step Two.   The second step is to select the evaluation criteria. First of all, the criteria should be measurable, feasible, and defendable. The criteria should be adequate to the evaluated alternatives and of course should include all aspects important to the decision-making body as well as to the community (travelers, business owners, and residents). In the previously mentioned Cincinnati case, the 59 criteria shown in Table 4.1 were formulated.

TABLE 4.1   Evaluation Criteria in the Central Area Loop Study for Cincinnati, Ohio

Cat.	Evaluation criterion
Cost-Effectiveness	To what extent does this transportation system represent a cost-effective investment?
	Are there front-end costs and time associated with this transportation system to ready it for implementation?
	How severe are the secondary costs (utilities, street changes) associated with placing this transportation system and its structures in likely locations?
	What is the technical life expectancy of this technology?
	To what extent does this transportation system imply a reasonable level of annual costs?
	Are there any extraordinary power requirements associated with this technology?
	Is this transportation system labor-intensive to operate and maintain?
	What relative degree of vehicle failure or downtime is likely with this technology?
	What level of vehicle spares seems indicated as prudent?
Equity	Will the transportation system distribute costs and benefits equally to all segments of the population within the service area?
	Will the transportation system serve a variety of population segments?
	Will the transportation system provide affordable transportation to low-income individuals?
Safety/Access	Is this transportation system ADA-compliant?
	Does the transportation system meet fire/life safety requirements?
	Is there a perception of personal safety within the vehicle and at the stations?
	Will there be difficulties meeting building code requirements?
	Does the transportation system provide convenient access to all users?
	Does the transportation system present a safety hazard to nonusers?
Effectiveness	Does the transportation system have acceptable point-to-point travel times including station dwell times?
	Does the system provide reliable service levels?
	Does the transportation system provide adequate service to the study area destinations in terms of frequency of service and geographic coverage?
	Will the transportation system adequately serve the projected ridership and/or attract sufficient ridership to justify the investment?
	How reliable is the transportation system in maintaining its schedule?
	Does this transportation system facilitate intermodal transfer movements among public transit service providers?

TABLE 4.2   Summary Results of the Cincinnati Central Loop Study

Estimation items	No-build	TSM (enhanced bus service)	Streetcar	Personal rapid transit
		Alternative		
Capital costs (with contingencies)	$0	$2,9000,000	$215,000,000	$450,000,000
Design life (average for components)	10.0	12.5	31.6	34.3
Annualization factor	0.142	0.123	0.079	0.078
Annualized capital costs per year	$0	$350,000	$17,000,000	$35,000,000
Operating and maintenance costs	$1,770,000	$5,710,000	$4,200,000	$13,900,000
Passengers per year	468,000	1,812,600	2,098,200	7,951,200
Cost per passenger	$3.78	$3.34	$10.10	$6,15

Source: As published in (OKI 2001).

Step Three.   The third step is to evaluate alternatives using the developed criteria. This step involves two tasks. First, it is important to determine the preferences of the decision makers in the context of the chosen evaluation criteria. There are several methods to achieve this, such as equal or direct weighting (Dodgson et al. 2001), the Delphi technique (Dalkey and Helmer 1963), the Gamble method (Anderson, Hobbs, and Bell 2009), and pairwise comparison of the criteria (Saaty 1977). Second, it is essential to establish a common scale, so that all the evaluation criteria are expressed in commensurate units. This enables direct comparison or combination of the criteria (e.g., with the use of utility, value, or cost-effectiveness functions; see Sinha and Labi 2007).

This is the stage where demand and traffic flows are predicted and travel quality estimated as one of important evaluation criteria. In addition, the economic and environmental impacts are quantified. Transportation modeling is extensively used in this phase, and the outcome includes tables summarizing fulfillment of the criteria by the alternatives. Table 4.2 shows example results for the four alternatives for Cincinnati as published in the final study report (OKI 2001). According to the rebuttal to the draft final report by the Sky Loop Committee and Taxi 2000 Corporation (2001), the final report had quadrupled the Personal Rapid Transit's cost and had seriously underestimated its ridership. The total cost per new rider was thus grossly distorted according to the rebuttal.

Step Four.   The fourth step is to select the best alternative by applying the selected criteria. Although an evaluation analysis assist in the selection process, the final selection belongs to a decision-maker.

In the Cincinnati case, the Ohio-Kentucky-Indiana Regional Council of Governments selected the improved bus service. The Personal Rapid Transit alternative was rejected as not proven technologically (OKI 2001). Other concerns included prohibitive capital costs, removing commuters from businesses at the street level, and poor aesthetics.

4.6.2   Conditions of Truthful Evaluation

The Central Area Loop Study for Cincinnati involved controversies around the modeling particulars. The draft final report prepared by a hired consultant was denied by one of the participating parties (Sky Loop Committee and Taxi 2000 Corporation 2001) as biased against one of the studied alternatives. It is not an unusual situation because the parties involved in the evaluation process have their own preferences, opinions, and agenda. That is why it is important that the evaluation results are defendable. Several unquestioned technical conditions of truthful evaluation could be pointed out.

1. *Properly identified transportation problem:* Incorrectly identified transportation problems immediately defeat the purpose of the evaluation study. Current transportation problems are typically obvious and are identified through data routinely collected and analyzed by transportation agencies or through special studies triggered by travelers' complaints. Identification of future transportation problems requires modeling and is more challenging, and it includes predicting

future conditions for the no-build scenario. A transportation analyst must decide where prediction of future demand is needed with all or some of the constraints removed (e.g., road capacity). Conditions 3 to 7 discussed below also apply here.

2. *Adequately and exhaustively developed alternatives:* Alternative projects are developed to solve the identified transportation problem. The proposed alternatives must cover a broad spectrum of possibilities.

3. *Input data of the best quality achievable:* The amount of data can be large and its collection may be costly. It is difficult to explain why existing needed data are not used if it is available. Collection of new data depends on the budget and time frame.

4. *Honest assumptions where data is lacking:* It is not possible to have all the required data available. In such cases, assumption may be the only means, but these assumptions must be justified with analogous cases, common knowledge, or at least a consensus of the participating parties. The last option seems to be most difficult.

5. *Adequate model calibrated for local conditions:* Model selection and calibration issues were discussed previously. It is obvious that an inadequate or inadequately calibrated model may generate inaccurate and even useless results. Model selection should consider the resources required for model calibration to avoid a situation where there is no time or funds to calibrate the selected model.

6. *Adequate and exhaustive evaluation criteria:* Evaluation criteria must reflect the preferences of the stakeholders and should consider all foreseeable benefits and losses caused by the alternatives. Therefore, input from the public is often sought if the subject transportation system is public or considerably influences the community.

7. *Proper documentation of the evaluation process and presentation of the results:* The final report must clearly state all the data sources, the assumptions made, and the tools used, including the procedure for reaching the conclusions. The results should be presented in layman's terms to help the public understand the results, conclusions, and justification for the conclusions. All the critical results must be included along with a discussion of the possible limitations of the study. Modern presentation techniques allow for smart and appealing layouts.

4.7   CONCLUSION

The high costs of transportation investments require sound decisions. Evaluation of transportation systems is needed and required before a decision can be made. This evaluation can be complex if its subject is complex. Modeling makes the evaluation task manageable.

Good modeling of transportation systems is resource-demanding. The costs can be reduced by using an adequate modeling tool that does not require redundant data and extensive calibration. Planning models and micro-simulation are typically data demanding. Combining these two creates large computational and input requirements. It should be kept in mind that an inappropriate model or one lacking or having incorrect calibration may produce results that are inaccurate.

Ethical issues are frequently present in transportation modeling and evaluation. The high costs of good modeling, the shortage of time and data, and the quality control difficulties create conditions where the best modeling and evaluation practice may be replaced with a substitute process where the tools and outcomes are accepted because the results are timely available and seemingly usable.

This chapter discussed in general terms the conditions of technically sound modeling and evaluation of transportation systems. The reader may visit the *Online TDM Encyclopedia* at http://www.vtpi.org/tdm/ created and maintained by the Victoria Transport Policy Institute. This site offers a practical compendium of transportation demand management methods and issues. The United States Department of Transportation sponsors a clearinghouse for information on travel demand forecasting and dynamic traffic assignment at http://tmip.fhwa.dot.gov/. This site contains a large number well-organized and classified documents.

REFERENCES

Adamus, P., L. Stockwell, E. Clairain, M. Morrow, L. Rozas, and R. Smith. 1991. *Wetland Evaluation Technique (WET)*, Vol. I, *Literature Review and Evaluation Rationale*. Tech. Rep. WRP-DE-2, U.S. Army Corps of Engineers, Waterways Experiment Station, Vicksburg, MS.

Anastasopoulos, P. and F. Mannering. 2009. "A Note on Modeling Vehicle-accident Frequencies with Random-parameters Count Models." *Accident Analysis and Prevention* 41(1):153–159.

Anastasopoulos, P., A. Tarko, and F. Mannering. 2008. "Tobit Analysis of Vehicle Accident Rates on Interstate Highways." *Accident Analysis and Prevention* 40(2):768–775.

Anderson, R., B. Hobbs, and M. Bell. 2009. "Multi-Objective Decision-Making in Negotiation and Conflict Resolution." In *Conflict Resolution*, Editor: Keith William Hipel, Vol. 2, EOLSS E-book Library, http://www.eolss.net/ebooklib/. Accessed in Nov. 2010, 203–228.

Barton-Aschman Associates, Inc., and Cambridge Systematics, Inc. 1997. *Model Validation and Reasonableness Checking Manual*, Prepared for Travel Model Improvement Program Federal Highway Administration, http://www.ctre.iastate.edu/educweb/ce451/LECTURES/Validation/finalval.pdf, Accessed Nov. 2009.

Ben-Akiva, M., M. Bierlaire, J. Bottom, H. Koutsopoulos, and R. Mishalani. 1997. Development of a Route Guidance Generation System For Real-Time Application. Proceedings of the *IFAC Transportation Systems 97 Conference*, Chania, Greece.

Ben-Akiva, M., M. Bierlaire, H. Koutsopoulos, and R. Mishalani. 1998 DynaMIT: A Simulation-Based System for Traffic Prediction. *Proc. DACCORD Short Term Forecasting Workshop*, Delft, The Netherlands, http://citeseerx.ist.psu.edu/viewdoc/download?doi=10.1.1.37.1549&rep=rep1&type=pdf. Accessed Nov. 2009.

Benson, P. 1988. "Development and Verification of the California Line Source Dispersion Model." *Transportation Research Record* 1176:69–77.

Booz Allen & Hamilton Inc. 1999. *The California Life-Cycle Benefit/Cost Analysis Model (Cal-B/C)*. Technical Supplement to User's Guide. California Department of Transportation, Sacramento, CA, http://www.dot.ca.gov/hq/tpp/offices/ote/benefit_files/tech_supp.pdf. Accessed Oct. 2009.

Bureau of Public Roads. 1964. *Traffic Assignment Manual*. U.S. Dept. of Commerce, Urban Planning Division, Washington D.C.

Cambridge Systematics. 2000. *Surface Transportation Efficiency Analysis Model (STEAM 2.0): User Manual*. Federal Highway Administration, U.S. Department of Transportation, Washington D.C. http://www.fhwa.dot.gov/steam/20manual.htm. Accessed Oct. 2009.

Carruthers, D., R. Holroyd, J. Hunt, W. S. Weng, A. Robins, D. Apsley, D. Thompson, and F. Smith. 1994. "UK-ADMS: A New Approach to Modeling Dispersion in the Earth's Atmospheric Boundary Layer." *Journal of Wind Engineering Ind. Aerodynamics* 52:139–153.

Chiu, C. Y. and C. M. Walton. 2002. *An Integrated Simulation Method to Evaluate the Impact of New Large Aircraft on Passenger Flows at Airport Terminals*, paper presented at the 2002 Annual Meeting of the Transportation Research Board, Washington, D.C.

Dalkey, N. and O. Helmer. 1963. "An Experimental Application of the Delphi Method to the Use of Experts." *Management Science* 9(3):458–467.

Daniels, G. and W. Stockton. 2000. "Cost-Effectiveness of High-Occupancy Vehicle Lanes in Texas." *Transportation Research Record* 1711:1–5.

Dodgson, J., M. Spackman, A. Pearman, and L. Phillips. 2001. *Multicriteria Analysis: A Manual*. London: Department for Environment, Food and Rural Affairs.

Evink, G. 2002. *Interaction Between Roadways and Wildlife Ecology*. NCHRP Synth. Rep. 305, National Cooperative Highway Research Program, Transportation Research Board, National Research Council, Washington D.C.

FAA. 2006. "Airport Noise Compatibility Planning Toolkit." Federal Aviation Administration, Washington D.C. http://www.faa.gov/about/office_org/headquarters_offices/aep/planning_toolkit/. Accessed Nov. 2009.

Fellendorf, M. 1994. "VisSim: A Microscopic Simulation Tool to Evaluate Actuated Signal Control including Bus Priority." *Proc. 64th ITE Annual Meeting*, 1994, Dallas, Texas.

FHWA. 2002. *Highway Economic Requirements System*. Tech. Rep., Federal Highway Administration, U.S. Department of Transportation, Washington D.C. http://www.fhwa.dot.gov/infrastructure/asstmgmt/hersindex.cfm. Accessed Oct. 2009.

FHWA. 2003. IHSDM Preview CD-ROM, Version 2.0. FHWA-SA-03-005, Federal Highway Administration, U.S. Department of Transportation, Washington D.C.

Fredendall, L. D., and E. Hill. 2000. *Basics of Supply Chain Management*. St. Lucie Press, CRC Press LLC, United States of America.

Goetz, A., J. Szyliowicz, T. Vowles, and S. Taylor. 2007. "Assessing Intermodal Transportation Planning at State Departments of Transportation." *World Review of Intermodal Transportation Research* 1(2):119–145.

Gomes, G., A. May, and R. Horowitz. 2004. "Congested Freeway Microsimulation Model Using VISSIM." *Transportation Research Record* 1876:71–81.

Greene, W., 2007. *Econometric Analysis* (6th ed.). Prentice Hall.

Guenslar, R., S. Washington, and W. Bachman. 1998. Overview of the MEASURE Modeling Framework. *Proc. ASCE Conference on Transportation Planning and Air Quality III*, Lake Tahoe, CA.

Harwood, D. W., F. M. Council, E. Hauer, W. E. Hughes, and A. Vogt. 2000. *Prediction of the Expected Safety Performance of Rural Two-Lane Highways*, Federal Highway Administration, Rep. FHWA-RD-99-207, http://www.fhwa.dot.gov/ihsdm/. Accessed Oct. 2009.

Hawas, Y. E., H. S. Mahmassani, R. Taylor, A. Ziliaskopoulos, G. L. Chang, and S. Peeta. 1997. *Development of DYNASMART-X Software for Real-Time Dynamic Traffic Assignment*. Technical Report ST067-85-TASK E, Center for Transportation Research, University of Texas, Austin, TX.

Hollander, Y. and R. Liu. 2008. "The Principles of Calibrating Traffic Microsimulation Models." *Transportation* 35(3):347–362.

Jack Faucett Associates, 1994. *Evaluation of MOBILE Vehicle Emission Model*, U.S. Department of Transportation, National Transportation Library, http://ntl.bts.gov/DOCS/mob.html. Accessed Nov. 2009.

Johnston, R. A., C. J. Rodier, J. E. Abraham, J. D. Hunt, and G. J. Tonkin, 2001. *Applying an Integrated Model to the Evaluation of Travel Demand Management Policies in the Sacramento Region*, MTI REPORT 01-03, The Mineta Transportation Institute, College of Business, San Jose State University, San Jose, California. http://www.transweb.sjsu.edu/mtiportal/research/publications/documents/01-03.pdf. Accessed Oct. 2009.

Jones, R., 1999. Statewide Travel Demand Forecasting Process in California. Tech. Rep. E-C011, California Department of Transportation, Irvine, CA, http://onlinepubs.trb.org/onlinepubs/circulars/ec011/jones.pdf. Accessed Oct. 2009.

Los Alamos National Laboratory. 2002. *TRansportation ANalysis SIMulation System (TRANSIMS), Portland Study Reports*, Vol. 0-7.

Mahmassani, H., T. Hu, and R. Jayakrishnan. 1995. "Dynamic Traffic Assignment and Simulation for Advanced Network Informatics (DYNASMART)." In N. H. Gartner and G. Improta, Editors, *Urban Traffic Networks: Dynamic Flow Modeling and Control*. Springer, Berlin/New York.

Mannering, F., S. Washburn, and W. Kilareski. 2009. *Principles of Highway Engineering and Traffic Analysis* (4th ed.). New York: John Wiley and Sons.

McFarland, W., J. Memmott, M. Chui, M. Richter, and A. Castano-Pardo. 1993a. *MicroBENCOST User's Manual*. NCHRP Project 7-12, Texas Transportation Institute, Texas A&M University. College Station, TX.

McFalrand, W., J. Memmott, and Chui, M., 1993b. *Microcomputer Evaluation of Highway User Benefits*. NCHRP Rep. 7-12, Transportation Research Board, National Research Council, Washington D.C.

McTrans, HCS+, *On-line Products Catalog*, at http://mctrans.ce.ufl.edu/index.htm. Accessed Nov. 2010.

Menge, C., C. Rossano, G. Anderson, and C. Bajdek. 1998. *FHWA Traffic Noise Model Version 1.0, Technical Manual*, FHWA-PD-96-010 and DOT-VNTSC-FHWA-98-2, John Volpe National Transportation Systems Center, Cambridge, MA.

Meyer, M., and E. Miller. 2001. *Urban Transportation Planning: A Decision-Oriented Approach*. New York: McGraw-Hill.

Moulton, H. 1990. *Air Force Procedure for Predicting Aircraft Noise Around Airbases: Noise Exposure Model (NOISEMAP) User's Manual*. AAMRL-TR-90-011, Armstrong Laboratory, Wright-Patterson AFB, OH.

OKI Regional Council of Governments. 2001. *Central Area Loop Study*. Final Report. Prepared by Parsons Brinckerhoff with Lea+Elliott, Inc., JKH Mobility Services, and Infrastructure Services, Inc. at http://www.oki.org/transportation/centralarea.html. Accessed Nov. 2010.

Park, B. and J. Schneeberger. 2003. "Microscopic Simulation Model Calibration and Validation: Case Study of VISSIM Simulation Model for a Coordinated Actuated Signal System." *Transportation Research Record* 185–192.

Patriksson, M. 1994. *The Traffic Assignment Problem: Models and Methods (Topics in Transportation Series)*. V.S.P. Intl Science.

Peeta, S. and A. Ziliaskopolous. 2001. "Foundations of Dynamic Traffic Assignment: The Past, the Presence, and the Future." *Networks and Spatial Economics* 1:233–265.

Raney, B., A. Voellmy, N. Cetin, M. Vrtic, and K. Nagel. 2002. "Towards a Microscopic Traffic Simulation of All of Switzerland." In *Lecture Notes in Computer Science*—Proc. International Conference on Computational Science-Part I, 2329, pp. 371–380.

Rathi, A. and Santiago, A. 1990. "Urban Network Traffic Simulation: TRAF-NETSIM Program." *Journal of Transportation Engineering* 116(6):734–743.

Robertson, H. D. (Editor). 2000. *Manual of Transportation Engineering Studies*. Washington, D.C.: Institute of Transportation Engineers.

Rouphail, N., C. Frey, J. Colyar, and A. Unal, A. 2001. "Vehicle Emissions and Traffic Measures: Exploratory Analysis of Field Observations at Signalized Arterials." *Proc. Transportation Research Board 80th Annual Meeting*, Washington D.C.

Saaty, T., 1977. "A Scaling Method for Priorities in Hierarchical Structures." *Journal of Mathematical Psychology* 15(9):234–281.

Savolainen, P. and F. Mannering. 2007. "Probabilistic Models of Motorcyclists' Injury Severities in Single- and Multi-Vehicle Crashes." *Accident Analysis and Prevention* 39(5):955–963.

Schamberger, M. and W. Krohn. 1982. "Status of the Habitat Evaluation Procedures." *Proc. Transactions of the Forty-Seventh North American Wildlife and Natural Resources Conference*, ed. Kenneth Sabol, Washington, D.C. http://digitalcommons.unl.edu/cgi/viewcontent.cgi?article=1047&context=usfwspubs. Accessed Nov. 2009.

Schintler, L. A. and M. A. Farooque, 2001. *Partners In Motion and Traffic Congestion in the Washington, D.C. Metropolitan Area*, Center for Transportation Policy and Logistics, School of Public Policy, George Mason University, prepared for Federal Highway Administration, Virginia Department of Transportation, School of Public Policy, George Mason University, at http://www.itsdocs.fhwa.dot.gov/JPODOCS/REPTS_TE/13500.html. Accessed Oct. 2009.

Sheffi, Y. 1985. *Urban Transportation Networks: Equilibrium Analysis with Mathematical Programming Methods*. Englewood Cliffs, NJ: Prentice-Hall.

Simon, H. 1959. "Theories of Decision-Making in Economics and Behavioral Science." *American Review* 49:253–283.

Sinha, K. and S. Labi. 2007. *Transportation Decision-Making: Principles of Project Evaluation and Programming*. John Wiley & Sons Inc. New Jersey.

Sky Loop Committee and Taxi 2000 Corporation. 2001. *A Rebuttal to the Central Area Loop Study Draft Final Report*, Cincinnati, Ohio, at http://www.jpods.com/JPods/004Studies/T2K-Rebuttal-to-CALS.pdf. Accessed Nov. 2010.

Smith, L., R. Beckman, D. Anson, K. Nagel, and M. Williams. 1995. TRANSIMS: Transportation Analysis and Simulation System. *Proc. National Transportation Planning Methods Applications Conference*, Seattle, WA.

System Application International. 2002. *User's Guide to HYROAD: The Hybrid Roadway Intersection Model*. Tech. Rep. SYSAPP-02-073d. National Cooperative Highway Research Program, National Research Council, Washington D.C.

Transportation Research Board. 2000. *Highway Capacity Manual*. Washington, D.C.: National Research Council.

University of Birmingham. 2005. *HDM Technical User Guide*. UB, Birmingham, UK. http://www.hdmglobal.com/AboutHDM4.htm. Accessed Oct. 2009.

U.S. Department of Transportation. 1998. *Transportation Case Studies in GIS—Case Study 4: Maine Department of Transportation Statewide Travel Demand Model*, Federal Highway Administration, Office of Environment and Planning, Washington, DC 20590, http://tmip.tamu.edu/clearinghouse/docs/gis/maine/maine.pdf. Accessed Nov. 2009.

USEPA. 1998. *Assessing the Emissions and Fuel Consumption Impacts of Intelligent Transportation Systems (ITS)*. EPA 231-R98-007, U.S. Environmental Protection Agency, Washington D.C.

USEPA. 2002. *User's Guide to MOBILE6.0 Mobile Source Emission Factor Model*. EPA 420-R-02-001, U.S. Environmental Protection Agency, Washington D.C.

USEPA. 2006. Green Vehicle Guide. U.S. Environmental Protection Agency, Washington D.C. http://www.epa.gov/greenvehicles/Index.do. Accessed Nov. 2009.

M. Van Aerde & Assoc., Ltd. 2005. *INTEGRATION© Release 2.30 For Windows: User's Guide—Volumes I and II*. Kingston, Ontario, at http://filebox.vt.edu/users/hrakha/Software. Accessed Nov. 2010.

Venugopal, S. and A. Tarko. 2000. "Safety Models for Rural Freeway Work Zones." *Transportation Research Record*, Transportation Research Board. 1715:1–9.

Volpe Center. 2002. *Integrated Noise Model (INM) Version 6.0 Technical Manual*, U.S. Department of Transportation, Federal Aviation Administration, FAA-AEE-97-04.

Wade, A. R., D. Harwood, J. P. Chrstos, W. D. Glauz. 2000. *The Capacity and Enhancement of VDANL and TWOPAS for Analyzing Vehicle Performance on Upgrades and Downgrades within IHSDM*, Federal Highway Administration, Report No. FHWA-RD-00-078, http://www.tfhrc.gov/safety/00-078.pdf. Accessed Oct. 2009.

Wang, Y. and P. D. Prevedouros, 1998. "Comparison of INTEGRATION, TSIS/CORSIM and WATSim in Replicating Volumes and Speeds on Three Small Networks," *Paper presented at the 1998 Annual Meeting of the Transportation Research Board*, Washington, D.C.

Washington, S., M. Karlaftis, and F. Mannering. 2003. *Statistical and Econometric Methods for Transportation Data Analysis*. Chapman and Hall/CRC, Boca Raton, FL, First edition 439 pages, 2003.

CHAPTER 5
SOFTWARE SYSTEMS AND SIMULATION FOR TRANSPORTATION APPLICATIONS

Elena Shenk Prassas
Department of Civil Engineering
Polytechnic University
Brooklyn, New York

The time has long passed when signal optimization was done by physical time-space boards or when traffic assignment models were executed overnight on a mainframe computer. Software systems are now the *basic* components of a professional transportation engineer's arsenal and just about every professional transportation study uses software in its analysis.

Thanks to massive advances in computing and communications technology, we now live in an era in which much data can be downloaded electronically, transmitted wirelessly, and presented in summary reports and visuals by automated processes. In traffic control, data availability in real time is a reality. Sophisticated computer programs aid computation, signal optimization, and network assignments. This includes dynamic traffic assignments, responsive to incidents and events. Networks are routinely modeled and simulated at the microscopic level. The emphasis on intelligent transportation systems (ITS) has raised the level of both technological capability and public expectation. Global positioning systems (GPS), geographic information systems (GIS), pervasive cellular telephones, and electronic toll collection systems are no longer considered new technologies but common tools.

At the same time, much of the sensing is still done by road tubes, albeit linked to sophisticated devices, and integration of these varied technological capabilities into a seamless system of information is now less of a challenge. The very power and sophistication of some computational tools has led to a new generation of planning-level estimators, consistent with the newer operational tools.

The purpose of this chapter is to provide some information on the span of software systems available to practicing engineers and planners. Because of the abundance of software applications in all areas of transportation, most of the emphasis here is on traffic operations tools, and particularly simulation.

An excellent reference concerning the tools available for traffic analysis is the FHWA Traffic Analysis Toolbox[1-9], a nine-volume reference set on the various types of traffic analysis tools. One of the first and most daunting question a transportation professional faces is deciding what software type to use, so Volumes I and II describe the purpose, capabilities, and use of the different tools available to transportation professionals.[1,2] Volumes III and IV are specific to microscopic simulation and give guidelines for its use, in general,[3] and for the use of the CORSIM[10] simulation model, in particular.[4] Volume V presents real-world case studies that show which tools are applied and how they can be used effectively.[5] Volume VI discusses measures of effectiveness and how to interpret results.[6] Volume VII discusses how to predict performance using traffic analysis tools,[7] and Volumes VIII and IX are specific to using simulation in work zones.[8,9]

5.1   COMPUTATIONAL SOFTWARE VERSUS SIMULATION SOFTWARE

It is important to understand the difference between the different types of software tools and their purpose, and the difference between computational tools versus simulation. If an equation or algorithm is well-defined but being executed by computer for speed or efficiency, it is a *computational* tool; if subsystems are modeled, randomness introduced, and the subsystems linked, then it is generally a *simulation* tool.

In this spirit, the *Highway Capacity Manual (HCM)*[11] is a collection of computational procedures that are being realized in a set of computational tools, such as HCS[12] and Signal 2000.[13] The HCM is a collection of deterministic procedures for predicting capacity, delay, speed, queues, and other measures of effectiveness (MOE) for the various types of transportation facilities, such as freeways, signalized and unsignalized intersections, arterials, and roundabouts. The *HCM* procedures are best used for light to moderately congested facilities. They are relatively quick and easy to use, but many of the procedures can no longer be done by hand. For instance, the *HCM* 2010 signalized intersection methodology, and arterial methodology now incorporates iterative models that preclude hand calculations.

A number of signal optimization programs, including TRANSYT-7F[14] and PASSER,[15] are also computational tools because they are based on deterministic models and will always produce the same result, that is, there is no randomness introduced. Some optimization programs, such as TRANSYT-7F, also include a macroscopic simulation model for predicting platoon dispersion, queue spillback, and actuated control simulation.[16] Some tools combine capacity (*HCM*-based models) and signal optimization, such as SIG/Cinema,[17] Synchro/SimTraffic,[18] and TEAPAC *Complete*.[19]

One of the earliest simulation tools was the UTCS-1,[20] prepared for the Federal Highway Administration (FHWA) as part of the Urban Traffic Control System testbed in Washington, DC, dating to the late 1960s. This was a true microsimulation, in that such subsystems as turning, speed selection, car-following, and lane selection were modeled stochastically, calibrated individually, and then incorporated into a system that simulated traffic movements on arterials and in networks at a microscopic (i.e., individual vehicle) level.

This tool later evolved into NETSIM,[21] and was validated over the years at the macro level of overall network flows, speeds, and delay. An analogous tool was developed for traffic on freeways, and aptly named FRESIM.[22] These were both integrated under the direction of FHWA into a corridor tool, CORSIM,[10] and combined with a user-friendly interface and an animation program for viewing the results, into one package called TSIS.[23] Other corridor tools, such as WATSIM,[24] have been produced as competing products, as have such traffic simulators as VISSIM,[25] Paramics,[26] and Aimsun.[27]

The TEAPAC *Complete*[19] package has, in addition, organized the links between various computational and simulation models. It includes pre- and postprocessors for many of the software programs listed above, such as TRANSYT-7F, CORSIM, and VISSIM, thus making it easy to run the various levels of models with one input data set. Figure 5.1 shows a diagram of the TEAPAC *Complete* system, and the relation to other tools.

5.2   IMPORTANT AREAS AND TOOLS NOT COVERED

This chapter does not address a host of transportation computational tools and software available in the profession: transit scheduling and routing algorithms; travel demand models, such as TransCAD,[28] which combines GIS and transportation modeling; commercial vehicle route selection; or dynamic routing, just to name a few.

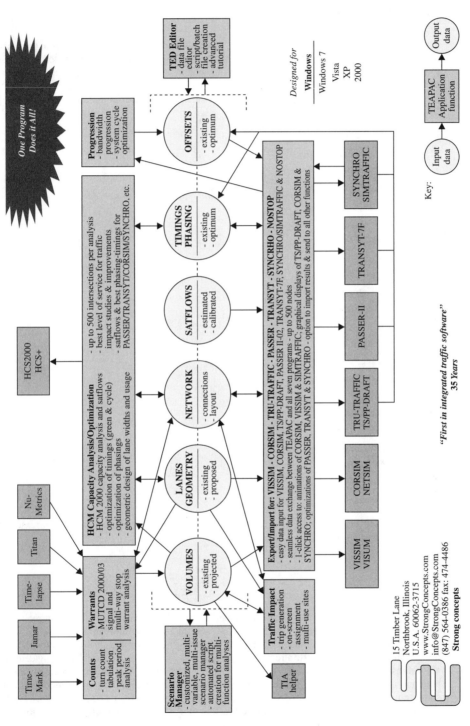

FIGURE 5.1 TEAPAC *Complete* and its linkages for an integrated system.

5.3   *TRAFFIC SOFTWARE SYSTEMS*

There are national FHWA-designated software distribution centers, notably McTrans at the University of Florida (http://mctrans.ce.ufl.edu) and PC-TRANS at Kansas University (www.kutc.ku.edu/pctrans). Originally funded by FHWA during their startup phase, they now exist based upon software sales and related services.

To appreciate the range of software products available to the user community, note the categories listed on the two Web sites just cited:

- Highway engineering
- Construction management
- Demand modeling
- Environmental engineering
- General traffic
- Highway design
- Highway hydraulics
- Highway surveying
- Mapping and GIS
- Network assignment
- Pavements/maintenance
- Planning data
- Project management
- Safety and accidents
- Signal timing/signal warrants
- Site analysis
- Structural engineering
- Traffic data
- Traffic maintenance
- Traffic simulation
- Transit operations
- Transit planning

Of this list, this chapter limits its attention primarily to traffic simulation, and to some extent, signal timing, and capacity analysis.

5.4   *BASICS OF SIMULATION*

Simulation is used extensively in traffic engineering. Simulation can be divided into three categories: macroscopic simulation, mesoscopic simulation, and microscopic simulation.

Macroscopic simulation models, unlike microscopic ones, do not replicate the movements of the individual vehicles in the traffic stream. Rather, platoons of vehicles are modeled, using deterministic algorithms for the speed-flow-density relationship. Macroscopic simulation (sometimes also called continuous flow simulation) is somewhat less data intensive than microscopic models. Figure 5.2 shows an output from the macroscopic simulation model in TRANSYT-7F, which shows a detailed flow profile for a specific node and traffic movement. A similar flow profile can be shown for any

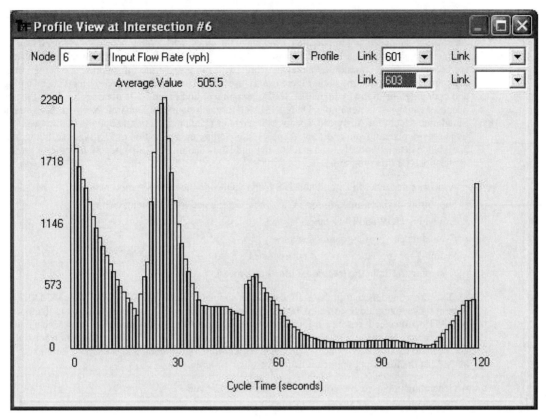

FIGURE 5.2 Macroscopic simulation of traffic flow. [*Source*: http://mctrans.ce.ufl.edu/featured/TRANSYT-7F/Release11/MacroSim.htm]

traffic movement and time period; and the software could also show queue length profiles for any traffic movement and time period. This type of output could not be obtained from a computational tool, such as the *HCM*.

Microscopic simulation models replicate as closely as possible the movements of the individual vehicles in the traffic stream. There are submodels for the various driver behaviors for responding to external stimuli that are linked together to reproduce the traffic stream. There are no deterministic algorithms for predicting the measures of effectiveness. Rather, data is gathered as the vehicles move along the facility, given that the program knows exactly where each vehicle is at each second (or subsecond). The models and capabilities of microsimulation will be discussed in detail later in the chapter. Microsimulation models can, in addition, show realistic animations of the movement of vehicles on the facility or network, since only with microsimulation do you have all the detailed information on the position of each individual vehicle at each second. Some microscopic simulation programs also replicate pedestrian behavior and the interaction between pedestrians and vehicles.

Mesoscopic simulation combines features from both macroscopic and microscopic simulation. Mesoscopic models do simulate the individual vehicle, however with aggregated behavior on the travel links, so there will be a simpler car following and lane changing model. Then, rather than moving the simulation clock ahead each second (or subsecond) and knowing where each individual vehicle is every second, the model will work off of key events and move the simulation clock ahead to the next key event. Mesoscopic models are useful for very large networks; the problem or focus areas can be simulated at a microscopic level while the other areas can be more aggregated. Aimsun 6, for instance, incorporates both a microscopic and mesoscopic simulator. Mesoscopic simulators are becoming more common as larger networks with more real-time information are being simulated and evaluated. As the nation increasingly uses intelligent transportation systems for planning and operating our highway networks, traffic assignment and planning is done at the same time as operations analysis. For example, DYNASMART-P, which stands for Dynamic Network Assignment Simulation Model for Advanced Roadway Telematics (Planning version), supports both transportation network planning and operational analyses, including the evaluation of ITS systems, but it does not include detailed models for car-following and lane-changing.[30] Reference 31 lists the following applications for this program:

- Assessing impacts of ITS and non-ITS technologies on the transportation network
- Supporting decision-making for work-zone planning and traffic management
- Evaluating HOV and HOT lanes
- Evaluating congestion pricing schemes
- Planning for special events and emergencies
- Assigning traffic in the traditional four-step process

Two other models that include a mesoscopic simulator are DynaMIT[32] and DYNASMART-X, both of which integrate historical databases and real-time inputs. These tools were designed to support ATIS (Advanced Traveler Information Systems) and ATMS (Advanced Traveler Management Systems) applications at a traffic management center (TMC). They provide information for traffic engineers to make proactive decisions based on real-time information and scenarios. References 31 and 32 list the following applications for these two programs:

- Generating unbiased and consistent information to drivers
- Efficient operating of Variable Message Signs
- Off-line evaluation of real-time incident management strategies
- Real-time incident management and control
- Evaluating alternative traffic signal and ramp metering operational strategies
- Coordinating evacuation and real-time emergencies that could block highway links
- Optimizing the operation of TMCs with real-time predictions
- Generating real-time signal control strategies
- ATIS applications
- Generating congestion pricing schemes for varying locations, times, and prevailing network state

These types of dynamic traffic management systems are the future of transportation. They use real-time information and origin-destination information, as well as simulation models, to predict network conditions and thus make it possible to analyze the effects of various traffic management decisions or traffic control measures. Such dynamic systems can also provide better traveler information, which can then optimize route choice and reduce congestion and delays because they use real-time information obtained from loop detectors and/or roadside sensors and from GPS-equipped vehicles.[31]

Table 5.1 is a portion of a table taken from Volume II of the FHWA Traffic Analysis Toolbox. It shows the types of facilities that can generally be analyzed with the different types of simulation models.

TABLE 5.1 Relevance of Tool with Respect to Facility Type

Facility type	Macroscopic simulation	Mesoscopic simulation	Microscopic simulation
Isolated intersection	•	•	•
Roundabout	Φ	O	Φ
Arterial	•	•	•
Highway	•	•	•
Freeway	•	•	•
HOV lane	•	•	•
HOV bypass lane	Φ	Φ	•
Ramp	•	•	•
Auxiliary lane	•	•	•
Reversible lane	O	O	Φ
Truck lane	Φ	O	•
Bus lane	Φ	O	•
Toll plaza	O	O	•
Light-rail lane	O	O	•

O The tool does not generally address the specific context.
Φ Some tools do and some do not address the specific context.
• The tool does generally address the specific context.

Source: From Traffic Analysis Toolbox Volume II: Decision Support Methodology for Selecting Traffic Analysis Tools, June 2004, http://ops.fhwa.dot.gov/trafficanalysistools/tat_vol2/index.html

5.5 *MICROSCOPIC SIMULATION MODELS*

As mentioned earlier, microsimulation involves modeling the behavior of individual vehicles, but what this really means is modeling the behavior of people, by replicating the responses of drivers to stimuli in the real world. People make decisions differently, they drive differently from one another, and this variability in driving characteristics is modeled.

In microsimulation, the detailed movements of each vehicle in the traffic stream are represented, that is, the model calculates the detailed trajectory of each vehicle, defining its speed and location at each second (or subsecond). Microsimulators model how the driver responds to the vehicle that is directly ahead with a car following model. There is a lane-changing model that replicates the driver's decision whether to change lanes and how to do it safely. All the ways that the driver responds to perceived stimuli are modeled, and since the behavior of any two drivers is never identical, the model must access statistical distributions to quantify specific human responses. For example, in CORSIM, when each vehicle enters the network a random number is generated that defines the driver as being somewhere on a distribution that ranges from very timid to very aggressive. This vehicle (driver) will carry this designation for its entire trip. Then when this vehicle is faced with a decision to make a permitted left turn, for instance, depending upon the driver characteristic that it was assigned, it will accept a small gap if it is an aggressive driver and will wait for a larger gap if it is a timid driver.[33]

An important benefit of simulation is that it can bring the real-world environment into the office. Because of the level of detail, simulation can provide insights that are not available from any other source. It is a tool that can be used to test various ideas and designs relatively inexpensively; because of the ability to show the results in animation, it also can be used to display these ideas to the client.

5.5.1 Random Number Generation

The most basic feature of simulation is randomness, and yet digital computers are deterministic by their very nature. How then does one generate random numbers on a computer?

The answer is that one creates a computer code that generates a sequence of numbers that *appear* to be random. In fact, there is a pattern, but it repeats only after millions of numbers. Sequential numbers in the chain do not appear to be correlated to each other even with rather sophisticated statistical techniques. These "pseudorandom" number generators are now used extensively and are rather sophisticated in masking the underlying relation. Indeed, they are so routine that spreadsheets now incorporate random-number-generating capabilities.

In many applications, the user can specify a starting or "seed" number for the chain. By specifying a different number, the user picks up the chain in a different place for each choice.

The user can also specify the same number, and be assured that exactly the same numbers are generated in exactly the same pattern. This is extremely useful when one wishes to see how two different control policies affect *exactly* the same traffic, something that is impossible in the real world.

From a statistical point of view, it also allows paired t-tests to be run on the performance data from a set of N replications under two different control policies, if the same set of N different seed numbers is used in the N runs for each policy.

It is rather natural to think of moving ahead in discrete steps of time, say one second into the future. With knowledge of vehicle speeds and positions, a car-following relation, a lane-changing rule, and certain other rules, we can then simulate the next positions (and speeds) of individual vehicles. We can also estimate whether a new vehicle entered the system in each entry lane on each link. Further, we can gather information needed for the performance indices and vehicle trajectories. This approach is called *discrete time-based* simulation.

5.5.2   Modeling the Mechanisms

The first issue is actually *enumerating* the mechanisms of the proposed simulation model and defining their interactions. When this is done, it is then necessary to model the individual mechanisms, link them, use them, and refine them in some iterative process that introduces reality. Consider the simple case of traffic traveling on an arterial. The primary mechanisms might be:

- Vehicle arrives
- Vehicle travels in lane at desired speed
- Vehicle interacts with others (car-following)
- Vehicle changes lane, or not (Overtaking? Turning? When? Why?)
- Signal indication influences vehicle
- Vehicle decelerates
- Vehicle accelerates

This simplified list does not include the ways in which the vehicle may interact with pedestrians, buses, geometrics (grade of road, for instance), weather, or sun glare. Nor does the above list include the ways in which cross traffic is considered, or many other factors.

Nonetheless, having defined these basic mechanisms, the immediate challenge is then to establish the details of the mechanisms, the links to other mechanisms (because they are generally interdependent), and the calibration/validation data needs.

Moreover, there must be a guiding principle that *anticipates the uses* to which the model will be put. This is often elusive, because users find new applications and because the state of the art changes. Nonetheless, the need exists.

Consider the simple decision to affix an identifying tag to each vehicle as it enters the network, which it retains as it moves through the network. This allows: (1) future applications to assign a route to that particular vehicle, and to update that route periodically; (2) desired speed and accel/decel patterns to be linked to that vehicle; (3) location data to be stored at each increment of time, allowing trajectories to be reconstructed. Without such an identity tag, the same vehicle might bounce through the network by Brownian motion, subject to a series of random turn decisions (with the turn percentage being a characteristic of the intersection), with no meaningful path information recoverable.

Some of the mechanisms that a simulation model can replicate include:

1. The interactions of buses with the general traffic and delay that his causes
2. The interaction between left-turning vehicles and traffic in the oncoming direction
3. The delay that is caused by turning vehicles interacting with pedestrians in crosswalks
4. Oversaturated conditions can be modeled including the queues that spillback into the upstream intersection and cause delay; also the delay to the side street is modeled due to spillback
5. Right turn on red is modeled realistically by looking for acceptable gaps in the conflicting traffic
6. Parking maneuvers causing delay to vehicles in the lane adjacent to the parking
7. The effect of short-term or long-term events, such as taxis or work zones that block one or more lanes
8. All controller types (pretimed, semiactuated, and fully actuated), as well as stop and yield signs
9. Traffic patterns that vary over time

5.5.3 Simulation of Pedestrian Environments

Many simulators (e.g., VISSIM and Aimsun) have now added a module for realistically simulating pedestrian behavior. As with the vehicles, the pedestrians are simulated individually and can model the interaction between vehicles and pedestrians (at crosswalks) as well as be used for the planning and design of urban spaces, planning layouts of public transportation terminals, designing pedestrian crossing facilities, safety analysis, and more.[34,35] The Urban Analytics Framework (UAF),[35] which is a pedestrian simulation model that uses Quadstone Paramics' traffic microsimulator combined with a "free space agent model," allows the simulated people to "move freely within the constraints of their defined space." Unlike a traditional traffic model the free space model has no links/lane/node etc. that dictate the agents movement."[36] Figure 5.3 is a graphics output from the UAF model. The model can also show Level of Service (LOS) using colored bands to show where the hot spots for pedestrian LOS are, where the definition of LOS can be defined by the user. See Figure 5.4, which is in black and white so that the colored bands cannot be seen, but the LOS definitions can be seen in the figure.

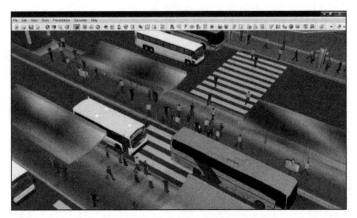

UAF presentation graphics

FIGURE 5.3 Graphic output of UAF pedestrian simulation model. [*Source*: http:// www.paramics-online.com/downloads/info_sheets/UAF_brochure.ped]

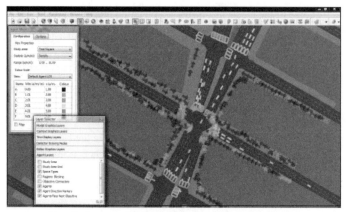

Level of service density and agent direction markers

FIGURE 5.4 Pedestrian levels of service shown on facility. [*Source*: http://www.paramics-online.com/downloads/info_sheets/UAF_brochure.ped]

5.5.4 Calibrating the Mechanisms and/or the Model

Calibration is done to adjust the model parameters used in the various mathematical relationships so that they better match to local conditions.

Consider the left-turn mechanism under permissive signalization, which is basically a gap acceptance rule that considers the size of the available gap, the speed of the approaching vehicle, and perhaps the type of opposing vehicle. It might also consider the type of left-turning vehicle, the "pressure" from queued vehicles behind the turner, the number of lanes to be crossed, and other factors.

The challenge is to define the mechanism in a way that is both realistic and capable of calibration with an affordable amount of data. The model outputs are compared against your real-world data.

At this level of detail, it may be that the model as an entity is not calibrated as such, but that it is run for "reality checks" on whether the overall performance is credible.

5.5.5 Validating the Mechanisms and/or the Model

Validation is a distinct operation from calibration, and requires data reserved or collected for the purpose. Calibration data cannot be used to validate a model, simply because it is then a self-fulfilling prophecy.

Validation can be done on two levels: the microscopic checking of individual mechanisms and the macroscopic checking of aggregate performance measures. The latter category may verify that arterial travel times, average speeds, and delay conform to real world observations when observed traffic data (volumes, composition, etc.) are fed into the model. Because this is a statistical test, usually operation with a null hypothesis that the model and real-world results are the same, a validation may require a significant amount of data.

With current technology, calibration and/or validation can more accurately be done using GPS-GIS integrated systems to collect and process actual vehicle movement information. Research done at Louisiana State University has shown that car following data, for instance, can be collected and processed using GPS with significant benefits when comparing the cost to the accuracy.[37]

5.6 ISSUES IN TRAFFIC SIMULATION

A number of issues in planning a traffic simulation deserve special attention.

5.6.1 Number of Replications

Assuming that the traffic simulation model is as random as the real world, at least in its internal mechanisms, the resulting performance measures will be *samples* or *observations* from a set of possible outcomes. That is, each performance measure is a random variable, with a mean and variance (among other properties).

Because of this, several runs (or "replications") of the situations are needed, each with its own set of seed random numbers. If one desires a certain confidence bound, a considerable number of replications might be required. If one can only afford a limited number of replications, then the resultant confidence bound might be disappointing.

Some practitioners cite a "rule of thumb" that three replications are the minimum. Many have translated that to mean that three replications will suffice. *This is without foundation, and confidence bounds must be considered in each case.*

Consider the nine replications in the table, in which the average speed is read from simulation results.

Run	Average speed (mph)
1	45.2
2	52.5
3	43.7
4	48.4
5	47.3
6	53.2
7	46.7
8	42.9
9	50.1

From the results in the table, one can estimate a mean of 47.8 mph and a standard deviation of 3.6 mph. Further, the 95 percent confidence bounds on the mean are ±2.4 mph. If ±1.0 mph were desired, then 51 replications would have been required, rather than 9.

If one could only afford 9 replications (or had time for only 9), then the 95 percent confidence bound on the mean is unavoidably ±2.4 mph.

Some situations are not as bleak as this particular example. There is also some advantage to be gained when two control policies are to be compared, in terms of their effect on a performance measure (in this case, the mean speed).

Consider the case in which nine replications are done with each of two control policies, using the same nine sets of seed numbers for each policy (that is, the identical traffic, so to speak). This "pairs" specific sets of runs and allows us to construct the following table.

Run	Control policy one, average speed (mph)	Control policy two, average speed (mph)	Paired difference (mph), CP2-CP1
1	45.2	48.1	2.9
2	52.5	54.3	1.8
3	43.7	44.3	0.6
4	48.4	51.4	3.0
5	47.3	50.2	2.9
6	53.2	55.8	2.6
7	46.7	48.3	1.6
8	42.9	45.2	2.3
9	50.1	53.5	3.4

If a test were done on this data without pairing, under the hypothesis of "the two means are the same", the variability in the data would lead us to *not reject* the hypothesis of equal means. *Even if such a difference existed, it could not be detected.*

However, by pairing the runs according to the seed numbers (i.e., the exact traffic patterns), the differences in each pair can be computed, as shown in the last column in the table.

In this particular case (which we arranged to be so dramatic), *all* of the differences are positive, with a mean of 2.3 mph and a standard deviation of 0.9 mph. Indeed, the 95 percent confidence bounds on the mean are ± 0.6 mph, and a hypothesis of "zero difference" is easily rejected. While not all cases are so dramatic, the reader should understand the advantages of a well-planned set of runs.

5.6.2 Length of Run

Early on in the practice of simulation, another rule of thumb emerged: Runs should be (at least) 15 minutes long. Again, many users interpreted this to be that they can all be 15 minutes long.

Rather than follow such a rule of thumb, the user should focus on the defining events, determine how often they occur, and select the run duration so that a reasonable number of these cases occur during the observation period.

Consider the case in which the productivity of an approach is being considered, and there are two defining events—the vehicle at the head of the queue is a left-turner who traps everyone else, or it isn't. If the first situation occurs only 10 percent of the time, and by definition can occur only once every cycle length, then a 15-minute period with a 90-second cycle length will have *no* such blockages 35 percent of the runs, and only one such blockage another 39 percent of the runs. Lengthening the run duration will dramatically lower these probabilities, if that is desired.

It is good practice to use runs of 1 to 2 hours of simulated time when a number of such "rare" events should be included in the typical period. Another situation which may lead to longer runs is taking the effect of buses into account, because the interarrival times on even a very busy route is often at most 3 to 5 minutes.

5.6.3 Specific Detailed Mechanisms

Consider the case of bus traffic in an urban area. If the simulation model specifies a dwell time distribution but does *not* link it to the interarrival time between buses, an important mechanism contributing to platooning of buses can be overlooked. This is because the delayed buses *in the model* do not have more people waiting for them, and early-arriving buses do not have fewer. If a user is looking at overall performance, this may not be a problem. But if the user is trying to study bus platooning specifically, this can be a major problem, and the mechanism might have to be created and the model revised.

5.6.4 Avoiding Use of the Model Beyond Its Limits

This is simply another aspect of the point just made, but is a particular challenge when a user is trying a new application. A working knowledge of the model is needed, on a level that only the developers and a few others might have.

5.6.5 Selecting Performance Measures for the Model

The traffic engineering profession has a number of well-established measures (volume, flow rate, speed, delay) and a number of evolving ones, particularly as related to multimodal considerations. Even so, the measures defined in some simulation models may not conform to the standard definitions and usage. For instance, there are many forms of "speed" used in both practice and in traffic models; the delay reported in a simulation model is not necessarily the same delay reported in the *HCM*.

5.6.6 User-Friendly Input and Output

Whereas the preceding items focus on some of the important issues in applying models, the most important user issues tend to be in how user-friendly the input/output is, and how efficiently it can link to existing data bases, if at all. For instance, TEAPAC, Synchro/SimTraffic, and to some extent CORSIM make major points on how clear the input process is, and/or how much they link to other models.

5.7 VISUALIZATION

There are two aspects of visualization that deserve attention: (1) the displays that tell system managers and the public the condition of the network, and (2) graphic representations of simulator outputs that provide powerful visual images of the situation.

5.7.1 Visualization of Simulation Results

Computer models are measured by their ability to effectively communicate the results, in two-dimensional or—preferably—three-dimensional representations.

Figure 5.5 shows the output of the VISSIM model in which the two-dimensional simulation output is shown in the context of aerial photographs of the area; this is a standard feature available in all microsimulation software. Figure 5.6 shows another VISSIM output, with a three-dimensional perspective. The PARAMICS and AIMSUN models also have three-dimensional displays.

Other models provide three-dimensional animations as part of the standard output, or as specialty displays for specific needs. Figure 5.7 shows an AIMSUN output, and Figures 5.8 and 5.9 show a WATSIM output tailored to an acute operational problem in a very constrained urban environment.

FIGURE 5.5 VISSIM animation of output, with aerial photographs for context. [*Source: VISSIM web site.*[25]]

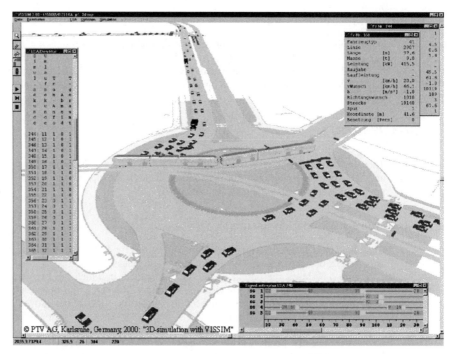

FIGURE 5.6   VISSIM three-dimensional animation of output, including an LRT line. [*Source: VISSIM web site.*[25]]

FIGURE 5.7   AIMSUN three-dimensional display of animated output. [*Source: TSS web site.*[27]]

FIGURE 5.8 WATSIM three-dimensional display, tailored to specific constrained urban setting. [*Source: KLD Associates materials,[24] used with permission.*]

FIGURE 5.9 Three-dimensional display of acute traffic problem, prepared to consider signal and design changes. [*Source: KLD Associates materials,[24] used with permission.*]

5.8   WHY SIMULATE? ADVANCED TRAFFIC CONTROL SYSTEMS, AND OTHER

The most basic reasons for using simulation extensively are straightforward:

1. Users can experiment with various control policies without disrupting real traffic and making that traffic part of an "experiment."

2. The same can be said of alternative designs for remedies to existing or future conditions, such as lane additions, rerouting of traffic, changes in direction of links, and so forth.

3. Above all, these alternative control implementations and designs can be considered rapidly, without major capital investment.

4. Whenever desired, various alternatives can be considered with exactly the same traffic load and conditions, something that is simply not feasible in the real world.

The potential place of traffic simulators in real-time control is another important reason. Historically, simulators and signal optimization programs dependent upon simulators could really only be used off-line. However, now if one aspires to truly real-time control, the time period available for the computations is rather short.

5.9   TRAFFIC SIMULATORS

At the beginning of this chapter, a number of simulators and computational tools that imbed or use simulations were enumerated. At the risk of overlooking a key model, it will simply be noted that the available and widely used tools (nationally and internationally) include CORSIM, WATSIM, VISSIM, PARAMICS, AIMSUN, and TEAPAC *Complete*. There are other tools that incorporate some simulation, such as HCM/Cinema (for intersection animation), SYNCHRO/SimTraffic for signal optimization and simulation; and others that handle signal optimization and/or capacity, including SIGNAL2000 and HCS.

Because there are a number of models in the commercial market, each with its own satisfied user base, this chapter will not attempt to rank or rate the alternative models. Indeed, the literature contains little information on extensive *comparative* testing of various models, relative to each other or to base cases with real field data.

Several of the models emphasize their suitability for intelligent transportation systems (ITS) applications. One example application is the modeling of toll plazas with a mix of exact change, cash, and electronic toll collection (ETC).

As the use of these technologies spread even more widely, the potential for a new range—and quality—of network data has become available. This shaped what simulators can *do, and how* they *are designed, to address problems that heretofore were deemed infeasible.*

5.10   REAL-TIME TRAFFIC MANAGEMENT

Simulation is now being used to support real-time decision making for traffic management centers. Simulation tools have become so fast and integrated that large systems can use macro- to meso- to micromodeling in a very short period of time. Aimsun 6, for example, intergrates static and dynamic models and includes a static traffic assignment model, a mesoscopic simulator, and a microsimulator.[38] In the Greater Toronto Area, for example, Aimsun's macro-, meso-, and micro-simulation framework is being used on the 400-series highway network and the major arterials, which covers its busiest freeways, HOV and collector-express lanes, and a fully electronic tolled highway.[39]

5.11 PLANNING TOOLS

This chapter has focused on computational models and simulation models that are becoming more powerful over time, and more capable of addressing sophisticated problems. This is proper, logical, and accurate. The profession has to some extent moved to a set of tools that simply cannot be executed by hand. For a document such as the *Highway Capacity Manual (HCM)*[1] to recognize and accept this principle for operational analyses is truly ground-breaking. The 2010 *HCM*, for the first time, does not have worksheets for the signalized intersection nor urban streets chapter, and the methods can no longer be done by hand. They are still deterministic models, but they are iterative and more complex than ever before.

At the same time, there is a growing demand for *planning* techniques that are fast, simple, and easy to use and communicate. It is generally acknowledged that these must be consistent with operational techniques but capable of giving preliminary guidance on the capability of facilities in terms of AADT and quality of flow. The Florida DOT has been a pioneer in this development of modern simplified tools with software such as the multimodal ARTPLAN[40] for arterial planning and FREEPLAN[41] for freeway planning, to handle both the early stages of planning (generalized planning) for "in the ball park" estimates of LOS, and for preliminary engineering, which covers conceptual planning, preliminary design, and project development, as well as some aggregate assessment of operational impacts.

This is given special note because these models contribute to the development of a spectrum of tools available to the professional, suited to the precision and data availability needed at different stages; in this continuum, it is the simulation models that often represent the most sophisticated of the approaches.

5.12 RESOURCES FOR FURTHER INFORMATION

To a large extent, the references for this chapter have been listed as a set of Web sites, covering most of the well-used models and some software distribution centers. These Web sites contain much information on the availability and performance of software systems used for transportation.

REFERENCES

1. Traffic Analysis Toolbox Volume 1: Traffic Analysis Tools Primer, June 2004, http://ops.fhwa.dot.gov/trafficanalysistools/tat_vol1/index.html

2. Traffic Analysis Toolbox Volume II: Decision Support Methodology for Selecting Traffic Analysis Tools, June 2004, http://ops.fhwa.dot.gov/trafficanalysistools/tat_vol2/index.html

3. Traffic Analysis Toolbox Volume III: Guidelines for Applying Traffic Micro-simulation Modeling Software, June 2004, http://ops.fhwa.dot.gov/trafficanalysistools/tat_vol3/index.html

4. Traffic Analysis Toolbox Volume IV: Guidelines for Applying CORSIM Micro-simulation Modeling Software, January 2007, http://ops.fhwa.dot.gov/trafficanalysistools/tat_vol4/index.html

5. Traffic Analysis Toolbox Volume V: Traffic Analysis Toolbox Case Studies—Benefits and Applications, November 2004, http://ops.fhwa.dot.gov/trafficanalysistools/tat_vol5/index.html

6. Traffic Analysis Toolbox Volume VI: Definition, Interpretation, and Calculation of Traffic Analysis Tools Measures of Effectiveness, January 2007, http://ops.fhwa.dot.gov/publications/fhwahop08054/index.htm

7. Traffic Analysis Toolbox Volume VII: Predicting Performance with Traffic Analysis Tools, March 2008, http://ops.fhwa.dot.gov/publications/fhwahop08055/index.htm

8. Traffic Analysis Toolbox Volume VIII: Work Zone Modeling and Simulation—A Guide for Decision-Makers, August 2008, http://ops.fhwa.dot.gov/wz/trafficanalysis/tatv8_wz/index.htm

9. Traffic Analysis Toolbox Volume IX: Work Zone Modeling and Simulation—A Guide for Analysts, March 2009, http://ops.fhwa.dot.gov/wz/trafficanalysis/tatv9_wz/index.htm

10. CORSIM: Microscopic Traffic Simulation Model, Version 5.1, http://mctrans.ce.ufl.edu/featured/tsis/Version5/corsim.htm

11. Highway Capacity Manual, Transportation Research Board, 2004, http://www.trb.org/Main/Blurbs/Highway_Capacity_Manual_2000_152169.aspx

12. Highway Capacity Software (HCS+) at McTrans Web site, http://mctrans.ce.ufl.edu/hcs/

13. SIGNAL*2000* at Strong Concepts Web site, http://www.strongconcepts.com/

14. TRANSYT-7F at McTrans Web site, http://mctrans.ce.ufl.edu/featured/TRANSYT-7F/index.htm/

15. PASSER II-02 at McTrans Web site, http://mctrans.ce.ufl.edu/store/description.asp?itemID=29

16. *General Features of TRANSYT-7F*, http://mctrans.ce.ufl.edu/featured/TRANSYT-7F/Release11/MacroSim.htm

17. SIG/Cinema at KLD Associates Web site, www.kldassociates.com

18. Synchro/SimTraffic at Trafficware Web site, http://www.trafficware.com/

19. TEAPAC *Complete* at Strong Concepts Web site, http://www.strongconcepts.com/

20. *The Urban Traffic Control System in Washington DC*, USDOT, FHWA, Washington DC, Sept 1974; see also "Logical design and demonstration of UTCS-1 network simulation model" by Lieberman, Worrall, and Bruggeman in *Transportation Research Record* 409, (1972).

21. Andrews, B. et al. (1989). "The NETSIM Graphics System." *Transportation Research Record* 1112, Transportation Research Board.

22. *FRESIM User Guide, Version 4.5*, Turner Fairbank Highway Research Center, USDOT, FHWA, Washington DC, April 1994.

23. *TSIS User's Guide, Version 4.0 beta, CORSIM User's Guide Version 1.0 beta*, Kaman Sciences Corporation, January 1996.

24. WATSIM at www.kldassociates.com

25. VISSIM Web site, www.atacenter.org/tst/Vissim.html

26. PARAMICS Web site, http://www.paramics-online.com/index2.htm

27. AIMSUN Web site, www.aimsun.com/site/content/category/1/32/57/

28. TransCAD, Caliper Corporation, www.caliper.com/tcovu.htm

29. Priority, Market-Ready Technologies and Innovations List, Federal Highway Administration, http://www.fhwa.dot.gov/crt/lifecycle/dynasmart.cfm

30. DYNASMART-P, https://mctrans.ce.ufl.edu/store/description.asp?itemID=780

31. What are Dynamic Traffic Management Systems? http://eng.odu.edu/transportation/itsdecision/newitsd/ITS_Serv_Tech/traff_manag/dyn_trafficmangmt_summary.html

32. DynaMIT, http://mit.edu/its/dynamit.html

33. TSIS-CORSIM Documentation, www.mctrans.ce.ufl.edu/featured/tsis

34. Pedestrian Simulation with VISSIM, www.vissim.de/software/transportation-planning-traffic-engineering/software-system-solutions/vissim/pedestrian-simulation/

35. Pedestrian Simulation The Urban Analytics Framework, http://www.pedestrian-simulation.com/

36. UAF, Many Problems One Solution, http://www.pedestrian-simulation.com/reason_integrated_approach.php

37. Sivaram, C. and M. Kulkarni, M., "GPS-GIS integrated systems for transportation engineering." http://www.gisdevelopment.net/technology/gps/techgp0008pf.htm

38. Aimsun 6, The Integrated Transport Modeling Software, http://www.aimsun.com/site/content/category/1/32/53/

39. "MTO Chooses Aimsun as the Modeling Software for Ambitious Proof-of-Concept." http://www.docstoc.com/docs/25117464/

40. ARTPLAN, Florida Department of Transportation, part of LOSPLAN software package, http://www.dot.state.fl.us/planning/systems/sm/los/los_sw2m2.shtm

41. FREEPLAN, Florida Department of Transportation, part of LOSPLAN software package, http://www.dot.state.fl.us/planning/systems/sm/los/los_sw2m2.shtm

CHAPTER 6

APPLICATIONS OF GIS IN TRANSPORTATION

Gary S. Spring

Department of Civil Engineering, Merrimack College
North Andover, Massachusetts

6.1 INTRODUCTION

The concepts that drive geographic information systems (GIS) have become commonplace in today's everyday activities—witness the use of guidance systems in our cars, our use of on-line mapping programs (such as Google maps, Yahoo maps, etc.) to explore destinations. All of these are based upon basic GIS functions, namely, linking data to maps. The application of geographic information systems in the transportation industry (GIS-T) has consequently become widespread in the 20 years or so. Indeed, 8 years ago, Miller and Shaw (2001) stated that GIS-T had "arrived" and had become one of the most important applications of GIS. In the intervening years, the use of GIS has expanded dramatically. All fifty states have institutionalized their GIS activities and forty five of them provide clearinghouses for their geospatial data (AASHTO, 2008). For a full list of GIS acronyms and jargon visit the National Center for Geographic Information and Analysis Web site (Padmanabhan et al., 1992), and download report number 92-13.

In the United States, several key pieces of federal legislation, passed since the early 1990s—from the Intermodal Surface Transportation Efficiency Act of 1991 through to the latest $73.3 billion budget request to Congress (FY2010), and legislation by states that mandate the development of transportation programs to reduce traffic impacts—contain explicit requirements for local and state governments to consider transportation systems through their interdependence with other natural, social and economic systems. There exists a need, therefore, for enhanced approaches to store, manipulate and analyze data spanning multiple themes—for example, highway infrastructure, traffic flow, transit characteristics, demographics and air quality.

Later in this chapter, several applications from each of these areas, and others, are described. GIS offers a data management and modeling platform capable of integrating a vast array of data from various sources, captured at different resolutions, and on seemingly unrelated themes. The objectives for this chapter are to define and describe the basic elements of GIS-T; provide a review of GIS-T application areas, and to present issues involved in implementing these systems. The discussion begins with some background to GIS and GIS-T providing definitions and some discussion regarding the "why" of using them.

6.2 BACKGROUND

The GIS is a computerized database management system that provides graphic access (capture, storage retrieval, analysis, and display) to spatial data. GIS software provides a map display that allows thematic mapping of data and its graphic output overlain onto a map image. Figure 6.1 depicts the generic framework used by virtually all current GIS software packages.

The key element that distinguishes GIS from other data systems is the manner in which geographic data are stored and accessed. The types of GIS packages used for transportation applications store geographic data using topological data structures (objects' locations relative to other objects are explicitly stored and therefore are accessible) that allow analyses to be performed that are impossible using traditional data structures. The addition of this spatial dimension to the data base system is, of course, the source of power of GIS. Without this dimension, the GIS is merely a database management engine. Linked with the spatial dimension, its database management features enable GIS to capture spatial and topological relationships among georeferenced entities even when these relationships are not predefined. Standard GIS functions include thematic mapping, statistics, charting,

1:500,000

(a)

1:20,000

(b)

FIGURE 6.1 Effects of scale.

matrix manipulation, decision support systems, modeling algorithms, and simultaneous access to several data bases.

In contrast to other GIS applications, GIS-T has as its central focus a transportation network that has an intrinsic complexity due to its varying legal jurisdictions and multimodal nature. The digital representation of such a network is nontrivial and will be discussed in the following section.

6.3 BASIC PRINCIPLES OF GIS-T

6.3.1 GIS-T Data

Data and the modeling of those data are crucial elements of a successful system. The GIS derives information from raw data—one of its primary strengths. Although data and information are often used interchangeably, they can be quite different. Data consist of facts or numbers representing facts whereas information derives from data and gives meaning to those facts. This Section describes the nature, types and sources of data, and the modeling techniques and data models which are used to convert the raw data to information.

The Nature of Data.   There are two types of data in general—spatial data and attribute data. The former describes the physical geography represented in a data base whereas the latter, attribute data, linked to the spatial data, describe the "attributes" of the spatial objects. For example, a point could represent a tree location and its attributes might be species, diameter, or height. If a line feature represents a road segment, attributes might include pavement type, number of lanes, or speed limit. Area attributes might include soil type, vegetation cover, or land use. An attribute is the generic descriptor for a feature, while each feature has a specific value. In the point example above, likely values for the species attribute would be Pine, Fir, and Aspen. Attributes may be described as the questions that would be asked and values as the answers. Attributes to be collected would be determined in the software planning process, while specific values would be entered in the field.

The two most common ways of representing spatial data are the vector and raster formats. A familiar analogy for these formats is bit-mapped photos (such as jpeg) corresponding to the raster format and vector-based ones (such as tiff) representing the vector format. Each of these approaches has advantages and disadvantages and the choice between them lies in the nature of the application for which the data are to be used. Data structure, that is choice between raster and vector, is, in general, one of the first major decisions to be made in establishing a GIS. With regard to GIS-T, this decision is fairly straightforward, however. Transportation applications typically are concerned, as was pointed out earlier, with networks, political boundaries and so on. These are all well-defined line systems and thus lend themselves better to the vector data structure. This notwithstanding, there are situations in which the raster format is used for transportation applications and, so, the next few paragraphs provide a description of both structures including some discussion of their relative advantages and disadvantages.

The Raster Data Structure.   Raster data consist of rows of uniform cells coded according to data values. For example, a landscape scene would be gridded, and each cell in the grid would be given a single landscape identity, usually a code number that refers to a specific attribute measure (e.g., a particular land use or type of land cover). The number might also be an actual measurement value, such as an amount of rainfall. These cells are akin to pixels on a computer monitor. Indeed, carrying the analogy further, pixels have values associated with them as well—color, hue, and shades of grey—and the degree of resolution provided relates to the size of the pixel. Similarly, the degree of approximation in a landscape relates to the size of the raster cell. A grid comprised of small cells will follow the true location of a boundary line more closely, for example. However, as with photo resolution, there is an overhead cost, namely, the size of the raster file containing the data will increase as cell size decreases. In general, raster provides a simple data model and fast processing speed at the expense of the excellent precision provided by the vector model with its higher data needs. The nature of raster data lends itself to natural resources applications whose data elements tend to be continuous in nature—such as soil type.

The Vector Data Structure. Vector data consist of points, lines and closed polygons or areas, much the same as the drawing elements in computer-aided drafting (CAD) programs. The lines are continuous and are not broken into a grid structure. In the vector model, information about points, lines and polygons is encoded and stored as a collection of *x-y* coordinates. The location of a point feature, such as a manhole, is described by a single *x-y* coordinate. Linear features, such as roads, are stored as a string of point coordinates. Area features, such as census tracts or traffic analysis zones, are stored as a closed loop set of coordinates. The vector model represents discrete features like buildings better than continuous features like soil type. The vector format, in general, provides a more precise description of the location of map features, eliminates the redundancy afforded by the raster model and therefore reduces mass storage needs, and allows for network-based models. Vector models are, however, often computationally intensive—much more so than their raster counterparts.

Vector models are most often used for manmade infrastructure applications—such as transportation. Indeed, for transportation applications, vector-based GIS is, by far, the most commonly used data structure. GIS-T, in representing guideways (or roadways) uses a coordinate system to store and display primitive feature elements of points, lines, and areas. The roadway is represented as a "line" feature.

A combined collection of graphical links form a roadway network, but this representation alone is considered as having no "intelligence"—that is, connectivity (also known as topology) does not exist. Unlike purely graphical software applications, however, GIS can build and manage topology. A "road network" consists of a series of roadway "line" features that have the same defined attributes, for example, road name. This connectivity of line features is critical for use in routing and network data modeling. With respect to the latter, GIS network algorithms may be used to determine optimal travel routes, which roadway or other facility is closest, or what is near a particular site. Appropriate topology may be used most effectively if built into the GIS data.

Topology describes how graphical objects connect to one another by defining relative positions of points, lines and areas. For example, topology allows queries about which street lines are adjacent to a census tract area, or what intersection node points form the end points of a street segment (link). The latter information is essential for routing applications. Only vector-based data include topological information which is in part why these data may have higher costs associated with them—the presence of topology implies the maintenance of additional databases whose purpose is to store information on the connectedness of points, lines and areas in the database.

Scale and Accuracy. Modeling the road network as a spatial or graphical layer in GIS must reflect the needs and requirements of an agency. Scale and accuracy are important data considerations, especially when using more than one data source, which is often the case in transportation applications. Generally, the source material and the standards of data development determine both the scale and precision of spatial datasets.

Simply defined, scale is the relationship between distance on the map and distance on the ground. A map scale usually is given as a fraction or a ratio—1/10,000 or 1:10,000. These "representative fraction" scales mean that 1 unit of measurement on the map—1 in or 1 cm—represents 10,000 of the same units on the ground. Small-scale maps, have smaller representative fractions than large-scale maps; for example, 1 in 500,000 versus 1 in 24,000, respectively, that is, large is small. As scale size increases, the ability to depict detail also increases. For example, a map with scale 1 in 500,000 would necessarily represent only main roads and only by centerline whereas with a scale of 1 in 20,000, details such as ramps, collectors, and direction of travel could be shown as well. See Figure 6.1 for examples. On much larger scales, such as 1 in 600 (1 in = 50 ft), features such as pavement markings, actual lane designations, and specific design elements can be depicted. The latter scale is the level of resolution often used in roadway design work.

GIS employs a wide range of data sources reflecting the varied goals of the systems themselves. Since GIS-T may involve applications as varied as archeological analysis, marketing research, and urban planning, the source materials can be difficult to inventory and classify comprehensively. The fact that many different scales may be encountered adds to the complexity of the problem. Even within a single GIS project, the range of materials employed can be daunting. If multiple datasets are contemplated, and they do not have common scales, the process of conflation (the fusion or marrying together of data) may be used in some cases to spatially integrate the datasets to create a new master coverage with the best spatial and attribute qualities. In situations such as this, the analyst should use extreme care in the use of these conflated data.

Geocoding. The linkage between attribute data and spatial data as stated previously is the key advantage of using GIS. The process by which this is accomplished is called geocoding. There exist a great many tools to accomplish this process. Some involve matching a non-map-based database, such as a list of names and addresses, with a map-based database, such as one containing zip code centroids. Using a common field, and the relational model (described in the next section), coordinate information is attached to the name and address.

Other common geocoding tools involve the use of distances along reference lines and offsets from these lines to place objects on a map, thus requiring some sort of linear referencing system (LRS), which will also be described in a later section. For example, going back to the non-map-based names and addresses database, one could use a street centerline file which includes address ranges on its links to geocode addresses by estimating the address locations based on the length of the segment and the address range assigned to the segment. For example, an address of 41 Elm Street would plot as shown in Figure 6.2.

Linear Referencing Systems. Address ranges as linear references are rarely adequate for maintenance of information on transportation infrastructure. Consequently, transportation agencies and planning organizations (e.g., U.S. Department of Transportation, state departments of transportation, and metropolitan planning organizations) have developed more sophisticated linear referencing systems (LRS) for their transportation facilities. These are similar to the address range example mentioned previously but use transportation network related measures, such as mile post, to locate objects. Baker and Blessing (1974) define LRS as "a set of office and field procedures that include a highway location reference method." Linear referencing is a means of identifying a location on a

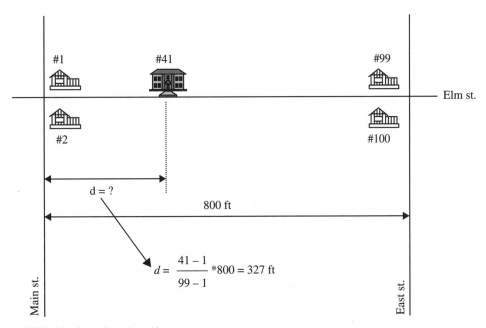

FIGURE 6.2 Geocoding using address ranges.

linear feature, such as a road or railroad. LRS support the storage and maintenance of information on events that occur within a transportation network—such as pavement quality, crashes, traffic flow and appurtenances (traffic signs and signals for example). They provide the mechanism by which these data elements can be located in a spatial database. LRS has traditionally been accomplished using either references to the roadway or references to monuments along the roadway. This method is known as "linear referencing." Several variations of linear location referencing systems exist, the most common of which fall into the five categories listed and described below (Adams, et al., 2000).

- Route-Milepost (RMP)
- Route-Reference Post (RRP)
- Link-Node (LN)
- Route-Street Reference (RSR)
- Geographic Coordinates

The Route-Milepost (RMP) system is, perhaps, the most common linear referencing method used, particularly at the State DOT level. It uses measured distance from a known point, such as a route's beginning or where it crosses a state or county line, to a referenced location. The distance is usually specified to the nearest hundredth of a mile, although, in some cases, other less precise measures may be used. The point of interest (such as the location of a crash) is always offset in a positive direction from the zero mile point, and is not referenced to other intermediate points along the route. An advantage of this system is that reference posts or signs need not be maintained in the field. However, it requires that field personnel know where the route begins and the primary direction of the referencing system. It also means that, with each realignment along the route, the referenced distances change.

The Route-Reference Post (RRP) method uses signs posted in the field to indicate known locations. These signs, known as "reference posts," may or may not reflect mileposts. All road feature data collected in the field are referenced to these markers in terms of distance and direction. The features can later be converted to corresponding mileposts using cross-referencing tables and maps. The advantage of this system over RMP is that the problems associated with changes in route length due to realignment are eliminated. This method overcomes the disadvantages of the mile point method but adds the burden of maintaining signage in the field.

In a Link-Node (LN) system, specific physical features, such as intersections, are identified as nodes. Each node is considered unique and is assigned some sort of unique identifier such as a node number. Links are the logical connection between nodes and may vary in length. Links also have unique identifiers that are often derived from the associated pair of node numbers. All features are measured as an offset distance from the nearest or lowest node number along a link.

The Route-Street Reference (RSR) system is more commonly used in municipalities and relies on a local system of streets to locate physical features. In this system, an event is typically recorded as occurring on one street at a specified distance and direction from a reference street. A variation of this system is the use of two reference streets and no distance measurement. For example, a crash may be coded as occurring on Street A between Streets 22 and 23. This option results in a loss of detail with regard to precise location, but still provides enough information for some uses—such as identifying road sections that have a high crash frequency.

The use of coordinates as a reference system is becoming more common with the advent of technologies such as global positioning systems (GPS). Cartesian coordinates use an x-y coordinate plane to represent location. Geographic coordinates use latitude and longitude that are measured in degrees along the axes of the sphere of the Earth. Transportation authorities at the state and local levels generally use the former system, in the form of state plane coordinates, to measure (in meters or feet) the distance east and west or north and south along a State origin or datum.

The need to manage, analyze, and understand transportation information in a linear context will persist because transportation systems are essentially linear phenomena. The current trend is to collect data using two- or three-dimensional devices, such as GPS, which allows for new economies in the field and the continued use of existing legacy while supporting network models, applications, and displays (Fletcher et al., 1995). Advances in the use of GPS and their inexpensive data collection and navigation tools hold promise in replacing the LRS. For example, Oklahoma DOT has developed and implemented a GPS-based inventory of its rail crossings (Greager, 2009).

Data Processing. For the processing and management of data the two paradigms most commonly used are the relational and object oriented models. It should be noted that the database is the operation center of the GIS where much of the primary work is done. Graphics may or not be necessary for an application but, almost without exception, the database is a key part of any analysis.

Relational database management systems (RDBMS) are well suited for ad hoc user queries—an important aspect of GIS analysis. The relational model uses tables of data arranged as columns (categories of data) and rows (each observation entry). Columns are called fields, and rows are called records—as shown in Figure 6.3. Consider, for example, the table of courses shown in Figure 6.4. Note that each course's attributes are read across, on a row. Queries may be made from these tables by specifying the table name, the fields of interest and conditions (for example, course credit hours greater than 3). The rows that meet those conditions are returned for display and analysis. The power of RDBMS lies in its ability to link tables together via a "unique identifier" which is simply a field common to both tables and whose values appear only once. Thus, using the current example, one could, in addition to querying about courses from Figure 6.4, query about which students take courses from which professor, number of students in each class who hail from Missouri, geographic distribution of students, and so on. This is done via the common field, student ID, which links the two data tables, as shown in Figure 6.5. Within GIS, this capability to link tables is used to link spatial objects with tables containing information relating to those objects—as shown in Figure 6.6.

Object-oriented database management systems (ODBMS) offer the ability to integrate the GIS database with object-oriented programming languages such as Java and C++. Martin and Odell (1992) state that the object-oriented approach: "models the world in terms of objects that have properties and behaviors, and events that trigger the operations that change the state of the objects. Objects interact formally with other objects." In short, the database has a set of objects with attached attributes. At this simplest level, ODBMS is analogous to the RDBMS model in that its objects represent the latter's "rows" in tables, and its attributes the RDBMS fields. The ODBMS model is much more powerful than this simple definition implies, however. Three concepts are crucial to understanding the ODBMS: abstraction, encapsulation, and inheritance.

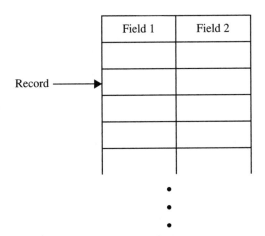

Field 1	Field 2

Record ———▶

•
•
•

FIGURE 6.3 DBMS—The relational model.

Professor	Course	Hours	Course ID
Adams	CE341	3	44356
Burken	CE441	3	77877
Nanni	CE223	4	33645
Spring	CE210	3	65645
Zhang	CE446	4	48655

FIGURE 6.4 Table in the RDBMS.

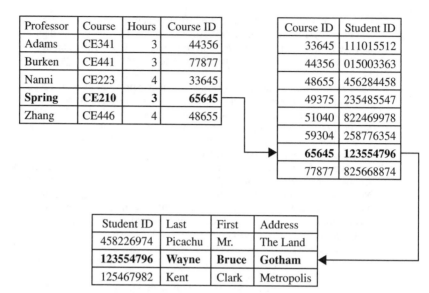

Professor	Course	Hours	Course ID
Adams	CE341	3	44356
Burken	CE441	3	77877
Nanni	CE223	4	33645
Spring	**CE210**	**3**	**65645**
Zhang	CE446	4	48655

Course ID	Student ID
33645	111015512
44356	015003363
48655	456284458
49375	235485547
51040	822469978
59304	258776354
65645	**123554796**
77877	825668874

Student ID	Last	First	Address
458226974	Picachu	Mr.	The Land
123554796	**Wayne**	**Bruce**	**Gotham**
125467982	Kent	Clark	Metropolis

FIGURE 6.5   Linking tables in the RDBMS.

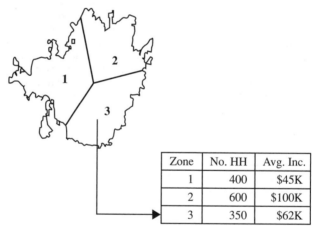

Zone	No. HH	Avg. Inc.
1	400	$45K
2	600	$100K
3	350	$62K

FIGURE 6.6   RDBMS in GIS.

Data abstraction is the process of distilling data down to its essentials. The level of abstraction indicates what level of detail is needed to accomplish a purpose. For example, on a small scale, a road network may be presented as a series of line segments, but on a large scale, the road network could include medians, edges of roadways, and roadway fixtures (Rumbaugh, 1991). Encapsulation includes procedures with the object data. In other words, code and data are packaged together. The code performs some behavior that an object can exhibit—for example, calculating the present serviceability index of a section of road. Thus, encapsulation allows the representation of an object to be changed without affecting the applications that use it. Modern software programming provides a familiar example of this concept. The application Excel may be called as an object in a Visual

Basic routine. The routine would then have access to the application's full functionality through its attributes (built-in functions).

Finally, inheritance provides a means to define one class of object in terms of another. An object class represents a group of objects with common operations, attributes, and relationships (Fletcher et al., 1995). An object is a specific instance of a class. Each object can have attributes and operations (or code, as explained above). For example, a conifer is a type of tree. There are certain characteristics that are true for all trees, yet there are specific characteristics for conifers. An attribute is a data value held by objects in a class (Rumbaugh et al., 1991)—for example, for the attribute "tree color" the value might be "green." An operation is a function that may be applied to or by objects in a class. Objects and classes of objects may be connected to other objects. That is, they may have a structural relationship called an association with these other objects. The primary purpose for inheritance is software reuse and consistency (Lewis and Loftus, 1998).

ODBMS hold great promise as GIS tools. Their power, with respect to their abilities to interface with object oriented programming languages and to provide "smart" data, and their utility when applied to extremely complex databases all support their eventual preeminence in the GIS industry. Indeed, in the late 1980's most GIS vendors were moving from RDBMS to ODBMS models (Sutton, 1996). However, today most states continue to use the RDBMS models to manage their geospatial data.

Data Sources.   Digital transportation data may be obtained in a variety of ways and from a large number of sources, all of which fall into one of the following four categories:

- Digitizing source materials
- Remote sensing
- GPS for road centerline data
- Existing digital sources

Digitizing Source Materials.   Digitizing source materials involves digitizing data elements from aerial photographs (considered to be primary source material) or from hardcopy maps (considered to be secondary source material). The process involves collecting the *x-y* coordinate values of the line features by tracing over each one using a digitizing tablet with a cursor or puck as the input device to locate and input map features into the computer from a paper map. This may also be done using a mouse with a digital raster image map. This latter approach is called "heads up" digitizing. This type of manual production requires planning, source material preparation, and production setup, in addition to postprocessing of data (Davis, 1996). The costs for this type of data acquisition often represent the majority of system startup costs. Semiautomated methods using map scanning and line tracing technologies are sometimes used to lower the cost and improve the accuracy of the digitizing process.

Remote Sensing.   Remote sensing is the science (and to some extent, art) of acquiring information about the Earth's surface without actually being in contact with it. This is done by sensing and recording reflected or emitted energy and processing, analyzing, and applying that information. Remote sensing provides a convenient and efficient means of identifying and presenting planimetric data. Imagery is available in varying scales to meet the requirements of many different users. In general, there are two sources for remotely sensed data: photogrammetry, which uses cameras on-board airplanes, and satellite imagery. The former produces photographs that may either be in paper form, in which case they must be digitized, as was described in the previous section, or in digital form which may then be used directly as raster images. Cameras and their use for aerial photography are the simplest and oldest of sensors used for remote sensing of the Earth's surface. Cameras are framing systems that acquire a near-instantaneous "snapshot" of an area of the surface. Camera systems are passive optical sensors that use a lens (or system of lenses collectively referred to as the optics) to form an image at the focal plane, the plane at which an image is sharply defined. Aerial photographs are most useful when fine spatial detail is more critical than spectral information, as their spectral resolution is generally coarse when compared to data captured with electronic sensing devices. The geometry of vertical photographs is well understood and it is possible to make very accurate measurements from them, for a variety of different applications (geology, forestry, mapping, etc.). The use of satellites involves an interaction between incident radiation and the targets of interest. Note, however, that remote sensing also involves the sensing of emitted energy and the use of nonimaging sensors. With the advent of online mapping software (as mentioned before, such as Google and Yahoo provide), satellite imagery for the entire United States and most countries around the world is available. These maps are fairly small scale and are therefore are mainly useful for planning purposes only.

Global Positioning Systems. These systems use a constellation of 31 actively broadcasting satellites and spherical geometry calculations to determine geo-positions (Wiki, 2009a). These data, in the form of coordinates, are transformed into path-oriented or linear locations. These measurements are referenced to a standard ellipsoid model instead of base maps or linear field monuments. This allows for new economies in the field and the continued use of existing legacy, and it supports path and network models, applications, and displays.

Existing Digital Data from Other Sources. A cost-effective alternative to digitizing is to acquire digital data from a third-party source, such as those described below.

In the recent past, most GIS projects have had to rely almost exclusively upon data available only in printed or "paper" form. Much of these data are now available in digital form while continuing to be published on paper. The ever-increasing pace of this transformation from paper to digital sources has many repercussions for GIS. Inexpensive, and in many cases free, access to high-quality data will enhance the use of GIS in the coming years. The Internet and Worldwide Web are being used more and more to distribute data and information, thus requiring users to know where to look and how to search networks.

All data sources have strengths and limitations. Digital sources are no different. It is important to understand their characteristics, costs, and benefits before using them. Learning a little about commonly employed digital formats will save much work in the long run. Although the types of materials will vary greatly from project to project, GIS practitioners know something of the characteristics and limitations of the most commonly available data sources. These are materials collected and published by a variety of government agencies and commercial interests and are used quite widely.

Local, state, and federal government agencies are major suppliers of digital data. Better than 90 percent of states maintain databases with state and county route centerlines, and 98 percent of these also maintain other geo-spatial databases, such as other transportation networks and features (rail lines, airports, etc.). The majority of the states (62 percent) distribute these data free of charge and the others have policies in place that allow data to be shared with other public agencies. Finding the data appropriate for a given system may often require significant on-line research. This is perhaps less the case at the federal level mainly because certain key agencies such as the Bureau of the Census, United State Geological Survey, Soil Conservation Service, National Aeronautics and Space Administration (NASA), and Federal Emergency Management Agency, provide standard sorts of information for the entire nation.

Key federal data sources and helpful indexes to those sources include:

The Bureau of Transportation Statistics (BTS, n.d.) provides links to state GIS resources, national transportation atlas data files, which consist of transportation facilities, networks and services of national significance throughout the United States. The files are in shapefile format, which is the proprietary format for ESRI data. There are also links to DynaMap/1000 files which contain centerline data and features for nearly every street in the Nation as well as the National Highway Planning Network which is a comprehensive network database of the nation's major highway system. It consists of over 400,000 miles of the nation's highways comprised of Rural Arterials, Urban Principal Arterials and all National Highway System routes. The data set covers the 48 contiguous States plus the District of Columbia, Alaska, Hawaii, and Puerto Rico. The nominal scale of the data set is 1:100,000 with a maximal positional error of ±80 m. The network was developed for national level planning and analysis of the U.S. highway network. It is now used to keep a map-based record of the National Highway System and the Strategic Highway Corridor Network.

The United States Geological Survey (USGS) (USGS, n.d.) produces topographic maps for the nation, as well as land use and land cover maps, which include information about ownership and political boundaries, transportation, and hydrography. For access to these data as well as an extensive index of other federal on-line data sources see the USGS publications Web site. The index includes links to the Global Land Information System and tends to concentrate on data created through the USGS.

The Bureau of the Census (Census, n.d.) provides socioeconomic and demographic data, census tract boundary files and street centerline networks (TIGER files) for the entire nation. TIGER comes from the acronym **T**opologically **I**ntegrated **G**eographic **E**ncoding and **R**eferencing and was developed by the U.S. Census Bureau to support its mapping needs for the Decennial Census and other Bureau programs. The topological structure of the TIGER database defines the location and relationship of streets, rivers, railroads, and other features to each other and to the numerous geographic entities for which the Census Bureau tabulates data from its censuses and sample surveys. It has a scale of approximately 1 in equals 100 ft and is designed to assure no duplication of these features or areas.

The National Aeronautics and Space Administration (NASA, n.d.) provides remotely sensed data from all over the world. These data are typically in raster format and are 30-m resolution.

The National Technical Information Service (NTIS) established FedWorld, a clearinghouse for digital data, which is an excellent resource for on-line federal information of any kind.

As mentioned above, 45 states provide clearinghouses for their geo-spatial data, significantly easing the task of finding the data desired. Additionally, every state now has a contact person who takes responsibility for the state's GIS. Most data of this type are now available in standard formats, come with adequate documentation and data quality reports, and but still may need to be checked as to origin and quality. Furthermore, the spatial data holdings themselves have limitations. As mentioned earlier different data sources may have different scales associated with them and require conflation to integrate them. Conflation, however, may not yield the information required. For example, the most widely held data, the USGS 7.5-min quadrangle data (1:24,000 scale), are increasingly becoming unsuitable for the uses of a transportation agency. For example, when collecting driveway data with a GPS unit at 1-meter accuracy, centerlines based on 40-ft accuracy do not work well. Updating base maps with greater accuracy has the potential to be an expensive issue.

A primary advantage offered by government data sets is that most are in the public record and can be used for free or for a small processing fee. Some of the agencies that do not provide their data free of charge have found that this practice seriously impedes data sharing.

There also exist many private sources of information. Commercial mapmaking firms are among the largest providers, but other firms have for years supplied detailed demographic and economic information, such as data on retail trade and marketing trends. Some of this information can be quite expensive to purchase. Also, it is important to check on restrictions that might apply to the use of commercially provided data.

Many software vendors repackage and sell data in proprietary formats as well. These data are usually checked and corrected during the repackaging process. The use of these converted datasets can save time. Some firms will also build datasets to a user's specifications. These are often termed "conversion" firms. They are usually contracted to build special purpose datasets for utility companies and some government agencies. These datasets are often of such special purpose that they cannot be assembled from existing publicly available sources, say when an electric utility wishes to digitize its maps of its service area.

6.3.2 Data Quality

It has been said that an undocumented dataset is a worthless dataset. If a data set's pedigree and quality are unknown, for example, the user must spend time and resources checking the data. Vendors should provide a data dictionary that provides a description of exactly what is in the file (data types and formats), how the information was compiled (and from what sources), and how the data were checked. The documentation for some products is quite extensive and much of the detailed information may be published separately, as it is for USGS digital products.

Some characteristics to consider when evaluating data sets are:

- Age
- Origin
- Areal coverage
- Map scale to which the data were digitized
- Projection, coordinate system, and datum used
- Accuracy of positional and attribute information
- Logic and consistency
- Format
- Reliability of the provider

In summary, sometimes the costs of using and converting publicly and commercially available digital files outweigh their value. No matter how much data becomes available publicly, there is no guarantee that it will contain exactly the sorts of information necessary for specific projects.

Data Standards. For a variety of reasons, GIS users often want to share or exchange data, usually having to do with overlapping geographic interests. For example, a municipality may want to share data with local utility companies and vice versa. GIS software developed by commercial companies store spatial data in a proprietary format protected by copyright laws. Although vendors often attempt to provide converters, in general, one company's product may not read the data stored in another's format. Applications of digital geospatial data vary greatly, but generally users have a recurring need for a few common themes of data. These themes include transportation, hydrography (rivers and lakes), geodetic control, digital imagery, government boundaries, elevation and bathymetry, and land ownership (or cadastral) information. A lack of investment, common standards, and coordination has created many situations in which these needs are not being met. As a result in some cases, important information may not be available and in others, data sets may be duplicated. A means to maintain and manage the common information being collected by the public and private sector does not exist. This results in increased costs and reduced efficiency for all involved.

Thus, data standardization is therefore a fundamental consideration in developing GIS for integration with existing databases. All users of GIS data depend upon the establishment of standards that should address simple integration and processing of data within the GIS. Standards should include spatial data modeling and scale, accuracy, resolution, and generalization, and datum and projection mapping. As illustrated by the internet and its interoperability, hardware platforms, operating systems, network environments, database systems, and applications software are generally *not* an issue. The standards for data definitions are much more important than these latter items for providing reliable and portable systems and applications. The GIS software standards, however, can add to the complexity of sharing data sets, since not all GIS share a common route system that is easily transferred from one vendor-specific application to another.

During the 1980s the USGS worked with academic, industrial, and federal, state, and local government users of computer mapping and GIS to develop a standard for transfer and exchange of spatial data. In 1992, after 12 years of developing, reviewing, revising, and testing, the resulting standard—SDTS—was approved as Federal Information Processing Standard (FIPS) Publication 173-1 (NIST, 1994). The SDTS requires a two-step process for transferring data from one platform to another. The source data are exported by the first GIS to the SDTS format, then the second GIS imports the transfer file, creating a target dataset in its own file format. This approach, while extremely useful in enabling data sharing, is cumbersome at best. Recognizing the criticality of data sharing The National Spatial Data Infrastructure (NSDI) was established by President Clinton's Executive Order 12906 on April 11, 1994 to implement the recommendations of the National Performance Review published by his Administration in the Fall of 1993 [superseded in 1998 by ANSI NCITS 320-1998 (ANSI, 1998)].

The Order states that:

*Geographic Information is critical to promote economic development, improve our stewardship of natural resources, and protect the environment.... National Performance Review has recommended that the Executive Branch develop, in cooperation with state, local, tribal governments, and the **private sector**, a coordinated National Spatial Data Infrastructure to support public and private sector applications of geospatial data...*

The concept of this infrastructure was developed by representatives of county, regional, State, Federal, and other organizations under the auspices of the Federal Geographic Data Committee (FGDC, 2009). Rechartered in August 2002, the Committee, an interagency committee, promotes the coordinated development, use, sharing, and dissemination of geospatial data on a national basis. This nationwide data publishing effort is known as the National Spatial Data Infrastructure (NSDI).

The private sector is also actively engaged in building the "National Spatial Data Infrastructure" to meet market place needs defined as business opportunities. The Open GIS Consortium of 388 private sector, public sector, universities, and not-for-profit organizations representing technology users and providers is addressing the issue of easy access to spatial information in mainstream computing. OGC is working to develop open software approaches that facilitate the development and use of location dependent software applications using spatial data to increase farm productivity, identify disease and health threats, assist police and law enforcement in identifying crime patterns and many more. In other words, GIS users will be able to access one another's spatial data across a network even if they are using different GIS software programs (OGC, 2009).

6.3.3 Spatial Modeling

The benefits of GIS-T are well-established. It provides:

- The capability of storing and maintaining large data sets of spatial and tabular information
- Display and analytical capabilities that model the physical proximity of spatial features
- Flexibility in modeling spatial objects to suit the particular needs of the user or application
- Database integration
- Image overlay capabilities
- Network analyses (e.g., shortest path routing).

These capabilities have developed as the technology has matured. Over the past 10 years, GIS-T has adapted to accommodate linear referenced data. Crash and roadway inventory data are examples of this type of linear data and can now be brought into GIS for display and analysis. This feature makes it easier to understand the spatial relationships within data that are not found in other information systems. In addition, GIS-T provides a programming environment that allows users to develop specific analysis programs or customize existing programs. All functions for display and analysis can be employed in a single-system design using common programming languages, such as Visual Basic, C++, and Java (Smith et al. 2001). With the advent of object oriented programming, GIS-T can be integrated into more mainstream enterprise applications, as well as Web-based client applications. GIS-T provides abilities broader than simply mapping data and includes several types of analytical capabilities that may be broadly categorized into five groups:

- Display/query analysis
- Spatial analysis
- Network analysis
- Cell-based modeling
- Dynamic segmentation

Display/Query Analysis.   The primary appeal of GIS to many is its graphical capabilities as the adage "a picture is worth a thousand words" bespeaks. Maps are the pictures GIS uses to communicate complex spatial relationships that the human eye and mind are capable of understanding. The computer makes this possible, but still, it is the GIS user that determines what data and spatial relationships will be analyzed and portrayed, or how the data will be thematically presented to its intended audience. In short, the GIS allows analyses at a higher level of abstraction than do standard database tools. Using the database capabilities of GIS, the analyst can query the database and have the results displayed graphically. This query analysis, when spoken in everyday conversation, takes on the form of a "show me" question, such as "Can you show me sections of road that are in poor condition?" However, query analysis in GIS can also be used for other purposes, such as database automation, which might be used for error checking and quality control of coded data. As an example, the GIS roadway database could be queried automatically during the crash data entry process to verify the accuracy of speed limit and other crash report variables coded by an officer.

Spatial Analysis.   Several analytical techniques, grouped under the general heading "overlay analysis," are available in GIS for spatial analysis and data integration. GIS provides tools to combine data, identify overlaps across data, and join the attributes of data sets using feature location and extent as the selection criteria. For example, the number of acres of wetlands impacted by a proposed highway corridor could be obtained by this overlay process. Overlay techniques may also be used in combining data features by adding (or by applying some other function) one data set to another, or by updating or replacing portions of one data set with another—thus, creating a new spatial data set. For example, the analyst could use these techniques to combine number of households and the average number of school age children with pedestrian-related crashes, in order to derive risk factors for the total number of pedestrian-related crashes relative to the total number of school age children per road segment, for pedestrian-to-school safety analysis.

Proximity analysis represents the fundamental difference of GIS from all other information systems. Buffering is a means of performing this practical spatial query to determine the proximity of neighboring features. It is used to locate all features within a prescribed distance from a point, line, or area, such as determining the number of road crashes occurring within one half mile from an intersection, or, the number of households that fall within 100 ft of a highway's layout line.

Network Analysis.   Unlike proximity analysis that searches in all directions from a point, line, or area, network analysis is restricted to searching along a line, such as a route, or throughout a network of linear features, such as the road network. Network analysis can be used to define or identify route corridors and determine travel paths, travel distances, and response times. For example, network analysis may be used to assess the traffic volume impact of a road closure on adjacent roadways. GIS networking capabilities can also be used for the selection of optimal paths or routes - for example, finding shortest paths between zonal centroids. The network may include turning points, avoid improper turns onto one-way streets, represent posted traffic control restrictions, and include impedance factors to travel (such as mean travel speeds, number of travel lanes, and traffic volumes) to enhance the network analysis. It is this capability that one sees in the on-board navigation tools currently available in some models of automobile.

Cell-Based Modeling.   Cell-based modeling, also referred to as "grid-based" analysis, uses a grid or cells to aggregate spatial data for discrete distribution. Although similar to raster-based systems, these are vector-based. In cell-based modeling, the spatial data are developed as tiles of a given dimension, or points of a uniform distribution, as defined by the user, for display and analysis. Cell-based modeling is effective in displaying patterns over larger areas, such as representing the sum total of crashes that are located within a cell. This capability provides a quick means to view spatial clustering of crash data. Another example of this modeling approach is the representation of ground water contamination levels using statistical models as the mathematical base. Since cell-based modeling aggregates data at a specified grid resolution, it *would not* be appropriate for site-specific spatial analysis. In cell-based modeling, special tools are available to merge grid data for overlay analysis. Cell-based overlay analysis is similar to the GIS overlay analysis previously discussed; however, the techniques and functions available in cell-based modeling are somewhat different. When the cells of different data sets have been developed using the same spatial dimensions, they can be merged on a cell-by-cell basis to produce a resulting data set. The functions and processes used in cell-based modeling to merge grid data are referred to as "map algebra," because the grid data sets in cell-based modeling are merged using arithmetic and Boolean operators called "spatial operators."

Dynamic Segmentation.   Dynamic segmentation (dynseg) is a process that allows the association of multiple sets of attributes to a portion of a linear feature without having to modify feature geometry or topology. Although implementation of dynamic segmentation varies by GIS vendor, GIS uses dynamic segmentation to locate and display linear features along a route and/or to segment the route itself. The process consists of interpolating the distance along the measured line of the GIS route from the beginning measure to the ending measure of the line using attributes as the interpolation criteria. In short, the dynseg process allows temporary modification of feature geometry "on the fly" based upon the attribute data being considered. Consider for example a vector in a database representing a segment of road. As explained earlier, this line object is actively linked to a row in a table of attribute data—say road condition. The issue addressed by dynamic segmentation is the inability of the GIS to represent features that do not all begin at the beginning of the line and end at its end. In other words, using the road condition example, a road segment's geometry may run from station 0+00 to station 5+00 but the road condition may (and probably does) change somewhere in between. Figure 6.7 depicts a situation where pavement condition and traffic volume levels are shown along the line segment and demonstrates the difficulty in determining various combinations of pavement condition and traffic volume using fixed segmentation. As the exhibit shows, dynamic segmentation allows representation of these nonuniform attributes without requiring the physical segmentation of the link—that is, no modification of geometry is necessary.

This robust method to represent model links allows the dynamic management of many-to-one relationships between GIS segments attribute features. The method offers several advantages over fixed segmentation. First, the editing of the routes is preferable to editing the underlying base map. Second, the routes may be stored in simple correspondence tables that can be easily read by planning models, for example. Third, depicting the transportation network as linear features (or "linear events") in the GIS means that route tables and associated transit line tables can be managed automatically through dynamic segmentation. Dueker and Vrana (1992) describe dynamic segmentation in great detail.

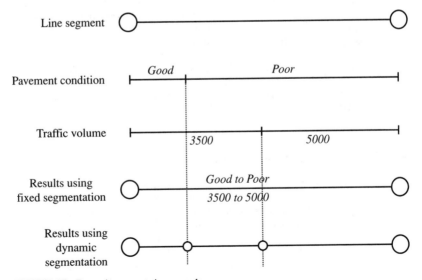

FIGURE 6.7   Dynamic segmentation example.

6.4 *GIS APPLICATIONS IN TRANSPORTATION*

All state departments of transportation (DOT) have now initiated some GIS activity and more than 40 now have at least one full-time staff person with a geography or cartography background as a member of the GIS unit. GIS-T applications have evolved dramatically in the past 5 to 10 years. Just 10 years ago, GIS was primarily used in single applications. Now it has evolved into serving enterprise wide information through multiplatform applications. Additionally, GIS application development has grown in the past few years significantly. More than one third of states now spend in excess of a quarter of a million dollars solely on GIS application development, almost half of which is outsourced. There has been a significant increase in recent years in the number of state DOTs developing more accurate road base maps using digital imagery and kinematic GPS. Indeed, more than thirty states now report that the spatial resolution of their road centerline databases is 1:12,000 or better. There has also been significant growth in Web-based GIS applications and the use of GIS for managing road feature inventories (AASHTO, 2009).

Certainly the legislation described at the beginning of this chapter has played a major role in the increasing use of GIS-T. But, perhaps the more compelling reasons contributing to the advent of this technology are the technical ones such as the increased ability to integrate diverse databases, better integration with modeling tools, and linkage with global positioning systems and with advanced imaging systems such as Light Detection and Ranging (LiDAR). As indicated earlier, transportation agencies across the nation have invested significant resources in developing GIS data sets and transportation models leading to the large number of clearinghouses that they now maintain. These have facilitated better data integration and display of a wide array of transportation network files and databases. Almost ten years ago, Dueker and Butler (2000) addressed the difficult issue of data integration. Different database formats supporting diverse applications lead to inconsistencies, inaccuracies and duplication in updating. They proposed a framework for sharing transportation data that attempts to clarify roles among participants, data producers, data integrators, and data users. In a follow up study, Dueker et al. (2001) proposed a clearinghouse approach to collecting and disseminating GIS-T data. The approach allows users to control the selection and segmentation of transportation features in updating existing GIS-T databases that meet the needs of various private and public applications and has since been implemented, as stated earlier, in 45 states.

Certainly wherever there exists a need to analyze, manage or view spatial data, GIS can play a powerful role. There exist a host of application examples in traditional transportation areas such as planning, traffic analysis, network analysis and routing, and emergency response. In the past five years, however, state DOTs have overwhelmingly (and consistently) identified five key areas of development for their GIS: asset management, enterprise deployment, improved location referencing systems, web-based applications, and, most recently, Light Detection and Ranging (LiDAR)—a remote sensing system used to collect topographic data. In 2009, all but the last item appear in the top-five GIS activity categories for state DOTs. This section describes current GIS-T applications to common transportation problems and provides examples in these other key areas of development as well.

6.4.1 Enterprise GIS/Asset Management

The days of state DOT enterprises being data rich and information poor are ending. As indicated above, the use of GIS as a resource for business intelligence, management and accurate product and marketing information to clients and customers. Most states acknowledge that GIS as a tool holds a high priority for most state DOTs. Indeed, as the ESRI enterprise report points out (ESRI, 2007), GIS technology is fast becoming one of the linchpins of their command centers. A study conducted by the Florida Department of Transportation (Cambridge Systematics, 2008) supports this statement. It found that a broad array of transportation agencies have moved away from the traditional, so-called, stovepipe business model (narrow and rigid responsibilities, output, and feedback) to an integrated approach to implementing GIS.

The concept of Enterprise GIS has been defined in many different ways based on the different objectives of particular projects. While some have defined it as a centralized repository of data, others called it a package of applications and tools. Based upon the findings of its Phase I report, the Florida Department of Transportation using the following definition (Cambridge Systematics, 2008):

> *Enterprise Geographic Information Systems (GIS) is an organization wide framework for Department communication and collaboration of shared geospatial data and GIS resources that enhances existing business processes and provides an efficient way to plan, analyze, and manage transportation infrastructure and related elements.*

Enterprise data integration continues to be cited by most states as yielding the greatest benefit (70 percent) and the most difficult and costly to implement. (AASHTO, 2009 summary). It is a data-intensive process whose resulting data integration improves integrated decision making. The evolution of geographic information systems (GIS) and spatial technologies provides powerful mechanisms for developing enterprise GIS/asset management decision-making products.

Benefits of such systems include (Hall, 2006):

- Visualization for DOT personnel, managers, legislature
- Enabling more complex analysis of data for decisions—e.g., program development, bundling of projects, intermodal
- More sophisticated analysis—travel flow—modal direction, ramps
- Enterprise integration of data sources for asset and modal evaluation
- Intranet/Internet delivery of information
- E-government capabilities—public access
- Integration of external data

Many agencies struggle with the development of these spatial products on an enterprise basis, however. Reported difficulties with data integration and quality include:

Wide variety of legacy systems and platforms

Data integration problems

Multiple linear referencing methods

Data quality, timeliness, accuracy problems—common and potentially critical issue.

Problems in integrating more precise information, especially with future data collection

Need for more details in the data, e.g., specific bridge spans

Access to historical data that are typically not spatially enabled within agencies

Technical and organizational difficulties in implementing an enterprise resource

Field data collection, validating, and updating—expensive, especially with large geographic coverage

Updating—validating road centerline

Integrating databases internal and external—different identifiers, timeframes, precisions

Limited established transportation models for analysis

Right-of-way and parcel map inaccuracies

Training and cultural issues of users, managers

Lack of consistent internal standards

6.4.2 Transportation Planning

Decision support is the primary goal of transportation planning and consequently of GIS-T applied to transportation planning. Several models and datasets are available to planners for modeling trip generation for impact studies, land use projection, population and employment projection, four step traffic projection for network traffic assignment and corridor studies, air quality assessments of mobile sources of pollution, alternatives analyses, noise analysis, and energy studies.

Urban planning models have long been available to assist planners in recommending decisions regarding transportation investment. Today, many planning agencies also use GIS for inventory and data presentation. Many studies have suggested the utility of GIS in the context of transportation forecasting and scenario analysis, and many agencies have begun integrating these two commonly used computer packages.

There exist several commercially available transportation modeling and/or GIS programs that incorporate both modeling and GIS capabilities. Indeed, a recent study found that more than 54% of state DOTs and MPOs surveyed have integrated these functions (Boxill, 2005). For example, TransCAD (Caliper Corporation), UFOSNET (RST International Inc.), ESRI's "Arc" products (more than 60% of users report using the latter in the Boxill 2005 study), GRASS and FreeGIS (both are open source software projects) and Intergraph are all GIS-based systems that include transportation forecasting procedures (FHWA, 2009). Transportation planning software packages that have added GIS-capable interfaces, such as Tranplan, also exist (CTRE, 2009).

6.4.3 Traffic Engineering

Application of GIS to traffic engineering is a fairly recent trend. GIS-T is especially well suited to maintain, analyze, and present city-wide spatial data useful in traffic engineering programs. The ability of GIS to integrate advanced technologies such as GPS, surveillance cameras and vehicle location devices is now being applied to address traffic systems issues in large, congested urbanized areas (Easa and Chan, 2000).

Safety management systems lend themselves extremely well to analysis using GIS tools. There have been a host of efforts to apply GIS to safety needs ranging from integrating GIS with existing safety analysis tools to the integration of computational intelligence tools with GIS. Vonderohe et al. (1993) documented the development of a conceptual structure, implementation plan, and management guidelines for a GIS-T capable of supplying the highway safety information requirements for most transportation agencies. Indeed, several agencies have begun to implement GIS-T within their safety management systems. Many have used global positioning systems in concert with GIS to develop point topology (Miller and Shaw, 2001). Crash data, once geocoded, may be integrated with road attributes, and sign and signal inventories to enhance safety analyses using routing, dynamic segmentation and buffering—all of which are built-in GIS functions. Much of the effort in such systems, though, has been focused on the very difficult problem of getting the data into the database—relying on built-in GIS functions for applications. In 1999, responding to the need for specialized tools, specifically designed for GIS applications, FHWA released its GIS Safety Analysis Tools CD which expands the analytical features of the Highway Safety Information System (HSIS) by integrating them with GIS functions. The tools provide practitioners with programs to perform spot/intersection analysis, cluster analysis, strip analysis, sliding-scale evaluations, and corridor analysis. The 2001 update includes additional pedestrian and bicycle safety tools to select safe routes to schools, assess the bicycle compatibility of roadways, and define high pedestrian crash zones (Smith et al., 2001).

Given the ability to program within many GIS packages, researchers have attempted to expand GIS functionality by integrating advanced technologies. Spring (2002) developed a prototype knowledge-based GIS for the identification and analysis of hazardous highway locations. The system examined the use of these tools to address the problems of "false positives" intrinsic to most existing methods for identifying hazardous locations, and examined how knowledge-based systems could best interface with GIS to analyze these so-called "black spots." An earlier study by Carreker and Bachman (2000) used GIS to address problems with crash data quality by providing better access to and improving multiple public and private road databases. Panchanathan et al. (1995) developed knowledge-based GIS for managing and analyzing safety-related information for rail-highway grade crossings. They developed an integrated system for the management of safety information. The system integrated rail-highway grade crossing safety data from the Federal Rail Administration and Delaware Department of Transportation (DelDOT), and an accident prediction model developed by USDOT.

The GIS environment makes it possible to develop interrelated maps, databases and traffic management applications. Several studies have examined the use of GIS for incident management in particular. Siegfried and Vaidya (1993) developed a GIS-based automated incident management plan for the Houston area. The study evaluates the use of GIS to relate incident locations with the transportation network and to make decisions and calculations for incident management. Their prototype system does alternate routing, incident response and resource management, and it allows for network and incident information queries. Ozbay et al. (2000) implemented a web-based real-time decision support system for wide-area incident management that can be used by all the involved agencies. The system is based upon an earlier effort described by Kachroo et al. (1997), which developed an integrated expert system/GIS for this purpose. Ozbay and Mukherjee (2000) implemented the system on the World Wide Web using a collection of Java-based tools and describe implementation and development issues, such as communication speed, security issues, technological obsolescence, use of heuristics to estimate key parameters (for example, incident duration), that they faced.

6.4.4 LiDAR

Light Detection and Ranging (LiDAR) is a remote sensing system used to collect topographic data. The National Oceanic and Atmospheric Administration (NOAA) and NASA scientists use this technology to document topographic changes along shorelines. The data are collected with aircraft-mounted lasers capable of recording elevation measurements at a rate of 2,000 to 5,000 pulses per second and have a vertical precision of 15 cm (6 in) (NOAA, 2009). The primary difference between LiDAR and radar is that with LiDAR, much shorter wavelengths of the electromagnetic spectrum are used, typically in the ultraviolet, visible, or near infrared. In general it is possible to image a feature or object only about the same size as the wavelength, or larger (Wiki, 2009b).

The state of Iowa has recently acquired and is in the process of implementing a LiDAR-based system using the Quick Terrain Modeler software (produced by Applied Imagery; http://appliedimagery.com/) in which it will collect terrain data to be used for urban and land use planning, emergency response, transportation and hydrological applications (Abrams, 2009). Trimble's Geospatial Division has developed tools for the automatic detection of road signs, poles, and pavement marking, as well as the automatic calculation of road geometry. The tools are based upon advanced photogrammetric techniques combined with an array of integrated sensors provides an efficient, safe and cost effective solution for large scale infrastructure that can have multiple uses. The most recent development is integrating LiDAR (laser) technology for automation (Brown, 2009).

6.5 IMPLEMENTATION ISSUES

Implementing GIS-T is not as simple as merely installing a new piece of software into an existing system (such as a pavement management system) and then operating it. In addition to the information technologies intrinsic to GIS-T, successful GIS implementation involves elements, such as personnel and their GIS skills, the organizational structure within which they work, and the institutional relationships that govern the management of information flow (Hall, 2006).

GIS is an enabling technology and serves as a platform for integrating various types of data, systems, and technologies. It may be applied to a host of different applications, as demonstrated by the preceding limited review of current applications. Among these are systems that allow for near real-time assessment of conditions, simultaneous sharing of large databases and integration, and enhancement of existing transportation models—all complex and challenging applications. The fact that the adoption rate of GIS technologies has grown exponentially makes successful implementation of these technologies even more challenging. Key areas that must be addressed when implementing GIS include data, people and organizations, and technology.

6.5.1 Data

The collection, maintenance, and use of data are challenging issues that must be addressed early in the implementation process. Who will collect and maintain data, who "owns" the data and who will serve as its custodian are all questions that must be answered.

The main reason that these questions are important is that they are key in determining if one can take advantage of GIS's main strength, namely, its ability to share data. That is, as stated previously, GIS serves as a logical and consistent platform in a common location reference system and allows diverse databases to be integrated and shared among different divisions of a department, for example. Integration standards should therefore be established to integrate different databases, some spatial, some not. Answers to the questions also address barriers, such as institutional and organizational arrangements, to the implementation process.

Technical questions, such as what is the nature of the data, how is it to be spatially referenced and what is its accuracy must be considered—as well as the means by which data from various sources and applications can be integrated. To understand the various data elements in a complex system, metadata (information about data, such as when collected, by whom, scale) should be part of a system. Additionally, spatial data requires establishment of some sort of linear referencing system as was described earlier.

All GIS data, spatial data and associated attribute data, suffer from inaccuracy, imprecision and error to some extent. Data quality assurance and quality control rules ensure the delivery of high quality data. Use of a data steward to collect, maintain and disseminate GIS data would facilitate this.

6.5.2 Technology

While GIS applications no longer bear the stigma of "new technology" and have been accepted into the mainstream of professional practice, there nevertheless exist technological issues that must be addressed when implementing them. Among these are the identification of critical technologies to be used, and keeping abreast of technological innovations. These include technologies that support interoperability (sharing of data and processes across application/system boundaries), Web-based GIS (given its significant growth in the past few years), and a rapidly expanding range of GIS applications. Of course, chosen technologies must match the architecture environment as well. The fact that industry standards are in their infancy exacerbates the problem of choosing the appropriate technologies for a given system. It is crucial to identify limiting technologies, statements by software vendors notwithstanding.

In making this decision one should look to the future, but not too far into the future. One should also keep in mind that there exists an increasing pressure to make georeferenced transportation data more accessible (and understandable) to the general public.

The introduction of new information technologies is necessarily accompanied by a change in organizational structures and institutional arrangements.

6.5.3 Organizations and People

Organizational and people issues are perhaps the most difficult to address. They continue to be more critical and more difficult to solve than technological issues. Personnel at both management and technical levels must be involved in implementing the system. Yet, this has been identified consistently as a problem area—especially maintaining support of upper management. Convincing

1. Integration of information for better decisions
2. Decreased risk of poor decisions due to incomplete data
3. Access to more accurate data
4. Reduction of duplicate data
5. Ability to quickly visualize interrelationships of various data and projects
6. Ability to develop public information maps
7. Assist with revenue initiatives
8. Quick response to internal and external queries
9. Easy access to information and maps
10. Significant time savings for information access
11. Unprecedented analysis of information
12. Identification of trends
13. Assist in public relations
14. Better cooperation with regulatory agencies
15. Increased safety for personnel
16. Ability to monitor department commitments

FIGURE 6.8   Intangible benefits of GIS.

decision-makers to accept the idea of GIS is key to a system's success. A top-down, rather then bottom-up GIS management strategy should be adopted for GIS planning and implementation.

Of primary interest to decision-makers are two questions: Who pays (for the hardware, software and personnel required) and from where will the resources be drawn? As with any investment decision, GIS must have economic justification. Hall et al. (2000) conducted a benefit cost analysis of GIS implementation for the state of Illinois. They calculated a ratio of slightly less than one but also identified several intangible benefits for GIS project, shown in Figure 6.8, which indicate that GIS benefits outweigh implementation costs.

Champions (at both the management and "grass-roots" levels) are critical to a system's success as well. They can facilitate a positive decision to purchase and enhance the chances for system success.

With regard to the latter, qualified GIS personnel must be available. What training and education are necessary for minimum acceptable qualifications are based upon the level of GIS knowledge needed. There are likely to be three levels of GIS staff and users: local GIS users, local GIS specialists, and a GIS application/data steward. Training for the GIS support personnel is extremely important to the success of GIS.

Early implementation of GIS is often more dependent on vendor supplied training. However, in the long term, the GIS support group should develop specifications for in-house training. Several GIS entities (Zhang et al., 2001) have proposed that a certification for GIS professionals be established to address this important issue. Getting and retaining adequately trained GIS staff continue to be problems. Indeed, there will be increased reliance on outside experts (consultants) for more complex GIS analyses (because of the difficulty in finding and keeping expert staff on internal payroll).

Additional problems of which to be aware relate to the nature of GIS data, namely, that, to be successful, it must be shared. This leads to turf battles and questions about who should have access and how much to the GIS.

6.6 CONCLUSIONS

GIS-T technologies hold much promise in improving how transportation does business and in making our transportation systems safer, cheaper, and more efficient. The transportation community has an unprecedented opportunity over the next few years to obtain, use, and distribute spatial data using these technologies. If used intelligently, the data will provide a wealth of information about transportation and its relationship to the quality of life on global and local scales.

Emerging issues and applications at the forefront of current attention include:

- Data visualization
- GIS-T customers
- Linear referencing systems
- LiDAR
- Enterprise GIS

6.6.1 Data Visualization

Data visualization continues to be an emerging issue for the GIS-T community. The growing popularity of commercial visualization tools like Microsoft's Virtual Earth and Google Earth have put pressure on GIS-T practitioners to build similar tools. The "look and feel" and speed of the commercial tools are sought by executives and the public. Traditional GIS-T tools, by themselves, do not have these attributes. Particular functions of interest are:

- Data and network security concerns
- Additional and new licensing, maintenance, and development costs
- Executive and management expectations
- Mash-up mapping techniques
- Three-dimensional views

It is an exciting time to be involved in GIS-T with the advance of geospatial technology into the mainstream. However, expectations for the technology have risen above the resources available to deliver on them. Commercial technology integration with traditional GIS-T tools will require follow-up discussions concerning development, implementation, and lessons learned.

6.6.2 GIS-T Customers

The more traditional areas of pavement, bridge, traffic, maintenance, and planning and programming are being imposed upon by emergency management, incident management, winter operations, Intelligent Transportation Systems (ITS), grade crossings, and automated vehicle locating. This increased desire to include geospatial technologies in diversified business processes has put new demands on GIS-T technologies and technologists.

State DOT's have resisted developing and maintaining an address geocoding centerline. It is a large investment with small return for the DOT. However, other state and local agencies use geocoding to solve many business problems on public and private roads.

6.6.3 Linear Referencing System (LRS)

Linear referencing continues to be leading topic of concern. There exists a critical need for multi-level linear referencing, that is, different linear referencing methods (LRMs) to be used on a single set of centerlines. Several business problems drive this concern:

- Address geocoding
- Adapting historic linear referencing methods (LRMs)
- Accommodating active multiple LRM's

- Enabling data exchange
- Centerline maintenance

The National Cooperative Highway Research Program (NCHRP) 20-27 framework, described earlier (Adams et al., 2000), is the most widely known data model for multilevel linear referencing. AASHTO's Technology Implementation Group is focused on introducing the concepts to other states.

6.6.4 Better Understanding of LiDAR

Geospatial technologists need to develop proficiency using and interpreting these data. Additionally, planners and engineers need to be educated on all of the possibilities LiDAR can offer them in decision support and efficiency. This remains a largely untapped data resource.

6.6.5 Enterprise GIS/Asset Management

Although a commonly used theme, this remains poorly defined to the GIS-T community. Without a proper definition, attainment can not be determined nor measured, and the role of GIS remains obscure (AASHTO, 2009).

This chapter has provided a brief overview of GIS-T, defined terms, reviewed general principles and analysis technologies, identified issues and described major application areas within the GIS-T arena. This material can only serve as a superficial introduction to the applications of GIS to transport problems but it is hoped that the reader will gain valuable insight into how these systems work, where they have been applied and what difficulties GIS users face when implementing such systems. It is further hoped that, with the "veil of ignorance" removed, users will more readily consider this important class of tools for use.

REFERENCES

AASHTO, Summary Report for 2008 Geographic Information Systems in Transportation Symposium, http://www.gis-t.org/, American Association of State Highway and Transportation Officials, 2008.

AASHTO, Summary Report for 2009 Geographic Information Systems in Transportation Symposium, http://www.gis-t.org/, American Association of State Highway and Transportation Officials, 2009.

Abrams, E. 2009. "Iowa DOT LiDAR," AASHTO GIS-T Conference, Session 1.4.1, Oklahoma City, OK.

Adams, T. M., N. A. Koncz, and A. P. Vonderohe. 2000. "Functional Requirements for a Comprehensive Transportation Location Referencing System," Proceedings of the North American Travel Monitoring Exhibition and Conference, Middleton, WI.

American National Standards Institute. 1998. ANSI, Spatial Data Transfer Standard, ANSI NCITS 320-1998.

Baker, W. and W. Blessing. 1974. "Highway Linear Reference Methods, Synthesis of Highway Practice 21," National Cooperative Highway Research Program. Washington, DC: National Academy Press.

Boxill, S.A. 2005. "The Integration of GIS and Transportation Modeling: A State-of-the-Practice Review," U.S. Department of Transportation, University Transportation Centers Program, Report No. SWUTC/05/473700-00043-1, September 2005.

Brown, D. 2009. "New Tools and Technologies of Mobile Mapping," AASHTO GIS-T Conference, Session 1.4.2, Oklahoma City, OK.

Cambridge Systematics. 2008. "Phase I of an Enterprise Geographical Information System for Transportation," Florida Department of Transportation, Final Report for Contract No. BDI40. 2008.

Carreker, L. E. and W. Bachman. 2000. "Geographic Information System Procedures to Improve Speed and Accuracy in Locating Crashes," *Transportation Research Record*. 1719, pp. 215–218, Transportation Research Board.

Center for Transportation Research and Education (CTRE), "GIS-T Tranplan Enhancement, http://www.ctre.iastate.edu/Research/enhance/, accessed August 2009.

Davis, B. 1996. *GIS: A Visual Approach*. New Mexico: OnWord Press.

Dueker, K. J. and R. Vrana. 1992. "Dynamic Segmentation Revisited: A Milepoint Linear Data Model." AASHTO GIS-T Symposium, Portland, OR.

Dueker, K., J. A. Butler, P. Bender, and J. Zhang. 2001. "Clearinghouse Approach to Sharing Geographic Information Systems—Transportation Data." Transportation *Research Record*, no. 1768, pp. 203–209, Transportation Research Board.

Dueker, K. and J. A. Butler. 2000. "A Geographic Information System Framework for Transportation Data Sharing," *Transportation Research Part C: Emerging Technologies*. 8(1):13–36.

Easa, S. and Y. Chan (eds.). 2000. *Urban Planning and Development Applications of GIS*. American Society of Civil Engineers, Reston, VA.

ESRI, "Enterprise GIS," http://www.esri.com, accessed January 2007.

Federal Geographic Data Committee (FGDC), "Data Clearing House," http://www.fgdc.gov/, accessed August 2009.

Federal Highway Administration (FHWA), "GIS in Transportation," http://www.gis.fhwa.dot.gov/, accessed August 2009.

Fletcher, D., T. Henderson, and J. Espinoza. 1995. "Geographic Information Systems—Transportation ISTEA Management Systems, Server—Net Prototype Pooled Fund Study Phase B Summary." Sandia National Laboratory, Albuquerque, New Mexico.

Greager, B., "Managing a Statewide Transportation Inventory," AASHTO GIS-T Conference, Session 3.4.3, 2009, Oklahoma City, OK.

Hall, J. P., T. J. Kim, and M. I. Darter. 2000. "Cost-Benefit Analysis of Geographic Information System Implementation: Illinois Department of Transportation." *Transportation Research Record*, Issue Number 1719, pp. 219–226, Transportation Research Board.

Hall, J. P. 2006. "Geospatial Information Technologies for Asset Management," Transportation Research Circular E-C108, Transportation Research Board.

Kachroo, P., K. Ozbay, and Y. Zhang. 1997. *Development of Wide Area Incident Management Support System*, FHWA Report DTHF71-DP86-VA-20.

Lewis, J. and W. Loftus. 1998. *Java Software Solutions: Foundations of Program Design*. Reading, MA: Addison-Wesley.

Martin, J. and J. Odell. 1992. *Object-Oriented Analysis and Design*. Englewood Cliffs, NJ: Prentice-Hall, Inc.

Miller, H. J. and S. L. Shaw. 2001. *Geographic Information Systems for Transportation: Principles and Application*. Oxford: Oxford University Press.

NIST, Spatial Data Transfer Standard, Federal Information Processing Standards Publication 173-1, National Institute of Standards and Technology, 1994.

NOAA, "LiDAR project," National Oceanic and Atmospheric Administration, http://www.csc.noaa.gov/products/sccoasts/html/tutlid.htm—accessed August, 2009.

Open GIS Consortium (OGC), http://www.opengeospatial.org/, accessed 2009.

Ozbay, K. and S. Mukherjee. 2000. "Web-based Expert Geographical Information System for Advanced Transportation Management Systems." *Transportation Research Record* n 1719, pp. 200–208, Transportation Research Board.

Padmanabhan, G., M. R. Leipnik, and J. Yoon. 1992. A Glossary of GIS Terminology, National Center for Geographic Information and Analysis, Report No. 92-13, ftp://ftp.ncgia.ucsb.edu/pub/Publications/Tech_Reports/92/.

Panchanathan, S. and A. Faghri. 1995. "Knowledge-Based Geographic Information System for Safety Analysis at Rail-Highway Grade Crossings," *Transportation Research Record* n 1497, pp. 91–100, Transportation Research Board.

Rumbaugh, J., M. Blaha, W. Premerlani, F. Eddy, and W. Lorensen. 1991. *Object-Oriented Modeling and Design*. Englewood Cliffs, NJ: Prentice-Hall, Inc.

Siegfried, R. H. and N. Vaidya. 1993. "Automated Incident Management Plan Using Geographic Information Systems Technology for Traffic Management Centers," Research Report 1928-1F, Texas Transportation Institute, College Station.

Smith, R. C., D. L. Harkey, and R. Harris. 2001. "Implementation of GIS-Based Highway Safety Analyses: Bridging the Gap," FHWA-RD-01-039, Federal Highway Administration.

Spring, G. S. 2002. "Knowledge-based GIS for the Identification, Analysis and Correction of Hazardous Highway Locations." Proceedings of the 7th International Conference on Applications of Advanced Technologies in Transportation Engineering (AATT-7), pp. 932–941, American Society of Civil Engineers, Boston, MA.

Sutton, J. C. 1996. "Role of Geographic Information Systems in Regional Transportation Planning," *Transportation Research Record*. n 1518, pp. 25–31, Transportation Research Board.

Vonderohe, Alan. 1993. "Adaptation of Geographic Information Systems for Transportation," NCHRP Report 20-27, National Research Council, Washington, DC.

Wikipedia, Global Positioning Systems, http://en.wikipedia.org/wiki/Global_Positioning_System, accessed, 2009a.

Wikipedia, Light Detection and Ranging, http://en.wikipedia.org/wiki/LiDAR, accessed, 2009b.

Zhang, Z., S.G. Smith, and R.W. Hudson. 2001. "Geographic Information System Implementation Plan for Pavement Management Information System: Texas Department of Transportation." *Transportation Research Record* n 1769, pp. 46–50, Transportation Research Board.

CHAPTER 7
TRAVEL DEMAND FORECASTING FOR URBAN TRANSPORTATION PLANNING*

Arun Chatterjee
Department of Civil and Environmental Engineering
The University of Tennessee, Knoxville, Tennessee

Mohan M. Venigalla
Civil, Environmental, and Infrastructure Engineering Department
George Mason University, Fairfax, Virginia

7.1 INTRODUCTION

7.1.1 The Need for Determining Travel Demand: Existing and Future

The basic purpose of transportation planning and management is to match transportation supply with travel demand, which represents the need for transportation infrastructure. A thorough understanding of existing travel pattern is necessary for identifying and analyzing existing traffic-related problems. Detailed data on current travel pattern and traffic volumes are also needed for developing travel forecasting/prediction models. The prediction of future travel demand is an essential task of the long-range transportation planning process for determining strategies for accommodating future needs. These strategies may include land use policies, pricing programs, and expansion of transportation supply—highways and transit service.

7.1.2 Scope of Analysis and Levels of Planning

There are different levels of planning, directed to different types of problems. The terminology for these levels of planning and analysis varies according to the context. For example, the expressions "micro," "meso," and "macro" are sometimes used to describe the level of detail or the size of an area used for an analysis. Similarly, the expressions "site-specific," "corridor," and "area-wide" or "metropolitan" are used to describe variations in the scope of a problem. The approach and techniques for analyzing and forecasting travel would vary according to the level of analysis. Even for a particular level of analysis, the techniques may have to be adjusted to match the constraints of available data and human resources.

An example of a microlevel or site-specific analysis is the case of a congested road intersection. In this case traffic engineers would be interested in detailed traffic flow characteristics, including

---

*Reprinted from the First Edition.

turning movements of vehicles along each approach and pedestrian volumes across each approach. Management strategies in this case would involve traffic operation and roadway design-oriented techniques. A corridor-level analysis, on the other hand, would cover a larger area, say 10 miles long and 2 miles wide. A major highway with severe congestion problem may require a corridor analysis. The origin and destination of trips and modal choice of travelers would be of interest in this case. Station-to-station movements of passengers may have to be estimated in the case of a rapid transit service along the corridor. At the macro level the concern may be total energy consumption by the transportation sector or the total emission of an air pollutant; for these cases, information on total vehicle-miles traveled (VMT) on each functional class of roads will be needed.

It is important to recognize that the nature of problems to be examined dictates the level of planning to be used as well as the technique for travel demand analysis. The discussion of this chapter will be oriented mostly to mesoscale or area-wide travel demand analysis that is commonly performed in urban transportation planning studies. Even for this type of analysis for an urban area at the mesoscale, the approach and details of techniques and models to be used would depend on the size of the area as well as the resources available for carrying out the work. For example, a small urban area may not have the manpower or funding needed for carrying out large-scale surveys and developing advanced mathematical models. The need for customizing the planning and modeling approaches based on specific situations was discussed in detail by Grecco et al. (1976).

7.2 CHARACTERISTICS OF TRAVEL

Certain special characteristics of travel demand require recognition for planning and design purposes. These are discussed below.

7.2.1 Spatial and Temporal Variations

The total magnitude of travel demand alone is not sufficient for detailed planning and management purposes. The spatial and temporal distributions of travel are also important items of information to be considered in determining supply strategies. The peaking of travel at certain time periods requires a level of transportation supply that is not needed at other times. However, due to the nature of supply, which cannot be adjusted easily, large investments have to be made to provide roadway or transit service capacities to accommodate peak period travel, and this capacity is not utilized efficiently at other times. An imbalance in the directional distribution of travel also creates similar inefficiencies.

The spatial orientation of trips has important influence on supply requirements and costs. A few typical spatial distribution patterns of trips in urban areas are listed below:

- Travel along dense corridors, which are usually radial connecting suburbs to central business district (CBD)
- Diffused travel pattern caused by urban sprawl
- Suburb to suburb or circumferential travel
- Travel within large activity centers in CBD and suburbs

Different modes of transportation may be needed to serve these different travel patterns. For example, fixed-route public transit service usually is efficient for concentrated travel along a dense corridor, but it is not ideally suited to serve a diffused travel pattern in a cost-effective manner.

Choice of domicile and workplace, lifestyles and different travel needs of individuals and families make the comprehension of trip-making characteristics of a large metro area very complex. These complexities may be illustrated through trips made by a typical suburban U.S. household

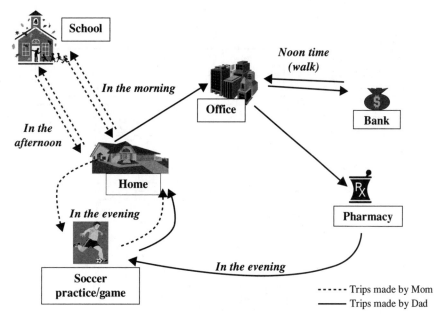

FIGURE 7.1   The complexity of trip making at a typical suburban household.

on a given weekday (Figure 7.1). Assume that this household has four members, including two children who go to grade school, and two cars. It can be seen that there are at least 11 trips made by this household at different times of day. Most of the trips are auto trips, and two trips are taken in the walk mode. Travel demand modeling attempts to capture such spatial and temporal variations in travel at an aggregate level, such as a zone, in which a number of households, businesses, and offices exist.

7.2.2   Classification of Travel by Trip Purpose and Market Segments

In addition to the spatial and temporal characteristics of travel demand, several other aspects of travel demand must be recognized. Trip purposes such as work, shopping, and social-recreation, and trip-maker's characteristics such as income and car ownership, are important factors influencing the elasticity of demand, reflecting its sensitivity with respect to travel time and cost. For example, work trips may be more likely to use public transit for a given level of service than trips of other trip purposes.

For a metropolitan study, it is useful to classify travel according to spatial orientation and trip purpose as shown in Figure 7.2. The concept of market segmentation is applicable to the classification of travel based on trip purpose, trip-makers' characteristics, and spatial-temporal concentration. This concept is used in the field of marketing for developing different types of consumer products targeted to match different tastes and preferences of potential users/buyers of these products. The concept of market segmentation is applicable to public transportation planning. A single type of transit service is not suitable for all transit market segments. For example, express buses may be needed for a commuter market segment. Taxicabs serve a different market segment. Woodruff et al. (1981) examined this subject in depth.

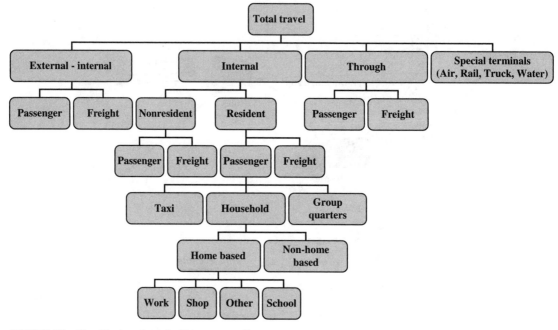

FIGURE 7.2 Classification of travel within a metropolitan area.

7.3 UNITS FOR MEASURING TRAVEL DEMAND

Travel demand is measured and expressed in different ways for different types of analysis. Examples of different units of measurement are:

1. Trip (between two areas)
2. Trip end (in a given area)
3. Traffic volume (on a road segment)
4. Person trip and vehicle trip
5. Passenger vehicle and freight vehicle
6. Person-miles traveled and vehicle-miles traveled

The definition of each of these units should be understood clearly and an appropriate unit of measurement should be used to match the case being analyzed. For example, for a parking study, trip end is the appropriate unit for expressing parking demand. For estimating the number of lanes to be provided in a road segment, the demand should be expressed in terms of traffic volume. As pointed out earlier, the appropriate unit of travel for estimating fuel consumption and/or air pollution attributable to transportation is vehicle-miles traveled (VMT).

7.4 MEASURING EXISTING TRAVEL

Detailed information on existing travel is needed for two purposes: analyzing existing problems and developing mathematical models for forecasting travel. A variety of surveys can be performed for gathering information related to existing travel demand. However, travel surveys are expensive, and therefore care must be taken to identify the types of information that really would be useful for specific purposes, and then the most suitable procedures should be selected for gathering the information. Sampling techniques are useful, and adequate time and care must be devoted to developing sampling procedures. There are several different types of survey techniques, some of which are suitable for automobile travel, some for transit travel, and some for general passenger movement. Survey procedures for freight vehicles and commodity movements may be very different in certain respects from those of passenger travel. Good references for travel demand-related survey techniques are Dresser and Pearson (1994), Stopher and Metcalf (1996), and *Travel Survey Manual* (Cambridge Systematics, Inc. and Barton-Aschman Associates, Inc. 1996).

7.4.1 Time Frame for Travel Surveys

Since travel demand varies during a given year according to the season (or month of year) and day of week, a decision must be made carefully to select a specific time frame or reference for surveys. For urban transportation studies it is a common practice to develop travel demand information for an average weekday in the fall. However, the time can be different based on the nature of the problem to be analyzed. For example, in the case of a tourist-oriented urban area the major concern may be traffic problems during weekend days or holidays, and surveys may be done to capture information for one of those days. A few major types of surveys are discussed in the following sections.

7.4.2 Origin-Destination Surveys

The classification of trips into the three classes of internal, external-internal (and vice versa), and through trips is useful for mesoscale and metropolitan-level as well as small-area studies. This classification scheme is useful for developing forecasting procedures/models as well as policies and strategies for accommodating travel because strategies for each of these classes of travel would be different. For example, through trips may require a bypass facility. External-internal trips may have to be intercepted before they reach a heavily congested area such as the central business district (CBD).

The origins and destinations (O-D) of trips along with some other characteristics such as trip purpose and mode of travel can be determined in different ways:

1. Home interviews (for internal travel)
2. Roadside interviews at cordon stations (for external-internal and through trips)
3. On-board survey on transit vehicles

All three of these techniques involve sampling and require careful planning before their execution. The Federal Highway Administration (FHWA 1975b) and Urban Transportation Systems Associates (1972) developed detailed guidelines for O-D survey procedures. The reliability of the results of an O-D survey depends on its sampling scheme and sample size, and this issue was examined by Makowski, Chatterjee, and Sinha (1974).

Full-scale origin-destination surveys were widely used during the 1960s and 1970s to develop a variety of information, including "desire lines" of travel. Their use has decreased because of the cost and also due to the use of synthetic or borrowed disaggregate travel models, which require less survey data.

7.4.3 Traffic Volume and Passenger Counts

For determining the use of various roadway facilities and assessing their level of service, vehicle counts are taken at selected locations along roadways. Short-count techniques are useful provided appropriate expansion factors are developed based on previous or ongoing research on fluctuations of traffic by hour, weekday, and month. All state Departments of Transportation (DOTs) have extensive programs for gathering traffic volume data on an annual basis. These vehicle counts usually are taken with machines.

For urban transportation studies screen lines and cut-lines are established in the study area to select traffic count locations and take counts in an organized manner so that the major travel movements can be measured and analyzed. These counts are also used for checking the results of travel forecasting models. Similarly, traffic counts are taken at special traffic generators such as airports and large colleges and universities to capture their unique travel-generating characteristics.

For analyzing the use of a transit service, passenger counts are taken on-board transit vehicles and/or at selected stops or stations. These passenger counts usually are taken by observers who are assigned to specific transit vehicles and/or transit stops/stations according to a survey plan.

7.5   FORECASTING FUTURE DEMAND

The need for travel demand forecasts arises in various contexts of planning, short-range as well as long-range. Travel forecasting is one of the most important and difficult tasks of transportation planning. There àre different types of travel prediction techniques, and the one to be used in a particular case must be compatible with the nature of the problem and scope of planning. Constraints of available time and resource also influence the selection of a technique.

7.5.1   Predicting Response to Service Changes Using Elasticity Coefficients

For short-range planning or a transportation systems management (TSM) study, it is often necessary to predict the effect of a proposed change in transportation service that can be implemented in the near future. For example, a planner may be asked to evaluate the impact on transit ridership of improving transit service in a travel corridor by providing increased frequency and/or limited stop service. The impact of changing the fare structure on transit ridership may also be of interest. In these cases demand elasticity coefficients, if available from past studies, would be useful. Typically an elasticity coefficient is developed with respect to a specific factor such as travel time or fare based on actual observation. The coefficient should exclude the effect of other factors that also may be influencing demand at the same time. Mayworm, Lago, and McEnroe (1980) give information on demand elasticity models. More information on the elasticity of transit use with respect fare may be found in Parody and Brand (1979) and Hamberger and Chatterjee (1987).

7.5.2   Stated Preference Surveys and Concept Tests for Forecasting

For transit planning it is sometimes necessary to ask people about their preferences and their likes and dislikes for various service characteristics. These surveys are used for determining how to improve an existing service and/or designing a new service, and also for forecasting ridership on a new service. These attitudinal and stated-preference surveys need sound statistical design for selecting the sample and analyzing the results. A discussion on stated preference survey may be found in the *Travel Survey Manual*. In the field of marketing, concept tests are performed for estimating the potential demand for a new consumer product, and this approach can be extended to ridership forecasts for new/innovative transit services.

Hartgen and Keck (1976) describe a survey-based method of forecasting ridership on a new dial-a-bus service. The interpretation of results of opinion-based surveys must be done carefully in order to account for any bias reflected in apparent results. Chatterjee, McAdams, and Wegmann (1983) present a case study involving noncommitment bias in public opinion on the anticipated usage of a new transit service.

7.5.3 Forecasting Future Travel on Road Segments and/or Transit Lines

A variety of forecasting procedures are available, ranging from the extrapolation of past trends to complex mathematical models involving several steps. A transportation planner must recognize the advantages and disadvantages of each procedure. Two procedures are examined for illustration.

Direct Estimation of Traffic Volume by Trend Analysis.   If traffic volume data are available for a road segment or a transit line of interest for several years in the past, the historical trend can be identified and extrapolated to estimate future volumes. This approach, of course, is appropriate if the trend is expected to continue, which commonly is true for short-range forecasts. Trend-based forecasts are appropriate also for aggregate values such as total VMT or transit rides in an urban area. However, major changes in the land development pattern and/or transportation network can cause substantial changes in the travel pattern, and if such changes are likely then trend extrapolation will not be appropriate. Therefore, for long-range forecasts of traffic volumes on individual segments of a road network or the number of passenger trips on individual transit routes, trend analysis is not used.

Stepwise/Sequential Procedure.   A widely used travel estimation procedure for long-range forecasts of traffic volumes on a highway network uses several steps in a sequence, as shown in the flowchart of Figure 7.3. Each step requires a particular type of model or procedure, and there are different choices of models at each step. One of the major advantages of this procedure is its ability to reflect several types of changes that may occur in the future:

1. Changes in trip-making rates
2. Changes in development pattern, resulting in altered travel pattern
3. Changes in transportation mode usage
4. Changes in transportation network

Another advantage of the stepwise, or sequential, procedure is that it generates several types of useful information at the end of various steps. The disadvantage of the procedure is that it needs a large amount of data for model development. It also requires a sound knowledge of one of the available

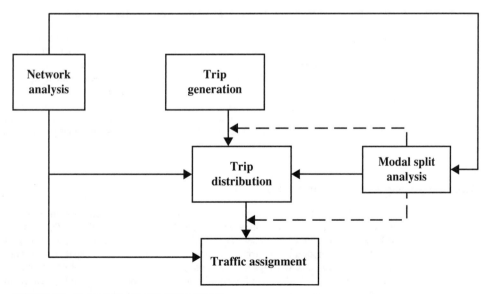

FIGURE 7.3   Sequential travel estimation procedure.

computer software models that is specially designed for developing and applying these models. A great deal of research has been performed and is still being continued to improve this stepwise modeling procedure. It should be acknowledged that the staff of the former Bureau of Public Works and later the staff of the Federal Highway Administration and Urban Mass Transportation Administration made tremendous contributions to the development of various procedures and computer software. An historical overview of the development of planning and modeling procedures in the United States is presented by Weiner (1992).

The stepwise procedure is popularly known as the four-step modeling process because it includes four major steps: trip generation, trip distribution, mode choice, and traffic assignment. Additionally, network analysis must be done to develop a few types of information that are needed for the other steps. These steps and procedures involved with each are discussed in detail in the following sections.

7.6 TRIP GENERATION

Trip generation is a very important step because it sets up not only the framework for the following tasks but also some of the controlling values, such as the total number of trips generated in the study area by location and trip purpose. The commonly used units for trip generation analysis usually include a household, a dwelling unit (DU), and a business establishment. However, the results of a trip generation analysis for a study area are aggregated based on larger areas known as traffic zones.

A typical classification scheme of trips used for trip generation analysis is presented in Figure 7.2. A detailed discussion of this classification scheme is presented by Chatterjee, Martinson, and Sinha (1977). A thorough analysis of all these types of trips shown in the figure requires a large amount of data. These data are collected by using origin-destination (O-D) surveys, discussed briefly in an earlier section. This section will focus primarily on trip generation models for internal passenger trips made by households.

7.6.1 Models for Internal Passenger Trips: Aggregate and Disaggregate Models

The goal of trip generation models for internal passenger trips is to estimate the total number of trip ends for each purpose generated in each traffic zone based on socioeconomic and/or land use data for the respective zones. This task can be accomplished with either aggregate or disaggregate models. For aggregate models the total number of trips (trip ends) generated in a zone is used as the dependent variable, whereas for disaggregate models trips made by a household (or a business establishment) are used as the dependent variable. When disaggregate models are used, the trip ends generated by households and/or any other trip generating units such as business establishments in a zone are combined to produce the zonal (total) value. Both disaggregate and aggregate trip generation models are used in planning studies.

7.6.2 Trip Generation by Households

Household-generated trips comprise more than 80 percent of all trips in an urban area. Trips by nonresidents and a variety of other vehicles, including commercial vehicles such as taxis and trucks, and public utility and public service vehicles comprise the remaining portion of total travel. For the purpose of modeling, the trips generated by households are classified as home-based and non-home-based. Home-based trips have one end, either origin or destination, located at the home zone of the trip maker. If both ends of a trip are located in zones where the trip maker does not live, it is considered a non-home-based trip.

Definitions of Productions and Attractions and Trip Purpose. Because of the predominance of home-based trips in an urban area, the model development is simplified if it is assumed that the home end of a trip is a production (P) in the zone of the trip maker's residence irrespective of whether it represents the origin or destination of the trip. According to this approach, the non-home end of a home-based trip is considered to be attraction (A). For a non-home-based trip, which has neither its origin nor its destination at the trip maker's residence, production and attraction are synonymous with origin and destination, respectively. This definition of productions (P's) and attractions (A's) is depicted in Figure 7.4. It should be noted that for home-based trips the activity at the non-home end determines the trip purpose, and that non-home-based trips usually are not further stratified by purpose.

7.6.3 Cross-Classification or Category Models for Household Trip Generation

Household trip rates have been found to vary significantly according to certain socioeconomic characteristics and the size of a household. Household characteristics that have been found to be significant for trip generation and are commonly used in trip generation models are:

1. Household size
2. Auto ownership
3. Income

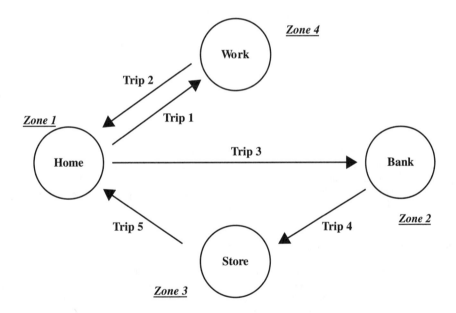

Zone	Production	Attraction (Purpose)
1	4 (2 work, 1 personal business, 1 shop)	0
2	1 (Non-home-based)	1 (Personal business)
3	0	2 (1 Non-home-based, 1 shop)
4	0	2 (work)
Total	5	5

FIGURE 7.4 Definitions of productions and attractions.

TABLE 7.1   A Hypothetical Category Model for Total Person Trips per Household

		Car ownership		
		No car	One car	Multicar
Household size	1 to 2	3.2	4.9	6.1
(persons per	3 to 4	5.2	7.1	8.7
household)	5 or more	7.2	9.6	12.0

A hypothetical example of a trip generation model for households is presented in Table 7.1, which includes trip production rates for different types or categories of households defined in terms of different combinations of household size and auto ownership. This type of model is referred to as a cross-classification or category models, and such models are used widely for estimating trip production by households.

Alternative techniques for a statistical analysis of these models are discussed by Chatterjee and Khasnabis (1973). When the households in a traffic zone are cross-classified by size and auto ownership, total trips made by households in a zone for a specific purpose (P) = Summation of (No. of households of a category) × (Trip rate for households of that category and for that specific purpose). In mathematical notation this relationship is shown below:

$$P = \sum_{kl=1}^{n} (HH_{kl})(TR_{kl}) \tag{7.1}$$

where HH = households
TR = trip rates
kl = a particular combination of household size k and auto ownership l
n = the number of combinations or categories

The choice of household characteristics to be used for developing the various categories for trip production rates may vary from one study to another. One advantage of disaggregate models is that for developing these models a full-scale O-D survey is not needed. A carefully selected small sample of households may be used for developing trip production rates as long as the number of cases for each category, or cell of the matrix, is statistically adequate.

7.6.4   Models for Trip Attractions

It is a common practice to use aggregate models in the form of regression equations for trip attractions. The dependent variable for these aggregate models is the total number of trip attractions for a specific trip purpose in a traffic zone. The independent variables usually are employment-related, and they represent zonal total values. Hypothetical examples of trip-attraction (A) models are presented below:

$$(HBW\ A)_j = 1.5\ (Total\ employment)_j$$

$$(HBNW\ A)_j = 8.5\ (Retail\ employment)_j + 1.0\ (Nonretail\ employment)_j + 0.9\ (Dwelling\ units)_j$$

$$(NHB\ A)_j = 3.0\ (Retail\ employment)_j + 2.0\ (Nonretail\ employment)_j + 0.8\ (Dwelling\ units)_j$$

where $(HBW\ A)_j$ = home-based work attractions in zone j
$(HBNW\ A)_j$ = home-based nonwork attractions in zone j
$(NHB\ A)_j$ = non-home-based attractions in zone j

The development of aggregate models usually requires a full-scale O-D survey. The coefficients of the regression equations will vary from area to area. The choice of independent variables and trip-purpose categories also may vary from one study to another.

7.6.5 Balancing of Productions and Attractions

Due to the definition of productions and attractions, home-based productions in a zone may not be equal to the corresponding attractions in the same zone. Non-home-based productions in a zone should be equal to corresponding attractions in the same zone. However, area-wide (total) productions (P's) of any trip purpose—home-based or non-home-based—should be equal to the corresponding area-wide (total) attractions. Thus,

$$\Sigma(\text{HBW P})_i = \Sigma(\text{HBW A})_j \tag{7.2}$$

$$\Sigma(\text{HBNW P})_i = \Sigma(\text{HBNW A})_j \tag{7.3}$$

$$\Sigma(\text{NHB P})_j = \Sigma(\text{NHB P})_j \tag{7.4}$$

When synthetic or borrowed models are used, the estimated area-wide (total) productions would not be equal to the estimated area-wide (total) attractions. Therefore, to achieve a balance, zonal attractions are adjusted proportionately such that the adjusted area-wide attractions equal area-wide productions. Adjustment or scaling factors for attractions are calculated as follows:

$$\text{Adjustment factor for HBW } A_j\text{'s} = \frac{\Sigma(\text{HBW P})_i}{\Sigma(\text{HBW A})_j} \tag{7.5}$$

$$\text{Adjustment factor for NBNW } A_j\text{'s} = \frac{\Sigma(\text{HBNW P})_i}{\Sigma(\text{HBNW A})_j} \tag{7.6}$$

$$\text{Adjustment factor for NHB } A_j\text{'s} = \frac{\text{Total NHB productions}}{\Sigma(\text{NHB A})_j} \tag{7.7}$$

7.6.6 Commercial Vehicle Traffic in an Urban Area

It should be pointed out that although internal trips made by residents in passenger vehicles account for a large proportion of total trips in an urban area, the other categories of trips must not be overlooked. The classification scheme presented in Figure 7.2 shows the other categories. The proportion of each category of trips varies according to the size and other characteristics of an urban area. For example, the proportion of through trips is usually larger in smaller size areas. In some cases, trips of one or more of these other categories may be the cause of major problems and thus will require special attention. For example, through traffic may be the major issue in the case of a small or medium-sized urban area, and the planners may have to analyze these trips thoroughly. Similarly, the movement of large trucks may be of major interest in some urban areas. A comprehensive study should pay attention to travel demand of all categories, although the level of detail may vary.

The analysis of commercial vehicle travel has been neglected in most urban transportation studies. These vehicles are garaged in nonresidential locations and include trucks of all sizes, taxicabs, rental cars, service vehicles of plumbers and electricians, etc. There are a few useful references on how to estimate truck traffic in urban areas, including an article by Chatterjee et al. (1979), Ogden (1992), and Fischer (2001).

7.6.7 Forecasting Variables Used in Trip Generation Models

In developing trip generation models the availability of data for the independent variables of the models is an important issue that can influence the selection of a variable. Usually the availability of data for the base year is less problematic than that for future years. Of course, if data for an independent variable are not available for the base year, they cannot be used in model development. However, what the model developer must recognize before building and adopting a model is whether the independent variables used in the model can be forecast by the responsible planning agency, and if such forecasts would be very difficult then it may be desirable to avoid using those variables in the model. Sometimes transportation planners have to develop a procedure or model to be used for making such forecasts.

Usually aggregate values of socioeconomic parameters used in trip generation models are not very difficult to forecast with the existing state of the art. The difficulty usually involves the task of disaggregating socioeconomic data at the zonal level. For example, it may not be very difficult to predict the total number of households in each zone along with their average size and auto ownership. However, it would be difficult to cross-classify the predicted number of households in a zone according to specific categories based on household size and auto ownership. Similarly, predicting the average income of households in individual traffic zones may not be very difficult, but developing a breakdown of the households in every zone by income groups would be difficult. The disaggregate trip generation models thus present a challenge to planners for making detailed forecasts of socioeconomic characteristics for future years. In order to provide assistance for making forecasts in a disaggregate form, a few procedures have been developed, and examples of such household stratification models can be found in Chatterjee, Khasnabis, and Slade (1977) and FHWA (1975a).

7.7   TRIP DISTRIBUTION

The purpose of the trip distribution step of the stepwise travel-modeling procedure is to estimate zone-to-zone movements, i.e., trip interchanges. This step usually follows trip generation analysis. In some cases, but not commonly, trip distribution may come after trip generation and modal split analysis. The inputs to a trip distribution model are the zonal productions (P_i) and attractions (A_j). The model strives to link the productions and attractions based on certain hypotheses/concepts.

When the trip distribution phase precedes modal split analysis, productions and attractions include trips by all modes and the distribution model should be multimodal in nature. In actual practice, however, multimodal trip distribution models are uncommon, and in most cases highway-oriented models have been used to distribute trips of all modes. It should be noted that in the rare case where the trip distribution phase follows modal split analysis, mode-specific distribution models are needed. It is generally believed that ideally trip distribution should be combined with modal split analysis because decisions related to destination and travel mode are usually made simultaneously.

In this section a widely used trip distribution technique, the gravity model, will be discussed in detail, followed by a brief overview of other types of models. A good review of commonly used trip distribution models can be found in Easa (1993).

7.7.1   Formulation of a Gravity Model

The basic hypothesis underlying a gravity model is that the probability that a trip of a particular purpose k produced at zone i will be attracted to zone j is proportional to the attractiveness or pull of zone j, which depends on two factors: the magnitude of activities related to the trip purpose k in zone j, and the spatial separation of the zones i and j. The magnitude of activities related to trip purpose k in a zone j can be expressed by the number of zonal trip attractions of the same purpose, and the effect of spatial separation between zones i and j can be expressed by a friction factor, F_{ij}^k, which is inversely proportional to an appropriate measure of impedance, usually travel time. The attractiveness or pull of zone j with respect to zone i is proportional to $A_j^k F_{ij}^k$. The magnitude of trips of purpose k produced in zone i and attracted to zone j, T_{ij}^k, of course, also depends on the number of trips being produced at zone i, P_i^k. This can be expressed mathematically as follows:

$$T_{ij}^k = f(P_i^k, A_j^k, F_{ij}^k) \tag{7.8}$$

The above formulation is not sufficient for estimating the T_{ij}^k values because it yet does not reflect any considerations for other zones that are competing as alternative destinations for the trips P_i^k. Actually, the effective attractiveness of a zone is relative to others, and it can be expressed as the ratio of its own attractiveness with respect to the total. Thus the relative attractiveness of a zone for trips of purpose k being produced in zone i is expressed by the ratio $A_j^k F_{ij}^k / \Sigma_{j=1}^n A_j^k F_{ij}^k$. Dropping the subscript k, the trip distribution model can be written as follows:

$$T_{ij} = \frac{P_i A_j F_{ij}}{\sum_{j=i}^{n} A_j F_{ij}} \tag{7.9}$$

7.7.2 Application of Gravity Model Concept

The application of the gravity model concept for the trip distribution step of the stepwise travel forecasting procedure was introduced by Voorhees (1955). The classic example of an application of a gravity model as presented by Voorhees is shown in Figure 7.5. This application is based on the assumption that the effect of spatial separation with respect to trip making is proportional to the inverse of the square of travel time between the respective pairs of zones. The calculations presented below shows how the accessibility and attractiveness of zone 4 changed due to a new expressway resulting in an increased number of trip attractions. The number of shopping trips attracted to zone 4 from zone 1 was 28 without an expressway and increased to 80 as a result of a new expressway. The increase in trips attracted to zone 4 resulted in a decrease of trips attracted to the other zones.

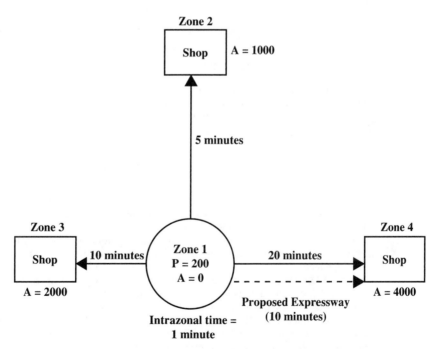

FIGURE 7.5 Example of gravity model concept as introduced by Alan M. Voorhees.

I. Situation without Expressway

Existing pulls from	Percent pull	No. of trips from zone 1 to
Zone $1 = \dfrac{0}{1^2} = 0$	$0/70 = 0\%$	$200 \times 0 = 0$
Zone $2 = \dfrac{1000}{5^2} = 40$	$40/70 = 57\%$	$200 \times 0.57 = 114$
Zone $3 = \dfrac{2000}{10^2} = 20$	$20/70 = 29\%$	$200 \times 0.29 = 58$
Zone $4 = \dfrac{4000}{20^2} = 10$	$10/70 = 14\%$	$200 \times 0.14 = 28$
Total pull $= 70$	100	200

II. Situation with Expressway

Existing pulls from	Percent pull	No. of trips from zone 1 to
Zone $1 = \dfrac{0}{1^2} = 0$	$0/100 = 0\%$	$200 \times 0 = 0$
Zone $2 = \dfrac{1000}{5^2} = 40$	$40/100 = 40\%$	$200 \times 0.40 = 80$
Zone $3 = \dfrac{2000}{10^2} = 20$	$20/100 = 20\%$	$200 \times 0.20 = 40$
Zone $4 = \dfrac{4000}{10^2} = 40$	$40/100 = 40\%$	$200 \times 0.40 = 80$
Total pull $= 100$	100	200

7.7.3 Calibration of Gravity Model Parameters

The three basic parameters of a gravity model are zonal trip productions and attractions, P_i's and A_j's, and friction factors F_{ij}'s. Whereas the P_i's and A_j's are estimated by trip generation models, friction factors must be derived as a part of the trip distribution phase. The basic concept of the gravity model implies the following form for friction factors:

$$F_{ij} = \frac{1}{(\text{Travel time between } i \text{ and } j)^x} \tag{7.10}$$

In the early application of the gravity concept, the exponent of the travel time was assumed to be a constant. However, further empirical analysis suggested that the exponent varies over the range of values for travel time. The actual values of friction factors are derived by a trial-and-error procedure and vary according to trip purpose.

For each trip purpose an arbitrary set of friction factors for a range of travel time values at an increment of 1 minute is assumed at the beginning of the calibration process, and the results of this initial application of the model are evaluated with respect to the actual trip distribution obtained from an O-D survey. The evaluation is made by comparing the trip-length frequencies generated by the model with those derived from the O-D survey, and if the results are not similar the friction factors are adjusted and the gravity model is applied again with the new factors. This trial-and-error procedure is continued until the trip-length frequencies of a model appear similar to those of the O-D survey. It may be noted that the absolute values of these factors have no special implications and that it is the relative weight with respect to each other that is important. The respective set of these frictions factors can be scaled up or down by a constant factor.

Balancing a Gravity Model.   It must be pointed out that due to the basic nature of the gravity model formulation, the zonal productions obtained from the model application must equal the values of P_i's originally used as inputs to the model. However, the same is not true for the zonal attractions, and the model results must be compared with the original A_j's. In the cases when model-generated A_j's do not match closely with original A_j's, a balancing procedure is used by adjusting the input values of A_j's until the model results are satisfactory.

7.8   OTHER TYPES OF TRIP DISTRIBUTION MODELS

The gravity model is by far the most widely used trip distribution technique, but there are also other techniques used in urban transportation planning. One technique utilizes growth factors for each traffic zone and uses an iterative balancing procedure to project a base year trip matrix to the future year. The most popular of the growth factor techniques is that introduced by Fratar (1954) and known as the Fratar technique. The limitation of a growth factor procedure is that they are basically extrapolation techniques and cannot be used to synthesize movements between zone pairs if the base year trips are zero. However, the Fratar technique is utilized regularly for projecting through trips in an urban area and sometimes even for external-internal trips.

A somewhat complex trip distribution technique that was used by the Chicago Area Transportation Study, the transportation planning agency for Chicago, is the intervening opportunities model. The trip distribution theory underlying this model states that the probability that a trip originating in a zone i will find a destination in another zone j is proportional to possible destinations in zone j and also the possible destinations in zones closer to the origin of the trip. This model is rarely used by any agency today.

7.9   MODAL SPLIT

One widely researched step/phase of the sequential travel-modeling procedure for urban transportation planning is the modal split analysis, which involves the allocation of total person trips (by all modes) to the respective modes of travel, primarily automobile and public transit. It should be noted, however, that many studies for small and medium-sized urban areas omit this step by developing and using models for automobile trips only. This omission is justified in areas where transit trips constitute a very small fraction of total trips and are made primarily by captive riders.

Modal split models basically relate the probability of transit usage to explanatory variables or factors in a mathematical form. The empirical data necessary to develop these models usually are obtained from comprehensive O-D surveys in specific urban areas. In applying these models to predict the future transit usage, one must make the implicit assumption that the variables which explain the present level of transit usage will do so in much the same manner in the future.

7.9.1   Factors Affecting Mode Choice

Factors that may explain a trip maker's choosing a specific mode of transportation for a trip are commonly grouped as follows:

Trip Maker Characteristics

Income
Car ownership
Car availability
Age

Trip Characteristics

Trip purpose—work, shop, recreation, etc.

Destination orientation—CBD versus non-CBD

Trip length

Transportation Systems Characteristics

Waiting time

Speed

Cost

Comfort and convenience

Access to terminal or transfer location

7.9.2 Categories of Modal Split Models

The possible sequence of different types of modal split models with respect to the other steps of travel-modeling procedure is shown in Figure 7.3.

Predistribution (or Trip End) Models. This type of modal split model is used to separate the trip productions in each zone into the different modes to be distributed by mode-specific trip distribution models. The primary disadvantage of these models is that they cannot include variables related to transportation system characteristics. Predistribution models are not commonly used.

Postdistribution (or Trip Interchange) Models. This type of modal split model is very popular because it can include variables of all types. However, conceptually it requires the use of a multimodal trip distribution model and currently such distribution models are not used commonly. Figure 7.6 illustrates the sequence of application of a postdistribution model.

Simultaneous Trip Distribution and Modal Split Models. This type of model strives to estimate the number of trips between two zones by specific modes in one step directly following the trip generation phase. Conceptually and theoretically this type of a model has a sound basis, but it is not commonly used at this time.

7.9.3 Developing a Modal Split Model

Modal split models are developed from observed data on trip making available from home-interview surveys. The analysis involves the processing of a variety of data for both demand and supply.

Aggregate Model. Modal split models of the 1960s and early 1970s in most cases were based on an aggregate approach, which examined the mode choice of trip makers and their trips in groups based on similar socioeconomic and/or trip characteristics. These mode choice models usually involved two modes only: auto and transit. A detailed stratification scheme was used, and the share of each mode was determined for each stratified group of trips, which was then correlated with selected independent variables. The dependent variable was percent transit applicable to a group of trips of similar characteristics made by similar trip makers. Commonly used independent variables included the ratio of travel time by transit to that by automobile; the ratio of travel cost by transit to that by automobile; and the ratio of accessibility by transit to that by automobile. The relationship of the dependent variable, percent transit, with the independent variable, say ratio of travel times, commonly was expressed by a set of curves. These curves sometimes were referred to as modal diversion curves.

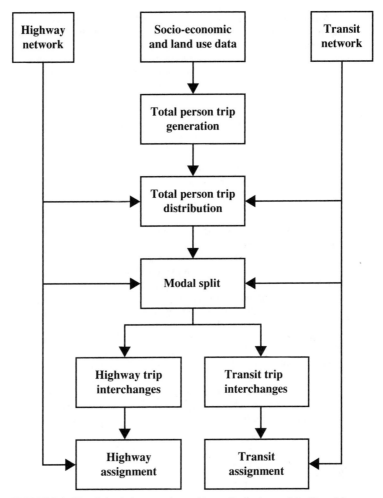

FIGURE 7.6   Travel simulation procedures with postdistribution modal split models.

The development of aggregate modal split models requires a large amount of data. Discussion of procedures used for developing different types of aggregate modal split models along with examples of these models can be found in Weiner (1969) and Chatterjee and Sinha (1975).

Disaggregate Behavioral Logit Models.   In late 1970s a new approach known as disaggregate behavioral method was developed and refined by a number of researchers. This approach recognized each individual's choice of mode for each trip instead of combining the trips in homogeneous groups. The underlying premise of this modeling approach is that an individual trip maker's choice of a mode of travel is based on a principle called utility maximization. Another premise is that the utility of using one mode of travel for a trip can be estimated using a mathematical function referred to as the utility function, which generates a numerical utility value/score based on several attributes of the mode (for the trip) as well as the characteristics of the trip maker. Examples of a mode's attributes for a trip include travel time and costs. The utilities of alternative modes can also be calculated in a similar manner. A trip maker chooses the mode from all alternatives that has the highest utility value for him or her.

A mathematical function that was used to represent the correlation of the probability of a trip maker's choosing a specific mode for a specific trip with a set of utility values is known as the logit function, and thus these models are also referred to as logit models. Binomial logit models deal with two modes, whereas multinomial logit models can deal with more than two modes. An example of the mathematical formulation of a multinomial logit model is given below:

$$p(k) = \frac{e^{U_k}}{\sum_{x=1}^{n} e^{U_x}}$$

(7.11)

where $p(k)$ = probability of using mode k
U_k = utility of using mode k
U_x = utility of using any particular mode x
n = number of modes to choose from

A special statistical procedure known as the maximum likelihood technique is used to derive an equation that combines different variable/factors in a meaningful way to calculate a utility (or disutility) value. The coefficients of each variable included in the utility (or disutility) function reflect certain behavioral aspects of a trip maker. Usually transportation-related variables used for a utility function include such items as access (or egress) time to (or from) transit stops/stations, wait time, line-haul time, and out-of pocket costs, and the coefficients of these variables are negative. Thus, the combined utility value comes out to be negative, which indicates disutility of using a mode. A trip maker's characteristics such as income are also built into the utility function.

One of the advantages of disaggregate mode choice models is that they do not need a full-scale O-D survey with household samples from every traffic zone. A carefully selected sample of 1,500 to 2,000 households would be adequate for developing these models. The mathematical theory related to multinomial logit models for mode choice analysis is fairly complex and beyond the scope of this chapter. Numerous articles and reports have been published on the subject of behavioral logit models, including Reichman and Stopher (1971) and McFadden (1978). Horowitz, Koppelman, and Lerman (1986) also have detailed information about disaggregate mode choice modeling.

7.10   TRAFFIC ASSIGNMENT

The task of the traffic assignment process is to develop the loadings, or user volumes, on each segment of a transportation network as well as the turning movements at intersections of the network. The user volumes may be the number of vehicles, the number of total persons, the number of transit riders, or any other units of travel demand that can be described by an origin and destination. For highway networks, user volumes are in terms of the number of vehicles, whereas for transit assignment the numbers of riders/passengers represent volumes. The relationship of the traffic assignment phase with respect to the other phases of the sequential travel simulation procedure is shown in Figure 7.3.

7.10.1   Inputs to Traffic Assignment Process

The two basic inputs to the assignment process are the transportation network and the zone-to-zone trip interchanges. The transportation network of automobiles, trucks, and taxis are analyzed separately from that of public transit systems, and usually traffic assignments are made separately for highway and transit systems. The typical inputs of a highway traffic assignment are shown in Figure 7.7. Transit network assignments are limited to internal person trips only.

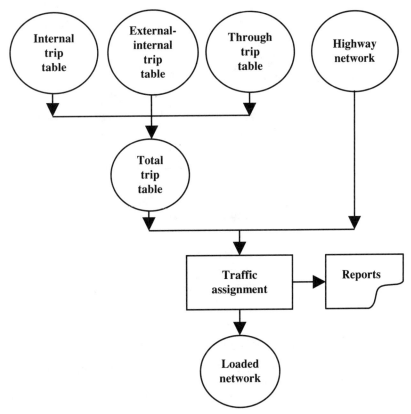

FIGURE 7.7   Inputs and outputs of traffic assignment.

7.10.2   Highway Network Analysis

For the purpose of computer analysis a highway network is represented by links and nodes and the traffic zones are represented by centroids, which are connected to the network. The characteristics of each link, such as the distance, speed, capacity, turn prohibitions, and functional classification, are coded. One of the primary tasks of network analysis is to determine the minimum time routes between each pair of centroids, and this task is performed utilizing Moore's algorithm. This algorithm does not require all possible routes between an origin and destination to be individually investigated to find the shortest route. Rather, a minimum tree is developed by fanning out from the origin to all other accessible nodes in increasing order of their impedance summation from the origin.

A tree is defined as the set of shortest routes from an origin to all other points in a network. An example of a path tree is shown in Figure 7.8. The travel time between a pair of zones is obtained by adding up the times on the individual links comprising the minimum time route, and this is repeated for every pair of zones. A skim tree usually refers to the interzonal travel time matrix.

It should be pointed out that the coding of a network for analysis with a computer-based algorithm requires a great deal of care and experience. There are many detailed issues and questions that come up with reference to such items as centroid connectors, representation of interchange ramps, whether to include certain roads or not, etc. Coding errors also can cause problems, and there are certain checks that can be done to minimize errors. Easa (1991) and Ismart (1990) discuss some of these issues and techniques for coding a network.

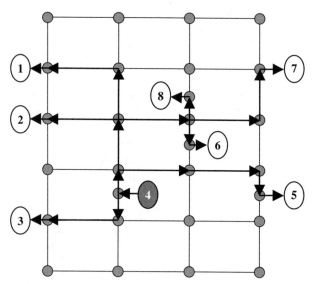

FIGURE 7.8   An example of a minimum path tree from home zone 4.

7.10.3   Alternative Techniques for Highway Traffic Assignment

A traffic assignment technique basically allocates the trips between each zone pair to the links comprising the most likely travel routes. The trips on each link are accumulated and the total trips on each link are reported at the end of the assignment process. Alternative assignment techniques vary in terms of the criteria for route selection.

All-or-Nothing Assignment (AON).   This procedure assigns all trips between a zone pair to its minimum time route. This is the most commonly used technique, although the realism of its basic hypothesis is questionable. It should be noted that other, more advanced techniques make use of this technique as a part of their more involved procedure.

Diversion Techniques.   A diversion technique allocates the trips between a zone pair to more than one route. The most commonly used diversion technique considers two routes. One of these routes uses freeways and the other is the quickest alternative nonfreeway arterial route. The procedure assumes that a proportion of trips as determined by a diversion curve will be diverted from an arterial route to a freeway route based on the ratio of time via freeway with respect to time via quickest alternate arterial route. This procedure is documented in the *Traffic Assignment Manual* (1964) (Bureau of Public Roads 1964).

Diversion techniques were widely used in the early 1960s. Their advantage is in getting a spread of traffic between competing routes, and these techniques appear to be more realistic than the all-or-nothing assignment. With the introduction of the capacity restraint assignment procedure, diversion techniques are rarely used today for network assignments, although their usefulness should not be overlooked for corridor-type applications.

Capacity Restraint Assignment.   The capacity restraint procedure explicitly recognizes that as traffic flow increases the speed of traffic decreases. In this procedure several assignments are made based on the all-or-nothing concept. At the end of each assignment, however, the assigned volume on each link is compared with the respective capacity and the travel time is adjusted according to a given formula. A new set of minimum time routes is computed for the next assignment.

The original capacity restraint procedure developed by the then Bureau of Public Roads, which is documented in the *Traffic Assignment Manual* (1964), assumed that the relationship between travel time and the volume peculiar to each link in a highway network can be expressed by the following equation:

$$t = t_0 \left[1 + 0.15 \left(\frac{V}{C_p} \right)^4 \right]$$

(7.12)

where t = travel time at which assigned volume can travel on the subject link
      t_0 = base travel time at zero volume = travel time at practical capacity × 0.87
      V = assigned volume
      C_p = practical capacity

This process may be continued for as much iteration as desired. Usually four iterations are adequate. The analyst has the choice to accept the results of any single iteration. Sometimes the link volumes obtained from all iterations are averaged to produce the final result. This procedure strives to bring the assigned volume, the capacity of a facility, and the related speed into a proper balance.

Equilibrium Assignment.   Traffic assignment has been the subject of intense research for many years, and the research has resulted in several alternatives to the all-or-nothing and capacity restraint techniques, which were widely used during the 1960s and 1970s. One example of an assignment technique that was developed after the capacity restraint technique is the probabilistic multi-path assignment technique, developed by Dial (1971). The most widely used procedure today is the user equilibrium assignment, which is based on the notion that traffic flows on network links are adjusted to an equilibrium state by the route-switching mechanism. That is, at equilibrium, the flows will be such that there is no incentive for route switching. As mentioned above, the travel time on each link changes with the flow, and therefore the travel time on several network paths changes as the link flows change. A stable condition is reached only when a traveler's travel times cannot be improved by unilaterally changing routes. This condition characterizes the user equilibrium (UE) condition.

The UE condition strives to optimize the utility of individual drivers. If the analysis is focused on optimizing a system-wide travel measure such as minimum aggregate travel time, then the problem is called a system optimal equilibrium (SOE) problem. Both UE and SOE problems rely on mathematical programming methods for developing the formulation and deriving a solution.

UE Problem Statement.   Given a generalized function S_a that relates arc/link costs to traffic volumes, find the equilibrium traffic volumes on each arc/link of a directed graph, $G(N, A)$, with N nodes, A arcs, and a total number of origin-destination zones Z. This user equilibrium traffic assignment problem may be formulated in the following nonlinear optimization form.

$$\text{Minimize} \quad f(x) = \sum_a \int_0^{V_a} S_a(x)\, dx$$

(7.13)

$$\text{subject to} \quad V_a = \sum_i \sum_j \sum_k \delta_{ij}^{ak} x_{ij}^k$$

(7.14)

$$\sum_k x_{ij}^k = T_{ij}$$

(7.15)

$$x_{ij}^k \geq 0$$

(7.16)

where i = subscript indicates an origin zone/node $i \in Z$
$\qquad$ j = subscript indicates a destination zone/node $j \in Z$
$\qquad$ k = indicates a path between the origin zone (root) i and the destination zone j
$\qquad$ a = subscript for link/arc, $a \in A$
$\qquad$ u = subscript for volume category, $u \in U$
$\qquad$ T_{ij} = number of trips (all modes) originated at i and destined to j
$\qquad$ C_a = capacity of arc a
$\qquad$ V_a = total volume in category u on arc a in current solution
$\qquad$ W_a = all-or-nothing volume of u trips on arc a in current solution
$\qquad$ $S_a(V_a)$ = generalized travel time (cost) function (also known as the link performance
$\qquad\qquad$ function) on link a which is determined by total flow V_a on each link
$\qquad$ x_{ij}^k = number of total trips from i to j assigned to path k
$\qquad$ $\delta_{ij}^{ak} = 1$ if link a belongs to path k from i to j, 0 otherwise

The most common form of the link performance function used for equilibrium traffic assignment problems is shown in equation (7.17).

$$t = t_0 \left[1 + a \left(\frac{V}{C_p} \right)^b \right] \qquad (7.17)$$

where a and b are constants (note that when $a = 0.15$ and $b = 4$, equation (7.17) reduces to the form of BPR function shown in equation (7.12)).

Even for very small networks, it is very difficult to obtain a mathematical solution to this nonlinear optimization problem. Frank and Wolfe (1956) developed a heuristic solution that decomposes the problem into a number of steps. The solution is popularly known as Frank-Wolfe decomposition of the user equilibrium problem. Presented in Figure 7.9 is a schematic of computer implementation of Frank-Wolfe decomposition.

After several iterations, the heuristic reaches a situation where the UE condition is satisfied. As can be seen in Figure 7.9, the user equilibrium assignment technique utilizes a convex combination method called the golden section search for direction finding. The mathematical concepts and optimization techniques underlying the user equilibrium assignment is fairly complex and beyond the scope of this chapter. More details about UE and other optimization solutions for network assignment problems can be found in Sheffi (1985).

7.11 MODEL ADJUSTMENTS AND USE OF SYNTHETIC/BORROWED MODELS

When the models at different steps of the sequential modeling process are developed based on detailed data collected in the study area using large samples as prescribed for O-D and other survey procedures, the results at each step can be verified against survey-generated data that represent true values of the dependent variables of the models, such as trip ends (productions and attractions), trip-length frequencies, mode choice proportions, and traffic volumes on network links. The availability of detailed data allows model developers to make adjustments at respective steps as needed. For example, if during trip generation analysis it is found that the trip ends at a certain zone cannot be estimated closely by trip production or attraction models, a special investigation may be done, and an off-model procedure may be used for that zone. A special generator analysis for a college campus or an airport is an example of this type of a case. For gravity models, also, a special adjustment factor has been used in some cases to reflect the impact of certain physical features such as a river crossing on the attractiveness (or pull) of a group of traffic zones. Traffic counts taken at selected screen-lines are useful for comparing model-generated travel pattern with actual volumes of traffic crossing screen-lines and making adjustments, if needed.

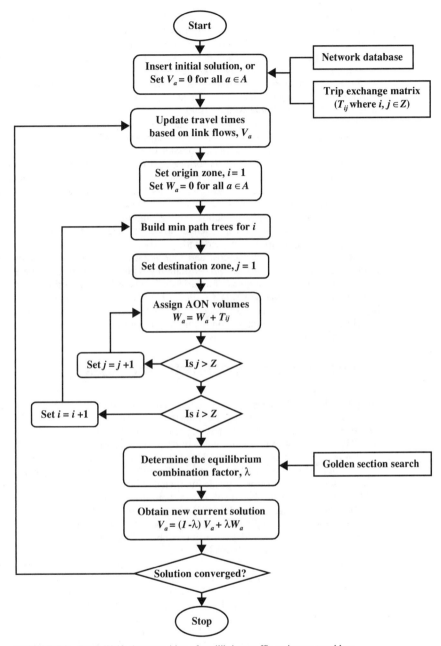

FIGURE 7.9 Frank-Wolfe decomposition of equilibrium traffic assignment problem.

At the traffic assignment step, network-related parameters such as travel speed and capacity of certain links may need adjustments in order to produce results that match closely existing traffic counts. For this purpose traffic counts taken at cut-lines are utilized for comparison with volumes generated by the traffic assignment model. For example, if it is found that along a freeway corridor the assigned volumes on freeway links are too high whereas those along a parallel arterial highway are too low in comparison to ground counts, the speeds and capacities of selected links along the corridor should be reexamined and adjusted. A good source of useful ideas for model adjustments including network coding is Ismart (1990).

Making adjustments to models to replicate the existing situation more closely is an important task of model development, and this requires some experience and sound understanding of how the models work. Adjustments to models assume a greater role in the case of synthetic or borrowed models. In this case a set of preexisting models for one area is transferred and adopted for another study area. Then the original values of key parameters such as trip generation rates and friction factors are adjusted, if necessary, to produce desired results. This process is referred to as model calibration. In the case of transferred/borrowed models, data from O-D survey are not available for checking the results at each step of the sequential modeling process, and the only data to check with are traffic volume counts on various road segments, which are to be replicated by the results of traffic assignment.

Synthetic models have been used widely, especially for small and medium-sized urban areas, in order to avoid or reduce the cost and time required for full-scale O-D surveys. Transportation planners in the North Carolina DOT's Planning and Research Branch have developed and used synthetic models for urban areas of a variety of sizes in North Carolina for many years. Chatterjee and Cribbins (1972), Bates (1974), Modlin (1974), Khasnabis and Poole (1975), and Chatterjee and Raja (1989) describe some these procedures. Sosslau et al. (1978) and Martin and McGuckin (1998) contain considerable information on the use of borrowed models and transferred parameters.

7.12 TRAVEL DEMAND MODELING SOFTWARE PACKAGES

The principles and the steps involved in the travel demand modeling process were first implemented in the form of computer software programs by the former Bureau of Public Roads and later refined by the Federal Highway Administration. This software package originally was called PLANPAC/BACKPAC, and it was primarily highway oriented. A report by FHWA (1974) contains the details of these programs. In early 1970s the Urban Transportation Planning System (UTPS) was developed by the former Urban Mass Transportation Administration to add the capability for transit planning. Later the scope of UTPS was expanded to include both highway and transit networks. UTPS was an IBM mainframe computer-based software system that has individual modules capable of performing a specific task. For example, the UMATRIX module performed matrix computations and the ASSIGN module performed AON or capacity restrained assignment. As with many computer programs in the 1960s and 1970s, the capabilities of early versions of UTPS were very limited. Performing a complete run of the four-step process took several days of work related to input preparation, debugging, and output analysis. Until the early 1990s several MPOs were still using the UTPS-based planning software.

With the advent and penetration of microcomputers in the early 1980s, different commercial versions of travel demand modeling (TDM) software were developed and marketed. Among the first was a software package called MINUTP, which was developed and marketed by COMSIS Corporation. MINUTP was an MS-DOS-based command-driven modeling package and similar to FHWA's PLANPAC software. Included among the command-driven TDM packages that were popular till the late 1990s and even in the early 2000s are TRANPLAN and QRS-II.

Since the advent of the Microsoft Windows operating system, travel demand modeling software landscape has changed even more dramatically. Current high-end TDM packages are not only capable of performing travel demand modeling, but are also compatible with geographic information systems (GIS). For example, TransCAD (Caliper Corporation) is a travel demand modeling package as well as a GIS software package. Other TDM packages include CUBE and TP+ by Citilabs, EMME/2, T-Model (Strong Concepts), and Saturn (UK).

7.13 APPLICATIONS OF TRAVEL MODELS

Depicted in Figure 7.10 is the traditional long-range planning process for a region (MPO) or sub-region. This process involves the identification of transportation-related problems followed by the determination of the future travel demand for a given situation. This in turn is followed by an attempt to find future transportation improvement that will meet the need of the future travel demand. Traditionally, in the planning process, the main criterion that is used objectively to evaluate alternative projects is congestion relief by capacity improvement, which typically involves building new highways, widening existing highways, and improving transit services. Land use-related alternatives also are examined. Travel forecasting models help assess the effectiveness of each alternative in reducing traffic congestion. Since traffic congestion of a serious nature usually occurs on major highways—primarily arterials—the travel forecasting procedure usually pays more attention to these highways, and this was reflected in network coding. Typically local roads and some minor collectors are not included in the network used for traffic assignment.

For the design of a new highway and/or the widening of an existing highway, the estimated traffic volume for the facility is the main item of interest that highway design engineers expect from travel forecasting models. For other related information, such as the proportion of design hourly volume with respect to average daily traffic, directional split, and truck percentage, highway design

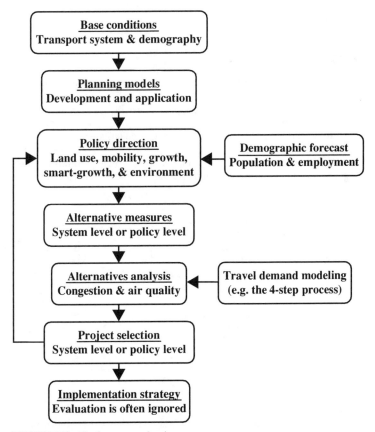

FIGURE 7.10 The long-range planning process.

engineers use other sources of information. However, this situation changed when more and more urban areas had to assess the air quality impact of highway networks as more detailed and accurate information related to travel was needed for air quality analysis.

7.14   *TRANSPORTATION AND AIR QUALITY PLANNING*

In recent years the planning process has given considerable emphasis on the assessment of the effect of transportation alternatives on the environmental consequences, especially air quality impacts. The United States Environmental Protection Agency (EPA) developed several versions of a model over time for estimating emission factors of air pollutants from mobile sources in terms of grams per mile. One of the recent versions of this model, MOBILE5, was used widely during the 1990s. The latest version, MOBILE6, is being used since early 2002. In the state of California a different emission factor model, called EMFAC, is used. These emission factor models need a variety of travel-related measures for the estimation of emissions from vehicular travel, and this need uncovered several deficiencies of the traditional travel forecasting models and led to various refinements and advancement of the modeling procedure. The integration of travel models with emission factors models is illustrated in Figure 7.11.

The travel-related inputs required for mobile source emissions estimation are discussed by Miller et al. (1992), and the deficiencies of the four-step models are examined by Stopher (1993). Another source of information on this subject is Chatterjee et al. (1997). A few examples of these weaknesses and refinements are discussed below.

7.14.1   Travel Speed

The amount of emissions released by vehicles when traveling varies considerably with speed, and therefore an accurate estimation of travel speed on each link of a highway network is important. However, the traffic assignment procedures usually focus on the accuracy of the assigned traffic volumes, and travel speeds are adjusted to produce better results for traffic volumes. In many cases the travel speeds generated by travel forecasting models are not accurate, and little effort was made in the past to improve the speed values because, as mentioned above, for the tasks of capacity-deficiency analysis and design of highway improvements, predicted traffic volume is the item of interest and there was no urgent need to improve the accuracy of speed estimates. During the 1990s, however,

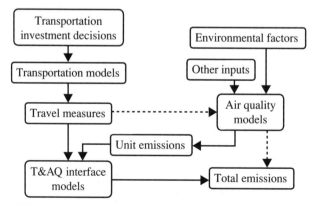

FIGURE 7.11   A schematic representation of TDM integrated with air quality modeling.

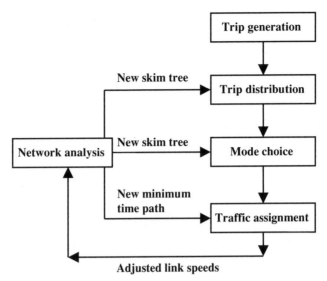

FIGURE 7.12   The planning process with feedback loops.

in response to the needs of emission factor models, considerable research was performed to improve the model-generated speed values. For this purpose feedback loops, as shown in Figure 7.12, and postprocessing procedures for speed calculations based on volume/capacity ratios were introduced, and these are widely used at this time. Alternative methods for feedback and iteration can be used, and there are also different convergence criteria to choose from. Boyce, Zhang, and Lupa (1994) discuss alternative methods of introducing feedback into the four-step procedure, and Dowling and Skabardonis (1992) provide good information on speed postprocessing.

7.14.2   VMT

For the calculation of total emissions from mobile sources in an urban area, a reliable estimate of total VMT for each class of roads is needed. The estimation of VMT generated on local and minor collector roads is important for this purpose. However, as discussed above, most of the local and minor roads usually are not included on the highway network used for travel demand modeling. Thus, network-based models are not capable of generating reliable estimates of VMT on local and minor collector roads. To fill this gap a few off-model procedures have been developed in recent years, and some of these are discussed by Chatterjee et al. (1997).

7.14.3   Vehicle Class Mix and Vehicle Age Distribution

For traditional transportation planning, which determines future needs of new and improved highways and transit services, detailed information regarding different types of vehicles operating on various highways is needed. Usually it is sufficient to use a rough estimate of the proportion of large trucks for capacity analysis. For air quality analysis, however, MOBILE5 provided separate emission rates for 8 different types of vehicles. The MOBILE6 moel requires input parameters for up to 28 different types of vehicles. The proportions of these classes of vehicles actually vary according to the functional class of roads and time of day. The data need becomes more demanding and complicated when it is recognized that emission rates depend not only on the size and weight of a vehicle but also on its age and mileage accumulation and that this information too has to be developed to take advantage of the MOBILE model's capabilities.

It is clear that commonly used travel forecasting models are not designed to develop detailed information on vehicle class and age and that other off-model procedures must be used for this purpose. However, the responsibility for developing this information lies with transportation planners who are working in the area of travel demand analysis.

7.14.4   Start versus Running Emissions

Earlier versions of MOBILE (version 5 and earlier) combined start emissions and running emissions into one composite emission factor. Owing to the realization that start emissions are much more pronounced than the running emission, the latest version of MOBILE (version 6) separates them from running emissions. Inputs on starts are provided in the form of start distributions by vehicle class.

7.14.5   Cold and Hot Soak Periods and Operating Modes

MOBILE 6 accounts for engine temperatures in the form of soak distributions (cold and hot soak distributions). The time lag between two successive trips of a vehicle determines the emissions at the beginning (start emissions) and end (evaporative emissions) of a trip. Inputs on this time lag are provided to the emission models in the form of soak period distributions. Venigalla and Pickrell (2002) describe procedures to obtain soak period inputs from large travel survey databases.

The operating modes of vehicles are related to engine temperature. Earlier versions of the MOBILE model classified operating modes into two broad categories—transient and hot stabilized modes. The transient mode is further categorized into two separate subcategories, cold start and hot start modes. EPA uses a few criteria based on engine soak period to determine the operating mode of a vehicle at a particular time. A few researchers used these time-based criteria in conjunction with travel surveys and innovative traffic assignment procedures to demonstrate how operating mode fractions can be estimated analytically (Venigalla, Miller, and Chatterjee 1995a, 1995b; Venigalla, Chatterjee, and Bronzini 1999; Chatterjee et al. 1996).

7.14.6   The Role of Transportation Planners in Air Quality Modeling

More and more the regulatory burdens of air quality modeling related to transporation projects are being shouldered by transportation planners. Air quality models are constantly being updated with new knowledge gained on transportation-related emissions. The current state of the art in emissions modeling requires more from the transportation planning community than ever before. For example, the concept of trip ends and trip chaining is easily extended to deriving travel-related inputs to the emission factor models (Chalumuri 2003). While model-improvement efforts undoubtedly improve the state of the practice, additional burdens are placed on the transportation modeling community to develop innovative methods to derive travel-related and vehicle activity inputs to the emissions model. In order to accommodate the needs of transportation-related air quality modeling in the foreseeable future, transportation planners are expected to develop new methods or adapt existing methods.

REFERENCES

Bates, J. W. 1974. "Synthetic Derivation of Internal Trips for Small Cities." *Transportation Research Record* 526, 93–103.

Boyce, D. E., Y. Zhang, and M. R. Lupa. 1994. "Introducing 'Feedback' into Four-Step Travel Forecasting Procedure versus Equilibrium Solution of Combined Model." *Transportation Research Record* 1443:65–74.

Bureau of Public Roads. 1964. *Traffic Assignment Manual.* U.S. Department of Commerce, Urban Planning Division, Washington, DC.

Cambridge Systematics, Inc., and Barton-Aschman Associates, Inc. 1996. *Travel Survey Manual*. Prepared for U.S. Department of Transportation and U.S. Environmental Protection Agency, Travel Model Improvement Program (TMIP), July.

Chalumuri, S. *Emersion Impacts of Personal Travel Variables*. A thesis submitted in partial fulfillment of Master of Science degree in Civil and Infrastructure Engineering Department, George Mason University, Fairfax, VA, August 2003.

Chatterjee, A., and P. D. Cribbins. 1972. "Forecasting Travel on Regional Highway Network." *Transportation Engineering Journal of ASCE* 98(TE2):209–24.

Chatterjee, A., and S. Khasnabis. 1973. "Category Models—A Case for Factorial Analysis." *Traffic Engineering*, October, 29–33.

Chatterjee, A., and M. Raja. "Synthetic Models for Through Trips in Small Urban Areas." *Journal of Transportation Engineering*, September, 537–55.

Chatterjee, A., and K. C. Sinha. 1975. "Mode Choice Estimation for Small Urban Areas." *Transportation Engineering Journal of ASCE* 101(TE2):265–78.

Chatterjee, A., S. Khasnabis, and L. J. Slade. 1977. "Household Stratification Models for Travel Estimation." *Transportation Engineering Journal of ASCE* 103(TE1):199–213.

Chatterjee, A., D. R. Martinson, and K. C. Sinha. 1977. "Trip Generation Analysis for Regional Studies." *Transportation Engineering Journal of ASCE* 103(TE6):825–41.

Chatterjee, A., M. A. McAdams, and F. J. Wegmann. 1983. "Non-Commitment Bias in Public Opinion on Transit Usage." *Transportation* 11:347–60.

Chatterjee, A., T. L. Miller, J. W. Philpot, T. F. Wholley, R. Guensler, D. Hartgen, R. A. Margiotta, and P. R. Stopher. 1997. *Improving Transportation Data for Mobile Source Emission Estimates*. Transportation Research Board, National Research Council, National Cooperative Highway Research Program Report 394.

Chatterjee, A., P. M. Reddy, M. M. Venigalla, and T. M. Miller. 1996. "Operating Mode Fractions on Urban Roads Derived by Traffic Assignment." *Transportation Research Record* 1520:97–103.

Chatterjee, A., F. J. Wegmann, J. D. Brogan, and K. Phiu-Nual. 1979. "Estimating Truck Traffic for Analyzing UGM Problems and Opportunities." *ITE Journal* 49(5):24–32.

Dial, R. B. 1971. "A Probabilistic Multipath Traffic Assignment Model Which Obviates Path Enumeration." *Transportation Research* 5(2):83–111.

Dowling, R., and A. Skabardonis. 1992. "Improving Average Travel Speeds Estimated by Planning Models." *Transportation Research Record* 1366:68–74.

Dresser, G. B., and D. F. Pearson. 1994. *Evaluation of Urban Travel Survey Methodologies*. Texas Transportation Institute Research Report 1235-10, Texas Department of Transportation, Austin.

Easa, S. M. 1991. "Traffic Assignment in Practice: Overview and Guidelines for Users." *Journal of Transportation Engineering* 117(6):602–23.

Easa, S. M. 1993. "Urban Trip Distribution in Practice. I: Conventional Analysis." *Journal of Transportation Engineering* 119(6):793–815.

Eash R. 1993. "Equilibrium Traffic Assignment with High-Occupancy Vehicle Lanes." *Transportation Research Record* 1394, 42–48.

Eash, R. W., B. N. Janson, and D. E. Boyce. 1979. "Equilibrium Trip Assignment: Advantages and Implications for Practice." *Transportation Research Record* 728:1–8.

Federal Highway Administration (FHWA). 1974. *FHWA Computer Programs for Urban Transportation Planning*. U.S. Department of Transportation, FHWA, July.

———. 1975a. *Trip Generation Analysis*. U.S. Department of Transportation, FHWA, August.

———. 1975b. *Urban Origin-Destination Surveys*. U.S. Department of Transportation, FHWA, reprinted July.

Fischer, M. J. 2001. *Truck Trip Generation Data*. NCHRP Synthesis 298, National Cooperative Highway Research Program, Transportation Research Board.

Frank, M., and P. Wolfe. 1956. "An Algorithm for Quadratic Programming." *Naval Research Logistics Quarterly* 3:95–110.

Fratar, T. J. 1954. "Forecasting Distribution of Inter-Zonal Vehicular Trips by Successive Approximations." *Proceedings of Highway Research Board* 33.

Gallo, G., and S. Pallottino. 1988. "Shortest Path Algorithms." *Annals of Operations Research* 13:3–45.

Grecco, W. L., J. Spencer, F. J. Wegmann, and A. Chatterjee. 1976. *Transportation Planning for Small Urban Areas*. National Cooperative Highway Research Program Report 167, Transportation Research Board.

Hamberger, C. B., and A. Chatterjee. 1987. "Effects of Fare and Other Factors on Express Bus Ridership in a Medium-Sized Urban Area." *Transportation Research Record* 1108:53–59.

Horowitz, J. L., F. S. Koppelman, and S. R. Lerman. 1986. *A Self-Instructing Course in Disaggregate Mode Choice Modeling.* Technology Sharing Program, U.S. Department of Transportation, December.

Ismart, D. 1990. *Calibration and Adjustment of System Planning Models.* U.S. Department of Transportation, Federal Highway Administration, Washington, DC, December.

Khasnabis, S., and M. R. Poole. 1975. "Synthesizing Travel Patterns for a Small Urban Area." *Traffic Engineering,* August, 28–30.

Makowski, G. G., A. Chatterjee, and K. C. Sinha. 1974. "Reliability Analysis of Origin-Destination Surveys and Determination of Optimal Sample Size." *Proceedings of the Fifteenth Annual Meeting of Transportation Research Forum* 4(1):166–76. [Also published by UMTA through the National Technical Information Service, PB 236-986/AS.]

Martin, W. A., and N. A. McGuckin. 1998. *Travel Estimation Techniques for Urban Planning.* NCHRP Report 365, Transportation Research Board.

Mayworm, P., A. M. Lago, and J. M. McEnroe. 1980. *Patronage Impacts of Changes in Transit Fares and Services.* Ecsometrics Inc., Prepared for U.S. Department of Transportation, UMTA Report RR 135-1, Washington, DC, September.

McFadden, D. L. 1978. "The Theory and Practice of Disaggregate Demand Forecasting for Various Modes of Urban Transportation." In *Proceedings of the Seminar on Emerging Transportation Planning Methods,* Office of University Research, U.S. Department of Transportation, Washington, DC, August.

Miller, T., A. Chatterjee, J. Everett, and C. McIlvaine. 1992. "Estimation of Travel Related Inputs to Air Quality Models." In *Transportation Planning and Air Quality,* Proceedings of the National Conference of the American Society of Civil Engineers, 100–25.

Modlin, D. G. 1974. "Synthetic Through Trip Patterns." *Transportation Engineering Journal of ASCE* 100(TE2):363–78.

Moore, E. F. 1959. "The Shortest Path through a Maze." In *Proceedings of the International Symposium on the Theory of Switching.* Harvard University Computation Laboratory Annals 30. Cambridge: Harvard University Press, 285–92.

Ogden, K. W. 1992. *Urban Goods Movement: A Guide to Policy and Planning.* Aldershot: Ashgate.

Parody, T. E., and D. Brand. 1979. "Forecasting Demand and Revenue for Transit Prepaid Pass and Fare Alternatives." *Transportation Research Record* 719:35–41.

Reichman, S., and P. R. Stopher. 1971. "Disaggregate Stochastic Models of Travel-Mode Choice." *Highway Research Record* 369:91–103.

Sheffi, Y. 1985. *Urban Transportation Networks: Equilibrium Analysis with Mathematical Programming Methods.* Englewood Cliffs: Prentice-Hall.

Sosslau, A. B., A. B. Hassan, M. M. Carter, and G. V. Wickstrom. 1978. *Quick Response Urban Travel Estimation Techniques and Transferable Parameters: User Guide.* NCHRP Report 187, Transportation Research Board, Washington, DC.

Stopher, P. R. 1993. "Deficiencies of Travel-Forecasting Methods Relative to Mobile Emissions." *Journal of Transportation Engineering* 119(5):723–41.

Stopher, P. R., and H. A. Metcalf. 1996. *Methods for Household Travel Surveys.* NCHRP Synthesis 236, Transportation Research Board, Washington, DC.

Urban Transportation Systems Associates, Inc. 1972. *Urban Mass Transportation Travel Surveys.* Prepared for U.S. Department of Transportation, August.

Venigalla, M. M., A. Chatterjee, and M. S. Bronzini. 1999. "A Specialized Assignment Algorithm for Air Quality Modeling." *Transportation Research—D,* January.

Venigalla, M. M., T. Miller, and A. Chatterjee. 1995a. "Alternative Operating Mode Fractions to the FTP Mode Mix for Mobile Source Emissions Modeling." *Transportation Research Record* 1462:35–44.

———. 1995b. "Start Modes of Trips for Mobile Source Emissions Modeling." *Transportation Research Record* 1472:26–34.

Venigalla, M. M., and D. H. Pickrell. 1997. "Implication of Transient Mode Duration for High Resolution Emission Inventory Studies." *Transportation Research Record* 1587.

———. 2002. "Soak Distribution Inputs to Mobile Source Emissions Modeling: Measurement and Transferability." *Transportation Research Record* 1815.

Voorhees, A. M. 1955. "A General Theory of Traffic Movement." *Proceedings of the Institute of Traffic Engineers* 46–56.

Weiner, E. 1969. "Modal Split Revisited." *Traffic Quarterly* 23(1):5–28.

———. 1992. *Urban Transportation Planning in the United States: An Historical Overview,* rev. ed. Washington, DC: Office of the Secretary of Transportation, November.

Woodruff, R. B., D. J. Barnaby, R. A. Mundy, and G. E. Hills. 1981. *Market Opportunity Analysis for Short-Range Public transportation Planning: Method and Demonstration.* National Cooperative Highway Research Program Report 212, Transportation Research Board, Washington, DC, September.

CHAPTER 8
TRANSPORTATION AND LAND USE

Kara Kockelman
Department of Civil, Architectural and Environmental Engineering
University of Texas at Austin

Bin (Brenda) Zhou
Department of Engineering
Central Connecticut State University, New Britain

8.1 INTRODUCTION

The calculations and recommendations of transportation engineers and planners regularly recognize the fundamental role that land use conditions play in the theatre of transportation. An obvious example is the profession's heavy reliance on trip generation rates, with regular reference to the Institute of Transportation Engineer's *Trip Generation Manual* (ITE 2008) for over 150 land use types, with rates affected by development size and site location within urban regions.

As a derived demand, travel ensures that persons can engage in various activities at multiple sites, while packages and products reach their intended distributors and end users. Whether they be homes or businesses, parks or croplands, the more separated in space these activity sites are, the longer the travel distances. Accompanying these distances comes a shift to faster modes, an infeasibility of non-motorized modes, a greater need for high-speed freeways and jet airplanes. Within a given transportation system, greater distances caused by greater populations or less intensely developed land will result in greater demands on system components and a higher likelihood of congested travel conditions, over land, over water, and in the air. It is important that community planners and system designers recognize such relationships, while pursuing plans that enhance land use-transportation interactions.

Many will agree that the U.S. transportation profession has for far too long emphasized mobility enhancements for the motoring public (e.g., new highways and higher speeds), rather than a more balanced view of accessibility improvements, reflecting transport options in concert with land use patterns (Bartholomew 2007; Litman 2003, 2007; Handy 1994). This is no doubt because of challenges in effectively managing both land use and transport, with state departments of transport pursuing major network improvements and city officers permitting land owners' improvements to existing parcels. Transportation engineers and planners should seek to recognize how their decisions can impact access to jobs, schools, services, and other key destinations via a variety of modes, along with longer-term land use changes. In reality, various highway improvements can degrade access for local travelers, including walk and bike modes, and quality of life for local residents and shop owners, while improving travel times for through travelers. Such myopic planning led to America's

Freeway Revolts of the 1960s and 1970s (Mohl 2004). European models of transportation planning and land use management look very different. (See, e.g., Knoflacher 2007 and Pucher et al. 2010.) Many Americans have become fans of the principles embodied in concepts of Smart Growth, New Urbanism, Neotraditional Design, Traditional Neighborhood Development, and Transit-Oriented Development (TOD), as ways to moderate reliance on personal vehicles while curbing other ills of relatively standard U.S. design and development practices (see, e.g., Litman 2010; Evans et al. 2007; Handy 2005; Duany et al. 2000; Calthorpe 1993).

Travel is a complex phenomenon; and travelers trade off alternative destinations and routes, much as they do modes, vehicle ownership levels, and their own home (and work and school) locations. Thus, regions with double the density of activity sites (proxied by work and population densities) generally will not experience half the amount of travel distance or travel-related energy consumption, even though transit and carpooling may become more viable alternatives.[*]

Works by Newman and Kenworthy (1996, 1999, 2006), Holtzclaw (1991, 1994), and Holtzclaw et al. (2002) are regularly cited on this score: One may expect an elasticity of regional vehicle miles traveled (VMT) with respect to regional density of about 25 to 30 percent. In other words, as density doubles, energy use and VMT tend to fall by 25 to 30 percent. Or, as density halves, energy use and VMT have been estimated to rise by over 30 percent—even after controlling for certain demographic attributes like income and household size (Holtzclaw et al. 2002). Nevertheless, a wide variety of other attributes—including parking costs, land use balance, infrastructure provision, demographics, and even topography—can be critical. All are at play in the land use-transport connection, and density in isolation is no panacea for congestion and many other transportation problems.

Just as land use decisions help shape travel choices and traffic conditions, network investment decisions and transportation policies play some role in location choices and land development decisions, along with property values and other variables of interest to a variety of stakeholders. This chapter begins by summarizing typical categories of land use and then moves on to discussions of how such land uses affect travel-related choices, how transport improvements and policies impact land use patterns, how integrated land use-transportation models work, and how a variety of other meaningful topics relate to this complicated yet critical arena for demand forecasting, policymaking, and system design and management.

8.2 LAND USE CATEGORIES

The term "land use" carries at least three distinct implications: land cover, use type, and intensity. The first refers to land coverage (such as forested, barren, developed, wetlands, and shrubland), or land use/land cover. It is widely used in the field of geography, where biodiversity and ecosystem conservation are key and remote-sensing technologies are regularly used to enable large-scale geospatial data retrieval and classification. Of course, land cover change can be attributed to human exploration and activity, with ties to transportation system coverage and capacity (see, e.g., Laurance et al. 2001; Walker 2004). The second dimension of land emphasizes developed land, in urban and exurban areas, and is often further classified into categories that are pertinent to individual studies. These include residential versus nonresidential, commercial, industrial, civic, educational and transportation uses. Such details are useful for city and regional planning purposes, but do not serve as direct inputs to travel demand models, which rely instead on land use intensity details (in tandem with job and household type information).

With the increasing popularity of geographic information systems (GIS), a variety of spatial data sets now exist, for transportation engineering and planning use. For example, appraisal districts, which are responsible for appraising real, taxable property, are increasingly linking their data to a GIS layers for parcel-level data. The tax-related codes provide a meaningful measure of land use

---

[*]Reduced trip chaining, greater activity participation rates, travel to more preferred destinations, and more uniformly distributed (rather than poly-nucleated) activity sites may emerge in denser environments.

TABLE 8.1   Land Use Categories

Classification	Description
Large-lot single-family	Single-family homes on lots greater than 10 acres
Single-family	Single-family detached, two-family attached
Mobile homes	Mobile homes
Multifamily	Three- and four-plex, apartments and condos, group quarters, retirement facilities
Commercial	Retail and general merchandise, apparel and accessories, furniture and home furnishings, grocery and food sales, eating and drinking, auto related, entertainment, personal services, lodgings, building services
Office	Administrative offices, financial services (banks), medical offices, research and development
Industrial	Manufacturing, warehousing, equipment sales and service, recycling and scrap, animal handling
Civic	Semi-institutional housing, hospital, government services, educational facilities, meeting and assembly facilities, cemeteries, day care facilities
Mining	Resource extraction, quarries
Open space	Parks, recreational facilities, golf courses, preserves and protected areas, water drainage areas and detention ponds
Utilities	Utility services, radio towers, communication service facilities, water/wastewater facilities
Undeveloped/rural	Rural uses, vacant land, land under construction
Water	Inundated areas, such as lakes and rivers
Transportation	Railroad facilities, transportation terminal, aviation facilities, parking facilities, right-of-way and traffic islands

Note: Table details come from City of Austin's Land Use Survey Methodology (2000).

classification. Table 8.1 shows a typical coding system and broad categories often suitable for land use studies and neighborhood characterizations. (See, e.g., Zhou and Kockelman 2008a and 2009a.)

When detailed spatial data on actual land use types are lacking, zonal employment and household density measures can be used to represent local land use conditions. Such density measures can also be used to generate area-type categories. For example, the Texas Department of Transportation (TxDOT) classifies central business district (CBD) zones as those having eight or more person-equivalents per gross acre, urban areas as having three or more person-equivalents per gross acre, suburban as one or more, and rural as anything less dense.[*]

Such categories, while coarse, are regularly found to be statistically and practically significant in models of travel behavior, property value, and other variables of interest, particularly when alternative attributes of urban form (and/or demographic and firmographic information) are lacking. Nevertheless, it is best to retain, and control for, the continuous underlying measures of density that generate such categories (as well as the type of jobs and households or persons they reflect). In addition, a simple distance-to-CBD variable (Euclidean or network-based) tends to be highly practically significant in a variety of contexts, serving as a solid surrogate for regional accessibility (particularly in monocentric regions). Where feasible, measures characterizing the diversity, mixing, and balance of land uses (based on parcel-level data, but computed at a more spatially aggregate level, like a 600-meter radius circle or traffic analysis zone) can also prove quite meaningful in prediction.

Of course, network-based attributes, such as the share of intersections that are four-way, average block size, and distance to nearest principal arterial, can be helpful in prediction—not just of travel behavior, but also land use conditions and land use change. When used in tandem, land use and travel cost or other types of access variables can provide accessibility measures, both local and regional, to different types of activities and actors (see, e.g., Srour et al. 2002). And these are often key to prediction, as described below.

---

[*]Equivalent population is simply zone population plus zone employment times the region's persons-per-job ratio.

8.3 LAND USE EFFECTS ON TRAVEL DEMAND

Land use choices essentially determine activity site locations, and thus opportunities for trip origins and destinations. From trip generation and attraction decisions come travel distances, and these tie into each travel mode's feasibility and cost, with the automobile dominating choice for longer intra-regional trips within most developed countries.

Low-density land use patterns have been cited as an important source of roadway congestion, energy depletion, air pollution, and greenhouse gas (GHG) emissions (see, e.g., Dunphy and Fisher 1996; Newman and Kenworthy 2006; Ewing et al. 2008); and many investigations have concluded that vehicle ownership levels, shares of motorized trips, and household VMT depend on various features of urban form in both practically (and statistically) significant ways. (See, e.g., Fang 2008; Holtzclaw et al. 2002; Ewing and Cervero 2001, 2010; Cervero and Kockelman 1997.)

As an example, Musti and Kockelman's (2009) regressions of vehicle ownership levels on demographic and land use attributes at the level of traffic analysis zones (TAZs) in Austin, Texas, signal a striking 30 percent elasticity with respect to local employment density, ceteris paribus, suggesting that jobs density (or the attributes for which it proxies, such as regional access, central location, and land use balance) can play a key role in energy and VMT savings, per capita. Moreover, as distance to the CBD falls in such regressions, vehicle ownership falls further, providing a type of "double dividend" (since many jobs tend to be centrally located). Since VMT per vehicle owned is relatively stable, regardless of vehicle ownership level (averaging 9,000 to 10,000 miles per year, in the United States, according to National Household Travel Survey data [Kockelman et al. 2009]), much of the VMT and energy savings that can come from land use changes probably stem from vehicle ownership decisions.

Of course, there are other ways to moderate congestion, energy, air quality, and climate change concerns, without altering land use patterns (which can be slow to take hold, though arguably more enduring and beneficial in other respects[*]). Among these are congestion pricing, gas taxes, fuel-economy regulations, vehicle purchase feebates, full and mild hybridization of vehicles, preheating catalytic converters, and so forth (see, e.g., Kockelman et al. 2009). And, of course, land use conditions are not the only factor impacting travelers' choices. Demographics tend to offer much greater predictive power (see, e.g., Bento et al. 2005; Schimek 1996).[†] Over the past two decades, a great deal of literature has emerged concerning the relationship between the physical features of urban landscapes and traveler behavior. Ewing and Cervero's (2001, 2010) comprehensive reviews of such studies essentially conclude that regional-level accessibility is a key predictor of per-capita VMT, while travelers' vehicle ownership levels and mode choices are most affected by neighborhood-level land use patterns. As Boarnet and Crane (2001) note, behavioral processes at play are complex, and studies that use different data sets and geographic scales and focus on different aspects of travel behavior draw somewhat distinctive conclusions. In general, early work has used more aggregate statistics, whereas later work has benefited from access to more disaggregate data and richer controls on demographics, neighborhood attributes (both at the origin and destination), travel costs (across alternatives), and other factors.

To disentangle the relationship of travel behavior and the built environment, researchers have relied on quasi-experimental designs (e.g., pairing matched neighborhoods, as in Cervero and Gorham [1995], Khattak and Rodriguez [2005], and Shay and Khattak [2005]), cross-sectional data and regression techniques (see, e.g., Kockelman 1997; Cervero and Kockelman 1997, Crane and Crepeau 1998; Salon 2006), and analysis of longitudinal data (see, e.g., Krizek 2003). When comparing travel behavior in matched neighborhoods, households in neotraditional and transit-oriented neighborhoods engage in fewer automobile trips, less VMT, and more work trips by transit,

---

[*]The building stock generally enjoys lifetimes that exceed those of vehicles and many policies, and more compact development can lead to savings on infrastructure and other public expenditures (such as school bussing) while enabling more healthful mode choices (such as walking). (See, e.g., Burchell et al. 2002.)

[†]Of course, it is difficult, if not impossible, for planners, engineers and others to appreciably affect demographics (like income, household sizes, and the age and presence of children).

as compared to counterparts living in more conventional neighborhoods. Thanks to a reliance on longitudinal data (of 6144 moving households' travel choice changes), Krizek's (2003) findings in this vein appear most compelling.

Among all potential control variables, priced parking and higher regional accessibility appear to provoke the greatest reductions in personal-vehicle use (see, e.g., Kockelman 1997; TRCP 2004; Ewing and Cervero 2001; Kockelman et al. 2009). The associations are felt to be strong enough and the literature robust enough that the U.S. Environment Protection Agency relies on average estimates provided by Ewing and Cervero's (2001) extensive review to inform its Smart Growth Index Model, a tool used by transportation planners in U.S. regions seeking emissions credits for various land use actions (U.S. EPA 2002, Kuzmyak et al. 2008).

In their international review of city data, Newman and Kenworthy's (2006) argue that 14 jobs or persons per acre can serve as a very meaningful threshold density, for per-capita transport energy use. Above this density they notice a sharp increase in walk, bike, and transit use. They also recognize that it is unrealistic for cities to simply add a rail line through the center and expect significant distance and mode shifts. Nevertheless, they do suggest that auto-oriented cities can and should be restructured as smaller, transit-oriented cities, to save energy and travel. Of course, different cities around the world enjoy very different histories, cultures, incomes, and transport systems. Moreover, the notion of regional density relationships holding at a local level is problematic. In reality, density is just one of many factors at play. Density is highly correlated (and causally associated) with a variety of other features. (See, e.g., Kuzmyak et al.'s [2003] and Litman's [2010] factor descriptions and literature summaries, and Ewing and Cervero's [2010] meta-analysis of impacts.)

8.3.1 The Impact of Self-Selection

In light of all of the empirical estimates, to what extent is self-selection at play in location and travel decisions? In other words, are people's home location and destination choices purposefully supportive of travel choices that they wish to pursue, regardless of location? While much of the work supports, to some degree, a meaningful role for urban form, controlling for attitudes, to approximately correct for self-selection bias (because of residential sorting), diminishes the estimated influence of the built environment (Mokhtarian and Cao 2008). This highlights the importance of the self-selection issue. Of course, attitudes are typically difficult to measure, and may be largely shaped by one's location. Unfortunately, highly controlled experimental designs (like moving randomly selected household to different sites for some period of time and measuring their travel distances) are infeasible.

Researchers have had to rely on special econometric techniques to appreciate the magnitude of self-selection effects. Zhou and Kockelman (2008b) used Heckman's latent index model (Heckman 1979; Heckman and Vytlaci 1999; and Heckman, Tobias, and Vytlaci 2001) to investigate daily VMT by households surveyed in Austin, Texas. In their study, the daily VMT of the average household living in an urban or CBD zone (i.e., at densities of at least 3 person-equivalents per gross acre) is 47.5 miles per day, as compared to 71.0 miles per day for households living below this density threshold. Their results suggest that at least half the differences in VMT observed between ostensibly equivalent households living in more urban versus less urban neighborhoods is attributable to the location itself, while self-selection of such locations (by households that wish to meet special travel needs and/or preferences) accounts for the remainder.

In general, better control of relevant attributes (e.g., income, household size, the presence and age of children, occupation and education of working adults) diminishes estimates of self-selection effects (because location preferences are regularly associated with socioeconomic and other characteristics). Bhat and Guo (2004) discuss such issues, while controlling for a variety of standard demographic and neighborhood factors in their Oakland, California (Alameda County) data set. Their specification allowed for error-term correlation between location and vehicle ownership choices, to reduce self-selection effects; and results still showed significant built environment effects.

8.4   TRANSPORT'S EFFECTS ON LAND USE

Transportation system improvements can affect regional economies and land development through increased mobility of persons and goods, along with improved access to customers, suppliers, labor, and amenities. Land values are regularly used as a proxy for the access benefits (e.g., implicit value of travel time savings) that come with system improvements, and different types of improvement can have very different impacts on these values. Moreover, impacts can vary noticeably across land use types and across regions. (See, e.g., Transportation Research Board 1995, and ten Siethoff and Kockelman 2002.) In general, transit projects tend to have positive effects on both residential and commercial property values (e.g., Weinstein and Clower 1999; Cervero and Duncan 2002; Armstrong and Rodríguez 2006; Hess and Almeida 2007), while highway projects offer more variable effects (e.g., ten Siethoff and Kockelman 2002; Mikelbank 2004; Iacono and Levinson 2009). Concerns relating to air pollution, noise, safety, and other issues can dampen valuation of residential properties near highway corridors, while added visibility and enhanced access cause commercial property valuations to rise.

As an example, Weinstein and Clower (1999) found that property values within one-quarter mile of Dallas' light rail stations increased about 3 percent more than those in control neighborhoods (i.e., those with similar neighborhood characteristics) over a four-year period (2 years before and 2 years after station opening). And Armstrong and Rodriguez (2006) found properties in cities with a commuter rail station to be valued as much as 10 percent higher than their counterparts. Cervero and Duncan (2002) estimated such differences for commercial land values to be on the order of 120 percent within 0.25 miles of a commuter rail station in California's San Jose area.

Mikelbank (2004) used spatially correlated hedonic models and found that highway projects have negative impacts on housing values during the preconstruction and construction phases. Ten Siethoff and Kockelman (2002) estimated the negative impacts of highway-upgrade construction to be $0.50 per square foot of structure per year using tax assessment values along U.S. 183, a corridor in Austin, Texas. They also estimated a sizable benefit of being within one-half mile of the corridor, at about $50,000 per acre of land and $3 per square foot of structure. In contrast, Iacono and Levinson (2009) generally did not find statistically significant impacts of new-highway construction or improvement projects on property value changes in three Minnesota case studies.

The impacts of transportation on land use are also evident in the land development process and location preference of households and firms, with commute times, highway access, and airport access playing important roles (see, e.g., Zhou and Kockelman 2008a, 2011; Bina and Kockelman 2009; Bina et al. 2006; Van Ommeren et al. 1999; Rouwendal and Meijer 2001; Clark et al. 2003; Tillema et al. 2006; De Bok and Bliemer 2006). Specific locations within a network (e.g., corner parcels) and transport project timing are also important considerations for developers (ten Siethoff and Kockelman 2002).

In terms of residential preferences, Bina and Kockelman (2009) estimated home prices of recent buyers in Austin, Texas to fall by $8,000 with every mile (further) from the CBD, and by $4,700 for every minute in added (one-way) commute time (*ceteris paribus**). They also found that higher-income households are more willing to pay for centrality, which is not surprising given value of time effects. Similarly, Bina et al. (2006) estimated apartment rents in Austin to fall $20 per month with every mile of added distance to the CBD and by $24 for every added commute minute. Interestingly, commute time to work ranked second, right after apartment price, in terms of attribute importance (as evaluated by survey respondents on a scale of 1 to 5). Other access attributes ranked fifth, sixth, and eighth in the lineup of 15 apartment-choice considerations. Nevertheless, access was less of a factor in recent home-buyer decisions (Bina and Kockelman 2009), and home owners comprise roughly two-thirds of the U.S. housing market (according to the 2000 Census of Population).

In general, the effects of land use on transport choices appears to be more direct and strong than the reverse. This is caused, in large part, by the important roles of trip generation and attraction,

---

*Control variables include variables like home or apartment size, age of dwelling, number of bedrooms, lot size, and bus-stop density.

whose spatial distribution largely determines distances traveled between activity sites (see, e.g., Zhao and Kockelman 2002). Nevertheless, the role of transport decisions on land use patterns seems quite evident in many data sets and modeled processes. As a result, many regions throughout the world seek to forecast both land use and transport futures, in tandem.

8.5 LAND USE MODELING

Significant environmental, traffic, and other impacts of urbanization, as well as federal and local regulation (e.g., the U.S. Clean Air Act Amendments of 1990 and Intermodal Surface Transportation Efficiency Act of 1991), effectively require that transportation planning plans and programs account for the interaction and feedbacks between transport and land use (Lyons 1995). And passage of the U.S. Safe, Accountable, Flexible, and Efficient Transportation Equity Act (SAFETEA-LU) in 2003 emphasized the coordination between transportation and land use planning at the state and metropolitan-area levels (Center for Environmental Excellence 2008). Such legislation directly and indirectly encourages the development and application of land use models that tie to models of travel demand. The State of Oregon has been pursuing such integrated modeling efforts in earnest for many years, as a result of its urban growth boundary requirements (based on 1970s legislation), and the State of California now has a law (Senate Bill 375) to reduce GHGs by limiting urban sprawl, with integrated land use-transport modeling a key tool for policy comparisons.

Land use models (LUMs) seek to predict a region's future spatial distribution of households and employment. Though not nearly as complex as the human systems they seek to mimic, such model systems are very complicated. The forces that drive land use change range from regional climate to topography, public policies to human preferences, and social structures to transportation infrastructure; and these factors interact in intricate ways.

Theories of land use can be traced to von Thünen's (1826) concept of agricultural rents and travel costs around a market center, followed by Wingo's (1961) and Alonso's (1964) urban examples. These early models treat land as homogeneous and continuous, and recognize only one employment center. They also neglect latent taste heterogeneity. Thanks to increasing computational power and theoretical advances, many operational LUMs have been developed, with most applying at spatially aggregate levels (such as traffic analysis zones [TAZs]). Key theoretical constructs underlying the majority of LUMs include gravity allocation, cellular automata, spatial input-output, general equilibrium, and discrete response simulation.

8.5.1 Land Use Model Specifications

Multiple researchers have summarized and compared such models (PBQ&D 1999; U.S. EPA 2000; Wegener 2004; Dowling et al. 2005; Iacono et al. 2008). The general consensus is that many limitations remain and the appropriateness and usefulness of any tool varies by context. For example, gravity models tend to use regional totals to adjust forecasts across all zones, and have been found to perform less well with disaggregate zone systems and/or sparse zone activity levels (PBQ&D 1999). Zhou et al. (2009) found that reasonable forecasts emerged only after imposing a variety of hard-coded rules (e.g., restricting excessive growth and declines in population and jobs at the zone level), suggesting that local knowledge and expert opinion may be needed to manually adjust gravity model forecasts. They also found that household and employment allocations were relatively insensitive to land use consumption levels, and standard equations may suffer from overspecification.

Cellular automata (CA) models are a class of artificial intelligence (AI) methods, with SLEUTH (Slope, Land use, Exclusion, Urban extent, Transportation and Hill shade) being the most widely applied CA LUM (e.g., Clarke et al. 1997; Silva and Clarke 2002; and Syphard et al. 2005). It represents a dynamic system in which discrete cellular states are updated according to a cell's own state, as well as that of its neighbors. While CA models may mimic many aspects of the dynamic and complex land use systems, they generally lack behavioral foundations to explain the process. Moreover, they emphasize land-cover type, not land use intensity, so postprocessing is needed to generate employment and household count patterns (which are, of course, critical to travel demand modeling).

Spatial input-output models are used to anticipate the economic and related interactions of employment and household sectors across zones, using discrete choice models for mode and input-origin choices. Production and demand functions consider transport disutility between zones, and people (and generally freight) move from one location to another in order to equilibrate supply and demand. Representative models include TRANUS (e.g., Johnston and de la Barra 2000), PECAS (e.g., Hunt and Abraham 2003), and RUBMRIO (e.g., Kockelman et al. 2004). Trade-based spatial input-output models are most suitable for larger spatial units (e.g., countries, regions, states and/or nations), so spatial resolution can be poor. Good trade and production data are also difficult to come by. It is worth noting that PECAS now includes a disaggregate submodel for space development, to anticipate developer actions at the level of parcels or grid cells (see, e.g., Production, Exchange, and Consumption Allocation System 2007, and Hunt et al. 2008). This advance results in a hybrid of spatial input-output (for activity allocation) and microsimulation.

General equilibrium models rely on a modeling framework that balances demand for built space and the supply of real estate. They generally require analytical solutions to obtain results that equilibrate real estate markets at zonal levels. MUSSA for Santiago, Chile (e.g., Martínez and Donoso 2001, 2006; Martinez and Henriquez 2007), now embedded in Cube-Land commercial software, and MetroScope for Portland, Oregon (Conder 2007), are two such models. In addition to household and firm behavioral data, information on the supply side of built space is essential to calibrating and applying such models. However, such data are generally quite difficult to obtain, resulting in often heroic, yet necessary, assumptions. Such models are built on the notion of balancing supply and demand for land and/or space at the level of zones, and rents are generally endogenously determined. Information on monetary metrics allows modelers to study the impacts of economic incentives (and disincentives) on land development, land use patterns, and agent welfare.

Random utility maximization for discrete choices (McFadden 1978) is the basis for most microsimulation models. While utility maximization is a reasonably defensible behavioral principle, numerous factors affect individual household and firm decisions, and these factors interact in complicated ways, generally demanding some form of dynamic disequilibration. For such reasons, opportunities for model improvement always exist. Two operational microsimulation models are Waddell's UrbanSim (e.g., Waddell 2002; Waddell et al. 2003; Waddell and Ulfarsson 2004; and Borning et al. 2007) and Gregor's LUSDR (Land Use Scenario DevelopeR). UrbanSim simulates location choices of individual households and jobs, while anticipating new development on the basis of such models, but prices are not explicitly derived from the interaction of supply and demand. LUSDR emphasizes very fast model runs and the stochastic nature of results, seeking a balance between model completeness and practicality (Gregor 2007). Zhou and Kockelman (2011) recently simulated market bidding and clearance for Austin, Texas, parcels and their associated buildings, in harmony with developers' (random profit-maximizing) decisions, demonstrating how microsimulation models may soon evolve in such a way that they are even more disaggregate and realistic in nature.

8.5.2   Land Use Model Applications

Of course, the objective for transportation planners and engineers is a realistic model that successfully integrates, and accurately forecasts, both transportation and land use changes (Miller et al. 1999). And many relevant variables will always lie outside the model components. Preferences evolve in uncertain ways, along with incomes, household sizes, transport and building technologies, energy prices, loan rates, and other factors of interest. A single forecast, assuming that development trends observed over the calibration period will continue and no new policies are imposed, is generally not of great value. Land use-transport models can better serve communities and their policymakers through multiple-scenario analyses, preferably with various uncertainties explicitly recognized and quantified. (See, for example, Sevcikova et al. 2007, Zhao and Kockelman 2002, Gregor 2007, Lemp and Kockelman 2009, Duthie et al. 2009, Krishnamurthy and Kockelman 2003, and Pradhan and Kockelman 2002.)

As one example, Zhou et al. (2009) forecasted year-2030 land use and travel conditions across the Austin-Round Rock metropolitan statistical area, by integrating a gravity-based land use model (G-LUM) with a standard travel demand model (TDM). Three scenarios were investigated, including a business-as-usual (BAU) scenario (i.e., development trends observed over the five-year calibration were assumed to continue, and no new policies were imposed), a congestion pricing-plus-carbon tax scenario (i.e., marginal delay costs were applied on all congested freeway segments in the network, and carbon tax of 4.55 ¢/mile was applied to all network links), and an urban growth boundary (UGB) scenario (where all new development was restricted to a zones with two or more job-equivalents per acre, plus their adjacent zones), centered on existing population centers. Documentation associated with Putman's ITLUP® model was used to design three submodel components, for residential location assignments (by household type, in RESLOC), job assignments (by category, in EMPLOC), and zone-level land consumption estimates (by use type, in LUDENSITY), as illustrated in Figure 8.1.

Year 2030 predictions were summarized in terms of vehicle-miles traveled (VMT), traffic flows, volume-to-capacity ratios, speeds, and downtown accessibility indices (to households and employment), as described at length in Zhou and Kockelman (2011). Of particular interest is the fact that the road pricing (roughly 5 ¢/mile on most links) had almost no discernable effect on land use predictions, yet resulted in the same predicted reduction in regional VMT (roughly 15 percent) as the UGB policy (which also greatly impacted land use patterns).

Tirumalachetty and Kockelman's (2010) design and five-scenario application of a detailed microsimulation model (Figure 8.2) to the same Texas region resulted in GHG emissions estimates that were lowest under this same style of UGB policy. Their VMT and GHG estimates were lower under this UGB policy than estimates based on a $3-per-gallon gas tax increase coupled with road tolls of 10 cents per vehicle-mile. In other words, certain land use policies may be expected to have significant land use and transport effects, even when traditional land use models are used, and even when compared to considerable road tolling strategies.

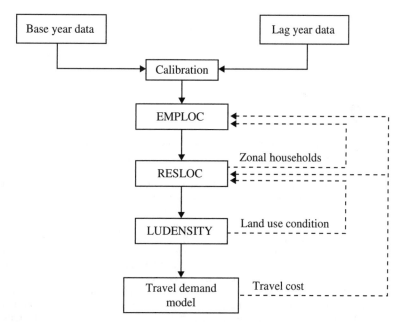

FIGURE 8.1 Gravity-based land use model example, in concert with a travel demand model. (from Zhou et al. 2009) (Note: Dashed lines represent one-period (t-1) lagged feedback of information. Each period is 5 years.)

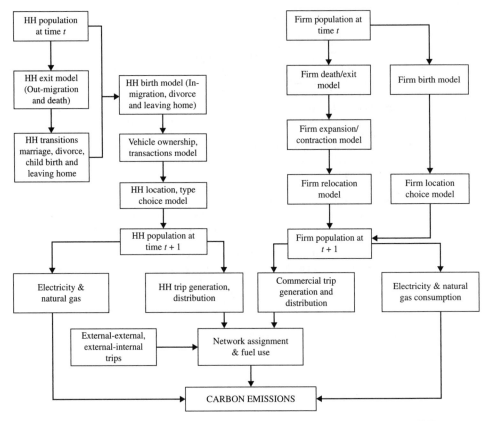

FIGURE 8.2   Example microsimulation model framework. (*Source:* Tirumalachetty and Kockelman 2010.)

8.6   OTHER TOPICS

Beyond the potential and perceived relationships between land use and transport, and their integrated modeling using mathematical algorithms, other topics are worthy of discussion in a chapter on land use and transport. For example, one wonders whether higher densities or more accessible neighborhoods produce significant local congestion. Kuzmyak et al. (2008) described Cox's (2003) argument that higher densities mean much higher traffic densities and congestion levels, since VMT per capita does not fall as fast as density rises. They recognized that, in reality, the VMT savings are so significant (using Cox's own numbers), and mode shifts so likely, at higher densities, that one cannot assume overall congestion will be worse. In fact, traveler delays are estimated to be noticeably higher in places like Dallas and Atlanta than in high-density locations like New York City and Boston, year after year, based on Schrank and Lomax's (2009) *Urban Mobility Report* for U.S. regions.

Another topic worthy of mention is access management. Management of property access, for safety, congestion and other reasons, is an important component of transportation engineers' toolkits, since zoning, setbacks, and other methods for moderating land use-transport interactions fall under the purview of local city planners, rather than regional or superregional transportation staff. The U.S. Transportation Research Board's *Access Management Manual* (TRB 2003) provides recommended practices for agencies struggling with the issues of preserving flow and protecting travelers while adequately serving property owners' interests. In general, access regulations (e.g., where and how to design driveways, relative to intersection locations and adjacent property site layouts) impact the layout, intensity, and type of business or residential property that can be sited along such corridors. In other words, something as basic as driveway (and parking lot) design can have a significant, though highly localized, land use impact.

Of course, air quality, GHG emissions, energy, noise exposure and other impacts also come to mind when contemplating the land use-transportation relationship, as alluded to earlier. Developed regions entail highly complex human interactions, over space and time, and land use and transport are key facets of these dynamic systems. Transportation engineers have major roles to play in balancing competing needs and interests, of diverse and dependent ecosystems, as they try to anticipate system responses to their proposed designs and policies. The significant literature relating to and modeling capabilities for land use-transport interactions provides ample opportunity for educated and enlightened decisions, as we navigate the paths that lie ahead.

REFERENCES

Alonso, W. 1964. *Location and Land Use*. Cambridge: Harvard University Press.

Armstrong, R. J., and Rodríguez, D. A. 2006. "An Evaluation of the Accessibility Benefits of Commuter Rail in Eastern Massachusetts Using Spatial Hedonic Price Functions." *Transportation* 33, 21–43.

Bartholomew, K. 2007. "The Machine, the Garden and the City: Towards an Access-Efficient Transportation Planning System." *The Environmental Law Reporter* 37(8):10593–10614.

Bhat, C. and Guo, J. 2004. "A Mixed Spatially Correlated Logit Model: Formulation and Application to Residential Choice Modeling," *Transportation Research Part B*, 38:147–168.

Bento, A., Cropper, M., Mobarak, A., and K Vinha, K. 2005. "The Effects of Urban Spatial Structure on Travel Demand in the United States." *The Review of Economics and Statistics* 87 (3):466–478.

Bina, M., and Kockelman, K. 2009. "Location Choice vis-a-vis Transportation: The Case of Recent Home Buyers," Chapter 28 in *The Expanding Sphere of Travel Behaviour Research*, Kitamura, R. and T. Yoshii (Eds.). Bingley, U.K.: Emerald Publishing, 597–619.

Bina, M., Kockelman, K., and Warburg, V. 2006. "Location Choice vis-à-vis Transportation: The Case of Apartment Dwellers." *Transportation Research Record* No. 1977, 93–102.

Boarnet, M., and Crane, R. 2001. *Travel by Design: The Influence of Urban Form on Travel*. Oxford, U.K.: Oxford University Press.

Borning, A., Waddell, P., and Förster, R. 2007. "UrbanSim: Using Simulation to Inform Public Deliberation and Decision-Making." In Traunmueller, R. et al. (Eds.), *Digital Government: Advanced Research and Case Studies*. New York: Springer-Verlag, 439–463.

Burchell, R., Lowenstein, G., Dolphin, W., Galley, C., Downs, A., Seskin, S., GrayStill, K., and Moore, T. 2002. *The Costs of Sprawl—2000*. Transit Cooperative Research Program (TCRP) Report 74. Transportation Research Board, Washington, D.C. Available at http://onlinepubs.trb.org/onlinepubs/tcrp/tcrp_rpt_74-a.pdf.

Calthorpe, P. 1993. *The Next American Metropolis: Ecology, Community, and the American Dream*. Princeton: Architectural Press.

Center for Environmental Excellence (CEE), 2008. American Association of State Highway and Transportation Officials (AASHTO). Retrieved in December 2008, http://environment.transportation.org/environmental_issues/land_use_sg/#bookmarkPolicyandGuidance.

Cervero, R. and Duncan, M. 2002. "Transit's Value-added: Effects of Light and Commuter Rail Services on Commercial Land Values." *Transportation Research Record* No. 1805, 8–15.

Cervero, R., and Gorham, R., 1995. "Commuting in Transit versus Automobile Neighborhoods." *Journal of the American Planning Association* 61(2):210–226.

Cervero, R., and Kockelman, K., 1997. "Travel Demand and the Three Ds: Density, Diversity, and Design." *Transportation Research Part D: Transport and Environment* 2(2):199–219.

City of Austin. 2000. Land Use Survey Methodology. Available at http://www.ci.austin.tx.us/landuse/survey.htm

Clark, W.A.V., Huang, Y., and Withers, S. 2003. "Does Commuting Distance Matter? Commuting Tolerance and Residential Change." *Regional Science and Urban Economics* 33:199–221.

Clarke, K. C., Hoppen, S., and Gaydos, L. 1997. "A Self-Modifying Cellular Automaton Model of Historical Urbanization in the San Francisco Bay Area." *Environment and Planning B: Planning and Design* 24:247–261.

Conder, S. 2007. Metroscope 3.0 Specifications as Implemented at Portland Metro. (Document received from Sonny Conder in November 2008.)

Cox, W. 2003. "How Higher Density Makes Traffic Worse." *Public Purpose* 57 (May). Available at http://www.publicpurpose.com/pp57-density.htm.

Crane, R., and Crepeau, R. 1998. "Does Neighborhood Design Influence Travel? A Behavioral Analysis of Travel Diary and GIS Data." *Transportation Research D*, 3(4):225–238.

DeBok, M., and Bliemer, M. 2006. "Infrastructure and Firm Dynamics: Calibration of Microsimulation Model for Firms in the Netherlands." *Transportation Research Record* 1977:132–144.

Dowling, R., Ireson, R., Skabardonis, A., Gillen, D., and Stopher, P. 2005. National Cooperative Highway Research Program NCHRP. Report 535: *Predicting Air Quality Effects of Traffic-Flow Improvements: Final Report and User's Guide*. Transportation Research Board, Washington, D.C.

Duany, A., Plater-Zyberk, E., and Speck, J. 2000. *Suburban Nation: The Rise of Sprawl and the Decline of the American Dream*. New York: North Point Press.

Dunphy R.T., and Fisher K. 1996. "Transportation, Congestion, and Density: New Insights. *Transportation Research Record* 1552:89–96.

Duthie, J., Voruganti, A., Kockelman, K., and Waller, S.T., 2009. "Uncertainty Analysis and its Impacts on Decision-Making in an Integrated Transportation and Gravity-based Land Use Model." Forthcoming in *Journal of Urban Planning and Development*.

Evans, J. E., Pratt, R. H., and Kuzmyak, J. R. 2007. "Transit Oriented Development." Chapter 17 in TCRP *Traveler Response to Transportation System Changes*. Transportation Cooperative Highway Research Program (TCRP) Report 95. Transportation Research Board, Washington D.C.

Ewing, R., and Cervero, R. 2001. "Travel and the Built Environment: A Synthesis." *Transportation Research Record: Journal of the Transportation Research Board*, No. 1780, 187–114.

Ewing, R., and Cervero, R. 2010. "Travel and the Built Environment: A Meta-Analysis." *Journal of the American Planning Association* 76(3):265–294.

Ewing, R., Bartholomew, K., Winkelman, S., Walters, J., and Anderson, G. 2008. "Urban Development and Climate Change." *Journal of Urbanism* 1(3):201–216.

Fang, H. A. 2008. "A Discrete-Continuous Model of Households' Vehicle Choice and Usage, with an Application to the Effects of Residential Density." *Transportation Research B* 42(9):736–758.

Gregor, B. 2007. "Land Use Scenario Developer: Practical Land Use Model Using a Stochastic Microsimulation Framework." *Transportation Research Record* 2003:93–102.

Handy, S. 1994. "Highway Blues: Nothing a Little Accessibility Can't Cure." *Access* 5, 3–7.

Handy, S. 2005. "Smart Growth and the Transportation-Land Use Connection: What Does the Research Tell Us?" *International Regional Science Review* 28(2):146–167.

Heckman, J. J. 1979. "Sample Selection Bias as a Specification Error." *Econometrica* 47(1):153–162.

Heckman, J. J., and Vytlaci, E. J. 1999. "Local Instrumental Variables and Latent Variable Models for Identifying and Bounding Treatment Effects." *Proceedings of the National Academy of Sciences* 96:4730–4734.

Heckman, J. J., Tobias, J. L., and Vytlaci, E. J. 2001. "Four Parameters of Interest in the Evaluation of Social Programs." *Southern Economic Journal* 68(2):210–223.

Hess, D. B., and Almeida, T. M. 2007. "Impact of Proximity to Light Rail Rapid Transit on Station-Area Property Values in Buffalo, New York." *Urban Studies* 44(5-6):1041–1068.

Holtzclaw, J. 1991. "Explaining Urban Density and Transit Impacts on Auto Use." San Francisco: Natural Resources Defense Council (California Energy Commission Docket No. 89-CR-90).

Holtzclaw, J. 1994. "Using Residential Patterns and Transit to Decrease Auto Dependence and Costs." San Francisco CA: Natural Resources Defense Council. Available at http://www.smartgrowth.org/library/cheers.html.

Holtzclaw, J., Clear, R., Dittmar, H., Goldstein, D., and Haas, P. 2002. "Location Efficiency: Neighborhood and Socio-Economic Characteristics Determine Auto Ownership and Use—Studies in Chicago, Los Angeles, and San Francisco." *Transportation Planning and Technology* 25(1):1–27.

Hunt J. D., and Abraham, J. E., 2003. Design and Application of the PECAS Land Use Modeling System. Presented at the *8th Computers in Urban Planning and Urban Management Conference*, Sendai, Japan.

Hunt J. D., Abraham, J. E., de Silva, D., Zhong, M., Bridges, J., and Mysko, J. 2008. Developing and Applying a Parcel-Level Simulation of Developer Actions in Baltimore. *Proceedings of the Transportation Research Board 87th Annual Meeting*, Washington, DC.

Iacono, M., and Levinson, D. 2009. The Economic Impact of Upgrading Roads: Draft Final Report. http://nexus.umn.edu/Projects/EIUR/EIUR.pdf.

Iacono, M., Levinson, D., and El-Geneidy, A. 2008. "Models of Transportation and Land Use Change: A Guide to the Territory." *Journal of Planning Literature* 22(4):323–340.

Institute of Transportation Engineers (ITE). 2008. *Trip Generation*, 8th ed. ITE, Washington, D.C.

Johnston, R. A., and de la Barra, T. 2000. "Comprehensive Regional Modeling for Long-Range Planning: Linking Integrated Urban Models and Geographic Information Systems." *Transportation Research Part A: Policy and Practice* 34(2):125–136.

Khattak, A. J., and Rodriguez, D. 2005. "Travel Behavior in Neo-Traditional Neighborhood Developments: A Case Study in USA." *Transportation Research Part A*, 39:481–500.

Knoflacher, H. 2007. "Success and Failures in Urban Transport Planning in Europe: Understanding the Transport System. *Sahana* 32(4):293–307.

Kockelman, K., 1997. "The Effects of Location Elements on Home Purchase Prices and Rents: Evidence from the San Francisco Bay Area." *Transportation Research Record* 1606:40–50.

Kockelman, K., Bomberg, M., Thompson, M., and Whitehead, C. 2009. GHG Emissions Control Options: Opportunities for Conservation. Paper commissioned by the National Academy of Sciences' Committee for the Study on the Relationships Among Development Patterns, VMT, and Energy Conservation, resulting in Special Report 298. Washington, D.C. Available at http://onlinepubs.trb.org/Onlinepubs/sr/sr298kockelman.pdf.

Kockelman, K., Jin, M., Zhao, Y., and Ruiz-Juri, N. 2004. "Tracking Land Use, Transport, and Industrial Production Using Random-Utility-Based Multizonal Input-Output Models: Applications for Texas Trade." *Journal of Transport Geography* 13(3):275–286.

Krishnamurthy, S., and Kockelman, K. 2003. "Propagation of Uncertainty in Transportation-Land Use Models: An Investigation of DRAM-EMPAL and UTPP Predictions in Austin, Texas." *Transportation Research Record* 1831:219–229.

Krizek, K. J. 2003. "Residential Relocation and Changes in Urban Travel: Does Neighborhood-Scale Urban form Matter?" *Journal of the American Planning Association* 69(3):265–281.

Kuzmyak, J., PolyTech Corporation, and Caliper Corporation. 2008. "Land Use and Traffic Congestion Study, Task 2 Report: Literature Review." Arizona DOT Research Project No. 619.

Kuzmyak, J. R., Pratt, R., Douglas, G. B., and Spielberg, F. 2003. "Land Use and Site Design." Chapter 15 in TCRP *Traveler Response to Transportation System Changes*. Transportation Cooperative Highway Research Program (TCRP) Report 95. Transportation Research Board, Washington D.C.

Laurance, W. F., Cochrane, M. A., Bergen, S., Fearnside, P. M., Delamônica, P., Barber, C., D'Angelo, S, and Fernandes, T. 2001. "The Future of the Brazilian Amazon." *Science* 291:438–439.

Lemp, J., and Kockelman, K. 2009. "Understanding & Accommodating Risk & Uncertainty in Toll Road Projects: A Review of the Literature." *Transportation Research Record* 2132:106–112.

Litman, T. 2003. "Measuring Transportation: Traffic, Mobility and Accessibility." *ITE Journal* 73(10):28–32.

Litman, T. 2007. Evaluating Accessibility for Transportation Planning. Victoria Transport Policy Institute. Available at http://www.vtpi.org/access.pdf.

Litman, T. 2010. "Land Use Impacts on Transport: How Land Use Factors Affect Travel Behavior." Victoria Transport Policy Institute. Available at http://www.vtpi.org/landtravel.pdf.

Lyons, W. M. 1995. "Policy Innovations of the US Intermodal Surface Transportation Efficiency Act and Clean Air Act Amendments." *Transportation* 22:217–240.

Martinez, F. J., and Donoso, P. 2001. "MUSSA: A Land Use Equilibrium Model with Location Externalities, Planning Regulations and Pricing Policies." Presented at the 7th International Conference on Computers in Urban Planning and Urban Management (CUPUM 2001), Honolulu, Hawaii.

Martinez, F. J., and Donoso, P. 2006. "MUSSA II: A Land Use Equilibrium Model Based on Constrained Idiosyncratic Behavior of All Agents in an Auction Market." Proceedings of the Transportation Research Board 86th Annual Meeting, Washington, D.C.

Martinez, F. J., and Henriquez, R. 2007. "A Random Bidding and Supply Land Use Equilibrium Model." *Transportation Research Part B,* 41:632–651.

McFadden, D. 1978. "Modeling the Choice of Residential Location." In Karlquist, A. et al. (Eds.), *Spatial Interaction Theory and Residential Location*. Amsterdam: North-Holland, 75–96.

Mikelbank, B. A. 2004. "Spatial Analysis of the Relationship between Housing Values and Investments in Transportation Infrastructure." *The Annals of Regional Science* 38:705–726.

Miller, E. J., Kriger, D. S., and Hunt, J. D. 1999. *TCRP Report 48: Integrated Urban Models for Simulation of Transit and Land Use Polices: Guidelines for Implementation and Use*. Transportation Research Board, National Research Council, Washington, D.C.

Mohl, R. A. 2004. "Stop the Road: Freeway Revolts in American Cities." *Journal of Urban History* 30(5): 674–706.

Mokhtarian, P. L., and Cao, X. 2008. "Examining the Impacts of Residential Self-Selection on Travel Behavior: A Focus on Methodologies." *Transportation Research Part B* 42:204–228.

Musti, S., and Kockelman, K. 2009. "Evolution of the Household Vehicle Fleet: Anticipating Fleet Composition and PHEV Adoption in Austin, Texas." Proceedings of the 12th Conference of the International Association for Travel Behaviour Research (IATBR), in Jaipur, India and under review for publication in *Transportation Research Part A*. Available at http://www.ce.utexas.edu/prof/kockelman/public_html/TRB10VehicleChoice.pdf.

Newman, P., and Kenworthy, J. 1996. "The Land Use-Transportation Connection: An Overview." *Land Use Policy* 13(1):1–22.

Newman, P., and Kenworthy, J. 1999. "Costs of Automobile Dependence: Global Survey of Cities." *Transportation Research Record* 1670:17–26.

Newman, P., and Kenworthy, J. 2006. "Urban Design to Reduce Automobile Dependence." *Opolis* 2(1):35–52.

Parsons Brinckerhoff Quade and Douglas (PBQ&D) 1999. National Cooperative Highway Research Program NCHRP. Report 423A: *Land-Use Impacts of Transportation: A Guidebook*. Transportation Research Board, Washington, D.C.

Production, Exchange, and Consumption Allocation System (PECAS). 2007. *Theoretical Formulation: System Documentation Technical Memorandum 1*. Received from John Abraham in March 2008.

Pradhan, A., and Kockelman, K. 2002. "Uncertainty Propagation in an Integrated Land Use-Transport Modeling Framework: Output Variation via UrbanSim." *Transportation Research Record* 1805:128–135.

Pucher, J., Dill, J., and Handy, S. 2010. "Infrastructure, Programs, and Policies to Increase Bicycling: An International Review. *Preventive Medicine* 48(2):106–125.

Putman, S. H. 1991. *Integrated Urban Models 2: New Research and Applications of Optimization and Dynamics*. London: Pion Press, London.

Rouwendal, J., and Meijer, E., 2001. "Preferences for Housing, Jobs, and Commuting: A Mixed Logit Analysis." *Journal of Regional Science* 41:475–505.

Salon, D., 2006. "Cars and the City: An Investigation of Transportation and Residential Location Choices in New York City." Dissertation for Doctorate in Agricultural and Resource Economics, The University of California at Davis.

Schimek, P. 1996. "Household Motor Vehicle Ownership and Use: How Much Does Residential Density Matter?" *Transportation Research Record* 1552:120–125.

Schrank, D., and Lomax, T. 2009. *Annual Urban Mobility Report*. Texas A&M University, Texas Transportation Institute.

Sevcikova, H., Raftery, A., and Waddell, P. 2007. "Assessing Uncertainty in Urban Simulations Using Bayesian Melding." *Transportation Research, Part B* 41(6):652–669.

Shay, E., and Khattak, A. J., 2005. "Automobile Ownership and Use in Neotraditional and Conventional Neighborhoods." *Transportation Research Record: Journal of the Transportation Research Board*, 1902:18–25.

Silva, E. A., and Clarke, K. C., 2002. "Calibration of the SLEUTH Urban Growth Model for Lisbon and Porto, Portugal." *Computers, Environment and Urban System* 26:525–552.

Srour, I., Kockelman, K., and Dunn, T. 2002. "Accessibility Indices: A Connection to Residential Land Prices and Location Choice." *Transportation Research Record* 1805:25–34.

Syphard, A. D., Clarke, K. C., and Franklin, J. 2005. "Using a Cellular Automaton Model to Forecast the Effects of Urban Growth on Habitat Pattern in Southern California." *Ecological Complexity* 2:185–203.

ten Siethoff, B., and Kockelman, K. 2002. "Property Values and Highway Expansions: An Investigation of Timing, Size, Location, and Use Effects. *Transportation Research Record* 1812:191–200.

Tillema, T., Ettema, D., and Van Wee, B. 2006. "Road Pricing and (Re)location Decisions of Households." Proceedings of the 85th Annual Meeting of the Transportation Research Board, Washington D.C.

Tirumalachetty, S., and Kockelman, K. 2010. "Forecasting Greenhouse Gas Emissions from Urban Regions: Microsimulation of Land Use and Transport Patterns in Austin, Texas." Proceedings of the 89th Annual Meeting of the Transportation Research Board (Paper #10-1338) under review for publication in the *Journal of Transport Geography*.

Transportation Research Board (TRB). 1995. *Expanding Metropolitan Highways*. Special Report Number 245. Washington, DC: National Academy Press.

TRANUS, 2008. TRANUS English. Modelistica. Retrieved in December 2008. http://www.modelistica.com/tranus_english.htm.

Transportation Research Board (TRB). 2003. *Access Management Manual*. Washington, DC: National Academy Press.

U.S. Environmental Protection Agency (U.S. EPA). 2000. *Projecting Land-Use Change: A Summary of Models for Assessing the Effects of Community Growth and Change on Land-Use Patterns*, Report EPA 600-R-00-098. Washington, D.C.

U.S. Environmental Protection Agency (U.S. EPA). 2002. Smart Growth Index (SGI) Model, U.S. Environmental Protection Agency. Available at www.epa.gov/livablecommunities/topics/sg_index.htm.

Van Ommeren, J., Rietveld, P., and Nijkamp, P. 1999. "Job Moving, Residential Moving, and Commuting: A Search Perspective." *Journal of Urban Economics* 46:230–253.

Von Thünen, J. H. (1826) *Der Isolierte Staat (The Isolated State)* in Beziehung auf Landwirtschaft und Nationalekonomie. Hamburg, Germany.

Waddell, P. 2002. "UrbanSim: Modeling Urban Development for Land Use, Transportation and Environmental Planning." *Journal of the American Planning Association* 68(3):297–314.

Waddell, P., Borning, A., Noth, M., Freier, N., Becke, M., and Ulfarsson, G. 2003. "Microsimulation of Urban Development and Location Choices: Design and Implementation of UrbanSim." *Networks and Spatial Economics* 3(1):43–67.

Waddell P., and Ulfarsson, G. F. 2004. "Introduction to Urban Simulation: Design and Development of Operational Models." In Haynes, K., Stopher, P., Button, K., and Hensher, D. (Eds.), *Handbook in Transport Volume 5: Transport Geography and Spatial Systems*. Oxford: Pergamon Press, 203–236.

Walker, R. 2004. "Theorizing Land-Cover and Land-Use Change: The Case of Tropical Deforestation." *International Regional Science Review* 27:247–270.

Wegener, M. 2004. "Overview of Land-Use Transport Models," in David A. Hensher and Kenneth Button (Eds.): *Transport Geography and Spatial Systems*. Handbook 5 of the *Handbook in Transport*. Kidlington, UK: Pergamon/Elsevier Science, 127–146.

Weinstein, B. L., and Clower, T. L. 1999. "The Initial Economic Impacts of the DART LRT System." Center for Economic Development and Research, University of North Texas. http://www.unt.edu/cedr/dart.pdf.

Wingo, L. 1961. *Transportation and Urban Land Use*. Baltimore: Johns Hopkins University Press.

Zhao, Yong, and Kockelman, K. 2002. "The Propagation of Uncertainty through Travel Demand Models: An Exploratory Analysis." *Annals of Regional Science* 36(1):145–163.

Zhou, B., and Kockelman. K. 2008a. "Neighborhood Impacts on Land Use Change: A Multinomial Logit Model of Spatial Relationships." *Annals of Regional Science* 42(2):321–340.

Zhou, B., and Kockelman, K. 2008b. "Self-Selection in Home Choice: Use of Treatment Effects in Evaluating the Relationship between the Built Environment and Travel Behavior." *Transportation Research Record* 2077:54–61.

Zhou, B., and Kockelman, K. 2009a. "Predicting the Distribution of Households and Employment: A Seemingly Unrelated Regression Model with Two Spatial Processes." *Journal of Transport Geography* 17:369–376.

Zhou, B., and Kockelman, K. 2011. "Land Use Change through Microsimulation of Market Dynamics: An Agent-Based Model of Land Development and Locator Bidding in Austin, Texas." Proceedings of the 90th Annual Meeting of the Transportation Research Board, Washington, D.C.

Zhou, B., Kockelman, K., and Lemp, J. 2009. "Applications of Integrated Transport and Gravity-Based Land Use Models for Policy Analysis." *Transportation Research Record* 2133:123–132.

CHAPTER 9
SUSTAINABLE TRANSPORTATION

Zhong-Ren Peng
Department of Urban and Regional Planning
University of Florida, USA; and
School of Transportation Engineering
Tongji University, Shanghai, China

Fei Yang
Department of Urban and Regional Planning
University of Florida
Gainesville, Florida

Sarah Perch
Department of Urban and Regional Planning
University of Florida
Gainesville, Florida

9.1 INTRODUCTION

Sustainability is a term used frequently, perhaps too frequently, by planners, engineers, government officials, and many others in related fields. However, the question still remains: What does this mean? How does sustainable transportation relate to the many fields and disciplines affected by sustainability? In the field of transportation, this concept is emerging as a key concern when thinking about transportation planning, engineering, operation, and management.

Previous literature provides many answers as to what sustainability means, and in turn, what sustainable transportation could mean. Perhaps the most widely used definition of sustainability comes from the World Bank, which describes it as a balance of economics, environmental and social concerns (World Bank 2010; Zegras 2006). Building on this definition, others have tried to elaborate on what sustainability really is. For example, Neumayer describes sustainability as being able to "maintain the capacity to provide nondeclining well-being over time" (Zegras 2006; Neumayer 2003). The Brundtland Report states that sustainability means meeting "the needs of the present without compromising the ability of future generations to meet their own needs" (UN WCED 1987; Zegras 2006, Mihyeon and Amekudzi 2005). While these definitions provide a framework that begins to address sustainability, along with some of the general principles behind the word, they do not provide a way to easily quantify or qualify what it means to be sustainable. It is something that is perhaps easier to see than to describe.

What then is sustainable transportation? How can it be measured?

Zegras (2006) asks a similar question, focusing on the many unknowns in quantifying what it means to be sustainable. He brings up the issue that sustainability, in addition to its unclear nature, is also inherently focused on a specific entity, or group of people, and therefore what is sustainable to someone might not be to another. Rather, it is a matter of priorities. This is an important consideration when thinking about what sustainability means and how to quantify it.

Ramani et al. (2009) introduce the idea of sustainable transportation as "an expression of sustainable development in the transportation sector" (p. 15), using a definition of sustainability as "the provision of safe, effective, and efficient access and mobility into the future while considering the economic, social and environmental needs of society" (p. 16). This clearly shows that there are some common themes when defining sustainability; namely, the World Bank's triangle of economic, social, and environmental concerns, as well as a thinking of the future.

Using this concept of sustainability, namely the three tenets (economic, social, and environmental), and a forward-thinking mindset, this chapter will discuss what it means to be sustainable, particularly in respect to sustainable transportation, and also the role that transportation plays in sustainable development. First, the chapter will more thoroughly discuss sustainable transportation, including its guiding principles and indicators. Then, sustainable planning strategies will be outlined and detailed. Following this, the chapter will introduce a few examples of cities or regions with sustainable transportation practices and discuss these as case studies for what sustainable transportation can look like.

9.1.1   Sustainable Transportation: Measurements

When thinking about sustainable transportation, there are a few main aspects that must be considered, namely the factors that are involved in transportation in respect to economic costs, environmental protection and social equity. For example, does practicing sustainable transportation mean lowering vehicle emissions? Spending less money on transportation investments? Lowering noise pollution? (Zegras 2006; Mihyeon and Amekudzi 2005)

Numerous indicators could be used when discussing sustainable transportation. For example, air quality could be an indicator and measure related to environmental protection. Accessibility is a measure of social equity. Cost is an economic concern. The list of potential indicators is vast and previous works have tried to consolidate and evaluate these many concerns (Litman 2009). Table 9.1 lists some examples of potential indicators. It is by no means a complete list, just examples of potential indicators.

Sustainable transportation practices must account for all kinds of transportation modes and therefore concerns related to transportation, transportation networks and development. A European Commission report declares that the goal of sustainable transportation is to have all transportation network elements functioning optimally, thereby better exploiting the network's capacity and enhancing strengths of each mode, which will therefore reduce overall congestion (European Commission 2009, p. 17). This strategy includes reducing air pollution and road accidents, reducing maritime accidents and pollution, as well as strengthening coastal infrastructure. In order to accomplish these goals, the European Commission believes that economic incentives, such as better priced signals, are necessary. Other measures being explored and taken include energy taxes; tolls and charges for infrastructure use; steering away from oil-reliance towards more electricity-based technology; and an increase in the use of ITS in electronic tolling system (ETC) and traffic management (European Commission 2009).

The measures currently being taken by the European Union provide a starting point when thinking about what tools can be implemented to promote sustainable transportation. They also indicate the underlying principles behind sustainable transportation practices: lowering energy use and increasing energy efficiency. The practices being implemented revolve around energy and efficiency. With the goal of promoting efficiency, there are several ways in which to approach measures that promote sustainable transportation. First, planning processes should encourage transit and alternative modes of transportation, besides single-occupancy vehicles (i.e., bicycles, walking). Other planning tools, such as zoning, may be used to control densities and optimize all modes of transportation. Second, environmental standards can be set, so that there are protective measures in place to prevent an excess of energy use and other environmentally-threatening activities. Third, economic strategies, such as taxation, may be implemented to promote travel habits that are more sustainable (Wittneben et al. 2009).

TABLE 9.1 Potential Sustainability Indicators

Category	Major concern	Subconcerns	Potential indicators
Environmental	Pollution	Emissions	Total vehicle emissions
		Air pollution exposure	Number of days of exposure per year
		Climate change	Climate change emissions (such as CO_2, CH_4, particulates etc.)
		Traffic noise	Exposure to traffic noise (decibel)
		Aircraft noise	Exposure to aircraft noise (decibel)
		Train noise	Exposure to train noise (decibel)
	Land use impacts	Sprawl	Per capita impervious surface area
		Transport land consumption	Changes in land devoted to transport facilities
		Ecological and cultural degradation	Habitat and cultural sites degraded by transportation facilities and at-risk areas studies
Equity	Affordability	Transportation costs	Portion of household budgets needed to provide for adequate transport
		Affordable housing	Affordable housing accessibility and availability
		Basic accessibility	Quality of accessibility for people with disabilities
	Overall accessibility	Mobility options	Availability of transportation options
		Land use accessibility	Quality of land use accessibility
		Mobility substitutes	Internet access and delivery service availability as alternatives to traditional travel
Economic	Economic measures	Transport costs	Consumer expenditures on personal transport
		Commute costs (time and money)	Ease of access to employment
		Infrastructure costs	Expenditures on roads, public transit, parking, ports, etc.
		Shipping costs	Freight expenditures
Transportation	Travel	Vehicles	Motor vehicle fleet size and composition
		Mobility	Motor vehicle travel trips and ease of travel
		Mode split	Portion of trips by auto, public transit, and nonmotorized modes
	Transport policy and planning	Pricing efficiency	Pricing strategies and evaluations
		Planning efficiency	Comprehensive planning documents
	Traffic risks	Crash casualties	Crash deaths and injuries
		Crashes	Crash report numbers
		Crash costs	Traffic crash economic costs

Source: Litman 2008.

Sustainable transportation not only involves promoting measures in the transportation sector, but it is inherently related to the concept of sustainable development. As transportation is the means by which people travel to destinations, there must be a connection between the two. Development contributes much towards greenhouse gas emissions, a major contributor toward the push to incorporate sustainable practices (United Nations 2010). The greater umbrella of sustainability includes development, transportation and their interconnectivity.

9.1.2   What Is Being Studied

A number of transportation practices support sustainable transportation planning, incorporating the strategies discussed above. These practices often involve both transportation and land use, as they are strongly connected. Sustainable transportation is heavily linked to sustainable development, so much so that it can be considered a subsidiary of development. The strategies and case studies discussed in this chapter incorporate both transportation and development.

Sustainable transportation planning strategies include multimodal transportation planning (MTP), land use integration, use of alternative energies, site design practices and travel demand management (TDM). Together, these strategies provide a context for understanding what can be done to implement policies and programs that are environmentally, economically and socially forward-thinking. This chapter will detail these strategies and then provide case studies of their implementation. The case studies vary from in scale, including the Portland Oregon Metropolitan Planning Organization (MPO); Vancouver, British Columbia, Canada; and the state of New Jersey. Each individual place has its own measures implemented that promote sustainable transportation planning. Each location has its unique set of needs and therefore has individual measures to address sustainable practices. The combination of multiple strategies provides a comprehensive approach to sustainable transportation and, together, each strengthens the other.

9.2   SUSTAINABLE TRANSPORTATION PLANNING STRATEGIES

It is widely known that sustainable transportation strategies are a principal component of sustainable development (Deakin 2001; Himanen et al. 2005). A variety of strategies that potentially increase transportation sustainability have been identified here, including multimodal transportation planning, land use integration, use of alternative energies, site design, travel demand management, and assessment of the effectiveness of sustainable transportation strategies. Previously, the implementation of these strategies was slow because of various socioeconomic barriers and therefore not considered thoroughly (Black 2000). However, a new interest in actively pursuing these strategies emerged in the early-21st century (Deakin 2001). This section describes these strategies and their effectiveness in implementation.

9.2.1   Multimodal Transportation Planning

General Definition.   Multimodal transportation planning (MTP) refers to planning and decision making that takes into account a wide range of transportation modes, and tries to optimize the function of each mode with an integrated network. Federal transportation legislation, starting with *Intermodal Surface Transportation Efficiency Act* (ISTEA) in 1991, highlights the importance of considering all transport modes when planning for improved mobility. The act was followed by the *Transportation Equity Act for the 21th Century* (TEA-21) (1998) and then in 2005, President Bush signed *Safe, Accountable, Flexible, Efficient Transportation Equity Act: A Legacy for Users* (SAFETEA-LU). These acts all call for alternative transport modes and help to populate the concept and practice of MTP in transportation organizations.

Characteristics of MTP. Unlike traditional transportation planning, which focuses almost exclusively on automobile travel, MTP favors diverse transportation options. It encourages transit, paratransit, ridersharing, biking and walking for personal travel, as well as rail as a substitute for truck and air freight. Lowering car dependency not only means less emission of pollutants, but also an increase in equal opportunities, since a transport system supporting diverse modes can provide equal opportunities for those who are disabled to drive or cannot afford a car. Therefore, MTP meets the requirements of sustainable development by protecting environment and improving social equity.

MTP integrates differing transport modes to form a new efficient system. This integration is not a simple combination of each mode; instead, it is the establishment of a network that integrates all modes together. Therefore, one can exploit all advantages of various transport modes and perform better than unimodal transportation in which only one mode, always motorized vehicles, is used. This network stresses the efficiency of the carriage of passengers and the distribution of goods, which gives rise to economic efficiency. MTP is more economically sustainable than traditional transportation planning that was heavily biased toward serving the automobile.

Furthermore, MTP encourages the use of innovative methods to determine road capacity (Comeau 2009; Litman 2009). "The traditional engineering method of measuring transportation capacity to handle new development assumes that a roadway or intersection has a theoretical design capacity to move vehicle traffic and then to measure traffic volumes or seconds of delay against the assigned design capacity of the arterial or intersection. The resulting ratio establishes the auto-centric operating level of service (LOS) typically during the highest demand period of a day" (Comeau 2009, p. 2). However, MTP calls for evolution from the LOS focusing on automobiles to multimodal LOS, which considers a wide range of transport modes and their impacts. In the case of Bellingham, Washington, for example, transportation planners defined their concurrency service areas (CSA) as 15 zones each of which has different land use/transportation characteristics (including land use patterns, transportation facilities, and services available). Bellingham adopted multimodal LOS as "person trips available by concurrency service area based on arterial and transit capacity for motorized modes and on the degree of network completeness for pedestrian and bicycle modes," (Comeau 2009, p. 8). LOS measures for each individual mode in the CSA are listed in the city's municipal code, see Table 9.2, below.

Transportation Modes and the Green Transportation Hierarchy. The various transportation modes embraced by MTP include walking, biking, public transit, and vehicle-sharing. However, each has different impacts on sustainable development practices. Thus, they do not have the same

TABLE 9.2   Bellingham's Level of Service Measures for Individual Transportation Modes

Motorized transportation modes	
Arterial streets	Peak hour LOS person trips available during weekday p.m. peak hour based on data collected at designated concurrency measurement points for each concurrency service area
Transit	Determine seated capacity, measure ridership, and equate to person trips available via public transit service during weekday p.m. peak hour based on data collected at designated concurrency measurement points for each concurrency service area
Nonmotorized transportation modes	
Bicycle	Credit person trips according to degree of bicycle network completeness for designated system facilities/routes for each concurrency service area
Pedestrian	Credit person trips according to degree of pedestrian network completeness for designated system facilities/routes for each concurrency service area
Trails	Credit person trips according to degree of bicycle and pedestrian network completeness, where trails serve a clear transportation function for a concurrency service area

Source: Cite from Comeau (2009, table 3).

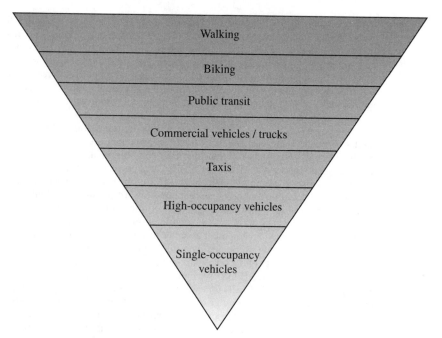

FIGURE 9.1 Green transportation hierarchy.

weight in sustainable transportation planning. In order to maximize sustainability, the Green Transportation Hierarchy system (Keegan 2001), displayed in Figure 9.1, was created to prioritize the transport modes that put less pressure on the environment. The hierarchy gives pedestrians and bicyclists highest priority because of their low cost and nonenvironmental impacts. Single-occupied vehicles are the lowest, because of their high cost, high environmental impacts, and low space efficiency. The essence of the green transportation hierarchy is that higher occupied vehicles enjoy higher priority. And this priority should be weaved into transportation planning and decision making, such as transportation grant provision and distribution of road space.

Recommendations for Multimodal Transportation Planning.   Although multimodal transportation has received intense attention across all levels of government, there are still many efforts needed to establish a successful practice. In his introduction article to MTP, Litman (2009) proposes valuable recommendations as to how to achieve best practice. These recommendations are summarized as follows:

- Consider a variety of transportation improvement options for a full range of transport modes, and also consider various combinations of these options.
- Consider all significant impacts, especially the long term, indirect and nonmarket impacts.
- Multimodal comparison should be comprehensive, and should account for factors such as transit system economies of scale and scope.
- Special attention should be given to the connectivity of different transport modes, and to the disadvantaged groups.
- Use comprehensive transportation models that consider multiple modes and the effects of various mobility management strategies.
- Transportation decision makers should take the lead to use the non-automobile transportation modes.

These recommendations are limited to technical aspects. They can be effective in promoting sustainable transportation systems. But the most important is the implementation. The decision makers should not only realize the importance to encourage alternative transport modes, but also commit transportation funding toward a multimodal transportation system.

9.2.2 Land Use and Transportation Integration

Interaction between Land Use and Transportation. Research on the interaction between land use and transportation dates back to the 1960s (Alonso 1964). Since then, the urban land use and transportation have become interrelated, and are often seen as such by planning researchers and practitioners (Wegener and Fürst 1999; Badoe and Miller 2000; Litman 2010). Any change in road networks impacts land use patterns. In turn, various land use patterns generate different trips and put different impacts on transportation system. Existing literature identifies two kinds of connections between land use and transformation. First, allocation of transportation funding and policy making influence development patterns (Handy 2005; Waddell et al. 2007). For example, commercial and business developed transport corridors and transport hubs; new suburbs emerge with the construction of new freeways. Second, land use patterns shape travel demand and behavior (Wegener and Fürst 1999; Shaw and Xin 2003; Handy 2005). A typical suburban land use pattern, with low density and scattered facilities, renders transit and walking a challenge. Furthermore, the concentration of employments is likely to generate more commuting trips. On the other hand, balanced workplaces and residences may reduce work trips. Therefore, incorporating land use with transportation can sustain economic development, eliminate car dependency with less environment pressure, and improve accessibility with equal opportunity to all. These can help communities to achieve sustainable transportation.

Important Land Use Factors Influencing Sustainable Transportation. In order to understand the land use/transportation system, one needs to first identify important land use factors that influence transportation, especially sustainable transportation. The dynamic nature of the land use/transportation system means that each element involved into the land use system will have both temporal and spatial influence on transportation, and vice versa. Therefore, the following discussion will be limited to major factors which play more intense roles in this relationship.

Land use density refers to both residential and employment density. Increased densities can reduce travel distances and the need for motorized vehicles to travel. This also tends to increase the use of alternative transport modes other than single occupied vehicles, because higher density means more concentrated capitals, which can afford more public facilities and services, such as sidewalks, better connected road networks, and better served public transit. As a result, the increased density is associated with lower car ownership and use, and increased use of alternative transport options (Wegener and Fürst 1999; Holtzclaw et al. 2002; Kuzmyak and Pratt 2003; Litman 2010).

Accessibility in the land use system measures the employment and public services available in a certain area. This greatly impacts travelers' perception of convenience on transportation system and therefore the use of personal vehicles. People living or working in suburbs are likely to drive much more often than people living closer to the urban center, as indicated by Kockelman (1996), who concluded that accessibility was the major predictor of household vehicle miles traveled.

Mixing land uses that are accessible by a wide range of transportation modes can facilitate a sustainable transport system in a community. Mixed use traditionally includes residential and commercial developments; public facilities, like recreation and education; and more. When homes are located within bicycling or walking distance to quality grocery stores, employment centers, and shopping, alternatives to driving become attractive, enabling more people to leave their automobiles and to help create a sustainable community.

Another important factor in land use and sustainable transportation is community design. Road network layout, pedestrian environment and bicycle lanes tend to lead to increases in walking and bicycling. In addition, attractive public spaces and various public services can induce more local trips (Wegener and Fürst 1999). This will be further detailed in the Site Design section of this chapter.

Coordination between Land Use and Transportation Planning. Attempts to search ideal approaches to coordinate land use and transportation system have yielded different results. Since the interaction between land use and transportation is quite complicated, the appropriate coordination of a certain land-use transport system depends on a large number of factors. However, there still exist some widely recognized strategies to incorporate land use and transportation planning. These strategies include mixed land use, compact building design, the creation of walkable communities, etc. They were all channeled into a new concept, smart growth.

Smart growth is a recent buzzword in urban planning. It was first used in connection with a Maryland state plan to control urban sprawl (Levy 2008). Although current smart growth concepts have developed to mean various things to different people (Downs 2001; Gearin 2004; Ye et al. 2005), the idea of smart growth has been widely accepted and is expressed by 10 principles (Smart Growth Networks and International City/County Association 2002).

1. Mix land uses.
2. Take advantage of compact building design.
3. Create a range of housing opportunities and choices.
4. Create walkable neighborhoods.
5. Foster distinctive, attractive communities with a strong sense of place.
6. Preserve open space, farmland, natural beauty, and critical environmental areas.
7. Strengthen and direct development towards existing communities.
8. Provide a variety of transportation choices.
9. Make development decisions predictable, fair, and cost effective.
10. Encourage community and stakeholder collaboration in development decisions.

These principles represent the logic behind numerous policies and strategies that strive to put Smart Growth into practice. That is why Handy (2005) claimed that "the connection between transportation and land use lies at the center of efforts in the United States to combat sprawl through smart growth strategies" (p. 147). In particular, transit-oriented development (TOD), the most cogent form of smart growth, has special value in terms of coordination between land use and transportation. Here, locations in and around transit stations can and do accommodate higher densities and mixed use. TOD can benefit society by increasing the ridership bonuses that can be found in both residential and office development (Cervero 2007; Cervero and Arrington 2008). Therefore, the policies and strategies supported by smart growth and transit oriented development can help coordinating land use and transportation planning.

9.2.3 Use of Alternative Energies

The Importance to Switch to Alternative Energies in Transportation Systems. Concern over the exhaustion of the world's energy resources, and the corresponding destruction of environment haunts development. In his famous work *Small is Beautiful*, Schumacher (1973) expressed this concern by arguing that people need to realize the difference between renewable and nonrenewable resources, and that the price of nonrenewable resources should take into account the needs of future generation. With the boom of automobiles and auto-dependent cities, the demand for fossil fuels has greatly increased. Transportation consumes many of these resources. In the United States, the transportation sector is a large component of the national economy, consuming more than a quarter of the nation's primary energy (Pro et al. 2005).

It is common knowledge that the fossil fuel reserve is finite. Black (1996) vividly described the future of fossil fuel as "The future of petroleum is that it has no future" (p. 152). Furthermore, burning fossil fuel is a major cause of air pollution and climate change. Therefore, a sustainable transportation should seek efficient approaches to collect and convert alternative energies other than fossil fuel.

Types of Alternative Fuels and Their Promise to Replace/or Complement Fossil Fuel. Alternative energies were defined by the *Energy Policy Act of 1992* as methanol, ethanol, natural gas, propane, hydrogen, coal-derived liquids, biological material, and electricity (Bull 1996, p. 1020). However, not all the alternative energies listed above have a promise to replace or complement fossil fuels. Because of the limitations of current technology and available funding, alternative energies in transportation will likely be limited to biofuel, hydrogen, and electricity in the near future.

Biofuels. Biofuels for transport include ethonal, biodiesel, and several other liquid and gaseous fuels. They have the potential to complement a significant amount of fossil fuels around the world in the near future, and that trend has begun (International Energy Agency 2004a). The benefits of using biofuels in transportation are substantial, including releasing less pollutants and greenhouse gases into the environment, improved vehicle performance, energy security, and enhanced rural economic development (International Energy Agency 2004a; Lund and Clark 2008). The barrier to spread the use of biofuels lies in its production costs, which are as much as three times that of fossil fuels. This is further highlighted because the benefits of biofuel are not easily quantified. Therefore, the free market favors petroleum. However, with the evolving international and national policies, the promise of biofuels is impending.

Hydrogen. Hydrogen is sometimes considered an ideal energy carrier in the foreseeable future (Kim and Moon 2008). It has high-density energy, which can be converted from a range of energy sources, such as solar, wind, and it emits no detrimental by-products to the environment. Hydrogen has recently become a source of fuel for transportation. The focus is mainly on cell application in buses (Cannon 1997). The research and production of hydrogen powered buses began in the early 1990s in the United States, followed by Germany, Italy and Belgium. As these efforts continued, more countries joined them. In 2003, the International Partnership on Hydrogen Economy was set up in Washington D.C. This international organization includes 14 most developed countries and 3 developing countries (China, India, and Brazil). Its member countries have paid great attention to weave hydrogen into their transportation systems and accelerate the transition to use hydrogen as alternative energy. In terms of private cars, several major automobile manufactures have tried to incorporate hydrogen into conventional combustion engines since the early 1980s. Hydrogen-driven vehicles have been built for demonstration. Current efforts are directed at commercialization of these new technologies, as well as more cost-effective vehicles (International Energy Agency 2004b). Therefore, hydrogen promises to be a major energy source for both private and public vehicles.

Electricity. Considering the potential negative impacts of the other alternative energies, such as insufficient CO_2 emission cuts and endangered food production with biofuels, and huge energy inputs and vast investments with hydrogen, it is sometimes argued that electricity is the most promising way of achieving sustainable transportation in the short term future (Netherlands Society for Nature and Environment 2009). Electricity is used in transportation by a rapid introduction of electrically powered transport, which involves electric vehicle (EV), hybrid electric vehicle (HEV), and plug-in hybrid electric vehicle (PHEV). The mass use of these vehicles requires innovative battery technology in terms of energy storage capacity and recharge efficiency. Under the joint efforts of both manufacturers and research institutes, there have been major breakthroughs in the past few years in respect to electricity. For example, a new type of battery, called carbon nanotube, is subject to North American testing. It is expected to deliver 380 miles range and can be recharged in less than 10 minutes (Next Alternative 2009). There are already electric cars on the market, however, the supporting infrastructures have yet to be developed. As the mass acceptance and production of electric vehicles becomes reality in the future, it is only a matter of time until supporting infrastructure is built.

In conclusion, there are already several alternative energies that could replace or at least complement fossil fuels in the transportation sector, namely biofuel, hydrogen, and electricity. However, each of them is not completely developed because of large infrastructure investment requirements and/or breakthroughs in current technology. There would be no guaranteed success in the market. But still, more efforts are being made in the research and development.

Weaving Alternative Energy into National Policy and Transportation Planning. Given the high costs of alternative energy-based transportation technologies, national policies and incentives are needed to promote alternative energy use in transportation and other energy-intensive industries. These policies and incentives could include but not limited to the following:

- Tax incentives on alternative fuels by reducing or eliminating fuel taxes.
- Tax incentives and rebates on alternative-energy-based vehicles.
- Incentives for alternative-energy-based infrastructure investments to support investments into new transportation facilities such as pipelines and recharging/refilling stations.
- Carbon taxes—in addition to incentives to develop alternative-energy-based vehicles and infrastructure, carbon taxes are fuel taxes based on the carbon content of the fuel, they tax the externality (carbon) directly. Furthermore, a new tax system should be developed to consider well-to-wheels emissions (emissions produced from a complete vehicle fuel cycle), not just tailpipe emissions (International Energy Agency 2004b).
- Fuel standards—some argue that raising the current fuel standard will help market development of alternative energies. This is arguably true because too high a standard may completely wipe out fossil fuels from the market. Therefore, if being used, this policy should be carefully evaluated so that allowing buffer time for fossil fuels.
- Free trading of alternative energies and related raw materials. There will be significant potential benefits from free international trade in alternative energies and their raw materials because of the wide range of production costs and raw material storage worldwide. Therefore, removing international trade barrier will improve the economic efficiency of alternative-energy markets.

In addition to public policies needed to encourage the research and development of alternative energies for transportation, there is a need for a national consensus in choosing which type of alternative energies to develop, as this decision will determine the types of refueling infrastructure needed. Currently, there is no consensus as to which type of alternative energies is the most promising one at the national level. This greatly limits the development of any alternative fuels, since no one will drive a vehicle without refueling networks outside the community.

9.2.4 Site Design

The Importance of Site Design to Sustainable Transportation. Site design itself is not a new strategy. It very much complements multimodal transportation planning and integrated land use. Site design aims to internalize travel needs and promote the use of sustainable transport modes, like walking, biking and public transit. At the same time, it integrates land uses to facilitate the use of public transit and improve the walking and biking environment. It differs from multimodal and land use planning in that site design is on the microscale, using design guidelines to "enhance the attractiveness, convenience and safety of walking, cycling and transit use, while not compromising the efficiency of travel by other modes (Canadian Institute of Transportation Engineers 2004, p. 9)."

Research indicated that site design has great impacts on travel behaviors. Moudon et al. (1997) studied 12 sites in Seattle for signs of travel behavior, showing that that people walked more often in places with smaller blocks and connected pedestrian facilities; a certain land uses, like apartments, schools and groceries generate pedestrian traffic. Similar findings were also reported by Kuzmyak and Pratt (2003), who found that where land uses are compatible and intermingled, and there is good transit access, there is a greater likelihood that people will travel less and have shorter trips, while using more nonmotorized transportation modes. This indicates that sustainable transportation needs to incorporate site design guidance into its guidelines.

Four Elements of Site Design to Promote Sustainable Transportation. Site design covers a wide range of design elements, including anything that can help to create a safe, convenient and comfort walking and bicycling environment. The Canadian Institute of Transportation Engineers (2004) categorized these elements into four categories: site organization, site layout, site infrastructure, and site amenity.

- **Site organization** refers to the elements influencing proximity and interconnectedness of buildings and key site features. It includes such elements as building placement, the location of building entrance and parking lot, and parking lot design (size, shape, and form). This will influence the walkability and accessibility of the site since these elements determine if the site is well connected to other transport network.

- **Site layout** refers to the elements influencing how people arrive at and get across a site. Elements include the layout of transport networks, such as the routes and stations of a wide range of modes, parking layout and configuration.

- **Site infrastructure** refers to design factors that involve the design of sidewalks and bike routes (e.g., width, materials, etc.), as well as their related signage. These elements influence accessibility of a site in terms of its physical features, such as signage. This determines if the site is friendly to pedestrians, bikers and public transit users.

- **Site amenity** refers to such as elements as street furniture, weather protected walkways, enclosed transit shelters, on-site change facilities, etc. These elements determines if a site looks appealing to people and functions well to its users. The attractiveness and functionality intend to encourage more people to choose alternative transport modes, namely, walking, biking and public transit.

All four categories of site design elements are important and should be integrated in the planning and implementation process. However, there is no "one shoe fits all" site design guideline, the following strategies have been widely used and proved to be beneficial to sustainable design.

Recommended Strategies.   The role of site design by itself is limited in promoting sustainable transportation without the support of other strategies, such as the multimodal transportation planning and integrated land use planning, to reinforce design measures.

Site design should be part of multimodal transportation planning. For example, cities can establish a mechanism to maintain foot and bicycle ways, or even make them a requirement to connect major transport corridors and transit hubs; create innovative walking and biking programs, like "Safe Routes to School" "Car Fee Day" and "Walk to Wellness."Supporting policies should be adopted to encourage public transit, such as requiring new developments to bring with them sufficient public transit services in the form of impact fee.

Site design should be integrated with land use planning. Without efforts to developing more compact, mixed-use, transit-supportive land use patterns, site design alone will do little to promote sustainable transportation (Canadian Institute of Transportation Engineers 2004). The Canadian Institute of Transportation Engineers also described four fundamental land use planning strategies that work simultaneously with site design to support sustainable development. These strategies include establishment of urban boundaries to combat urban sprawl; promotion of compact urban form and density; development of transit corridors and nodes; and layout of arterial and collector roads in a grid pattern.

9.2.5 Travel Demand Management

Introduction to TDM.   Travel demand management, or transportation demand management (TDM) refers to those transportation policies that attempt to marginally change travelers' behavior, so as to reduce travel demand, especially single occupied motorized vehicles, or to redistribute travel demand spatially and temporally (Ferguson 2000). TDM traces its origins to the 1970s and 1980s when, in response to the increasing oil price and worsening urban congestion, there emerged a desire to provide alternatives to single occupied commuting vehicles. Current TDM has expanded to embrace both commuting and noncommuting travel in order to create an efficient and sustainable transport system.

Reducing trips by single-occupancy vehicles (SOV) is the primary objective of TDM. The US Federal Highway Administration considers TDM to be "strategies to manage demand will be more critical to transportation than strategies to increase capacity (supply) of facilities" in the 21st century (U.S. DOT Federal Highway Administration 2004b).

Common Strategies Adopted in Current TDM and Their Effectiveness.   TDM includes a full range of strategies in terms of both temporal and spatial dimensions. Its major arena, which is previously concentrated on commuting trips, has shifted to encompass nonwork trips (shopping, entertainment), as well as some special events, such as the World Exposition and the Olympic Games. Victoria Transport Policy Institute (VTPI) produced an updated TDM encyclopedia which lists all the available effective TDM strategies. These strategies are categorized into four groups according to how they affect travel: incentives to use alternative modes and reduce driving, improved transport options, parking and land use management, policy and institutional reforms (www.vtpi.org/tdm). The effectiveness of these strategies depends on the level of overall trip reduction.

Using incentives is the most effective strategy to encourage alternative modes and reduce driving (U.S. DOT Federal Highway Administration 2004a). These incentives include road pricing, congestion pricing, parking pricing, carbon and fuel taxes. They can help manage congestion, motivate sustainable modal choices and also generate new revenues. For example, an European Union funded project, a real life experiment of road pricing in the greater Copenhagen region, found that there was significant decrease (up to 40 percent) in daily vehicle trips depending on the location and the pricing scheme (Schönfelder et al. 2007). When summarizing the effectiveness of TDM alternatives, Berman and Radow (1997) realized that the overall lesson from Seattle, Washington, the state of Maryland and other successful TDM programs in the United States, that when a TDM program provides incentives, such as time-saving and financial gains to the commuter, fewer people will drive alone during the peak hours.

Strategies used to improve alternative transport options are also effective TDM measures. These include favoring a more secure and attractive biking and walking environment, public transit, and high-occupancy vehicle trips. However, providing alternatives does not guarantee success. For instance, when summarized the effectiveness of TDM alternatives, Berman and Radow (1997, p. 1214) stated that "starting a new public transit service alone could not necessarily meet the trip reduction needs of an employment site." Neither could the vanpooling programs. Good service is only half the battle, the other part requires the users to capitalize on TDM alternatives by providing transit incentives to potential users. The majority of successful TDM programs offer some kinds of subsidies to those who would like to share a ride with others.

Incorporating parking and land use management into TDM is used to support alternative transportation modes and reduce vehicle trips. Those strategies overlap in many parts with integrated land use strategy, discussed above. Parking is an important factor influencing travel behavior (Shoup 2005). With unlimited parking, people are more intended to travel alone. Successful TDM programs often manage parking, perhaps providing preferential parking for ride-sharing vehicles, or restricting parking (Petiot 2004). Therefore, both parking and land use management are effective TDM strategies.

Recommendations to Achieve Successful TDM Programs.   The key to achieve successful TDM programs is to combine strategies and react to local conditions. U.S. DOT describes the successful TDM practices as: "Effective TDM employer programs usually employ a wide variety of TDM alternatives and strategies, each mutually supporting the overall objective of trip reduction" (U.S. DOT 1994, p. 5). In addition, the packaged TDM strategies should incorporate time and financial incentives and disincentives as discussed above. A truly effective TDM program is the one that promotes alternative travel modes over single occupied vehicles, reinforced by providing time and financial incentives, and uses rules, regulations and planning practices to ensure the successful implementation.

Furthermore, the success of TDM in the future is intrinsically related to the integration of advanced technologies, especially the advanced travel information technologies. The advanced travel information technologies enable travelers to make more timely and informed information about travel model, departure time and route choices with the use of web-based real-time traveler information, 3G network, dynamic message signs on roadsides and parking roads, etc. The easy-to-access, up-to-minute and detailed travel information is fundamentally changing the way we manage travel demand. In response, the focus of future TDM strategies should expand from transportation modes to timing, destination and route choices of trips (U.S. DOT 2005).

9.2.6   Assess the Effectiveness of Sustainable Transportation Strategies

Assessment of the effectiveness of sustainable transportation strategies is the key in monitoring the implementation of these strategies. This helps to assure that efforts are in line with sustainable transportation objectives, and that the expected goals are achieved. Furthermore, assessment can determine if the objectives are accomplished in a cost-effective manner. For example, an overwhelming number of sustainable transportation strategies focus on reducing motorized vehicle trips while encouraging the use of alternative transportation modes. However, each of these strategies requires unique costs in achieving objectives. Land use integration and site design mainly affect new development patterns and services, whereas travel demand management uses financial and policy tools to achieve trip reduction. Therefore, the effectiveness of these strategies needs to be assessed for each specific case in order to help communities make more informed decisions.

Methods and Criteria of Assessment.   The most difficult aspect of assess potential strategies is defining what sustainable transportation means to that community.

Since the publication of the Brundtland Report, also known as *Our Common Future* (UN WCED 1987), many proposals have been offered to measure sustainable development as well as sustainable transportation. It seems that indicators are, by far, the dominated and best way to measure sustainability (Ling 2005; Munier 2005; Jesinghaus 2007). The functions of indicators, according to Tunstall (1992), include:

- Assessing conditions and trends
- Comparing across places and situations
- Assessing conditions and trends in relation to goals and target
- Providing early warning information
- Anticipating future condition and trends

Similarly, transportation sustainability systems can also be assessed by individual performance indicators. Litman (2008b) categorized these perspectives into three groups: traffic, mobility and accessibility. Three additional criteria should be added to ensure a more complete list, including environmental impacts, cost-effectiveness, and behavior changes.

Traffic-Based Measurements.   The performance indicators of traffic-based measurements focus on the impacts on vehicle movement, including total number of trips, average traffic speeds, roadway level of service (LOS), congestion delays, available parking spaces, maintenance cost of transportation system. These measures are heavily oriented toward automobile travel and are less favorable to public transit and nonmotorized travel. From this perspective, the best transport system is a system that has sufficient road and parking capacity, operates at good speeds, satisfactory level of service, and with minimal costs and crashes.

Mobility-Based Measurements.   Mobility based measurements focus on the movement of people and goods, including person-miles, ton-miles, "trip" means person- or freight vehicle trip. These measurements favor increased travel mileage or speed. Although this perspective also favors automobiles the most that is similar to traffic-based measurements, it also put more weight on public transit, car/vanpooling, and cycling. Mobility based measures focus on a comprehensive transport system with strong connections among various transport modes, including also multimodal systems (Litman 2008a).

Accessibility-Based Measurements.   Accessibility-based measurements focus on the ability to reach desired destinations for goods, services and activities. This measurement focuses on the availability of activities within a certain travel distance and different ways to access these activities. Therefore, unlike the other two kinds of measurements, this one considers land use patterns, all modes of transportation, and mobility substitutes, such as telecommunication. It favors an integrated transportation and land use system that is multimodal, compact, and characterized by mixed-use, walkable communities. Accessibility is typically evaluated based on the accessibility index that takes into considerations of all these factors.

Environment-Based Measurements. Environment-based measurements focus on transportation's environmental impacts. This perspective assumes that the best transportation system is the one that minimizes the negative impacts on the environment. It not only considers all transportation modes, but more importantly, it favors alternative energies in the transportation system. Therefore, this type of measurement contains a comprehensive view to achieve sustainable transportation, including multimodal transportation, integrated land use, alternative energies and managing travel demand. The most commonly used assessment indicators involve air quality and energy use (Transportation Research Board 2009). Sometimes more complex model is employed to accommodate the temporal and spatial variability of transportation impacts on the environment, like sustainability footprint model (Amekudzi et al. 2008).

Cost-Benefit-Based Measurements. Cost-benefit measurements focus on the costs and benefits of implementing a certain strategy. The best sustainable transportation strategy should incur the least amount of costs while maximize the benefits to achieve certain objectives. The difficulty is in finding and evaluating the measurement of certain costs and benefits in monetary terms. For example, Florida Department of Transportation (FDOT) developed a standardized methodology for evaluating the costs and benefits of TDM. The result will then be used for comparative assessment of different TDM strategies. This methodology "models consumers' price responsiveness to various TDM strategies by embracing the most relevant trade-offs faced under income, mode cost and availability constraints (Florida Department of Transportation 2007, p. iii)." The decision makers can then incorporate the results into the decision-making process.

Behavior-Change-Based Measurements. Behavior-change-based measurement measures the effectiveness of certain sustainable transportation strategies on changing people's travel behaviors. This includes the elasticity and the amount of impacts of a certain transportation strategy. The elasticity measures the sensitivity of people to certain strategies, while the amount of impacts measures the number of people and the kinds of people who are sensitive to the strategies. Strategies that can more easily change behavior are considered to be more effective than those that cannot. If only small percentage of the population is sensitive to a particular strategy, especially when considered the economically disadvantaged, this methodology is not a reliable strategy. For example, one of the travel demand management (TDM) measures is to provide preferential parking to carpoolers and/or provide free transit pass to employees. But there are still not many people shifting to carpools or to transit, which means these strategies are not effective in changing people's behavior. People are not sensitive to these incentives. So, in theory, TDM is easy to implement and could have potentially large impacts. But in reality, it only has small impacts on changing people's behavior and achieving the goal of reducing single-occupancy vehicle driving. On the contrary, people are more sensitive to congestion pricing or higher parking charges, which suggests from the behavior change perspective, congestion and parking pricing are more effective in changing people's behavior.

In summary, these six types of measurements focus on varying ways of measuring the effectiveness of promoting transportation strategies. Ideally, sustainable strategies should help to increase the efficiency of the movements of both motorized and nonmotorized traffic; improve the accessibility of people to activities and services by all transport modes, reduce negative environmental impacts, cost-effective, and are able to change people's behavior in a large scale. However, some strategies are better than the others based on the six criteria. The following section describes our rough ranking of effectiveness of different strategies.

Rankings of Effectiveness of Sustainable Transportation Strategies. Table 9.3 shows our attempt to assess the effectiveness of certain common sustainable transportation strategies according to the six criteria discussed above. The assessment is based on the existing literatures on each strategy. If a

TABLE 9.3   The Rankings of Sustainable Transportation Strategies

Criteria Strategies	Traffic	Mobility	Accessibility	Environment	Cost-benefit	Travel behavior	Total
Multimodal transport	★	★	★	★		★	★★★★★
Travel demand management	★	★	★	★	★		★★★★★
Land use integration		★	★	★		★	★★★★
Site design			★	★		★	★★★
Alternative energies				★			★

strategy meets a certain criterion, then this strategy is given a star. As discussed in previous sections, the six criteria are major parts of the effectiveness of sustainable transportation. The more a strategy can meet these criteria, the more effective it tends to be.

However, it should be noted that this is a very rough and simple assessment. It does not take into consideration of the importance of the criteria and the quantity of the impacts. Therefore, the number of stars does not necessarily mean the absolute importance or effectiveness of the strategies. In realty, all the five strategies should work together and be weaved into long-term sustainable transportation planning and design process, because each of them performs better than others in a certain aspect of the sustainable transportation system.

9.3   EXAMPLES OF BEST PRACTICES OF SUSTAINABLE TRANSPORTATION PLANNING

The previous section introduced various strategies to achieve sustainable transportation. This section provides some examples about how governments at different levels adopt these strategies in their transportation planning process, and how transportation agencies review their policies and take actions to turn sustainable transportation from theory into practice.

There are numerous examples of successful practices of sustainable transportation around the world. Among them, United States, Canada and EU countries take the leading position. As more government agencies at different levels realize the need for an urgent transition to a sustainable transportation mode. For example, the United States federal government has set increasingly stricter environmental standards to mitigate the environmental impacts of transportation projects. The joint efforts of the U.S. Environmental Protection Agency, Department of Housing and Urban Development, Department of Transportation, etc. facilitate turning sustainable transportation practices into standards. This is likewise happening in many different places around the world, as concerns about sustainability increase.

Sustainable transportation has received increasing attention from state agencies within the United States. In compliance with *Intermodal Surface Transportation Efficiency Act* (ISTEA) requirements, state agencies must produce Master Plans and follow certain planning processes. Some pioneering states stand out among others in their efforts to follow sustainable transportation principles and practices. For example, the Maryland Department of Transportation (DOT) is committed to principles of sustainability and has introduced "A Sustainable Transportation Agenda" to fully incorporate sustainable transportation strategies. Michigan DOT has made similar commitments to sustainable transportation by weaving the rationale of sustainability into their long range plans. Oregon DOT is another sustainable transport pioneer, its 2006 Oregon Transportation Plan provides a framework to prioritize transportation investment to encourage state wide multimodal transportation. The New Jersey DOT has empowered its towns to partner with NJDOT and other state agencies to create balanced transportation and land use plans. These are only a few of the many states that have made efforts to follow transportation sustainability principles.

Cities are often main arena in implementing sustainable transportation practices. Many cities are trying to link sustainability and transportation policies together. Some cities start with simple effort, such as managing urban travel demand, while others act much more ambitiously to fully incorporate sustainable transport strategies, including multimodal transportation, land use integration, site design practices and the use of alternative energies such as Vancouver, British Columbia and New York City, New York.

In the United States, the Metropolitan Planning Organization (MPO), is often the in-between the state and city planning agency, playing an important role in achieving sustainable transportation. MPOs are federally mandated and funded transportation policy-making organizations that are composed of representatives from local governments and local transportation authorities. The MPO is responsible for allocating transportation investment, making plans according to the region's shared vision, facilitating collaboration of all stakeholders within the region. These are crucial elements in converting sustainability theory into practice.

The following section introduces examples of cities, states, and MPOs that are known for their sustainable transportation practices. The experience and lessons learned from these examples can serve as models and inspiration for other municipalities and governmental agencies.

9.3.1 State of New Jersey, United States

"Going Green" is one of the highest priorities of the current administration and for the State of New Jersey (New Jersey Department of Labor and Workforce Development 2009). The proposal for green development has been weaved into nearly all aspects of New Jersey development. As a major energy consumer and also a major source of air pollutants, transportation has been an intense focus for the idea of switching to sustainable transport. To achieve this, the New Jersey Department of Transportation (NJDOT) has made great efforts to incorporate sustainable transportation strategies and put these strategies into practice. As a result, NJDOT received the 2006 Transportation Planning Excellence Award. This biennial award program, sponsored by the Federal Highway Administration, Federal Transit Administration and the American Planning Association, recognizes innovative nation-wide transportation practices.

New Jersey Future in Transportation. In partnership with the Office of Smart Growth and other state agencies, NJDOT has created innovative ways to review the state development patterns and re-invest in its communities. These efforts became New Jersey Future in Transportation (NJFIT). The rationale of NJFIT is to use sustainable transportation strategies to achieve a smarter and greener transportation system. This program is considered as "a new paradigm for developing transportation projects in partnership with municipalities and other state agencies" (FHA and FTA 2006). The key desire of this program is to empower towns to partner with NJDOT, NJ Office of Smart Growth and other state agencies to create balanced transportation and land use plans. This planning process provided a unified vision for different levels of governments. Furthermore, it produces a vision that is shared by community members and can lead to wiser investments of public expense. These factors work together to ensure the implementation of proposed strategies.

Land Use Integration. New Jersey is an urban state. Though parts of the state are quite rural (the western part), much of the North, Central and Southern parts of the state are urban and have strong transit connections to major urban destinations (New York City and Philadelphia). The New Jersey Future in Transportation initiated and implemented a corridor study, the Integrated Land Use and Transportation Planning Studies (ILUTP), and a transit hub study, Transit Villages. Both of these studies highlight the importance to link land use to transportation planning. The philosophy behind them is described by NJDOT as "if you want to have a say in our transportation planning, we need to have a say in your land use planning" (NJDOT 2005b). The initiative of Transit Village is led by a partnership of 11 state agencies. It aims to prioritize the development of those municipalities that meet the requirement of land use changes. A granted Transit Village can receive a variety of assistance, including loans, fiscal subsidies, and technical support. These required land use changes include preservation of open space, redevelopment and revitalization of urban and suburban centers, mixed land uses, and compact urban forms.

Multimodal Transportation.   New Jersey has rich history in mass transit. There are already a few models to follow. For example, the Town of Westfield provides rail service from local train station to regional transit hubs, with a nonstop service during morning and evening rush hours. An express service to New York City is also available. Adding the convenient local bus service, the Town made public transportation highly desirable to commuters. New Jersey Future in Transportation (NJFIT) aims to continue this history and promotes alternative traveling modes to driving, including walking, biking and public transit. In order to encourage the use of public transit, NJFIT supports developments around transit hubs. In particular, residential developments around transit station provide residents with easy access to public transit. NJFIT also favors walking and biking by creating a well-connected street networks that emphasize accessibility and reduces the distance between destinations.

Site Design.   NJFIT uses design standards to naturally encourage walking, biking and transit ridership. It favors design roads in context. Rather than using the standardized solutions, NJFIT supports individualized solutions for individual places based on the context and the objectives sought. This can help achieve harmony with the community and preserve resources that otherwise might be lost or harmed. It incorporates design principles that limit the size of blocks and increase the number of intersections to encourage walking and biking; to reduce the width of streets to create safe environment; and to link together a series of memorable sites to create a pleasant walking and biking environment.

Travel Demand Management.   NJFIT encourages the creation of public-private mobility services to "reduce the use of single-occupancy vehicles (SOVs). Some of these services are subsidized through tax breaks or program funding, such as employers-provided paratransit vans, carpool and rideshare programs, transportation management associations, guaranteed ride home programs, and the commuter choices program which allows transit trips to be funded from pre-tax income" (NJDOT 2005a).

In summary, the State of New Jersey has made great efforts to pursue sustainable transportation. These efforts were epitomized by its state transportation plan, known as New Jersey Future in Transportation. The Plan employed various sustainable transportation strategies, including encouraging multimodal transportation, balancing land use and transportation developments, using site design and Travel Demand Management measures to create a less-car dependent and efficient transportation system. NJDOT has implemented these strategies in two major ways. On one hand, NJDOT has partnered with towns and other state agencies to redevelop state routes. Six routes are being studied to make them more environment-friendly, better access to more transport modes, and safer for cyclist and pedestrian. On the other hand, NJDOT has developed collaboration among other state agencies to fund Transit Villages. Since 1999, 16 communities have been designated as Transit Villages. These communities are in a powerful position to support their transit facilities, promote sustainable land use, and enhance their sense of place.

9.3.2   Vancouver, British Columbia, Canada

Vancouver is a coastal city located in British Columbia of Canada. It covers an area of 44.3 square miles and is the largest metropolitan area in Western Canada and also the eighth largest city in Canada with a population of 578,000 in 2006. The city is renowned for its innovative programs in search for sustainability and accessibility. It is also widely regarded as a leader in achieving sustainable transportation. Vancouver's efforts in search for sustainable transportation can be traced back to its 1997 Transportation Plan. The plan prioritized walking, cycling and public transit to reduce the use of private cars and was well implemented. As a result, Vancouver has become one of the most walkable cities in North America (Helmer and Gough 2010). In 2006, the City of Vancouver received Sustainable Community Awards: Sustainable Transportation for its Southeast False Creek Sustainable

Transportation Strategies (Transport Canada 2006). The Vancouver metropolitan area also paid special attention to sustainable transport. The metropolitan authority, known as Greater Vancouver Transportation Authority, received Sustainable Community Awards: Sustainable Transportation one year earlier than the City (Transport Canada 2005) for its student public transportation program, called U-Pass.

Regional Plans. The Greater Vancouver Regional District and Greater Vancouver Transportation Authority are responsible for designing regional plans and providing regional services. The strategies they propose therefore have a great impact on the City of Vancouver. Key elements of the strategies used by these agencies include: (1) integrating land use with transportation planning to create a more compact urban form and minimize travel times; (2) applying transportation demand management (TDM) to change travelers' behavior; (3) providing special facilities for high-occupancy vehicles (HOVs) and limit the use of single-occupancy vehicles (SOVs); (4) encouraging alternative travel modes, especially cycling and transit; and(5) supporting the use of alternative fuels by exempting their fuel tax until they reach 3.5 percent of British Columbia's motor fuel market.

Multimodal Transportation. The City of Vancouver strives to support multimodal transportation. To achieve this, the city focuses on sharing the existing road network, rather than expanding it. First, the 1997 Transportation Plan proposes allocating more road space to transit. Certain streets and sections of streets would assume a more clearly defined transit role, such as bus-only lanes. Second, the plan proposes improving truck access. An example is the proposal to complete the Port Road and to improve road conditions where there are concentrated sources of trucks. Third, the plan allocates particular space to bicyclists. The existing bikeway network and greenways program are expanded to guarantee well-connected bikeways can reach all areas around the city. Fourth, the plan proposes pedestrian comfort and safety improvements. Commercial centers would create pedestrian priority areas to make road crossing easier and safer, other measures include medians, corner bulges, and pedestrian-activated signals. Furthermore, the city's traffic-calming program was initiated to create safer neighborhoods.

Land Use Integration. The City of Vancouver made a great effort to link transportation and land use together. It supported land use that could reduce travel demand by motorized vehicles. The main strategy is to establish neighborhood centers which can create more jobs. Special attention has been paid to downtown. The city continues to develop new residential neighborhoods planned for downtown. The emphasis for transportation within downtown will focus on creating great streets which are fully integrated with the proposed downtown greenways and green links. Furthermore, the Downtown greenways will connect and build on the unique traditional characters of the passing neighborhoods, in favor of efficient ground floor uses, and serve as a stimulus for downtown revitalization.

Site Design. The city of Vancouver prefers using green design concept to make its streets more sustainable. This green design allows storm water runoff filter back into the ground rather than flowing into sewer system. It can also reduce air and water pollution, calm traffic, and create an enjoyable street for pedestrians and bicyclists. This green design is also expected to accommodate alternative transportation modes for different people. The green design concept was incorporated into the city's three main street standards and guidelines: S*treet Restoration Manual, Transportation Association of Canada Geometric Design Guide,* and the *Master Municipal Construction Document.* Under the guidance of these documents, the City has successfully developed several green projects. For example, its "Country Lane" program, which is a sustainable alternative to full width lane paving, removed all temporary surface and replaced it with a mixed surface including concrete, gravel, plastic mat, and grass grid. It provides vehicles with improved surface while maintaining the ambience of country living.

Travel Demand Management. Vancouver pays special attention to both car-sharing and ridesharing. It first provided on and off-street parking stalls for automobile cooperatives. Then, Vancouver contracted with the Cooperative Auto Network (CAN), a private car-sharing company, to locate additional low-emission CAN vehicles at City Hall to serve work-oriented travel needs. During work hours, these vehicles are used by city employees. At other times, the vehicles are available for car sharing by CAN members. The city has also increased the number of shared vehicles to encourage rapid growth of car sharing. Third, the city is cooperating with Bell Ride Foundation to establish ride-share matching database for city employees. This strategy greatly encourages commuters to use carpool, vanpool or rideshare.

Alternative Energies.   The city of Vancouver has recently started requiring all new single-family homes to install electric-car plug-ins and is evaluating the possibility in supplying outlets for multi-family residential buildings. The outlets to charge electric bicycles are required in off-street bicycle storage rooms. The city is also reforming its legislation to accommodate electric vehicles. The new by-law allows electric vehicles to be used on every street in Vancouver. This makes Vancouver the first major Canadian city to permit Neighborhood Zero Emission Vehicles on its streets. For more detailed information, please go to Vancouver Electric Vehicle Association Web site (www.veva.ca).

Plan Monitoring and Implementation.   The city specified the implementation schedule of its transportation plan based on the availability of funding and technology. It requires the Departments of Engineering and Planning to undertake regular monitoring and review of transportation services, in order to evaluate the progress of the implementation of current plans. The results of the evaluation, in turn, are the basis for recommendations for transportation components in the city's Capital Plan. This approach ensured that policies and principles can be translated into actions.

In summary, as one of the leading cities in striving to reach sustainable transportation goals, the city of Vancouver has initiated comprehensive and innovative programs to achieve the goal of sustainability. These programs cover land use integration, multimodal transportation, travel demand management, site design and alternative fuels promotion. In addition, Vancouver pays special attention to plan monitoring and implementation. Therefore, the city has been successful in achieving the desired results. As the Transportation Plan 10-year progress report pointed out: although the total trips in Vancouver has increased by 23 percent because of the increase of population and employment, vehicles entering and leaving the city have decreased by 10 percent over the same period. Since new trips are accommodated by public transit, bike and walk modes. Many of the mode share targets that were set for 2021 have already been achieved. The remaining initiatives will soon be completed (General Manager of Engineering Services and Director of City Plans 2006).

9.3.3 Portland Metro, Oregon

Metro is an elected regional government agency for the Oregon portion of the Portland-Vancouver metropolitan area, which is the only directly-elected Metropolitan Planning Organization (MPO) in the United States. It serves 25 cities in Clackamas, Multnomah and Washington Counties, covering 463 square miles with about 1.4 million residents, as of 2005 (Metro 2010). The main responsibilities of Metro are to design regional land use planning and plan for the metropolitan region's transportation system. Metro's efforts to pursue sustainable transportation measures make it one of the best examples of practice in North America (Marshall Macklin Monaghan Ltd 2007). They also won Metro honorable mentions in Environmental Excellence Awards in 2009 (Federal Highway Administration 2009).

Regional Transportation Plan.   The 2035 Regional Transportation Plan (RTP) updates the policies, projects and strategies that guide transportation planning and investments in the Portland metropolitan region for the year 2035. By 2035, the metropolitan region and surrounding counties are expected to grow by more than one million people. The employment sectors will create 0.5 million jobs. These increases are estimated to double current trips on transportation networks. These challenges resulting from this growth are addressed in 2035 RTP, which strives to create a sustainable regional transportation system.

The plan proposes to focus investment on the region's main streets, employment centers and major travel corridors to protect the community assets in which the region has already invested; reduce the distances that people travel to work, school and shop; and protect farm, forest and natural resource areas. Projects proposed in the 2035 plan include: fixing safety problems; address growing congestion; and providing viable options for walking, biking and transit; while making travel more affordable and reliable for everyone-including businesses and freight. In order to accomplish these goals, proposed projects include adding new sidewalks, adding new or expanded bicycle facilities and trails, using technology that makes travel safer and more efficient, adding new roads, expanding transit service and high-capacity transit connections, improved interchanges, and increasing capacity on the region's highway system.

Multimodal Transportation Planning. The 2035 RTP promotes multimodal transportation by considering expanding transportation choices as one of the planning goals. It states that "[m]ulti-modal transportation infrastructure and services provide all residents of the region with affordable and equitable options for accessing housing, jobs, services, shopping, educational, cultural and recreational opportunities, and facilitate competitive choices for goods movement for all businesses in the region" (Metro 2010, pp. 209). To achieve this goal, the plan proposes to increase the use of walking, biking, public transit and shared rides and to reduce reliance on automobiles and drive alone trips as well as vehicle miles traveled per capita. Furthermore, multimodal transportation also takes disadvantaged people into account. The plan tries to provide affordable and equitable access to travel choices including people with low income, children, elders and people with disabilities. In addition to the movement of people, the plan also considers multimodal transport for freight delivery. It supports multimodal freight transportation system that includes air cargo, pipeline, trucking, rail, and marine services to facilitate competitive choices for goods movement for business in the region. This comprehensive multimodal transport system will greatly decrease the use of single-occupancy vehicles (SOVs) and protect the public's interests as well.

Integrated Land Use and Transportation. In 1995, the Portland Metro area adopted the 2040 Growth Concept. This long-term plan aims to merge land use and transportation planning to rein-force the objective of both. The plan encourages mixed land use and high-density development in 38 urban activity centers, 33 light-rail communities around rail stations and 400-mile corridors connecting most activity centers. Then the plan developed strategies to connect Portland urban center with its 7 regional centers and 30 town centers by public transit. The 2035 RTP responds to the plan by considering transportation to be an integrated and interconnected system that must be completed over time. The effort of integrating land use and transportation also reflects in recogniz-ing the link between transportation and economic activities on certain land uses. The 2035 RTP and 2040 Growth Concept recognize investments that serve certain land use or transport facilities may return a greater profit. They also realize the importance of creating reliable and efficient connections between intermodal facilities and hubs, in order to reinforce the Portland's role as a gateway for business and tourism.

Site Design. The 2040 Growth Concept supports high-quality site design to improve transportation performance. In order to implement the plan, Metro staff has established livable streets programs, which aim to integrate site design with surrounding land uses to help prevent congestion as well as encouraging walking, biking and public transit. Metro has published three handbooks as a guide-line for site design, including *Creating Livable Streets: Street Design Guidelines for 2040, Green Streets: Innovative Solutions for Stormwater* and *Stream Crossings, and Tees for Green Streets: An Illustrated Guide*. These guidelines focus on creating safe, attractive spaces for neighborhood activi-ties. Street design elements such as sidewalks, bike lanes, crosswalks, ramps, street drainages, street trees and benches, street lighting and banners, and bus shelters are linked to improving attractiveness, slowing traffic and encouraging nonmotorized modes. Furthermore, the guidelines also focus on creating "green" streets. Design elements, including trees, greensward, and paving materials, can be used to alleviate the impact of storm water and protect aquatic ecosystems. The 2035 RTP proposed an update to these handbooks in 2010, which will integrate designs including low-volume bicycle boulevards, alternate designs for high volume arterial streets and regional trails. Updates should address the added design elements that are needed when these facilities serve as a bicycle parkway route, e.g., bicycle priority treatments and strategies for avoiding bike/pedestrian conflicts.

Transportation Demand Management. The Regional Transportation Systems Management and Operations (TSMO) Plan was developed in conjunction with the 2035 RTP to develop a focused strategy for investment in operations and demand management solutions. Three advisory committees, TransPort, Regional Travel Options, and a TSMO Policy Work Group, joined together efforts to create, review and revise the plan. TransPort, the operations subcommittee of Metro: Transportation Policy Alternatives Committee serves as the technical advisory committee for operations. The Regional Travel Options Subcommittee, the transportation demand management subcommittee of TPAC, provides guidance on TDM transportation demand management solutions. Lastly, a TSMO Policy Work Group was formed to provide high-level policy guidance for the plan.

Portland Metro TDM programs include rideshare matching services, technical assistance to employers. They also required major employers (with more than 50 employees) in the Metro region to report to a single work site to provide incentives for alternative commute modes to SOVs. This includes carpool matching, prioritized carpool parking, transit subsidies, etc.

In summary, Portland Metro manages regional transportation/land use system mainly through its Regional Transportation Plan. The newly updated Plan, known as 2035 Regional Transportation Plan, covers a wide range of sustainable transportation strategies, including multimodal transportation planning, integrated land use, site design and TDM. The plan identifies a 20-year list of future transportation projects. Metro established an outcomes-based approach to evaluate the effectiveness of the proposed strategies and monitor the implementation process. This approach takes a periodic overview of the transport data, which are related to vehicle level of services and mode share for walking, bicycling, transit use and share ride. Then, Metro refines its investment strategy to better serve the region in search for a sustainable transport system.

9.4 CONCLUSION

Sustainable transportation measures must be taken through a set of initiatives that promote the goals of sustainable transportation. By incorporating practices such as multimodal transportation planning, land use integration, alternative energy sources, site design and travel demand management, engineers and planners alike are able to best serve their community. There is great variety in what measures are most successful, depending greatly on the scale of implementation; however, only with coordination and a multifaceted approach can sustainable transportation truly be addressed.

Sustainable transportation policies are currently being implemented throughout North America and Europe. Cities, metropolitan regions and states are all thinking about what policies and practices will best meet sustainable transportation goals. This includes using a combination of practices that work together to promote multimodal transportation practices, lower dependence on the automobile and fossil fuels, and improve travel accessibility and mobility. These practices are a combination of technology, services and design that promote social, economic and environmental friendly behaviors.

Although the number of studied cases in this chapter is limited, there are some common and unique strategies that emerge. State governments are more focused on statewide policies and plan, as seen with New Jersey Future in Transportation, which coordinates among different levels of governments. Metropolitan Planning Organizations and similar regional authorities are more focused on regional land use policies, regional transportation system planning, allocation of transportation planning funding, and coordination and collaborations of different jurisdictions. Lastly, cities make more specified plans and related policies to turn the state and regional plans into reality. They are focused on land use transportation coordination, urban design, travel demand management, investments into multimodal transportation, and monitoring strategies to ensure the implementation of proposed strategies.

There is no "one shoe fits all" strategy. Sustainable transportation solutions depend on the integration of various strategies and involve different levels of agencies to work together. Unique circumstances that govern place and the individual laws and attitudes of each place govern how sustainable practices form and develop.

It should be noted, however, developing sound strategies is only one measure. Implementation is often the more challenging aspect of sustainable transportation. When accessing Canada's progress in achieving sustainable transport practices, Weller (2006) realized a serious, expanding gap between the promise to achieve sustainability and the performance to fulfill it. His findings illustrate the "say one thing, do another" mismatches. For example, if governments encourage more transit ridership, but spend more and more money on road expansion and maintenance, there is a disconnection. Local politicians favor walking and cycling but fail to guarantee the safety and quality of sidewalks and bike trails. These mismatch problems not only happen in Canada, they are also quite common in the U.S. and European countries (Banister et al. 2007; Hull 2005). Therefore, a rigorous implementation with funding priority oriented toward sustainable transportation plans and a continuous evaluation and adjustment are critical to our sustainable future of transportation and development.

ACKNOWLEDGMENTS

This research was supported in part by US National Science Foundation award BCS-0616957, National Natural Science Foundation of China award 50738004, China National High Technology Research and Development Program 863 award 2009AA11Z220. The authors would like to thank Mr. Jian Zhang for his assistance in preparing for the draft. Any opinions, findings, and conclusions or recommendations expressed in this chapter are those of the authors and do not necessarily reflect the views of the sponsors.

REFERENCES

Alonso, W. 1964. *Location and Land Use: Toward a General Theory of Land Rent.* Cambridge: Harvard University Press.

Amekudzi, A., Khisty, C., and Khayesi, M. 2008. "Using the Sustainability Footprint Model to Assess Development Impacts of Transportation Systems." *Transportation Research Part A* 43:339–348.

Badoe, D., and Miller, E. 2000. "Transportation-Land-Use Interaction: Empirical Findings in North America, and Their Implications for Modeling." *Transportation Research Part D* 5:235–263.

Banister, D., Pucher, J., and Lee-Gosselin, M. 2007. "Making Sustainable Transport Politically and Publicly Acceptable." In *Institutions and Sustainable Transport: Regulatory Reform in Advanced Economics.* Edited by Rietveld, P., and Stough, R. Cheltenham, England: Edward Elgar Publishing, 17–50.

Berman, W., and Radow, L. 1997. "Travel Demand Management in the USA: Context, Lessons Learned and Future Directions." *Energy Policies* 25:1213–1215.

Black, W. 1990. "Global Warming: Impacts on the Transportation Infrastructure." *Transportation Research News* 150:2–8.

Black, W. 1996. "Sustainable Transportation: A US Perspective." *Journal of Transport Geography* 4(3):151–159.

Black, W. 2000. "Socio-Economic Barriers to Sustainable Transport." *Journal of Transport Geography* 8(2):141–147.

Bull, S. 1996. *Renewable Energy Transportation Technologies.* Golden, Colorado: National Renewable Energy Laboratory. Accessed online at: http://cat.inist.fr/?aModele=afficheN&cpsidt=3154894, February 2010.

Canadian Institute of Transportation Engineers. 2004. "Promoting Sustainable Transportation through Site Design: An ITE Proposal Recommended Practice." Washington, D.C.: Canadian Institute of Transportation Engineers. Accessed online at: http://www.cite7.org/resources/documents/ITERP-PromotingSustainableTransp ortationThroughSiteDesign.pdf, March 2010.

Cannon, J. 1997. "Clean Hydrogen Transportation: A Market Opportunity for Renewable Energy." Washington, D.C.: Renewable Energy Policy Project Accessed online at http://www.repp.org/repp_pubs/pdf/issuebr7.pdf, February 2010.

Cervero, R. 2007. "Transit Oriented Development's Ridership Bonus: A Product of Self-Selection and Public Policies." *Environment and Planning A* 39(9):2068–2085.

Cervero, R., and Arrington, G. 2008. "Vehicle Trip Reduction Impacts of Transit-Oriented Housing." *Journal of Public Transportation* 11(3):1–17.

Comeau, C. 2009. "Moving beyond the Automobile: Multimodal Transportation Planning in Bellingham, Washington." Chicago: American Planning Association. Accessed online at http://www.planning.org/practicing-planner/ print/2009/fall/case.htm?print=true, February 2010.

Deakin, E. 2001. "Sustainable Development and Sustainable Transportation: Strategies for Economic Prosperity, Environmental Quality, and Equity." Institute of Urban and Regional Development, UC Berkeley. Accessed online: http://iurd.berkeley.edu/catalog/Working_Paper_Titles/Sustainable_Development_and_Sustainable_ Transportation_Strategies, February 2010.

Downs, A.2001. "What Does 'Smart Growth' Really Mean?" *Planning* April 2001 67(4):20.

"Driving and the Built Environment: The Effects of Compact Development on Motorized Travel, Energy Use and CO_2 Emissions. Committee for the Study on the Relationships among Development Patterns, Vehicle Miles Traveled and Energy Consumption." National Research Council. 2009. Transportation Research Board. Special Report 298.

European Commission. Directorate-General for Energy and Transport. *A Sustainable Future for Transport: Towards an Integrated, Technology-Led and User-Friendly System.* 2009. Luxembourg: Publications Office of the European Union.

Federal Highway Administration. 2004. "Travel Demand Management in a Brief Overview." Accessed online at http://www.ops.fhwa.dot.gov/tdm/index.htm, March 2010.

Federal Highway Administration (FHA) and Federal Transit Administration (FTA). 2006. 2006 Transportation Planning Excellence Awards. Washington, D.C.: FHA and FTA. Accessed online at http://www.fhwa.dot.gov/ planning/tpea/awards2006.htm, May 2010.

Federal Highway Administration. 2009. 2009 Environmental Excellence Awards. Washington, D.C.: Federal Highway Administration. Accessed online at http://www.fhwa.dot.gov/environment/eea2009/contents.htm, May 2010.

Ferguson, E. 2000. *Travel Demand Management and Public Policy.* Farnham, UK: Ashgate Publishing.

Florida Department of Transportation. 2007. Economics of Travel Demand Management: Comparative Cost Effectiveness and Public Investment. Florida: FDOT. Accessed online at http://www.nctr.usf.edu/abstracts/ abs77704.htm, June 2010.

Gearin, E. 2004. "Smart Growth or Smart Growth Machine? The Smart Growth Movement and Its Implications." In *Up Against the Sprawl.* Edited by Wolch, J., Pastor, M., and Dreier, P. Minneapolis: University of Minnesota Press, 279–309.

General Manager of Engineering Services and Director of City Plans. 2006. "Vancouver Transportation Plan, Progress Report." Vancouver: General Manager of Engineering Services and Director of City plans. Accessed online at http://vancouver.ca/ctyclerk/cclerk/20060530/documents/rr1a.pdf, May 2010.

Handy, S. 2005. "Smart Growth and the Transportation-Land Use Connection: What Does the Research Tell Us?" *International Regional Science Review* 28(2):146–167.

Helmer, J., and Gough, J. 2010. "Best Practices in Sustainable Transportation." *Institute of Transportation Engineers Journal* 80 (3):26–35.

Himanen, V., Lee-Gosselin, M., and Perrels, A. 2005. "Sustainability and the Interactions between External Effects of Transport." *Journal of Transport Geography* 13:23–28.

Holtzclaw, J., Clear, R., Dittmar, H., Goldstein, D., and Haas, P. 2002. Location Efficiency: Neighborhood and Socio-Economic Characteristics Determine Auto Ownership and Use?" *Transportation Planning and Technology* 25:1–27.

Hull, A. 2005. "Integrated Transport Planning in the UK: From Concept to Reality." *Journal of Transport Geography* 13(4):318–328.

Institute for Transportation & Development Policy. "New York City Wins 2009 Sustainable Transport Award." Accessed online at http://itdp.org/index.php/news_events/news_detail/new_york_city_wins_2009_sustainable_ transport_award/, May 2010.

International Energy Agency. 2004a. "Biofuels for Transport: An International Perspective." Paris: International Energy Agency. Accessed online at http://www.iea.org/textbase/nppdf/free/2004/biofuels2004.pdf, February 2010.

International Energy Agency. 2004b. "Hydrogen & Fuel Cells: Review of National R&D Programs." Paris: International Energy Agency. Accessed online at http://www.iea.org/textbase/nppdf/free/2004/hydrogen.pdf.

Jesinghaus, J. 2007. "Indicators: Boring Statistics or the Key to Sustainable Development." In *Sustainability Indicators: A Scientific Assessment*. Edited by Hak, T., and Moldan, B. Wasington, D.C.: Island Press, 83–96.

Keegan, K. 2001. "Reclaiming the Streets." *Transportation Alternatives* 7(2):8–10. Accessed online at http://transalt. org/ files/newsroom/magazine/012Spring.pdf, February 2010.

Kim, J., and Moon, Il. 2008. "The Role of Hydrogen in the Road Transportation Sector for a Sustainable Energy System: A Case Study of Korea." *International Journal of Hydrogen Energy* 33:7326–7337.

Kockelman, K. 1996. "Travel Behavior as a Function of Accessibility, Land Use Mixing, and Land Use Balance: Evidence from the San Francisco Bay Area." Thesis submitted to UC Berkeley, accessed online at http://www. ce.utexas.edu/prof/kockelman/public_html/mcpthesis.pdf, February 2010.

Kuzmyak, R., and Pratt, R. 2003. "TCRP Report Chapter 15: Land Use and Site Design-Traveler Response to Transport System Changes." Washington, D.C.: Transportation Research Board. Accessed online at http://gulliver.trb.org/publications/tcrp/tcrp_rpt_95c15.pdf, February 2010.

Levy, J. 2008. *Contemporary Urban Planning*. 8th ed. Upper Saddle River, NJ: Pearson Education.

Ling, O. 2005. *Sustainability and Cities: Concept and Assessment*. Singapore: World Scientific Printers, Ltd.

Litman, T. 2008a. "Measuring Transportation: Traffic, Mobility and Accessibility." Victoria, Canada: Victoria Transport Policy Institute. Accessed online at http://www.vtpi.org/measuring_transportation.pdf, March 2010.

Litman, T. 2008b. "Sustainable Transportation Indicators: A Recommended Research Program for Developing Sustainable Transportation Indicators and Data." TRB Sustainable Transportation Indicators.

Litman, T. 2009. "Introduction to Multi-Modal Transportation Planning: Principles and Practices." Victoria, Canada: Victoria Transport Policy Institute. Accessed online at http://www.vtpi.org/multimodal_planning.pdf, February 2010.

Litman, T. 2010. "Land Use Impacts on Transport: How Land Use Factors Affect Travel Behavior." Victoria, Canada: Victoria Transport Policy Institute. Accessed online at http://www.vtpi.org/landtravel.pdf, February 2010.

Lund, H., and Clark II, W. 2008. "Sustainable Energy and Transportation System Introduction and Overview." *Utilities Policy* 16:59–62.

Marshall Macklin Monaghan Ltd. 2007. "Best Practices Review: Sustainable Transportation Programs Across North America." Thornhill, Canada: Marshall Macklin Monaghan Ltd. Accessed online at http://www.york.ca/ NR/rdonlyres/mm574ixhu4x3nsnmetqbknywmkxkohl7dumtmvunvhdmdvarep3nysm3ffeyl5v5anlzx6s366egz4 p54gu2dikr5b/%28IP%29+TMP_Best+Practices_Apr07.pdf, May 2010.

Metro. 2010. "2035 Regional Transportation Plan: Final Draft Plan." Portland: Metro. Accessed online at http:// library.oregonmetro.gov/files//03_22_10_2035_rtp_final_web.pdf, March 2010.

Mihyeon Jeon, C., and Amekudzi, A. March 2005. "Addressing Sustainability in Transportation Systems: Definitions, Indicators, Metrics." *Journal of Infrastructure Systems* 11(1):31–51.

Moudon, A. V., Hess, P., Snyder, M. C., and Stanilov, K. 1997. "Effects of Site Design on Pedestrian Travel in Mixed-Use, Medium-Density Environments." Seattle: Washington State Transportation Center. Accessed online at: http://www.wsdot.wa.gov/Research/Reports/400/432.1.htm, March 2010.

Munier, N. 2005. *Introduction to Sustainability: Road to a Better Future*. Ottawa: Springer.

Netherlands Society for Nature and Environment. 2009. "Action Plan for Electric Transport: A Route to One Million Electric Cars in 2020!" Utrecht: Netherlands Society for Nature and Environment. Accessed online at http://www.snm.nl/pdf/090804_action_plan_electric_transport_en_def.pdf, February 2010.

Neumayer, E. 2003. *Weak versus Strong Sustainability: Exploring the Limits of Two Opposing Paradigms,* 2nd ed. Cheltenham, UK: Edward Elgar.

New Jersey Department of Transportation. 2005a. "Mobility Management." Trenton, NJ: NJ DOT. Access online at http://www.state.nj.us/transportation/works/njfit/partnership/, May 2010.

New Jersey Department of Transportation. 2005b. What Is the NJFIT Philosophy? Trenton, NJ: NJ DOT. Access online at http://www.state.nj.us/transportation/works/njfit/links/faq.shtm, May 2010.

New Jersey Department of Labor and Workforce development. 2009. "New Jersey Going Green." Accessed online at http://lwd.dol.state.nj.us/labor/lpa/pub/studyseries/njgreen.pdf, May 2010.

Next Alternative Inc. 2009. "Carbon Nanotube Lead Acid Battery." Ottawa, Ontario: Next Alternative Inc. Accessed online at http://www.next-alternative.com/product-batterya.html, June 2010.

Petiot, R. 2004. "Parking Enforcement and Travel Demand Management." *Transport Policy* 11: 399–411.

Pro, B., Hammerschlag, R., and Mazza, P. 2005. "Energy and Land Use Impacts of Sustainable Transportation Scenarios." *Journal of Cleaner Production* 13:1309–1319.

Ramani, T. L. , Zietsman, J., Eisele, W. L., Rosa, D., Spillane, D. L., and Bochner, B. S. 2009. "Developing Sustainable Performance Measures for TxDOT's Strategic Plan: Technical Report. 0-5541-1." Texas Transportation Institute, College Station, TX, April.

Shaw, S., and Xin, X. 2003. "Integrated Land Use and Transportation Interaction: A Temporal GIS Exploratory Data Analysis Approach." *Journal of Transport Geography* 11:103–115.

Schönfelder, S., Rich, J., Nielsen, O., Würtz, C., and Axhausen, K. 2007. "Road Pricing and Its Consequences for Individual Travel Patterns." *Mobilities* 2(1):75–98.

Shoup, D. 2005. *Parking Cash Out*. Chicago: American Planning Association.

Schumacher, E. 1973. *Small Is Beautiful: Economics as if People Mattered*. London: Blond and Briggs.

Smart Growth Network, International City/County Association. 2002. *Getting to Smart Growth: 100 Policies for Implementation*. Washington, D.C.: Smart Growth Network.

Transport Canada. 2006. Sustainable Community Awards: Sustainable Transportation 2006. Ottawa, Canada: Transport Canada http://www.tc.gc.ca/eng/programs/environment-utsp-awards-1065.htm, accessed online at May 2010.

Transportation Research Board. 2009. "Driving and the Built Environment: The Effects of Compact Development on Motorized Travel, Energy Use and CO_2 Emissions." Washington, D.C. special report 298. Accessed online at http://www.nap.edu/catalog/12747.html, February 2010.

Tunstall, D. 1992. *Developing Environmental Indicators: Definitions, Frameworks and Issues*. Background materials for the World Resources Institute, Workshop on Global Environmental Indicators, Washington D.C., December 7–8.

United Nations Environment Programme. Climate Change. http://www.unep.org/climatechange/.

United Nations World Commission on Environment and Development (UN WCED). 1987. *Our Common Future*. Oxford: Oxford University Press.

Urban Land Institute. 2009. "Transportation for a New Era: Growing More Sustainable Communities." Washington, D.C.

U.S. DOT. 1994. "Overview of Travel Demand Management Measures: Final Report." Washington, D.C.: U.S. DOT DOT-T-94-11 Accessed online at http://ntl.bts.gov/DOCS/273.html, March 2010.

U.S. DOT Federal Highway Administration. 2004a. "Mitigating Traffic Congestion: The Role of Demand-Side Strategies." Washington, D.C.: U.S. DOT Federal Highway Administration. Accessed online at http://www.ops.fhwa.dot.gov/ publications/mitig_traf_cong/index.htm, March 2010.

U.S. DOT Federal Highway Administration. 2004b. "Travel Demand Management." Washington, D.C.: U.S. Federal Highway Administration Accessed online at: http://ops.fhwa.dot.gov/aboutus/one_pagers/demand_mgmt.htm, February 2010.

U.S. DOT Federal Highway Administration. 2005. "Managing Demand trough Travel Information Services." Washington, D.C.: U.S. DOT Federal Highway Administration. Accessed online at http://www.ops.fhwa.dot.gov/publications/manag_demand_tis/travelinfo.htm, March 2010.

Ye, L., Mandpe, S., and Meyer, P. 2005. "What Is 'Smart Growth'—Really?" *Journal of Planning Literature* 19(3):301–317.

Waddell, P., Ulfarsson, G., Franklin, J., and Lobb, J. 2007. "Incorporating Land Use in Metropolitan Transportation Planning." *Transportation Research Part A* 41:382–410.

Wegener, M., and Fürst, F. 1999. "Land-Use Transport Interaction: State of Art." Dortmund: Institute for Raumplanung. Accessed online at http://129.3.20.41/eps/urb/papers/0409/0409005.pdf, February 2010.

Weller, B. 2006. "Sustainable Transport Practices in Canada: Exhortation Overwhelms Demonstration." Ottawa: Transport 2000 Canada. Accessed online at http://www.transport2000.ca/SustainableTransportPractices.pdf, May 2010.

Wittneben, B., Bongardt, D., Dalkmann, H., Sterk, W., and Baatz, C. January 2009. "Integrating Sustainable Transport Measures into the Clean Development Mechanism." *Transport Reviews* 29(1):91–113.

World Bank. 2010. "About Sustainable Development Network." http://go.worldbank.org/4G5284K5A0.

Zegras, C. July 2006. "Sustainable Transport Indicators and Assessment Methodologies." Background Paper for Plenary Session 4 at the Biannual Conference and Exhibit of the Clean Air Initiative for Latin American Cities.

OPERATIONS AND ECONOMICS

CHAPTER 10
TRANSPORTATION PLANNING AND MODELING

Konstadinos G. Goulias
University of California, Santa Barbara, California

10.1 INTRODUCTION

Transportation modeling and simulation aims at designing an efficient infrastructure and service to meet our needs for accessibility and mobility. At its heart is good understanding of human behavior that includes identification of the determinants of behavior and the change in human behavior when circumstances change either because of controls (e.g., taxation, land use controls, strategies to improve efficiency), trends (e.g., demographic change of resident population, immigration, workforce shifts, aging), or unexpectedly (e.g., natural and anthropogenic disasters). This understanding is a key ingredient for decisions in transportation planning and traffic operations. Since transportation systems are the backbone connecting the vital parts of a region, in-depth understanding of transportation-related human behavior is essential to the planning, design, and operational analysis of all the systems that make a region function.

Understanding human nature requires us to collect data and analyze and develop synthetic models of human agency in its most important dimensions and the most elemental constituent parts. This includes, and it is not limited to, understanding individual evolution along individual life cycle paths (e.g., from birth to entry in the labor force to retirement to death) and the complex interaction between an individual and the anthropogenic environment, natural environment, and the social environment. Travel behavior research is one aspect of analyzing human nature and aims at understanding how traveler values, norms, attitudes, and constraints lead to observed behavior. Traveler values and attitudes refer to motivational, cognitive, situational, and disposition factors determining human behavior. Travel behavior refers primarily to the modeling and analysis of travel demand, based on theories and analytical methods from a variety of scientific fields. These include, but are not limited to, the use of time and its allocation to activities and travel, methods to study this in a variety of time contexts and stages in the life of people, and the arrangement or artifacts and use of space at any level of social organization such as the individual, the household, the community, and other formal or informal groups (Goulias 2009a). This also includes the movement of goods and the provision of services having strong interfaces and relationships with the engagement in activities and the movement of persons.

Travel behavior analysis and synthesis can be examined from both objective (observed by an analyst) and subjective (perceived by the human) perspectives in an integrated manner among four dimensions of *time, geographic space, social space,* and *institutional context*. In a few occasions, the models reviewed here include and integrate time and space as conceived in science with perceptions of time and space by humans in their everyday life. Research that includes theory formation, data collection, modeling, inference, and simulation methods aims at the creation of decision support systems for policy assessment and evaluation combining different views of time and space.

Another objective of understanding individual and group travel behavior is conceptual integration. Explanation of facts from different perspectives can be considered jointly to form a comprehensive understanding of people and their groups and their interactions with the natural and built environments. In this way, we may see explanations of human behavior fusing into the same universal principles. These principles eventually will lead to testable hypotheses from different perspectives offering Wilson's, 1998, famous consilience among, for example, psychology, anthropology, economics, the natural sciences, geography, and engineering. Unavoidably this is a daunting task with many model propositions in the research domain and just a few ideas finding fertile ground in applications. The analysis-synthesis path in travel behavior offers methods that help us understand and predict behavior partially leaving many gaps (Timmermans 2003). However, policy questions are becoming increasingly impossible to address with old tools, a large pool of researchers is actively working on new methods, and many public agencies commenced a variety of tool development projects to fill the travel behavior analysis gaps. To capture these trends, we see modeling examples with ideas from a transdisciplinary viewpoint and contributors to modeling and simulation from a variety of merged backgrounds (e.g., see the evolution of ideas in a sequence of the International Association for Travel Behaviour Research conferences; www.public.asu.edu/~rpendyal/iatbr/iatbr_index.htm; accessed January 2010).

The impressive movement forward of transportation modeling and simulation emerges from three related but distinct sources. The first source is a fundamental change in planning practice that we could name *dynamic planning practice* to indicate the existence of bidirectional time (from the past to the future and from the future to today), as well as, assessment cycles and adjustments taking place within the short term, medium term, and long term horizons. These cycles are also bidirectional in time. A fundamental motivation that generates the substantive problems that we need to solve and the specific policies we need to examine is a vision for a *sustainable and green transport system*. Problems and solutions in this general area motivate and inspire contemporary substance and content of policies throughout the world. Three complementary and mutually strengthening directions are the *economy*, *environment*, and *society*—the three fundamental pillars of sustainability.

Planning today is heavily influenced by *dynamic thinking,* which means that time and change are intrinsic in the thought processes underlying planning activities. In the past, assumptions about the existence of a tenable and general equilibrium and our ability to build the infrastructure needed to meet demand did not require careful orchestration of actions. This was radically changed in the industrialized world to meet specific goals using available finite resources to maximize benefits. Together with our inability to build at will and a tendency to the preservation of nonrenewable resources (e.g., land and open space, fossil fuels, time) we are much more motivated to think strategically and to consider in a more careful way the performance of the overall anthropogenic system as we plan, design, operate, and manage transportation systems. Any action of this type, however, requires that we have a detailed and accurate picture of our facilities, their interconnectedness, their status within the hierarchy of movements, their conditions, and their evolving role. An accurate and more complete picture like this is called an *inventory*. This inventory includes the typical information about the resident population—demographics and employment, land availability and land uses, economic development and growth, and so forth. It contains data and relationships within the geographic area of interest (region) but also the region's relationship with other areas with which substantial flow of people, goods, and communication takes place. Inventories may also include data and information about cultural and historical factors. For example, statewide plans identify a variety of corridors as buffers of land and communities around major routes of the movement of people and goods. Some of these routes were created centuries ago when pioneers were still exploring uncharted lands. These routes experienced a major change when waterways were the main links among economic and military centers, and they are still evolving. Today these same routes contain as backbones railways, roadways, rivers, and often they surround major distribution locations such as ports and airports. Their nature is heavily influenced by their historical and cultural context and the path of their evolution.

Travel behavior analysts are familiar with inventories created for the regional long range plans, which subdivide the study area in traffic analysis zones with data from the Decennial Census suitably reformatted and packaged for use in a specific application (i.e., the long range regional transportation plan). Then, additional data are assigned to these same subdivisions to build a richer context for modeling and simulation. Thus, the inventory for a typical long range plan is an electronic map of where people live and work, the network(s) that connect different locations, availability of different modes on each segment of the network, as well as information about travel network performance (e.g., link capacities, speeds on links, congestion, and connectivity). The tool of choice for data storage and visualization is a Geographic Information System (GIS) (see Kwan 1994, 1997). One of the thorniest problems within this context is maintaining an up to date inventory (e.g., characteristics of the population in each zone, presence of certain types of businesses, location and characteristics of intermodal facilities). This is a particularly important issue for periods in between decennial censuses. Year-to-year updates are very often required to provide "fresh" data. Many of these updates are becoming widely available and much less expensive than in the past. For example, the inventory of the highway network, with suitable additions and improvements, is available from the same private providers of in-vehicle navigation systems. In a similar way, inventories of businesses and residences can also be purchased from vendors. Although the need for inventories is undoubtedly extremely important many important issues are yet to be resolved including levels of detail we should use in updating the data we have, treatment of errors in the data and model sensitivity to these errors, frequency of data updates and treatment of missing data, and questions about merging different databases. Investment on resolving some of these issues depends on the budget (time and money) available, consequences of errors in the data, and the use of models in decision making. Let us turn now to the core of the dynamic planning practice which is about strategy and performance.

Strategic planning and performance-based planning changed the way we plan for the future. This has been a 20-year-long process in the United States as its transportation policy at the federal, state, and metropolitan levels is shaped by three consecutive legislative initiatives (ISTEA, TEA-21, and SAFETEA-LU). Under all three legislative frameworks and independent of role, location, and perceived need for investment, the overall goal of funding allocation has been to maximize the performance of the transportation system in its entirety and avoid major new infrastructure building initiatives. As a result, planning practice at the federal, state, and local levels is becoming heavily performance-based and designed in a way that motivates the measurement of policy and program outcomes and judging these outcomes for funding allocation. Some states have created long range plans that are strategic and they measure transportation performance. Yearly evaluative updates are also used for a state's strategic transportation plan. After a comprehensive public involvement campaign a few themes capturing the desires of the resident population are first identified. To these themes, technical requirements based on planners and agency inputs are added, a large number of objectives are created, and then a variety of measures of performance are developed. These measures are given target levels that evolve over time to a desired future performance for the entire state and for a finite number of corridors of statewide significance. Yearly evaluations contain measures of target achievement and they should be used to guide an agency in its investments. The interface with regions is also included in this performance-based framework. Many infrastructure improvement projects in the United States are selected from lists of projects that regions (called Metropolitan Planning Organizations) submit to their state to be included in a list of projects in the Transportation Improvement Program (TIP) and become candidates for funding. Under statewide performance-based planning, these projects are evaluated with respect to their contribution in meeting the statewide performance measures and in some states the performance measures of the relevant corridor (NCHRP 446 2000). Although these examples are far-ranging in time and space, they contain operations components and yearly evaluations that: a) require data collection, modeling, and simulation at finer spatial and temporal scales than their counterpart planning feedbacks used in the long range transportation planning practice, and b) need methods that are able to coordinate the short-, medium-, and long-term impacts. Emerging from these considerations are questions about the types of consistency we need among geographic scales for planning and operations actions to perform evaluations, policy requirements for coordination among planning activities to ensure consistency, need for suitable methods to coordinate smaller projects in broader contexts (either of policy assessment or geographical area), development of tools required to perform measurement of impacts and program evaluation at the newly defined assessment cycles, and optimal planning activity with evaluation methods.

As illustrated later in this chapter, a new approach emerged in the past few years in which models of discrete choice are applied to individual decision makers that are then used to (micro)simulate most of the possible combinations of choices in a day. The result is, in essence, a synthetic generation of travel patterns. When the microsimulation also includes activities and duration at activity locations it becomes a synthetic schedule (called the activity-based approach to travel demand forecasting). In parallel, for forecasting purposes a synthetic population is first created for each land subdivision with all the relevant characteristics, and then models are applied to the residents of each subdivision to represent area-wide behavior. Changes are then imposed on each individual as a response to policies and predictive scenarios of policy impacts are thus developed. The evolution of individuals, their groups, and the entire study area can be used for trend analysis that includes details at the level of decision makers (either for passenger travel and/or for freight). In addition, progression in time happens from the present to the future and one could identify paths of change by individuals and groups if the application has been designed in the proper way (e.g., keeping detailed accounting of individuals as they move in time, using models that are designed for transitions over time and so forth). In a forecasting setting, progression in time follows calendar time, temporal resolution is most often a year, and the treatment of dynamics is a one-way causal stream to the future. Within the broader study of futures, forecasting is the method we use to develop *projective scenarios*. Performance-based planning, however, requires tools that can extrapolate from future performance targets the actions required today to reach them. In essence we also need *prospective studies* that start from a desirable future and move backwards to identify specific actions that will lead us to that prospect (see the review by Goulias 2009b).

10.2 POLICY AND PLANNING

Policy actions also view the world surrounding us as an integral ecosystem placing more emphasis on its overall survival by examining direct and indirect effects of individual policy actions and entire policy packages or programs (see the examples in Meyer and Miller 2001). This trend is not limited to transportation. Lomborg (2001) shows that a sustainable and green vision encompasses the entire range of human activity and the entirety of the ecosystem we live in. Although this is good news, because the approach enables analyses and policies that are consistent in their vision about futures, comprehensive views also reveal that the pace of economic growth and development clearly conflicts with the biological pace of evolution with unknown consequences (Tiezzi 2003), strengthening the view that more comprehensive analytical frameworks are required.

In fact, research syntheses that address the transportation and environment relationship by the Transportation Research Board of the National Academies (TRB 1999, 2002), expands the envelope to incorporate ecology and natural systems and addresses human health in a more comprehensive way than in the past reiterating the urgency to address unresolved issues about environmental damage. As a result, we also experience a clear shift to policy analysis approaches that have an expanded scope and domain and they are characterized by explicit recognition of transportation system complexity and uncertainty.

Reflecting all this, *sustainable transportation* is now often used to indicate a shift in the mentality of the community of transportation analysts to represent a vision of a transportation system that attempts to provide services that minimize harm to the environment. In fact, in one of the most comprehensive reviews of policies in North America, Meyer and Miller, 2001, contrast the nonsustainable to the sustainable approaches. They provide a compelling argument about the change in these policies and pathways toward a more sustainable path. In the United States the need to examine these new and more complex policy initiatives became increasingly pressing due to the passage of a series of legislative initiatives (Acts) and associated Federal and State regulations on transportation policy, planning, and programming. The multimodal character of the new legislation, its congestion management systems and air quality requirements for many U.S. regions motivated many new forecasting applications (Niemeier 2003). An added motivation is also lack of substantial funding for transportation improvement projects and a shift to charge the firms that benefit the most from transportation system improvements creating a need for impact fee-assessment for individual private developers. These assessments create the need for higher resolution in the three dimensions of geography (space), time (time of day), and social space (groups of people with common interests and missions, households, individuals) used in typical regional forecasting models but also the domain of jurisdictions where major decisions are made.

As Garrett and Wachs (1996) discuss in the context of a lawsuit against a regional planning agency in the Bay Area, traditional four-step regional simulation models (Creighton 1970; Hutchinson 1974; Ortuzar and Willumsen 2001) are outpaced by the same legislative stream of the past 20 years that defined many of the policies described above. Unlike the "energy crisis" of the 1970s, the urgency and timeliness of modeling and simulation is becoming more urgent, more complex, and requires an "integrated" approach. Under these initiatives, forecasting models, in addition to long-term land use trends and air quality impacts, need to also address issues related to technology use and information provision to travelers in the short and medium terms. The European Union focuses on similar issues as documented in van der Hoorn, 1997. Tables 10.1 and 10.2 provide an overview of policy tools that are loosely ordered from the longer term of land use and governance to medium and shorter-term operational improvements depending on the lag time required for their impacts to be realized.

These policy initiatives place more complex issues in the domain of regional policy analysis and forecasting and amplify the need for methods that produce forecasts at the individual traveler

TABLE 10.1 Typical Tools for the Planner and Operations Manager in Transportation

Type of policy tool	Brief description	Source of information*
Land use control	Legislation that controls for the growth of cities in sustainable paths	www.smartgrowth.org www.awcnet.org www.fhwa.dot.gov/planning/ppasg.htm www.planning.dot.gov/Documents/DomesticScan/domscan2.htm
	Land use design and attention to neighborhood design for nonmotorized travel	www.compassblueprint.org http://www.ecoiq.com/onlineresources/ center/
	City boundaries are divided into incorporated, within the sphere of influence, and external to manage growth	countypolicy.co.la.ca.us/BOSPolicyFrame.htm www.ite.org/activeliving/files/Jeff_Summary.pdf
Public Education and Involvement	Individualized Marketing	http://www.travelsmart.gov.au/ http://www.dft.gov.uk/pgr/sustainable/travelplans/
	Public involvement	www.fhwa.dot.gov/reports/pittd/contents.htm
	Health Promotion	www.activelivingbydesign.org
Emissions Control	Programs that shift taxation from traditional sources toward pollutant emissions and natural-resource depletion agents	www.fresh-energy.org/ www.fhwa.dot.gov/environment/ www.fightglobalwarming.com/
	Accelerated retirement of vehicles programs	http://www.cars.gov/
Pricing and Taxation	Congestion pricing and toll collection programs	www.vtpi.org/london.pdf
	Parking fee management to restrict access by space and time	http://onlinepubs.trb.org/onlinepubs/tcrp/tcrp_rpt_95c13.pdf

*Accessed January 2010.

TABLE 10.2 Typical Tools for the Planner and Operations Manager in Transportation

Type of policy tool	Brief description	Other source of information
Security	Preparedness through metropolitan planning processes	http://www.tsa.gov/what_we_do/index.shtm
Safety	Incorporate safety considerations in transportation planning	http://tsp.trb.org/
Time Shifting	Programs that change the workweek of individuals and firms	www.its.dot.gov/JPODOCS/REPTS_PR/13669/section05.htm
Operational Improvements	Goods movements (freight) programs to improve operations	ntl.bts.gov/DOCS/harvey.html
	Highway system improvements in traffic operations and flow	http://www.dot.ca.gov/dist11/operations/trfimprv.htm
	Intelligent Transportation Systems (ITS)	www.itsa.org/ www.ertico.com/ www.its.dot.gov/index.htm
	Special event planning and associated traffic management	http://ops.fhwa.dot.gov/publications/fhwahop07108/ pl_imp_op_eval.htm

*Accessed January 2010.

and her/his household levels instead of the traffic analysis zone level. In addition to the long-range planning activities and the typical traffic management activities, analysts and researchers in planning need to also evaluate traveler and transportation system manager information provision and use (e.g., location-based services, smart environments providing real-time information to travelers, vehicles, and operators), combinations of transportation management actions and their impacts (e.g., parking fee structures and city center restrictions, congestion pricing), and combinations of environmental policy actions (e.g., carbon taxes, health impacts of lack of bike and pedestrian facilities, and information campaigns about health effects of pollutants).

The tools to study and plan for these policies need to also have forecasting and backcasting/performance planning capabilities that are more accurate and detailed in space and time. In fact, planning initiatives are moving toward parcel by parcel analysis and yearly assessments. It is also conceivable that we need separate analyses for different seasons of a year and days of the week to capture seasonal and within a week variations of travel. Echoing all this and in the context of the Dutch reality Borgers et al. (1997) have identified five information need domains that the new envisioned policy analysis models will need to address and they are (in a modified format from the original list):

- Social and demographic trends that may produce a structural shift in the relationship between places and time allocation by individuals invalidating existing travel behavior model systems
- Increasing scheduling and location flexibility and degrees of freedom for individuals in conducting their everyday business leading to the need to consider additional choices (e.g., departure time from home, work at home, shopping by the internet, shifting activities to the weekend) in modeling travel behavior
- Changing quality and price of transport modes based on market dynamics and not on external to the travel behavior policies (e.g., the effect of deregulation in public transport)
- Shifting of attitudes and potential cycles in the population outlook about travel options
- Changing scales/jurisdictions (scale is the original term used to signify the different jurisdictions)– different policy actions in different sectors have direct and indirect effects on transportation and different policy actions in transportation have direct and indirect effects in the other sectors (typical example in the United States is the welfare to work program)

The first substantive implication of all these considerations is an expanded envelope of modeling and simulation. Many processes that were left outside the realm of transportation modeling and simulation need to be included as stages of the travel model system. One notable example is the inclusion of *residential location choice, work location choice, and school location choice* to capture the spatial distribution and relative location of important anchor points on travel behavior and to also capture the impact of transportation system availability and level of service on these choices. In this way when implemented policies lead to improved level of service and the relative attractiveness of locations change, shifts in residential location, work location, and possibly school location can be incorporated as impacts of transportation. A similar treatment is needed for *car ownership and car type choices* of households or *fleet sizes and composition* for firms. These choices are expressed as functions of parking availability, energy and other costs and level of service offered by the transportation system (highway and transit). To account for other resources and facilities available for household travel we also need to consider processes for *driver's licensing*, acquiring of *public transportation subscription (passes)*, and participation in *car-sharing programs*. In this way, variables of car availability and public transportation availability in households can be used as determinants of travel behavior. Similar treatment is required for policies that change attitudes, perceptions and knowledge about travel options.

To address some of the policies of Tables 10.1 and 10.2, we need to transition to a domain that contains a variety of outputs that include shares of program participation, sensitivity to accessibility and prices, and the usual indicators of travel on networks using input variables from the processes and behaviors discussed up to this point. Although the number of vehicles per hour per lane is the typical input of traffic operations software, a variety of other variables such as speeds on network links and types of vehicles are also needed for other models, such as emissions estimation.

Ideally, longer-term social, economic, demographic, and resource/facilities circumstances of people should be converted into yearly schedules identifying periods of vacation, workdays, special occasions, and so forth. These, in turn, should lead to weekly schedules separating days during which people stay at home from days during which people go to work and days during which they run errands and/or engage in other non-work and non-school-related activities. In this way, patterns of working days versus not working days can be derived in a natural (con)sequence. As we will see in a later section, a fundamental leap of faith intervenes in practice and converts all this background information into a representative day that is used to create a more or less complete sequence of activities and trips with their destinations and modes used.

In this way, decisions and choices people make are organized along the time scale in terms of the time it takes for these events to occur and their implications. For example, decisions about education, careers and occupation, and residential and job location are considered first and they condition everything that happens next. These should be formulated in terms of life-course long projects and not represented by a cross-sectional choice model. Similarly, decisions about yearly school and work schedules that determine work days and vacation days in a year are should also be modeled as a stream of interrelated choices. Conditional on all this are the daily schedules of individuals and the myriad of decisions determining a daily schedule, which are modeled in much more detail and paying closer attention to the mutual dependency among the different facets of a within a day schedule.

10.3 MODELING

Modeling made tremendous progress toward a comprehensive approach to build simulated worlds on computer enabling the study of complex policy scenarios (Henson et al. 2009). Although, passenger travel received the bulk of the attention, similar contributions to new research and technology are found in modeling the movement of goods (Southworth 2003; Stefan et al. 2005; Samimi et al. 2010). The emerging framework, although incomplete, is rich in the directions taken and has considerable potential for scientific discovery, policy analysis, and more comprehensive approaches in dealing with sustainability issues.

10.3.1   Modeling Dimensions

There are four dimensions characterizing simulation models. The first is the *geographic space* and its conditional continuity, the second is the *temporal scale* and calendar continuity, the third is inter-connectedness of *jurisdictions*, and the fourth and most important is the set of relationships in *social space* for individuals and their communities. The first dimension, *geographic space* here is intended as the physical space in which human action occurs. This dimension has played important roles in transportation planning and modeling because the first preoccupation of the transportation system designers has been to move persons from one location to another (i.e., overcoming spatial separa-tion). Initial applications considered the territory divided into large areas (traffic analysis zones), represented by a virtual center (centroid), and connected by facilities (higher level highways). The centroids were connected to the higher level facilities using a virtual connector summarizing the characteristics of all the local roads within the zone. As computational power increased and the types of policies/strategies required increased resolution, the zone became smaller and smaller. Today we expect software to handle zones that are as small as a parcel of land and transportation facilities that are as low in the hierarchy as a local road (the centroid becomes the building on a parcel and the centroid connector is the driveway of the unit, and they are no longer virtual).

In modeling and simulation we are interested in understanding human action. For this reason, in some applications, geographic space needs to consider more than just physical features (Golledge and Stimpson 1997, p. 387) moving us into the notion of place and social space. The second dimen-sion is *time* that is intended here as continuity of time, irreversibility of the temporal path, and the associated artificiality of the time period considered in many models. Models used in long-range planning applications use typical days (e.g., a summer day for air pollution). In many regional long-range models, the implied assumption is that we target a typical work weekday in developing models to assess policies. Households and their members, however, may not always (if at all) obey this strict definition of a typical weekday to schedule their activities and they may follow very different deci-sion making horizons in allocating time to activities within a day, spreading activities among many days including weekends, substituting out of home with in home activities in some days but doing exactly the opposite on others, and using telecommunications only selectively (e.g., on Fridays and Mondays more often than on other days). Obviously, taking into account these scheduling activities is by far more complex than what is allowed in existing transportation planning models. The third dimension is *jurisdictions* and their interconnectedness. The actions of each person are "regulated" by jurisdictions with different and overlapping domains such as federal agencies, state agencies, regional authorities, municipal governments, neighborhood associations, trade associations and societies, religious groups, and formal and informal networks of families and friends. In fact, the federal government defines many rules and regulations on environmental protection. These may end up being enforced by a local jurisdiction (e.g., a regional office of an agency within a city). On the one hand, we have an organized way of governance that clearly defines jurisdictions and policy domains (e.g., real estate tax collection in the United States). On the other hand, the relationships among jurisdictions and decision making about allocation of resources does not follow always this orderly governance principle of hierarchy. A somewhat different and more "bottom up" relationship is found in the social network and for this reason requires a different dimension that is the fourth and final dimension named *social space* and the relationships among persons within this space. For example, individuals from the same household living in a neighborhood may change their daily time allocation patterns and location visits to accommodate and/or take advantage of changes in the neighborhood and at the same time modify their activity and travel patterns because one member is actively engaging in another social network. Changes to the infrastructure and its management may motivate mutually canceling impacts because of the different networks. This may lead to the unintended consequence of policy failure.

One important domain and entity within this social space is the household. This has been a very popular unit of analysis in transportation planning recognizing that strong relationships within a household can be used to capture behavioral variation (e.g., the simplest method is to use a house-hold's characteristics as explanatory variables in a regression model of travel behavior). In this way, any changes in the household's characteristics (e.g., change in the composition because of birth, death, divorce, children leaving the nest, or adults moving into the household) can be used to predict changes in travel behavior. New model systems are created to study this interaction within a household looking at the patterns of using time in a day and the changes across days and years. It is therefore very important in modeling and simulation to incorporate in the models used for policy analysis interactions among these four fundamental dimensions, which bring us to the issue of *scale*.

The typical long-range planning analysis is defined for larger geographical areas (region, states, and countries) and addresses issues with horizons from 10 to 50 years. In many instances, we may find that large geographic scale means also longer timeframes applied to wider mosaics of social entities and including more diverse jurisdictions. On the other side of the spectrum, issues that are relevant to smaller geographic scales are most likely to be accompanied by shorter-term timeframes applied to a few social entities that are relatively homogeneous and subject to the rule of very few jurisdictions. This is one important organizing principle but also an indicator of the complex relationships we attempt to recreate in our computerized models for decision support. In developing the blueprints of these models, a variety of theories (e.g., neoclassical microeconomics) and conceptual representations of the real world can help us develop these models. At the heart of our understanding of how the world (as an organization, a household, a formal or informal group, or an individual human being) works are models of decision making and conceptual representations of relationships among entities making up this world.

Transportation planning applications are about judgment and decision making of individuals and their organizations. There are different settings of decision making that we want to understand. Three of these settings are the travelers and their social units from which motivations for and constraints to their behavior emerge, the transportation managers and their organizations that serve the travelers and their social units, and the decision makers surrounding goods movement and service provision that contain a few additional actors. These applications may also include land use markets (see www.urbanism.org). Travelers received considerable attention in transportation planning and the majority of the models in practice aim at capturing their decision making process. The remaining settings, e.g., airport and port travel, received much less attention and they are poorly understood and modeled.

Conceptual models of the interaction among agents are transformed into computerized models of a city, a region, or even a state in which we utilize components that are in turn models of human judgment and decision making, e.g., travelers moving around the transportation network and visiting locations where they can participate in activities. Models of this behavior are simplified versions of strategies used by travelers when they select among options that are directly related to their desired activities. In some of these models we also make assumptions about hierarchies of motivations, plans, actions, and consequences. Some of these assumptions are explicit, e.g., when deriving the functional forms of models as in the typical disaggregate choice models or the rules in a production system, these assumptions are implicit.

When designing transportation planning model interfaces for transportation planners and managers we also implicitly make assumptions about the managers' ability to understand the input, agent representation, internal functioning, and output of these computerized models. Our objective is therefore not only to understand travel behavior and build models that describe and predict human behavior but also to devise tools that allow transportation managers to understand the assumed behavior in the models, study scenarios of policy actions, and define and explain policy implications to others. This, in essence, implies that we, the model system designers, create a platform for a relationship between planners and travelers. A similar but more direct relationship also exists between travelers and transportation managers when we design the observation methods that provide the data for modeling but also the data used to measure attitudes and opinions such as travel surveys. In fact, this relationship is studied in much more detail in the survey design context and linked directly to the image of the agency conducting the survey and the positive or negative impression of the travelers about the sponsoring agency (Dillman 2000). Most transportation research for modeling and simulation, however, has emphasized traveler behavior when building surveys and their models neglecting the interface with the planners. The summary of theories below, however, applies to individuals traveling in a network but also to organizations and planners in the sense used by H.A. Simon in *Administrative Behavior* (1997).

10.3.2 Decision-Making Paradigms

Rational decision making is a label associated with human behavior that follows a strategy in identifying the best course of action. In summary, a decision maker solves an optimization problem and identifies the best existing solution to this problem. Within this more general strategy, when an operational model is needed and this operational model provides quantitative predictions about human behavior some kind of mathematical apparatus is needed to produce the predictions. One such machinery is the subjective expected utility (Savage 1954) formulation of human behavior. In developing alternative models to SEU, Simon (1983) defines four theoretical components:

- A person's decision is based on a utility function assigning a numerical value to each option—*existence and consideration of a cardinal utility function*

- The person defines an exhaustive set of alternative strategies among which just one will be selected—*ability to enumerate all strategies and their consequences*

- The person can build a probability distribution of all possible events and outcome for each alternate option—*infinite computational ability*

- The person selects the alternative that has the maximum utility—*maximizing utility behavior*

This behavioral paradigm served as the basis for a rich production of models in transportation that include the mode of travel, destinations to visit as well as the household residence (see the examples in the seminal textbook by Ben-Akiva and Lerman 1985). It served also as the theoretical framework for consumer choice models and for attempts to develop models for hypothetical situations (see the comprehensive book by Louviere, Hensher, and Swait 2000). It has also replaced the aggregate modeling approaches to travel demand analysis as the orthodoxy against which many old and new theories and applications are compared and compete with. SEU can be considered to be a model from within a somewhat larger family of models under the label of weighted additive rule (WADD) models (Payne, Bettman, and Johnson 1993). Real humans, however, may never behave according to SEU or related maximizing and infinitely computational capability models (Simon labels this the Olympian model, 1983). Based on exactly this argument, different researchers in psychology have proposed a variety of decision-making strategies (or heuristics). For example, Simon created alternate model paradigms under the label of *bounded rationality–the limited extent to which rational calculation can direct human behavior* (Simon 1983, 1997) to depict a sequence of a person's actions when searching for a suitable alternative. The modeled human is allowed to make mistakes in this search giving a more realistic description of observed behavior (see also Rubinstein 1998). Tversky is credited with another stream of decision making models starting with the *lexicographic approach* (1969), in which *a person first identifies the most important attribute, compares all alternatives on the value of this attribute, and chooses the alternative with the best value on this most important attribute*. Ties are resolved in a hierarchical system of attributes. Another Tversky model (1972) assumes *a person selects an attribute in a probabilistic way and influenced by the importance of the attribute, all alternatives that do not meet a minimum criterion value (cutoff point) are eliminated*. The process proceeds with all other attributes until just one alternative is left and that one is the chosen. This has been named the *elimination by aspects strategies* (EBA) model. Later, Kahneman and Tversky (1979) developed *prospect theory* and its subsequent version of *cumulative prospect theory* in Tversky and Kahneman (1992) in which a simplification step is first undertaken by the decision maker editing the alternatives. Then, a value is assigned to each outcome and *a decision is made based on the sum of values multiplying each by a decision weight*. Losses and gains are treated differently. All these alternatives to SEU paradigms did not go unnoticed in transportation research with early significant applications appearing in the late 1980s. In fact, more than a decade ago, a conference was organized attracting a few of the most notable research contributors to summarize the state of the art in behavior paradigms and documented in Garling, Laitila, and Westin (1998). One of the earlier examples using another of Simon's inventions, the *satisficing behavior—acceptance of viable choices the may not be optimal*—is a series of transportation-specific applications described in Mahmassani and Herman (1990). Subsequent contributions continue along the path of more realistic models and the most recent example, discussing a few models, by Avineri and Prashker (2003), uses cumulative prospect theory giving a preview of a movement toward more realistic travel behavior models. As Garling, Laitila, and Westin (1998) and Avineri and Prashker (2003) point out, these paradigms are not ready for practical applications, contrary to the Mahmassani and colleagues efforts that have been applied, and additional work is required to use them in a simulation framework for applications. In addition, Payne, Bettman, and Johnson (1993) provide an excellent review of these models, a summary of the differentiating aspects among the paradigms. They also provide evidence that decision makers *adapt* by switching between decision making paradigms *to the task and the context of their choices*. They also make mistakes and they may also fail to switch strategies. As Vause (1997) discusses to some length transportation applications are possible using multiple decision making heuristics within the same general framework and employing a production system approach (Newell and Simon 1972). A key consideration, however, that has received little attention in transportation is the definition of context within which decision making takes place. Recent production systems (Arentze and Timmermans 2000) are significant improvements over past simulation techniques. However, travelers are still assumed to be passive in shaping the environment within which they decide to act (action space). This action space is viewed as largely made by constraints and not by their active shaping of their context within wider time intervals. Goulias (2001, 2003, 2009a) reviews and builds another framework from human development and social ecology perspective that is designed to treat decision makers in their active and passive roles and explicitly accounts for mutual influence between an agent (active autonomous decision maker) and his or her environment in the life course.

10.3.3 Transportation Models

Transportation modeling and simulation for planning experienced a few tremendously innovative steps forward. Interestingly these key innovations are from nonengineering fields but very often transferred and applied to transportation systems analysis and simulation by engineers. One major push forward is the development of random utility models. At exactly the time that the Bay Area Rapid Transit system was studied and evaluated in the 1960s, Dan McFadden (the Year 2000 Nobel Laureate in Economics) and a team of researchers produced practical mode choice regression models at the level of an individual decision maker (see http://emlab.berkeley.edu/users/mcfadden/; accessed June 2007). The models are based on random utility maximization (of the SEU family) and their work opened up the possibility to predict mode choice rates more accurately than ever before. These models were initially named *behavioral travel-demand models* (Stopher and Meyburg 1976) and later the more appropriate term of *discrete choice models* (Ben-Akiva and Lerman 1985) prevailed. Although restrictive in their assumptions, these models are still under continuous improvement and they have become the standard tool in evaluating discrete choices. Some of the most notable and recent developments advancing the state of the art and practice are:

- Better understanding of the theoretical and particularly behavioral limitations of these models (Garling, Laitila, and Westin 1998; McFadden 1998; Golledge and Garling 2003).

- More flexible functional forms that resolve some of the problems raised in Williams and Ortuzar (1982) allowing for different choices to be correlated when using the most popular discrete choice regression models (Koppelman and Sethi 2000; Bhat 2000, 2003).

- Combination of revealed preference, stated choices by travelers, with stated preferences and intentions, answers to hypothetical questions by travelers, availability of data in the same choice framework to extract in a more informative way travelers willingness to use a mode and willingness to pay for a mode option (Ben-Akiva and Morikawa 1989; Louviere, Hensher, and Swait 2000). This latter "improvement" enables us to assess situations that are impossible to build in the real world.

- Computer-based interviewing and laboratory experimentation to study more complex choice situations and the transfer of the findings to the real world (Mahmassani and Jou 1998). This direction, however, is also accompanied by a wide variety of research studies aiming at more realistic behavioral models that go beyond mode choice and travel behavior (Golledge and Garling 2003) and the use of technologies to enhance our understanding of information use (Auld et al. 2009).

- Expansion of the discrete choice framework using ideas from *latent class models* with covariates that were first developed by Lazarsfeld in the 1950s and their estimation finalized by Goodman in the 1970s (see the review in Goodman 2002, and discrete choice applications in Bockenholdt 2002). This family of models was used in Goulias (1999) to study the dynamics of activity and travel behavior and in the study of choice in travel behavior (Ben-Akiva et al. 2002).

As mentioned earlier, the rational economic assumption of the maximum utility model framework (that underlies many but not all of the applied disaggregate models in transportation) is very restrictive and does not appear to be a descriptive behavioral model except for a few special circumstances when the framing of decisions is carefully designed (something we cannot expect to happen every time a person travels on the network). Its replacement, however, requires conceptual models that can provide the types of outputs needed in regional planning applications.

A few additional research paths, labeled as *studies of constraints*, are also functioning as gateways into alternate approaches to replace or complement the more restrictive utility-based models. A few of these models also consider knowledge and information provision to travelers. The first aspect we consider is about the choice set in discrete choice models. Choice set is the set of alternatives from which the decision maker selects one. These alternatives need to be mutually exclusive, exhaustive, and finite in number (Train 2003). Identification, counting, and issues related to the alternatives considered have motivated considerable research in choice set formation (Richardson 1982; Swait and Ben-Akiva 1987a, 1987b; Horowitz 1991; Horowitz and Louviere 1995). Key threat to misspecification of the choice set is the potential for incorrect predictions (Thill 1992). When this is an issue of considerable threat as in destination choice models where the alternatives are numerous, a model of choice set formation appears to be the additional burden (Haab and Hicks 1997). Other methods, however, also exist and they may provide additional information about the decision making processes. Models of the processes can be designed to match the study of specific policies in specific contexts. One such example and a more comprehensive approach defining the choice sets is the situational approach (Brög and Erl 1989). The method uses in depth information from survey respondents to derive sets of reasons for which alternatives are not considered for specific choice settings (individual trips). This allows separation of analyst-observed system availability from user-perceived system availability (e.g., caused by misinformation and willingness to consider information). This brings us to the duality between "objective choice attributes" and "subjective choice attributes." Most transportation applications, independently of the decision making paradigm adopted, assume the analysts (modelers) measured attributes and the travelers used attributes in deciding to be the same. Due to differential perception these attribute values may not be the same. Modeling the process of perceived constraints may be far more complex when one considers the influence of the context within which decisions are made. Golledge and Stimpson (1997, pp. 33–34) describe this within a conceptual model of decision making that has a cognitive feel to it. They also link the situational approach to the activity-based framework of travel extending the framework further (pp. 315–328) to also include the use of time.

Chapin's research (1974), providing one of the first comprehensive studies about time allocated to activity in space and time, is also credited for motivating the foundations of activity-based approaches to travel demand analysis. His focus has been on the propensity of individuals to participate in activities and travel linking their patterns to urban planning. At about the same time Becker also developed his theory of time allocation from a household production viewpoint (Becker 1976) applying economic theory in a nonmarket sector and demonstrating the possibility of formulating time allocation models using economics reasoning (i.e., activity choice). In parallel another approach was developing in geography and Hagerstrand's seminal publication on time space geography (1970) presents the foundations of the approach. The idea of constraints in the movement of persons was taken a step further by this time-geography school in Lund. In that framework, the movement of persons among locations can be viewed as their movement in space and time under external constraints. Movement in time is viewed as the one way (irreversible) movement in the path while space is viewed as a three dimensional domain. It provides the third base about *constraints* in human paths

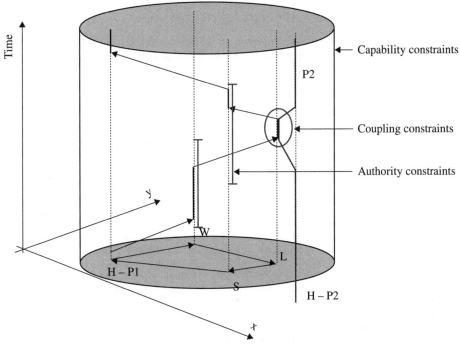

FIGURE 10.1 A two-person (P1 and P2) activity-travel pattern and the time and space limits imposed by constraints. H indicates home, W indicates work, L indicates leisure, and S indicates shopping (Source: Pribyl 2004).

in time and space for a variety of planning horizons. These are *capability constraints* (e.g., physical limitations such as speed); *coupling constraints* (e.g., requirements to be with other persons at the same time and place); and *authority constraints* (e.g., restrictions caused by institutional and regulatory contexts such as the opening and closing hours of stores). Figure 10.1 provides a pictorial representation in space and time of a typical activity-travel pattern of two persons (P1 and P2) and the three types of constraints.

Cullen and Godson (1975), also reviewed by Arentze and Timmermans (2000) and Golledge and Stimpson (1997), appear to be the first researchers attempting to bridge the gap between the motivational (Chapin) approach to activity participation and the constraints (Hagerstrand) approach by creating a model that depicts a routine and deliberated approach to activity analysis. The Cullen and Dobson study also defined many terms often used today in activity-based approaches. For example, each activity (stay-home, work, leisure, and shopping) is an episode characterized by start time, duration, and end time. Activities are also classified into fixed and flexible, and they can be engaged alone or with others. Moreover, they also analyzed sequencing of activities as well as pre-planned, routine, and on the spur of the moment activities. Within this overall theoretical framework is the idea of a project which according to Golledge and Stimpson (1997) *is a set of linked tasks that are undertaken somewhere at some time within a constraining environment* (pp. 268–269). This idea of the project underlies one of the most exciting developments in activity-based approaches to travel demand analysis and forecasting because seemingly unrelated activity and trip episodes can be viewed as part of a "big picture" and given meaning and purpose to complete models of human agency and explain resistance-to-change behavior.

Most subsequent contributions to the activity-based approach emerge in one way or another from these initial frameworks with important operational improvements (for reviews see Kitamura 1988; Bhat and Koppelman 1999; Arentze and Timmermans 2000; and McNally 2000). The basic ingredients of an activity based approach for travel demand analysis (Jones, Koppelman, and Orfeuil 1990 and Arentze and Timmermans 2000) are:

- Explicit treatment of travel as derived demand (Manheim 1979), i.e., participation in activities such as work, shop, and leisure motivate travel but travel could also be an activity as well (e.g., taking a drive). These activities are viewed as episodes (characterized by starting time, duration, and ending time) and they are arranged in a sequence forming a pattern of behavior that can be distinguished from other patterns (a sequence of activities in a chain of episodes). In addition, these events are not independent, and their interdependency is accounted for in the theoretical framework.

- The household is considered to be the fundamental social unit (decision making unit) and the interactions among household members are explicitly modeled to capture task allocation and roles within the household, relationships at one time point and change in these relationships as households move along their life-cycle stages and the individual's commitments and constraints change and these are depicted in the activity-based model.

- Explicit consideration of constraints by the spatial, temporal, and social dimensions of the environment is given. These constraints can be explicit models of time-space prisms (Lenntorp 1976; Pendyala 2003) or reflections of these constraints in the form of model parameters and/or rules in a production system format (Arentze and Timmermans 2000).

Input to these models are the typical regional model data of social, economic, and demographic information of potential travelers and land use information to create schedules followed by people in their everyday life. The output are detailed lists of activities pursued, times spent in each activity, and travel information from activity to activity (including travel time, mode used, and so forth). This output is very much like a "day-timer" for each person in a given region. Figure 10.2 provides an example of time allocation to different activities from an application that collected activity participation data (Alam 1998; Alam and Goulias 1999). It displays time allocation by one segment of the population showing the proportion of persons engaging in each activity by each hour of a day. Using this type of data and information about building allocation to uses, models are used to create the output shown in Figure 10.3 shows the output from a model that predicts the presence of persons in each building during each hour of a day engaging in each activity type. Combining an activity model with a typical travel demand model produces "volumes" of individuals at specific locations and on the network of a city as shown in Figure 10.4 (a more detailed description of this study can be found in Kuhnau and Goulias 2003 and Kuhnau 2001).

Many planning and modeling applications aim at forecasting. Inherent in forecasting are the time changes in the behavior of individuals and their households and their response to policy actions. At the heart of behavioral change are questions about the process followed in shifting from a given pattern of behavior to another. In addition to measuring change and the relationships among behavioral indicators that change in their values over time, we are also interested in the timing, sequencing, and staging of these changes. Moreover, we are interested in the triggers that may accelerate desirable or delay undesirable changes and the identification of social and demographic segments that may follow one time path versus another in systematic patterns. Knowledge about all this is required to collect data, design policies but it is also required to design better forecasting tools to assess future impacts of policies. Developments in exploring behavioral dynamics and advancing models for them have progressed in a few arenas. First, in the *data collection arena* with panel surveys, repeated observation of the same persons over time that are now giving us a considerable history in developing new ideas about data collection but also about data analysis (Golob, Kitamura, and Long 1997; Goulias and Kim 2003) and interactive and laboratory data collection techniques (Doherty 2003) that allow a more in-depth examination of behavioral processes. The second arena is in the *development of (microeconomic) dynamic formulations* for travel behavior that challenge conventional assumptions and offer alternative formulations (Kitamura 2000). The third arena, is in the behavior from a developmental viewpoint as a single stochastic process, a staged development process (Goulias 1999), or as the outcome from multiple processes operating at different levels (Goulias 2002). Experimentation with new theories from social psychology emphasizing

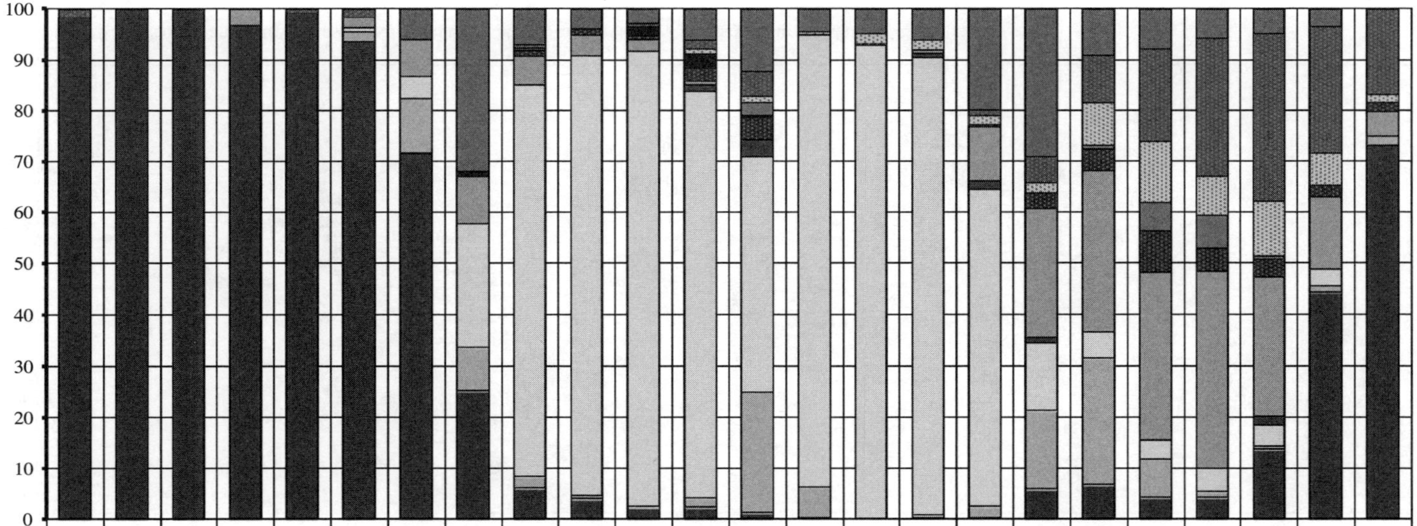

FIGURE 10.2 Time allocation to different activities in a day by staff members of a university.
A: Personal Needs (includes sleep), B: Eat meal, C: Paid work, D: Education, E: Household and family care, F: shopping, G: medical, H: Volunteering/Community, I: Socializing, J: Sports and Hobbies, K: Travel, L: All other. (Source: Alam 1998).

	1	2	3	4	5	6	7	8	9	10	11	12	13	14	15	16	17	18	19	20	21	22	23	24
L	0.0	0.0	0.0	0.0	0.0	0.0	0.0	0.0	0.0	0.0	0.0	0.0	0.0	0.0	0.0	0.0	0.0	0.0	0.0	0.0	0.0	0.0	0.0	0.0
K	0.0	0.0	0.0	0.0	0.0	0.0	0.0	31.9	7.0	4.0	2.9	6.3	12.3	4.6	5.1	6.3	20.0	29.2	9.2	8.0	5.9	4.9	3.5	0.1
J	1.6	0.0	0.0	0.0	0.8	1.6	0.0	0.0	0.8	0.0	0.7	1.6	4.9	0.0	0.0	0.0	1.1	5.2	9.4	18.2	27.0	33.0	25.0	16.8
I	0.0	0.0	0.0	0.0	0.0	0.0	0.0	0.0	0.1	0.0	0.0	1.0	1.2	0.4	1.6	1.6	1.6	1.8	8.2	11.9	7.8	10.7	6.1	1.6
H	0.0	0.0	0.0	0.0	0.0	0.0	0.1	0.3	0.3	0.0	0.0	0.1	2.6	0.0	0.0	0.3	0.0	0.0	1.2	5.9	6.4	0.8	0.0	0.0
G	0.0	0.0	0.0	0.0	0.0	0.0	0.0	0.8	0.3	0.3	1.8	2.9	0.5	0.0	0.0	0.0	0.0	0.0	0.0	0.0	0.0	0.0	0.0	0.0
F	0.0	0.0	0.0	0.0	0.0	0.0	0.0	0.0	0.7	1.0	0.8	2.4	4.4	0.0	0.4	0.0	0.5	3.2	3.7	8.0	4.5	3.3	2.5	1.6
E	0.0	0.0	0.0	3.3	0.0	2.0	7.1	9.4	5.7	4.0	2.3	0.8	0.0	0.0	0.0	0.7	10.7	25.0	31.7	32.5	38.3	27.0	14.2	4.9
D	0.0	0.0	0.0	0.0	0.0	0.0	0.0	0.0	0.0	0.0	0.0	1.2	0.0	0.3	0.0	0.7	1.6	1.2	0.0	0.0	0.0	1.6	0.0	0.0
C	0.0	0.0	0.0	0.0	0.0	0.8	4.5	24.1	76.9	86.6	89.3	79.9	46.2	88.7	92.9	89.9	62.2	13.3	5.3	4.1	4.9	4.6	3.5	0.0
B	0.0	0.0	0.0	0.0	0.0	2.0	10.8	8.6	2.5	0.5	0.1	1.8	23.8	6.0	0.0	0.5	2.0	15.4	24.7	7.5	1.4	0.6	1.2	2.0
A	98.4	100.0	100.0	96.7	99.2	93.4	71.5	24.9	5.7	3.7	2.0	2.0	0.8	0.0	0.0	0.0	0.1	5.7	6.5	4.0	3.8	13.4	44.0	72.8

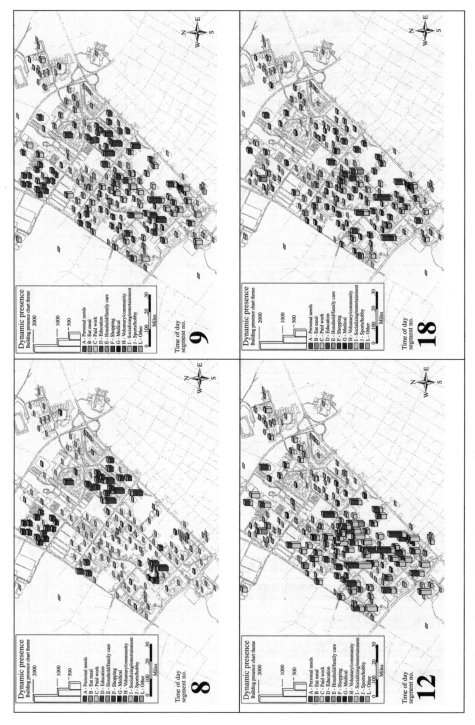

FIGURE 10.3 Persons and activities assigned to buildings at four different times in a day (Source: Alam 1998).

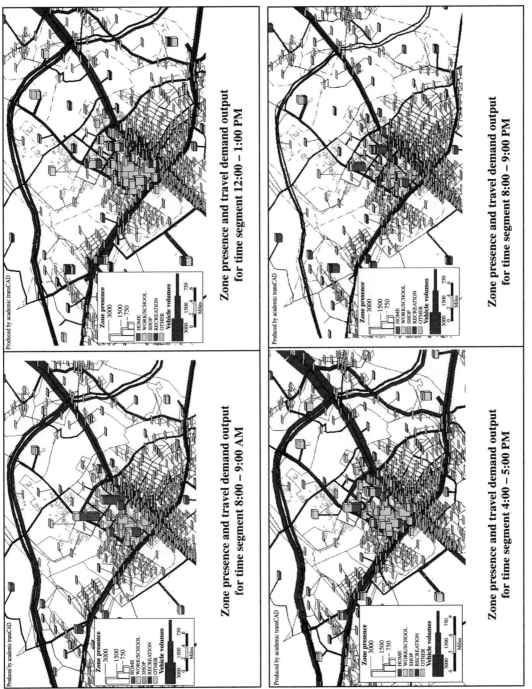

FIGURE 10.4 Persons and activities assigned to buildings and travel to the network in four different time period in a day (Source: Goulias, Zekkos, and Eom 2004).

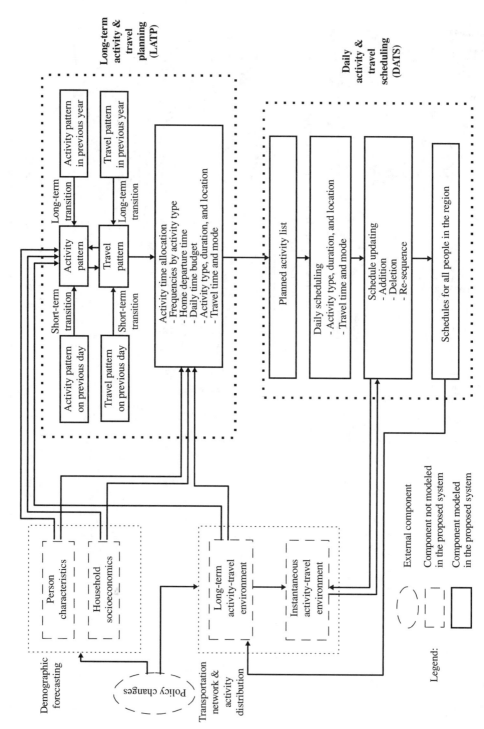

FIGURE 10.5 System structure of a dynamic travel model (Source: Ma 1997).

development dynamics is a potential fourth area that is just beginning to emerge (Goulias 2003; 2009a). Behavioral dynamics are also examined using more comprehensive analyses (Goulias et al. 2007) and models (Ramadurai and Srinivasan 2006).

An example of a dynamic model system is Figure 10.5, in which the model consists of two major components, each corresponding to a different decision stage. The first component is the long-term activity-travel planning (LATP). This component is based on identification of a few representative patterns of behavior from observed behavior. Using these *activity and travel pattern,* individuals are classified into relatively few homogenous groups. The probability of belonging to each group is a function other exogenous to the system sociodemographic characteristics that can be forecasted using microsimulation (see Goulias and Kitamura 1992 and Chung and Goulias 1997). Implied in this model system is that individuals are more likely to start their day by choosing a general pattern of behavior defined in approximate terms. For example, a person will decide whether to follow a typical working day pattern or a non-working day pattern without knowing the detailed timing of activities, destinations, modes to use, and so forth. An individual's choice of activity and travel patterns is considered to be a function of the person's contemporary characteristics, household socio-economics, and long-term activity-travel environment. The long-term activity-travel environment is introduced here to differentiate from the instantaneous activity-travel environment, e.g., congestion at specific locations and times of a day. The former is referred to a situation that the environment is evaluated based on a longer time interval (e.g., a day) while the latter on a finer time slice (e.g., an hour). A day's pattern choice also depends on the distant and immediate past. The distant past (e.g., one year in this scheme but expandable) captures behavioral inertia or habit, whereas the immediate past (i.e., one day) captures behavioral variation embedded in a weekly cycle. As the demand for travel is derived from the desire to participate in activities, travel pattern choice is considered to be a subsequent decision made after activity pattern choice. Thus, travel choice is dependent on, in addition to person characteristics, household sociodemographics, activity and travel environment, and the activity pattern choice in the past and at present. The use of this model system (LATP) up to this stage is suitable for long-range regional forecasting because of its ability to replicate policy responses in daily summary (synoptic measures of patterns) terms.

Given the predicted long-term activity and travel patterns, LATP then determines the expected daily activity and travel schedule for each person. This expected daily schedule may be viewed as an average schedule of all possible schedules a person may have had and/or considered. It should be noted here that the expected schedule is not a real schedule that can be expressed as an activity sequence with detailed timing. Rather, it is a summary of activity participation with a capability of linking all activities together. In other words, due to behavioral variation, a person's daily schedule will change from day to day, but it is always fluctuating around an average or typical pattern. This pattern is the expected schedule. The expected schedule is generated by a time allocation system that properly allocates activity times within a day. A person's expected daily schedule is a function of his characteristics, household socioeconomics, long-term activity-travel environment, and his long-term activity and travel pattern membership. The output of the expected schedule, is a group of potentially linked activities with timing and other information. This schedule constitutes activity frequencies of subsistence, maintenance, out-of-home leisure and returning home activities. It also contains information on when a person leaves home for an out-of-home activity (i.e., home departure time), total daily time the person is willing to spend on activity and travel including staying at home during the day (i.e., daily time budget), duration of each activity, and travel time of each trip.

The second component in the system structure is the daily activity and travel scheduling (DATS), which resembles the short-term choice in activity participation (see Figure 10.5). Conditional on the activity and travel pattern membership generated from LATP, DATS assumes that an individual makes his or her daily schedules according to the following three steps: (i) At the beginning of a given day, a person decides what types of activities he or she needs to pursue. These activities are called the planned activity list. (ii) Knowing this activity list, the person makes a planned schedule based on his or her experience. The individual may add some unplanned activities or delete some planned activities to best reflect his or her needs on that day. (iii) The person starts to execute the items one by one according to the planned schedule. Because of uncertainty involved in the decision-making process, at the end of each activity, the person evaluates his of her current situation against the plan and decides whether or not to update the planned schedule. These three steps lead to three subsystems of DATS: planned activity generating, daily scheduling, and schedule updating.

At the first step, the exact number of activities on the target day is determined. This represents the tasks a person plans to accomplish and is determined based on his or her expected schedule with a random term. The first expected schedule represents the person's general behavior while the random term allows for daily variation. The second step—daily scheduling—is to order all activities and travel on the planned activity list and determine their timing. Instead of searching for an optimal schedule from a set of combinatorial schedules, scheduling at this step is considered a stochastic process, which orders the activities in a probabilistic manner. Adding unplanned activities to and/or deleting activities from the planned activity list is part of the scheduling process. Yet, unlike the production system, which seeks a suboptimal solution through an interactive process, the scheduling process proposed here relies on the bounded rationality decision theory, i.e., a person will choose an alternative that meets his or her specific requirements. Therefore, if a generated activity schedule meets the time constraints in terms of total travel time or time to return home, then the schedule (called a feasible schedule) is considered to be the planned schedule and the scheduling process terminates. If, however, the schedule is infeasible (e.g., return-home time is later than that on the activity list), the scheduling process will generate another schedule until the requirements on the activity list are met or a predetermined number of generated schedules is reached. After obtaining the planned schedule, the person begins to execute these activities sequentially. Because of the uncertainty involved in the decision-making process and information gained during his or her activity participation, the actual activity duration and travel time may be different from the planned counterpart. For example, a shopping activity may be prolonged when a person has to spend more time on looking for some unusual items, or longer than expected travel time may be encountered when an accident occurs on the roadway. Moreover, individuals' habits or tastes are a source of variation in activity or travel time, even when the individuals may have exactly the same observed characteristics. These differences suggest that a more appropriate treatment of activity and travel time is to add a random value to the corresponding planned time. The random value accounts for the uncertainty in decision making and unexplained variation among individuals. As a result, the actual time becomes unexpected, which may cause a delay or advance in the planned schedule. This requires modifying the remaining activities on the schedule. In addition, while a person is executing an activity, the individual may gain some additional information that leads him or her to reschedule the remainder of the activities to best fit his or her need. For example, an overcrowded location (a result of many people choosing the same location for activity participation) may cause a person to change this expected location to a less-preferred place. This may also be viewed as the impact imposed by others through the instantaneous activity-travel environment. For this reason, a person is given an opportunity to modify the planned schedule when the person completes an activity and is about to start the next one. It should be noted that any schedule updating will affect the instantaneous activity-travel environment, which in turn, may induce further updating.

Descriptive measures of the long-term activity-travel environment obtained from these schedules may be fed back to the forecasting system. Totaling the actual schedules of all individuals is a rich source of data with which the actual number of people at a particular location at particular times can be obtained. This aggregation of patterns provides an origin-destination (OD) matrix for that time period, which is key to assessing time-sensitive policy analyses such as congestion pricing, zonal restrictions, changing work weeks, and access management. Assigning these OD matrices using a traffic assignment package would give the same types of, but more accurate information than, all of the current travel demand analysis models. The combination of LATP and DATS is repeatedly applied by simulating individuals and their characteristics as well as the context and environment within which they act. This is a typical example of stochastic dynamic microsimulation.

Another microsimulation that uses econometric models to simulate daily activity travel patterns for individuals and uses the time allocation-scheduling approach is the Comprehensive Econometric Microsimulator for Daily Activity-travel Patterns (CEMDAP) model (Bhat et al. 2003), which is also based on land use, sociodemographic, activity system, and level-of-service (LOS) attributes. Key distinctive element of CEMDAP is its reliance on hazard-based regression models to account for the continuous nature time of activity duration. Initially released in 2003, it is continually being expanded. The current version of CEMDAP complemented by a variety of components and name CEMUS includes population synthesis as well as the activity-pattern generation and scheduling of children, which is missing from many other simulators. CEMDAP is used currently to develop a new generation model system called SimAGENT for the entire Southern California Association of Governments is which 18 million persons reside. Another model that utilizes microsimulation is the Florida Activity Mobility Simulator (FAMOS) (Pendyala et al. 2005). FAMOS encompasses two modules: the Household Attributes Generation System (HAGS) and PCATS. Together, they comprise a system for modeling the activity patterns of individuals in Florida. The output is a series of activity-travel records. FAMOS is currently being further enhanced to include intra-household interactions and capture task allocation behavior among household members and is also integrated with other models as described later in this chapter. In a similar development path are the models AURORA by Arentze et al. (2006) FEATHERS (Forecasting Evolutionary Activity-Travel of Household and their Environmental Repercussions) that simulates activity-level scheduling decisions, within-a-day rescheduling, and learning processes in high resolutions of time and space, and from the same Dutch team also MERLIN and RAMBLAS (Veldhuisen, Timmermans, and Kapoen 2000).

Microsimulations have continued to gain in popularity in the activity-based modeling universe as they move from research applications to practice. Examples include the Portland Daily Activity Schedule Model, New York's "Best Practice" Model (2002) and the Mid-Ohio Regional Planning Commission (MORPC) Model (2003), both developed by Vovsha, Peterson, and Donnelly (2002, 2003), and the San Francisco model (Jonnalagadda et al. 2001) are currently being utilized by their respective MPO. The San Francisco model also includes destination choice models and is calibrated using household and census data. Four other models for Atlanta, Sacramento, the San Francisco Bay Area, and Denver are also microsimulation based (Bradley and Bowman 2006).

All these models focus on the paths of persons in space and time within a somewhat short time horizon such a day, week, or maybe a month. The consideration of behavioral dynamics has expanded the temporal horizons to a few years. However, regional simulation models are very often designed for long-range plans spanning 25 years or even longer time horizons. Within these longer horizons, changes in the spatial distribution of activity locations and residences (land use) are substantial, changes in the demographic composition and spatial distribution of demographic segments are also substantial, and changes in travel patterns, transport facilities, and quality of service offered can be extreme.

Past approaches in modeling and simulating the relationship among land use, demographics, and travel in a region attempted to disengage travel from land use and demographics and treat them as mutually exogenous in a cascading system. Interactions among them are scientifically more interesting and from a policy viewpoint pressing because of urban sprawl and suburban congestion. For these reasons, increasing attention is paid to their complex interdependencies. This has led to a variety of attempts to develop "integrated model systems" that enable the study of scenarios of change and mutual influence between land use and travel. An earlier review of these models with heavy emphasis on discrete choice models can be found in Anas (1982). Miller (2003) and Waddell and Ulfarsson (2003), 20 years later, provide two comprehensive reviews of models that have integrated many aspects in the interdependent triad of demographics-travel-land use models. Both reviews trace the history of some of the most notable developments and both link these models to the activity-based approach above. Both reviews also agree that a microeconomic and/or macroeconomic approach to modeling land and transportation interactions are not sufficient and more detailed simulation of the individuals and their organizations "acting" in an time-space domain need to be simulated in order to obtain the required output for informed decision making. They also introduce the idea of simulating interactive agents in a dynamic environment of other agents (multiagent simulation). The vast literature is reviewed by Timmermans 2003 and Miller 2006, from different viewpoints about progress made until now. However, they both agree that progress is rapidly made and that integration of land use and transportation models needs to move forward. Creation of integrated systems is further complicated by the emergence of an entire infrastructural system as another layer of human activity—telecommunication. Today telecommunication and transportation relationships are mostly absent from regional simulation planning and modeling as well from the most advanced land use and transportation integrated models. Considerable research findings, however, have been accumulating since the 1970s (Salomon 1986; Mokhtarian 1990; Mahmassani and Jou 1998; Marker and Goulias 2000; Weilland and Purser 2000; Patten and Goulias 2001; Golob 2001; Patten et al. 2003; Krizek and Johnson 2003; Goulias, Kim, and Pribyl 2003; and a framework of the ICT and travel interactions by Mokhtarian 2009).

Another type of technologies (named enabling herein) helped us move modeling and simulation further. A few of the most important enabling technologies are *stochastic simulation*, *production systems*, *geographic information systems*, *interactive and technology-aided data collection approaches*, and more *flexible data analysis techniques*. Stochastic microsimulation, as intended here and briefly described above, is an evolutionary engine software that is used to replicate the relationships among social, economic, and demographic factors with land use, time use, and travel by people. As discussed above, the causal links among these groups of entities are extremely complex, nonlinear, and in many instances unknown or incompletely specified. This is the reason that no closed form solution can be created for such a forecasting model system. An evolutionary engine, then, provides a realistic representation of person and household life histories (e.g., birth, death, marriages, divorces, birth of children, etc.), spatio-temporal activity opportunity evolution, and a variety of models that account for uncertainties in data, models, and behavioral variation (see Miller 2003 and Goulias 2002 for overviews and Sundararajan and Goulias 2003 for an application).

Production systems were first developed by Newell and Simon (1972) to explicitly depict the way humans go about solving problems. These are a series of condition-action (note the parallel with stimulus-response) statements in a sequence. From this viewpoint, they are search processes that may never reach an absolute optimum and they replicate (or at least attempt to) human thought and action. Models of this kind are called *computational process models* (CPM) and through the use of IF—THEN rules, they have made possible first the creation of a variety of new models and were used extensively in the ALBATROSS system mentioned above and the Kitamura and Fujii (1998) model systems. *Geographic information systems* are software systems that can be used to collect, store, analyze, modify, and display large amounts of geographic data. They include layers of data that are able to incorporate relations among the variables in each layer and allow to build relationships in data across layers. One can visualize a GIS as a live map that can display almost any kind of spatio-temporal information. Maps have been used by transportation planners and engineers for long time and they are a natural interface to use in modeling and simulation (Figures 10.3 and 10.4 are GIS examples). *Advanced data collection methods and devices* that are technologies that merit a note, although, not strictly developed for modeling, use internet (Doherty 2003) and geospatial technologies to build complex interviews that are interactive and dynamic (Wolf et al. 2001; Doherty et al. 2001; Auld et al. 2009). Very important development is also the emergence of devices that can record the bulk of environmental data surrounding a person movement, classify the environment in which the individual moves, and then ask simplified questions (Hato 2006). *Soft computing and nonparametric data analysis*. In the data analysis, we see greater strides in using data mining and artificial intelligence–borne techniques to extract travel behavior patterns (Teodorovic and Vukadinovic 1998; Pribyl and Goulias 2003) and advanced and less restrictive statistical methods to discover relationships in the travel behavior data (e.g., Kharoufeh and Goulias 2002). Soft computing is increasingly finding many applications in activity-based models (see www.imob.uhasselt.be). For a more recent and accessible review see Pribyl 2007, and the application in Pribyl and Goulias 2005. Henson and Goulias (2006, 2009) and Goulias (2009b) provide comprehensive reviews.

Microsimulation models also evolved in the interface between land use and travel behavior. The Integrated Land Use, Transportation and Environment (ILUTE) model (Salvini and Miller 2003) model is designed to simulate the evolution of people and their activity patterns, transportation networks, houses, commercial buildings, the economy, and the job market over time. Within this vision, Miller and Roorda (2003), developed the Toronto Area Scheduling model for Household Agents (TASHA) that uses *projects* to organize activity episodes into schedules of persons. Schedules for members in a household are simultaneously generated to allow for joint activities. Both ILUTE and TASHA utilize CPMs and econometric utility-based paradigms. Most recently, Ettema et al. (2006)

developed PUMA (Predicting Urbanization with Multiagents), a full-fledged multiagent system of urban processes that represents land use changes in a behaviorally realistic way. The evolution of new model development ideas continues in a variety of directions with one model system attempting to integrated many ideas in one system. This is the SimTRAVEL: *Sim*ulator of *T*ransport, *R*outes, *A*ctivities, *V*ehicles, *E*missions, and *L*and, which integrates long term choices employing Urbanism, activity and travel using an enhanced version of FAMOS, and a combination of highway and transit dynamic traffic assignment to produce estimates of emissions (http://urbanmodel.asu.edu/; accessed January 2010). These processes eventually will include the evolution of population, businesses, and land use as well as daily activity and travel patterns of people and the goods moving in a given area.

Similarities and differences among the implemented modeling ideas are:

- A hierarchy of decisions by households is assumed that identifies longer-term choices determining the shorter-term choices. In this way, different blocks of variables can be identified and their mutual correlation used to derive equations that are used in forecasting.

- Anchor points (home location—work location—school location) are inserted in the first choice level, and they define the overall spatial structure of activity scheduling in a day.

- Out-of-home activity purposes include work, school, shopping, meals, personal business, recreation, and escort. These expanded the original home-based and non-home-based purposes in the four-step models.

- In-home activities are explicitly modeled or allowed to enter the model structure as a "stay-at-home" choice with some models allowing for activity choice at home (work, maintenance and discretionary). In this way, limited substitution between at home and outside home can be reflected in the models.

- Stop frequencies and activities at stops are modeled at the day pattern and tour levels to distinguish between activities and trips that can be rescheduled with little additional efforts versus the activities and trips that cannot be rescheduled (e.g., school trips).

- Modes and destinations are modeled together. In this way the mutual influence, sequential, and/or simultaneous relationships can be reflected in the model structure.

- Time is included in a few instances in activity-based models. For example departure time for trips and tour time of day choice are modeled explicitly. Model time periods are anywhere between 30 minutes and second-by-second, and time windows are used to account for scheduling. This modeling component allows the incorporation of time-of-day in the modeling suites. It also allows the identification of windows of activity and travel opportunities. The presence of departure time also enables models to trip matrices for any desired periods in a day. In fact, output of time periods depends on traffic assignment needs and can be adjusted almost at will.

- Human interaction, although limited for now to the within-household interaction, is incorporated by relating the day pattern of one person to the day patterns of other persons within a household, their joint activities and trip making are explicitly modeled (joint recreation, escort trips), and allocation of activity-roles are also modeled.

- Spatial aspects of a region are accounted for using methods that produce spatially distributed synthetic populations using as external control totals averages and relative frequencies of population characteristics.

- Accessibility measures are used to capture spatial interaction among activity locations and the level of service offered by the transportation systems. These are also the indicators used to account for feedback among the lower level in the hierarchy decisions (e.g., activity location choices, routes followed, congestion) and the higher level such as residence location choice.

- Spatial resolution is heavily dependent on data availability and it reached already the level of a parcel and/or building at its most disaggregate level. Outputs of models are then aggregated to whatever level is required by traffic assignment, mode specific studies (nonmotorized and/or transit) and reporting needs and requirements.

10.4 FUTURE DIRECTIONS

The plethora of advances reviewed in this chapter includes models and experiments to create computerized virtual worlds and synthetic schedules at the most elementary level of decision making using microsimulation and computational process models; data collection methods and new methods to collect extreme details about behavior and to estimate, validate, and verify models using advanced hardware, software, and data analysis techniques; and integration of models from different domains to reflect additional interdependencies such as land use and telecommunications.

Much more work remains to be done in order to develop models that can answer more complex questions from policy analysis and for this reason a few steps are outlined here. In policy and program evaluation, transportation analysis appears to be narrowly applied to only one method of assessment that does not follow the ideal of a randomized controlled trial and does not explicitly define what experimental setting we are using for our assessments. Unfortunately this weakens our findings about policy analysis and planning activities. Although many laboratory experiments were done for intelligent transportation systems, we lack studies and guidelines to develop experimental and quasi-experimental procedures to guide us in policy development and large scale data collection.

In addition, many issues remain unresolved in the areas of coordination among scale in time and space and related issues. In addition very little is known about model sensitivity and data error tolerance and their mapping to strategy evaluations. This is partially because of the lack of tools that are able to make these assessments but also due to lack of scrutiny of these issues and their implications on impact assessment.

Regarding strategic planning and evaluation, we also lack models designed to be used in scenario-building exercises, such as backcasting and related assessments (Robinson 1982). The models about change are usually defined for forecasting and simple time inversion may not work to make them usable in backcasting. This area does not have the long tradition of modeling and simulation to help us develop suitable models, with a few exceptions, such as Miller and Demetsky 1999 and Sadek, Dessouki, and Ivan 2002.

In the new research and technology area, since we are dealing with the behavior of persons, it is unavoidable to consider perceptions of time and space. What role should perceptions of time and space (Golledge and Gärling 2004) play in behavioral models and what is the most appropriate use of these perceptions? The multiple dimensions of time such as tempo, duration, and clock time are neglected in behavioral models.

Human interaction is considered important and is receiving attention in research Golob and McNally 1997; Chandrasekharan and Goulias 1999; Simma and Axhausen 2001; Gliebe and Koppelman 2002; Goulias and Kim 2005; Zhang, Timmermans, and Borgers 2005; but only partially accounted for in applications as illustrated by Vovsha and Petersen (2005) and Pribyl (2004). Future applications will increasingly pay attention to motivations for human interactions and the nature of these interactions. This is particularly important in the relationship between land use policies and travel behavior (Yoon and Goulias 2009, 2010) and the important issue of perceptions of place and related urban form (Deutsch and Goulias 2010).

As we move closer to models that can satisfy the dynamic policy and planning need adopting a lifecourse perspective appears to be a better model framework (Goulias 2009a). This, however, requires data that are longitudinal and offer trajectories of individual paths. They also offer the opportunity to create richer and multidimensional theoretical frameworks. This type of modeling and simulation is also better suited for specifying models of everyday action that are connected with the lives of people as they progress in linked lives and their environment over time. Most important, however, is their explicit modeling of relations that should not be just a static explanation of the number of contacts and the amount of time people spend in activities and on travel. In order to capture the behavioral process underlying the outcomes we observe in typical household surveys we also need to identify relations and their evolution over time as well as their differing nature across different contexts.

REFERENCES

Alam, B. S. (1998). "Dynamic Emergency Evacuation Management System Using GIS and Spatio-Temporal Models of Behavior." MS Thesis. Department of Civil and Environmental Engineering, The Pennsylvania State University, University Park, PA.

Alam, B. S. and K. G. Goulias. (1999). "Dynamic Emergency Evacuation Management System Using GIS and Spatio-Temporal Models of Behavior." *Transportation Research Record* 1660:92–99.

Anas, A. (1982). *Residential Location Markets and Urban Transportation: Economic Theory, Econometrics and Policy Analysis with Discrete Choice Models.* New York: Academic Press.

Arentze, T. and H. Timmermans. (2000). *ALBATROSS–A Learning Based Transportation Oriented Simulation System.* European Institute of Retailing and Services Studies (EIRASS), Technical University of Eindhoven, Eindhoven, NL.

Arentze, T., H. Timmermans, D. Janssens, and G. Wets. (2006). Modeling Short-term Dynamics in Activity-Travel Patterns: From Aurora to Feathers. Presented at the Innovations in Travel Modeling Conference, Austin, TX. May 21–23.

Auld, J., C. Williams, A. Mohammadian, and P. Nelson. (2009). "An Automated GPS-Based Prompted Recall Survey with Learning Algorithms." *Transportation Letters* 1(1):59–79.

Avineri, E. and Y. Prashker. (2003). Sensitivity to Uncertainty: The Need for a Paradigm Shift. CD-TRB ROM Proceedings, Paper presented at the 82nd Annual Transportation Research Board Meeting, January 12–16, Washington D.C.

Becker, G. S. (1976). *The Economic Approach to Human Behavior.* Chicago: The University of Chicago Press.

Ben-Akiva, M. E. and S. R. Lerman. (1985). *Discrete Choice Analysis: Theory and Application to Travel Demand.* Cambridge, MA: MIT Press.

Ben-Akiva, M. E. and T. Morikawa. (1989). Estimation of Mode Switching Models from Revealed Preferences and Stated Intentions. Paper presented at the International Conference on Dynamic Travel Behavior at Kyoto University Hall, July, Kyoto, JP.

Ben-Akiva, M. E., J. Walker, A. T. Bernardino, D. A. Gopinath, T. Morikawa, and A. Polydoropoulou. (2002). "Integration of Choice and Latent Variable Models." In *In Perceptual Motion: Travel Behavior Research Opportunities and Application Challenges* (H. S. Mahmassani, Ed.), Amsterdam, NL: Pergamon. 431–470.

Bhat, C. R. (2000). "Flexible Model Structures for Discrete Choice Analysis." In *Handbook of Transport Modelling* (D. A. Hensher and K. J. Button, Eds.), Amsterdam, NL: Pergamon, 71–89.

Bhat, C. R. (2003). "Random Utility-Based Discrete Choice Models for Travel Demand Analysis." In *Transportation Systems Planning: Methods and Applications* (K. G. Goulias, Ed.), Boca Raton, FL: CRC Press, pp. 10–1 to 10–30.

Bhat, C. R., J. Guo, S. Srinivasan, and A. Sivakumar. (2003). "Activity-Based Travel Demand Modeling for Metropolitan Areas in Texas: Software-related Processes and Mechanisms for the Activity-Travel Pattern Generation Microsimulator." Research Report 4080-5, Center for Transportation Research, Austin, Texas.

Bhat, C. R. and F. Koppelman. (1999). "A Retrospective and Prospective Survey of Time-Use Research." *Transportation* 26(2):119–139.

Bockenholt, U. (2002). "Comparison and Choice: Analyzing Discrete Preference Data by Latent Class Scaling Models." In *Applied Latent Class Analysis* (J. A. Hagenaars and A. L. McCutcheon, Eds.), Cambridge, UK: Cambridge University Press, 163–182.

Borgers, A. W. J., F. Hofman, and H. J. P. Timmermans. (1997). "Activity-Based Modelling: Prospects." In *Activity-Based Approaches to Travel Analysis* (D. F. Ettema and H. J. P. Timmermans, Eds.). Oxford, UK: Pergamon, 339–351.

Bradley, M. and J. Bowman. (2006). "A Summary of Design Features of Activity-Based Microsimulation Models for U.S. MPOs." Conference on Innovations in Travel Demand Modeling, Austin, TX. May 21–23.

Brög, W. and E. Erl. (1989). "Interactive Measurement Methods—Theoretical Bases and Practical Applications." *Transportation Research Record* 765:1–11.

Chandrasekharan, B. and K. G. Goulias. (1999). "Exploratory Longitudinal Analysis of Solo and Joint Trip Making in the Puget Sound Transportation Panel." *Transportation Research Record* 1676:77–85.

Chapin, F. S. Jr. (1974). *Human Activity Patterns in the City: Things People Do in Time and Space.* New York: Wiley.

Chung, J. and K. G. Goulias. (1997). "Travel Demand Forecasting Using Microsimulation: Initial Results from a Case Study in Pennsylvania." *Transportation Research Record*, No. 1607:24–30.

Creighton, R. L. (1970). *Urban Transportation Planning.* Urbana, IL: University of Illinois Press.

Cullen, I. and V. Godson. (1975). "Urban Networks: The Structure of Activity Patterns." *Progress in Planning* 4(1):1–96.

Deutsch, K. E. and K. G. Goulias. (2010). "Exploring Sense-of-Place Attitudes as Indicators of Travel Behavior." Paper to be presented at the 89th Annual Transportation Research Board Meeting and included in the CD ROM proceedings (Paper 10-0070).

Dillman, D. A. (2000). *Mail and Internet Surveys: The Tailored Design Method* (2nd ed.). New York: Wiley.

Doherty, S. (2003). "Interactive Methods for Activity Scheduling Processes." In *Transportation Systems Planning: Methods and Applications* (K. G. Goulias, Ed.), Boca Raton, FL: CRC Press, pp. 7–1 to 7–25.

Doherty, S. T., N. Noel, M. Lee-Gosselin, C. Sirois, and M. Ueno. (2001). "Moving beyond Observed Outcomes: Global Positioning Systems and Interactive Computer-Based Travel Behavior Surveys." *Transportation Research Circular*, E-C026, March 2001, Transportation Research Board, Washington, DC.

Ettema, D., K. de Jong, H. Timmermans, and A. Bakema. (2006). *PUMA: Multi-Agent Modeling of Urban Systems.* 2006 Transportation Research Board CD-ROM.

Gärling, T., T. Laitila, and K. Westin. (1998). "Theoretical Foundations of Travel Choice Modeling: An Introduction." In *Theoretical Foundations of Travel Choice Modeling* (T. Garling, T. Laitila, and K. Westin, Eds.), Oxford, UK, Pergamon, pp. 1–30.

Garrett, M. and M. Wachs. (1996). *Transportation Planning on Trial. The Clean Air Act and Travel Forecasting.* Thousand Oaks, CA: Sage Publications.

Gliebe, J. P. and F. S. Koppelman. (2002). "A Model of Joint Activity Participation." *Transportation* 29:49–72.

Golledge, R. G. and T. Gärling. (2003). "Spatial Behavior in Transportation Modeling and Planning." In *Transportation Systems Planning: Methods and Applications* (K. G. Goulias, Ed.), Boca Raton, FL: CRC Press, Chapter 3, 1–27.

Golledge, R. G. and T. Gärling. (2004). "Cognitive Maps and Urban Travel." In *Handbook of Transport Geography and Spatial Systems*, Volume 5 (D. Hensher, K. Button, K. Haynes, and P. Stopher, Eds.), Amsterdam: Elsevier, 501–512.

Golledge, R. G. and R. J. Stimson. (1997). *Spatial Behavior: A Geographic Perspective.* New York: The Guilford Press.

Golob, T. F. (2001). "Travelbehaviour.com: Activity Approaches to Modeling the Effects of Information Technology on Personal Travel Behaviour." In *Travel Behavior Research, The Leading Edge* (D. Hensher, Ed.) Kidlington, Oxford: Elsevier Science/Pergamon, 145–184.

Golob, T. F., R. Kitamura, and L. Long, Eds. (1997). *Panels for Transportation Planning: Methods and Applications.* Boston, Kluwer.

Golob, T. F. and M. McNally. (1997). "A Model of Household Interactions in Activity Participation and the Derived Demand for Travel." *Transportation Research Board* 31:177–194.

Goodman, L. A. (2002). "Latent Class Analysis: The Empirical Study of Latent Types, Latent Variables, and Latent Structures." In *Applied Latent Class Analysis* (J. A. Hagenaars and A. L. McCutcheon, Eds.), Cambridge, UK: Cambridge University Press, 3–55.

Goulias, K. G. (1999). Longitudinal Analysis of Activity and Travel Pattern Dynamics Using Generalized Mixed Markov Latent Class Models." *Transportation Research* 33B:535–557.

Goulias, K. G. (2001). "A Longitudinal Integrated Forecasting Environment (LIFE) for Activity and Travel Forecasting." In *Ecosystems and Sustainable Development III* (Y. Villacampa, C. A. Brebbia, and J. L. Uso, Eds.), Southampton, UK: WIT Press, 811–820.

Goulias, K. G. (2002). "Multilevel Analysis of Daily Time Use and Time Allocation to Activity Types Accounting for Complex Covariance Structures Using Correlated Random Effects." *Transportation* Volume 10(1):31–48.

Goulias, K.G. (2003). "Transportation Systems Planning." In *Transportation Systems Planning: Methods and Applications* (K. G. Goulias, Ed.), Boca Raton, FL: CRC Press, 1–1 to 1–45.

Goulias, K. G. (2009a). "Travel Behavior Dynamics from a Lifespan Development Perspective." Paper presented at the 12th International Conference on Travel Behavior Research, December 13–18, Jaipur, India, and included in the CD ROM proceedings.

Goulias, K. G. (2009b). "Transportation Policy Analysis and Planning." In *The Expanding Sphere of Travel Behavior Research Selected Papers from the 11th International Conference on Travel Behaviour Research* (Eds. R. Kitamura, T. Yoshii, and T. Yamamoto), Emerald Group Publishing, Bingley, United Kingdom, 387–420.

Goulias, K. G. and T. Kim. (2003). A Longitudinal Analysis of the Relationship between Environmentally Friendly Modes, Weather Conditions, and Information-Telecommunications Technology Market Penetration." In *Ecosystems and Sustainable Development Volume 2* (E. Tiezzi, C. A. Brebbia, and J. L. Uso, Eds.), WIT Press, pp. 949–958.

Goulias, K. G. and T. Kim. (2005). "An Analysis of Activity Type Classification and Issues Related to the *With Whom* and *For Whom* Questions of an Activity Diary." In *Progress in Activity-Based Analysis* (H. Timmermans, Ed.), Amsterdam, Elsevier, 309–334.

Goulias, K. G. and R. Kitamura. (1992). "Travel Demand Analysis with Dynamic Microsimulation." *Transportation Research Record*, No.1607:8–18.

Goulias, K. G., T. Kim, and O. Pribyl. (2003). "A Longitudinal Analysis of Awareness and Use for Advanced Traveler Information Systems." *Journal of Intelligent Transportation Systems: Technology, Planning, and Operations,* 8(1):3–17.

Goulias, K. G., L. Blain, N. Kilgren, T. Michalowski, and E. Murakami. (2007). "Catching the Next Big Wave: Are the Observed Behavioral Dynamics of the Baby Boomers Forcing us to Rethink Regional Travel Demand Models?" Paper presented at the 86th Transportation Research Board Annual Meeting, January 21–25, 2007, Washington, D.C. and included in the CD ROM proceedings.

Haab, T. C. and R. L. Hicks. (1997). "Accounting for Choice Set Endogeneity in Random Utility Models of Recreation Demand." *Journal of Environmental Economics and Management* 34:127–147.

Hagerstrand, T. (1970). "What about People in Regional Science?" Papers of the Regional Science Association, 10, pp. 7–21.

Hato, E. (2006). "Development of Behavioral Context Addressable Loggers in the Shell for Travel Activity Analysis." Paper presented at the International Association of Travel Behavior Research conference, Kyoto, August 16–20, 2006, Japan.

Henson, K. and K. G. Goulias. (2006). "Preliminary Assessment of Activity and Modeling for Homeland Security Applications." *Transportation Research Record: Journal of the Transportation Research Board* 1942:23–30.

Henson, K., K. G. Goulias, and R. Golledge. (2009). "An Assessment of Activity-Based Modeling and Simulation for Applications in Operational Studies, Disaster Preparedness, and Homeland Security." *Transportation Letters* 1(1):19–39.

Horowitz, J. L. (1991). "Modeling the Choice of Choice Set in Discrete-Choice Random-Utility Models." *Environment and Planning A* 23:1237–1246.

Horowitz, J. L. and J. J. Louviere. (1995). "What Is the Role of Consideration Sets in Choice Modeling?" *International Journal of Research in Marketing* 12:39–54.

Hutchinson, B. G. (1974). *Principles of Urban Transport Systems Planning.* Washington, D.C.: Scripta.

Jones, P., F. Koppelman, and J. Orfeuil. (1990). "Activity Analysis: State-of-the-Art and Future Directions." In *Developments in Dynamic and Activity-Based Approaches to Travel Analysis. A Compendium of Papers from the 1989 Oxford Conference* (P. Jones, Ed.). Avebury, UK, 34–55.

Jonnalagadda, N., J. Freedman, W. A. Davidson, and J. D. Hunt. (2001). "Development of Microsimulation Activity-Based Model for San Francisco." *Transportation Research Record* 1777:25–35.

Kahneman, D., and A. Tversky. (1979). "Prospect Theory: An Analysis of Decisions under Risk." *Econometrica* 47(2):263–291.

Kharoufeh, J. P. and K. G. Goulias. (2002). "Nonparametric Identification of Daily Activity Durations Using Kernel Density Estimators." *Transportation Research, Part B Methodological* 36:59–82.

Kitamura, R. (1988). "An Evaluation of Activity-Based Travel Analysis." *Transportation* 15:9–34.

Kitamura, R. (2000). "Longitudinal Methods." In *Handbook of Transport Modelling* (D. A. Hensher and K. J. Button, Eds.), Amsterdam, NL: Pergamon, 113–128.

Kitamura, R. and S. Fujii. (1998). "Two Computational Process Models of Activity-Travel Choice." In *Theoretical Foundations of Travel Choice Modeling* (T. Garling, T. Laitila, K. Westin, Eds.), Amsterdam, Pergamon, 251–279.

Koppelman, F. S. and V. Sethi. (2000). "Closed-Form Discrete-Choice Models." In *Handbook of Transport Modelling* (D. A. Hensher and K. J. Button, Eds.), Amsterdam, NL: Pergamon, 211–225.

Krizek, K. J., and A. Johnson. (2003). "Mapping of the Terrain of Information and Communications Technology (ICT and Household Travel)." Transportation Research Board annual meeting CD-ROM, Washington, D.C., January.

Kuhnau, J. L. (2001). "Activity-Based Travel Demand Modeling Using Spatial and Temporal Models in the Urban Transportation Planning System." MS Thesis. Department of Civil and Environmental Engineering, The Pennsylvania State University, University Park, PA.

Kuhnau, J. L. and K. G. Goulias. (2003). "Centre SIM: First-Generation Model Design, Pragmatic Implementation, and Scenarios." In *Transportation Systems Planning: Methods and Applications* (K. G. Goulias, Ed.), Boca Raton, FL: CRC Press, pp. 16–1 to 16–14.

Kwan, M.-P. (1994). *A GIS-Based Model for Activity Scheduling in Intelligent Vehicle Highway Systems (IVHS)*. Unpublished Ph.D., Department of Geography, University of California Santa Barbara, Santa Barbara, CA.

Kwan, M.-P. (1997). "GISICAS: An Activity-Based Travel Decision Support System Using a GIS-Interfaced Computational-Process Model." In *Activity-Based Approaches to Travel Analysis* (D. F. Ettema and H. J. P. Timmermans, Eds.), New York; Elsevier Science, Inc., 263–282.

Lenntorp, B. (1976). "Paths in Space-Time Environment: A Time Geographic Study of Possibilities of Individuals." The Royal University of Lund, Department of Geography. Lund Studies in Geography, Series B. *Human Geography*, 44.

Lomborg, B. (2001). *The Skeptical Environmentalist: Measuring the Real State of the World*. Cambridge, UK: Cambridge University Press.

Louviere, J. J., D. A. Hensher, and J. D. Swait. (2000). *Stated Choice Methods: Analysis and Application*. Cambridge, UK: Cambridge University Press.

Ma, J. (1997). *An Activity-Based and Micro-Simulated Travel Forecasting System: A Pragmatic Synthetic Scheduling Approach*. Unpublished Ph.D. Dissertation, Department of Civil and Environmental Engineering, The Pennsylvania State University, University Park, PA.

Mahmassani, H. S. and R. Herman. (1990). "Interactive Experiments for the Study of Tripmaker Behaviour Dynamics in Congested Commuting Systems." In *Developments in Dynamic and Activity-Based Approaches to Travel Analysis*. A compendium of papers from the 1989 Oxford Conference. Avebury, UK. 272–298.

Mahmassani, H. S. and R. C. Jou. (1998). "Bounded Rationality in Commuter Decision Dynamics: Incorporating Trip Chainig in Departure Time and Route Switching Decisions." In *Theoretical Foundations of Travel Choice Modeling* (T. Garling, T. Laitila, K. Westin, Eds.), Amsterdam, Pergamon.

Manheim, M. L. (1979). *Fundamentals of Transportation Systems Analysis, Volume 1: Basic Concepts*. Cambridge, MA: MIT Press.

Marker, J. T. and Goulias, K. G. (2000). "Framework for the Analysis of Grocery Teleshopping." *Transportation Research Record* 1725:1–8.

McFadden, D. (1998). "Measuring Willingness-to-Pay for Transportation Improvements." In *Theoretical Foundations of Travel Choice Modeling* (T. Garling, T. Laitila, and K. Westin, Eds.), Amsterdam, Pergamon, 339–364.

McNally, M. G. (2000). "The Activity-Based Approach." In *Handbook of Transport Modelling* (D. A. Hensher and K. J. Button, Eds.). Amsterdam, NL: Pergamon, 113–128.

Meyer, M. D. and E. J. Miller. (2001). *Urban Transportation Planning* (2nd ed.). Boston, MA: McGraw-Hill.

Miller, E. J. (2003). "Land Use: Transportation Modeling." In *Transportation Systems Planning: Methods and Applications* (K. G. Goulias, Ed.), Boca Raton, FL: CRC Press, pp. 5–1 to 5–24.

Miller, E. J. (2006). "Resource Paper on Integrated Land Use-Transportation Models." Paper presented at the International Association of Travel Behavior Research conference, Kyoto, August 16-20,2006, Japan.

Miller, E. J. and M. J. Roorda. (2003). "A Prototype Model of Household Activity/Travel Scheduling." *Transportation Research Record* 1831:114–121.

Miller, J. S. and M. J. Demetsky. (1999). "Reversing the Direction of Transportation Planning Process." *ASCE Journal of Transportation Engineering* 125(3):231–237.

Mokhtarian, P. L. (1990). "A Typology of Relationships between Telecommunications and Transportation." *Transportation Research A* 24(3):231–242.

Mokhtarian, P. L. (2009). "If Telecommunications Is Such a Good Substitute for Travel, Why Does Congestion Continue to Get Worse?" *Transportation Letters* 1(1):1–17.

National Cooperative Highway Research Program. (2000). Report 446. Transportation Research Board, Washington D.C.

Newell, A. and H. A. Simon. (1972). *Human Problem Solving*. Englewood Cliffs, NJ: Prentice Hall.

Niemeier, D. A. (2003). "Mobile Source Emissions: An Overview of the Regulatory and Modeling Framework." Chapter 13 in *Transportation Systems Planning: Methods and Applications* (K. G. Goulias, Ed.), Boca raton, Fl: CRC Press, pp. 13–1 to 13–28.

Ortuzar, J. D. and l. G. Willumsen. (2001). *Modelling Transport* (3rd ed.). Chichester, UK: Wiley.

Patten, M. L. and K. G. Goulias. (2001). Test Plan: Motorist Survey—Evaluation of the Pennsylvania Turnpike Advanced Travelers Information System (ATIS) Project, Phase III PTI-2001-23-I. April 2001. University Park, PA.

Patten, M. L., M. P. Hallinan, O. Pribyl, and K. G. Goulias. (2003). Evaluation of the Smartraveler advanced traveler information system in the Philadelphia metropolitan area. Technical memorandum. PTI 2003-33. March 2003. University Park, PA.

Payne, J. W., J. R. Bettman, and E. J. Johnson. (1993). *The Adaptive Decision Maker.* Cambridge, UK: Cambridge University Press.

Pendyala, R. (2003). "Time Use and Travel Behavior in Space and Time." In *Transportation Systems Planning: Methods and Applications* (K. G. Goulias, Ed.), Boca Raton, FL: CRC Press, pp. 2–1 to 2–37.

Pendyala, R. M., R. Kitamura, A. Kikuchi, T. Yamamoto, and S. Fujii. (2005). "The Florida Activity Mobility Simulator (FAMOS): An Overview and Preliminary Validation Results." Presented at the Paper presented at the 84th Transportation Research Board Annual Meeting, January 9–13, 2005, Washington, D.C.

Pribyl, O. (2004). "A Microsimulation Model of Activity Patterns and within Household Interactions." Ph.D. Dissertation, Department of Civil and Environmental Engineering, The Pennsylvania State University, University Park, PA.

Pribyl, O. (2007). "Computational Intelligence in Transportation: Short User-Oriented Guide." In *Transport Science and Technology* (K. G. Goulias, Ed.), Amsterdam: Elsevier, 37–54.

Pribyl, O. and K. G. Goulias. (2003). On the Application of Adaptive Neuro-fuzzy Inference System (ANFIS) to Analyze Travel Behavior." Paper presented at the 82nd Transportation Research Board Meeting and included in the CD ROM proceedings and accepted for publication in the *Transportation Research Record*, Washington D.C., January.

Pribyl, O. and K. G. Goulias. (2005). "Simulation of Daily Activity Patterns." In *Progress in Activity-Based Analysis* (H. Timmermans, Ed.), Amsterdam, Elsevier Science, 43–65.

Ramadurai, G. and K. K. Srinivasan. (2006). "Dynamics and Variability in Within-Day Mode Choice Decisions. Role of State Dependence, Habit Persistence, and Unobserved Heterogeneity." *Transportation Research Record, Journal of the Transportation Research Board,* 1977:43–52.

Richardson, A. (1982). "Search Models and Choice Set Generation." *Transportation Research Part A,* 16A(5-6): 403–416.

Robinson, J. (1982). "Energy Backcasting: A Proposed Method of Policy Analysis." *Energy Policy* 10(4):337–344.

Rubinstein, A. (1998). *Modeling Bounded Rationality.* Cambridge, MA: The MIT Press.

Sadek, A. W., W. M. El Dessouki, and J. I. Ivan. (2002). "Deriving Land Use Limits as a Function of Infrastructure Capacity." Final Report, Project UVMR13-7, New England Region One University Transportation Center, MIT, Cambridge, MA.

Salomon, I. (1986). "Telecommunications and Travel Relationships: A Review." *Transportation Research A* 20A(3):223–238.

Salvini, P. and E. J. Miller. (2003). "ILUTE: An Operational Prototype of a Comprehensive Microsimulation Model of Urban Systems." Paper presented at the 10th International Conference on Travel Behaviour Research, Lucerne, August 2003.

Samimi, A., A. Mohammadian, and K. Kawamura. (2010). "A Behavioral Freight Movement Microsimulation Model: Method and Data." *Transportation Letters: The International Journal of Transportation Research* 2(1):53–62.

Savage, L. J. (1954). *The Foundations of Statistics.* Reprinted version in 1972 by Dover Publications, New York, NY.

Simma, A. and K. W. Axhausen. (2001). "Within-Household Allocation of Travel—The Case of Upper Austria." *Transportation Research Record: Journal of the Transportation Research Board,* No. 1752, TRB, National Research Council, Washington, D.C., 69–75.

Simon, H. A. (1997). *Administrative Behavior* (4th ed.). New York: The Free Press.

Simon, H. A. (1983). "Alternate Visions of Rationality." In *Reason in Human Affairs* (H. A. Simon, Ed.), Stanford, CA: Stanford University Press, 3–35.

Southworth, F. (2003). "Freight Transportation Planning: Models and Methods." In *Transportation Systems Planning: Methods and Applications* (K. G. Goulias, Ed.). CRC Press, Boca Raton, FL, 4.1–4.10.

Stefan, K. J., J. D. P. McMillan, and J. D. Hunt. (2005). "An Urban Commercial Vehicle Movement Model for Calgary." Paper presented at the 84th Transportation Research Board Annual Meeting, January 9–13, 2005, Washington, D.C.

Stopher, P. R. and A.H. Meyburg (Eds.) (1976). *Behavioral Travel-Demand Models.* Lexington, MA: Lexington Books.

Sundararajan, A. and K. G. Goulias. (2003). "Demographic Microsimulation with DEMOS 2000: Design, Validation, and Forecasting." In *Transportation Systems Planning: Methods and Applications* (K. G. Goulias, Ed.), Boca Raton, FL: CRC Press, pp. 14–1 to 14–23.

Swait, J. and M. Ben-Akiva. (1987a). "Incorporating Random Constraints in Discrete Models of Choice Set Generation." *Transportation Research Part B* 21B(2):91–102.

Swait, J. and M. Ben-Akiva. (1987b). "Empirical Test of a Constrained Choice Discrete Model: Mode Choice in Sao Paolo, Brazil." *Transportation Research Part B* 21B(2):103–115.

Teodorovic, D. and K. Vukadinovic. (1998). *Traffic Control and Transport Planning: A Fuzzy Sets and Neural Networks Approach.* Boston, MA: Kluwer.

Thill, J. (1992). "Choice Set Formation for Destination Choice Modeling." *Progress in Human Geography* 16(3):361–382.

Tiezzi, E. (2003). *The End of Time.* Southampton, UK: WIT Press.

Timmermans, H. (2003). The Saga of Integrated Land Use-Transport Modeling: How Many More Dreams before We Wake Up?" Conference keynote paper at the Moving through Net: The Physical and Social Dimensions of Travel. 10th International Conference on Travel Behaviour Research, Lucerne, 10–15, August. In *Proceedings of the meeting of the International Association for Travel Behavior Research (IATBR).* Lucerne, Switzerland.

Train, K. E. (2003). *Discrete Choice Methods with Simulation.* Cambridge, UK: Cambridge University Press.

Transportation Research Board. (1999). *Transportation, Energy, and Environment. Policies to Promote Sustainability.* Transportation Research Circular 492. TRB. Washington D.C.

Transportation Research Board. (2002). *Surface Transportation Environmental Research: A Long-Term Strategy.* Transportation Research Board, Washington, D.C.

Tversky, A. (1969). "Intransitivity of Preferences." *Psychological Review* 76:31–48.

Tversky, A. (1972). "Elimination by Aspects: A Theory of Choice." *Psychological Review* 79:281–299.

Tversky, A. and D. Kahneman. (1992). "Advances in Prospect Theory: Cumulative Representation of Uncertainty." *Journal of Risk and Uncertainty* 9:195–230.

van der Hoorn, T. (1997). "Practitioner's Future Needs." Paper presented at the Conference on Transport Surveys, Raising the Standard. Grainau, Germany, May 24–30.

Vause, M. (1997). "A Rule-Based Model of Activity Scheduling Behavior." In *Activity-Based Approaches to Travel Analysis* (D. F. Ettema and H. J. P. Timmermans, Eds.), New York: Elsevier Science, Inc., 73–88.

Veldhuisen, J., H. Timmermans, and L. Kapoen. (2000). "RAMBLAS: A Regional Planning Model Based on the Microsimulation of Daily Activity Travel Patterns." *Transportation Research A* 32:427–443.

Vovsha, P., Peterson, and R. Donnelly. (2002). "Microsimulation in Travel Demand Modeling: Lessons Learned from the New York Best Practice Mode." *Transportation Research Record* 1805:68–77.

Vovsha, P., Peterson, and R. Donnelly. (2003). "Explicit Modeling of Joint Travel by Household Members: Statistical Evidence and Applied Approach." *Transportation Research Record: Journal of the Transportation Research Board* 1831:1–10.

Vovsha, P. and E. Petersen. (2005). "Escorting Children to School: Statistical Analysis and Applied Modeling Approach." *Transportation Research Record: Journal of the Transportation Research Board* 1921:131–140.

Yoon, S. Y. and K. G. Goulias. (2009). "Of Individual Accessibility on Travel Behavior and Its Propagation through Intra-Household Interaction." Paper presented at the 12th International Conference on Travel Behavior Research, December 13–18, Jaipur, India, and included in the CD ROM proceedings.

Yoon, S. Y. and K. G. Goulias. (2010). "Constraint-Based Assessment of Intra-Household Bargaining on Time Allocation to Activities and Travel Using Individual Accessibility Measures." Paper to be presented at the 89th Annual Transportation Research Board Meeting and included in the CD ROM proceedings (Paper10-1820).

Waddell, P. and G. F. Ulfarsson. (2003). *"Dynamic Simulation of Real Estate Development and Land Prices within an Integrated Land Use and Transportation Model System."* Presented at the 82nd Annual Meeting of the Transportation Research Board, January 12–16, 2003, Washington, D.C. (also available in http://www.urban-ism.org/papers/; accessed April 2003).

Weiland, R. J. and L. B. Purser. (2000). "Intelligent Transportation Systems." In Transportation in the New Millennium. State of the Art and Future Directions. Perspectives from Transportation Research Board Standing Committees. Transportation Research Board. National Research Council. The National Academies, Washington, D.C., (2000). (Also in http://nationalacademies.org/trb/).

Wen, C.-H. and F. S. Koppelman. (2000). "A Conceptual and Methodological Framework for the Generation of Activity-Travel Patterns." *Transportation* 27:5–23.

Williams, H. C. W. L. and J. D. Ortuzar. (1982). "Behavioral Theories of Dispersion and the Mis-Specification of Travel Demand Models. *Transportation Research B* 16B(3):167–219.

Wilson, E. O. (1998). *Consilience, The Unity of Knowledge.* New York: Vintage Books.

Wolf, J., R. Guensler, S. Washington, and L. Frank. (2001). "Use of Electronic Travel Diaries and Vehicle Instrumentation Packages in the Year 2000." Atlanta Regional Household Travel Survey. Transportation Research Circular,E-C026, March 2001, Transportation Research Board, Washington, DC.

Zhang, J., H.J.P. Timmermans, and A.W.J. Borgers. (2005). "A Model of Household Task Allocation and Time Use." *Transportation. Research Board* 39:81–95.

CHAPTER 11
TRANSPORTATION ECONOMICS*

Anthony M. Pagano
Department of Managerial Studies
University of Illinois at Chicago
Chicago, Illinois

11.1 INTRODUCTION

Transportation economics is a very broad field. It includes the application of economic principles to pricing, cost analysis, and regulatory issues. It also includes the analysis of transportation impacts on land use, economic development, and the environment. The field of transportation economics also includes the analysis of the costs and benefits of transportation improvement and initial construction projects. It is this latter aspect of transportation economics that is of interest to transportation engineers and is the subject of this chapter.

The chapter begins with a discussion of the methodology of project appraisal. This includes cost analysis, methods to deal with uncertainty, methods of benefit quantification and evaluation, rate of return analysis, net present value analysis, the appropriate discount rate to use, benefit-cost ratios, net benefit analysis, and suboptimization. The chapter then discusses transportation user benefits, intangible costs and benefits, and the externalities associated with transportation projects.

11.2 THE METHODOLOGY OF PROJECT APPRAISAL

The methodology of project appraisal involves the ascertainment of the costs and benefits of a transportation project. It utilizes the methodology of benefit-cost analysis to analyze alternative projects and make decisions as to which course of action is best. Accordingly, we will focus on benefit-cost analysis techniques in this chapter.

The first question is: What is benefit-cost analysis? Prest and Turvey (1966) state that it is "a practical way of assessing the desirability of projects, where it is important to take a long view (in the sense of looking at repercussions in the further, as well as the 'nearer,' future) and a wide view (in the sense of allowing for side effects of many kinds on many persons, industries, regions, etc.); i.e., it implies the enumeration and evaluation of all the relevant costs and benefits." Quade (1965) says that "it is any analytic study designed to assist a decision-maker identify a preferred choice from among possible alternatives." In a sense, it is a way to look at a problem, analyze it, and arrive at some type of solution. It involves the comparison of various alternatives to achieve a specific objective and essentially consists of the following six steps:

1. The statement of the desired objectives
2. A complete specification of all the relevant alternatives

---

*Reprinted from the First Edition.

3. An estimation of all the costs involved

4. An enumeration of all the benefits

5. Development of a model, either verbally or mathematically

6. Development of criteria for choice among the relevant alternatives

It must be cautioned that these steps are so interrelated that any attempt to discuss them as mutually exclusive parts is surely doomed to failure. This chapter, then, examines each of the steps while keeping in mind their mutual interdependence.

11.3 OBJECTIVES OF PROPOSED INVESTMENTS

Any project appraisal must start with an enumeration of the objectives that the projects or programs are designed to attain. There may be one objective or many. What is important is that the objectives be enumerated clearly and completely. There is no room for ambiguity. The objective function cannot be misspecified. The importance of the objective function is taken up elsewhere in this chapter. At present, it should suffice to say that the objective function need not be formulated mathematically. It may be either verbally or mathematically explained, or both. What is necessary is that it be clear.

11.4 ALTERNATIVES

All the alternative systems or programs that, in the eyes of the analyst, may possibly attain the stated objectives should be enumerated exactly and in great detail. Sometimes all the alternatives are not obvious. The analyst must explore all avenues to obtain a list of all the alternatives.

Another important aspect concerning the specification of the alternatives is that if all the initially listed alternatives are examined and no one of them achieves the desired objectives adequately, the analyst may be forced to design new alternatives. This is one of the primary advantages of benefit-cost analysis: to force the examination and exploration of new avenues where alternatives may lie.

11.5 COSTS

A good benefit-cost study requires the complete enumeration of all the relevant costs of each alternative project. Even if all the costs are known and can be quantified, this is still not an easy task. The costs internal to the project must be broken down by type. One approach is to break them down into three categories—fixed or sunk costs, investment or capital, and maintenance costs. Sunk costs are money that has already been expended on plant and equipment or resources. These are costs that have been made in the past. Economic analysis maintains that because these costs were expended in the past they have no relevance to future systems. In other words, they should not be counted as a cost of the various systems that the analyst is examining. Only the difference among alternatives is what is relevant in their comparison. Inasmuch as sunk costs are the same for all alternatives, there is no need to consider it in an analysis.

Investment or capital costs are those costs that are outlays on plant and equipment for each alternative system. Maintenance costs are those costs that accrue over time and are expended to keep the various facilities at a suitable level of performance. These three categories, of course, can be broken down into various subcategories to enumerate more explicitly and completely the costs involved.

One approach is to calculate the net average annual cost of a transportation improvement project. This is the amount by which the annual cost of the improvement exceeds the cost if no improvements were made. An application to highway improvements would result in the following cost categories:

1. Right-of-way
2. Grading, drainage, and minor structures
3. Major structures
4. Pavement and appurtenances

The following equation can be used to calculate the net average annual highway improvement project cost:

$$\Delta H = (C_1 K_1 + C_2 K_2 + C_3 K_3 + C_4 K_4) + \Delta M$$

where
ΔH = net average annual highway improvement project cost
C_1 = capital cost of right-of-way
C_2 = capital cost of grading, drainage, and minor structures
C_3 = capital cost of major structures
C_4 = capital cost of pavement and appurtenances
ΔM = change in annual maintenance and operation cost for the project
K_1, K_2, K_3, K_4 = capital recovery factor for the known interest rate and service life of the respective item

If the service life of an improvement is T years and the known interest rate is r, the capital recovery factor is given by:

$$K = \frac{r(1+r)^T}{(1+r)^T - 1}$$

The capital recovery factor distributes the capital costs over the T years of service equally, taking into account the time value of money.

The use of the capital recovery factor is typical of analyses in transportation. The rationale behind its use is that the transportation agency must finance its improvements with borrowed money or that, even if it does not borrow money, the inclusion of interest is an indication of the investment opportunities forgone by the taxpayers when the agency spends tax money.

Another problem in cost estimation is establishing the planning or analysis period. The project life or planning period is usually highly subjective. It depends on personal judgments not only of the physical length of life of the project, but also of the likelihood of any changes that may make a particular project obsolete. The estimation of the planning period is further complicated in the analysis of transportation improvement projects due to varying physical lengths of life for each component of the improvement and the extreme uncertainty involved in the estimation of traffic growth in the more distant future. The estimation of traffic growth is an important part of an analysis of transportation improvements because the amount of benefit accruing from a given improvement is directly related to this factor. If the planning period is too long, the analyst may not be able to estimate this critical factor accurately. This can result in the wrong alternative being considered "best" or an unacceptable project being accepted.

There is not much agreement in transportation as to what constitutes the maximum length of time to use as the planning period. Periods of from 10 to 100 years have been mentioned by various authors as the maximum planning period. Some authors maintain, though, that the planning period should be either the physical length of life of the project or the useful life of the project, whichever is shorter.

Another problem in cost estimation is the estimation of the salvage value of the facility at the end of the analysis period. The salvage value can be positive in the case where the structures can be sold as scrap or negative in the case where there is a removal cost. Thus, the salvage value can either reduce or increase the cost of a given alternative and should be taken into account.

11.6 UNCERTAINTY

Rarely does the analyst have a chance to work on a problem where all the costs and benefits are known. Even when all the costs and benefits are quantifiable, the analyst usually is uncertain as to their actual values.

There are two types of uncertainty that need to be dealt with. The first is uncertainty about the world in the future. It includes items such as technological uncertainty and strategic uncertainty. This first type of uncertainty is difficult to deal with since these are unpredictable events. The second type is statistical uncertainty. It occurs because chance elements exist in the real world. This second type of uncertainty results from recurrent factors that can be modeled and estimated. Both types of uncertainty are usually present in long-run decision problems. Sensitivity analysis, contingency analysis, and a fortiori analysis are methods to treat the second type of uncertainty.

11.6.1 Contingency Analysis

Contingency analysis is most often used to evaluate military programs, and it may have some merit in transportation appraisal. The analysis must try to visualize the various changes that might occur that would have a significant impact on program outcomes and must take these contingencies into account in estimating the future worth of various investments.

11.6.2 A Fortiori Analysis

Another method to deal with uncertainty is a fortiori analysis. This method is used primarily in the military, but again may have some use in evaluating transportation improvement projects. To utilize this method, the analyst takes a set of circumstances unfavorable to a program and then compares that program with other possible programs. If the first program emerges as the winner in the unfavorable circumstances, the conclusion emerges, with high confidence, that that program is better than the others.

11.6.3 Sensitivity Analysis

Another approach to uncertainty is sensitivity analysis, which involves changing the values of certain parameters to see what effect this has on the final results. Given that the analyst is not sure exactly what the values of the parameters of various alternative systems will be, sensitivity analysis provides a means to examine the changes in the final results (i.e., which system should be accepted as best) if these parameters change. The analyst can vary each parameter and see how the final results change. If the final results are not appreciably affected by changes in certain parameters, it need not be considered further. If, on the other hand, the final result varies significantly with variations in other parameters, the analyst should spend more time, money, and energy on such parameters to try to achieve the best possible estimates of their actual values.

Sensitivity analysis can also involve finding a value for the uncertain parameter above or below which the optimal alternative may change. One example cited in the literature is the value of life. It can be helpful to know that below a certain value of life one project alternative is most desirable, but above that amount a different alternative would become optimal.

It might be worth mentioning that the finding of insensitivity is just as important as finding sensitivity in the estimates. If the results do not change much as key variables are changed, then the analyst can be more confident that the best project alternative is selected.

Sensitivity analysis, then, can be a helpful aid to the decision-maker either by indicating the risks involved in the decision or by reducing the number of parameters that need further consideration. It will not eliminate or even reduce the uncertainty surrounding the estimates of the various parameters. It will, however, indicate how uncertainty can or cannot affect the final results.

11.6.4 Monte Carlo Simulation

A final approach to dealing with uncertainty is to use Monte Carlo simulation. This involves specifying a probability distribution for each of the key parameters of the analysis. Each is varied according to the distribution and a distribution of outcomes developed for each alternative. Monte Carlo simulation has the advantage that all parameters and variables can be varied simultaneously. The disadvantage is that some knowledge of the probability distributions of the variables and parameters must be known.

11.7 BENEFITS—QUANTIFICATION AND EVALUATION

Probably the most conceptually difficult aspect of benefit-cost analysis is the measure of the benefits of each alternative project or system. The analyst would have a comparatively easy task if all the benefits were known and could be quantified. This is not the usual case. Most benefits are subject to a great deal of uncertainty, and the analyst may not be able to estimate even the likelihood that a benefit may be at any given level. Some benefits and costs are not even subject to quantification. How does the analyst measure the increased security and well-being of society? He cannot even measure, let alone place a value on, these benefits.

In view of these problems, the first point that should be noted is that all the benefits must be specified and enumerated. If not all benefits are taken into account, the wrong system may be considered best. This includes all benefits external as well as internal to the system. Also, the objectives should be kept in mind when the analyst decides what the appropriate measures of benefits or effectiveness will be. One example, cited in the literature, concerns the use of welfare payments as an alternative to training to help people out of poverty. One approach to evaluation is to discount the expected lifetime earnings of the trainees and match them against the cost of the program. If the earnings are greater than the costs, the training will be justified. Welfare payments could be justified as an alternative to training if the costs of training are greater than the discounted expected lifetime earnings of the trainees.

But the discounted expected lifetime earnings of the trainees substantially underestimate the benefits of the training programs. One of the major objectives should be to end poverty through the enhancement of personal opportunity. Clearly, the expected lifetime earnings of the trainees are not an adequate measure of the achievement of this objective. The alternative of welfare payments, if some measure of this personal opportunity objective were not included in the analysis, may be given more weight than it actually deserves. The analyst must examine carefully the objectives that he or she would like the alternatives to attain and use the measure or measures that most adequately indicate the level of attainment of the objectives. This is why the objectives must be clearly and distinctly defined. If the objectives are not precisely stated, it will be impossible to determine a suitable measure of the level of their attainment.

As mentioned previously, the most difficult task the analyst faces is what to do with benefits that cannot be quantified. There is a quantification fallacy that many analysts fall into, in which it is assumed that every factor pertinent to the analysis can be quantified. This cannot always be done. Certain factors, no matter how hard the analyst tries, cannot be quantified. If this is the case, the analyst should present his results to the decision-maker without the benefit of a quantitative analysis.

If the benefits can be quantified, the next step is to place some value on them. The usual analysis is to value benefits in a common measure so these can be compared. This common measure is in monetary units. Some authors have suggested the use of quasi- or shadow prices as a means to place some value on the various benefits. Shadow prices are those prices that would exist if a marketplace actually existed for these benefits. But some authors disagree with the use of shadow prices, arguing that the analyst should look elsewhere to place values on various benefits. For example, the use of the amount of money granted to individuals by courts of law is one means of placing value on the benefit received from reducing accidents or deaths.

11.8 EXPERT OPINION

One method of dealing with uncertainty and unquantifiables is to consult an expert or group of experts who are knowledgeable in the particular field of interest. One method is for the analyst to consult a single expert or groups of experts either individually or through direct fact-to-face confrontation of the experts. Another approach is to use the Delphi technique.

11.8.1 The Delphi Technique

The Delphi technique was developed by Olaf Helmer of the RAND Corporation in 1964. The Delphi technique eliminates the problems of group dynamics that are present in a face-to-face discussion and possibly permits a consensus. Direct discussion is replaced by a series of questionnaires, which can be administered through the mail or through e-mail.

As an example, suppose an analyst wants an estimate of the value of some number N. The analyst first asks each expert to place a value on N, independent of the others. The various responses are then arranged in order of magnitude and the quartiles Q_1, M, and Q_3 are determined.

N_1	N_2	N_3	N_4	N_5	N_6	N_7	N_8	N_9	N_{10}	N_{11}	N_{12}	N_{13}	N_{14}	N_{15}
1	1	1	1	1	1	1	1	1	1	1	1	1	1	1
				1			1				1			
				Q_1			M				Q_3			

Next, the values of Q_1, M, and Q_3 are given to the experts. Each is asked to revise the previous estimate, and if the revised estimate is outside the range Q_1 to Q_3, to explain why the estimate should be different than the 75 percent majority opinion. The results of the second round (that is, the revised values of Q_1, M, and Q_3) are given again to the experts along with the various reasons given in round 2 either to raise or lower the value of N. The reasons are given to the experts in such a way that the anonymity of the respondents is preserved. The experts are asked to evaluate the reasons given in round 2 and again revise their estimates. If their estimates still fall outside the range of Q_1 and Q_3 they are asked for the reasons why the estimate should be different. The estimation procedure can continue for an additional round if desired. The result is usually a consensus estimate that has less dispersion than the original estimates.

The various rounds reduce the dispersion of the estimates and increase the confidence that the analyst has in the estimate. The question remains, however, as to whether the resultant estimate is more accurate that the original.

The Delphi technique eliminates the problems of group dynamics that occur in face-to-face discussions, in which one or a few individuals can dominate the discussion. It also usually results in a consensus forecast that has less dispersion than the initial estimate. On the other hand, the Delphi technique can lead to ambiguous results, since a written questionnaire replaces face-to-face discussion. In addition, if there is too much time between rounds, the experts may have difficulty remembering the context in which the previous discussion was held. This last problem can be reduced through the use of e-mail.

11.9   THE MODEL

The next step in most benefit-cost studies is to design some sort of model to represent the system or systems that the analyst wishes to evaluate. This model is necessarily an abstraction, although it should be a reasonable representation of reality. The model can be either highly mathematical in form or merely a verbal description of reality. It can be a computer simulation or a written representation. The purpose of the model is to develop a *set of relationships* among the objectives, the alternatives available for attaining the objectives, the estimated cost of the alternatives, and the estimated benefits. The assumptions underlying the model should be made explicit.

After the model has been built, it should be checked to ascertain whether it is structured in such a way as to produce a reasonable representation of reality. Some possible questions that can be used to test the model are:

1. Can the model describe known facts and situations reasonably well?

2. When the principal parameters involved are varied, do the results remain consistent and plausible?

3. Can it handle special cases where we already have some indication as to what the outcome should be?

4. Can it assign causes to known effects?

11.10   DISCOUNTING

When the various costs and benefits appear as a stream over time, the analyst's job becomes more difficult. The costs or benefits may be larger in the first time period and decrease thereafter. Or the benefits and costs may grow larger through time, or the benefits may become larger while the costs become smaller. The important point is that both benefits and costs come in streams that are not necessarily equal, nor are they necessarily the same throughout the years.

If, in Figure 11.1, it is assumed that the two alternatives cost the same, which benefit stream is to be preferred? This is a question that the analyst often faces. Discounting can help answer it.

One of the more controversial aspects of benefit-cost analysis is the use of the discount rate. It is controversial because there is little agreement as to which rate to use. Discount rates from 2 percent all the way up to 25 percent have been suggested by various authors. Some argue that discounting

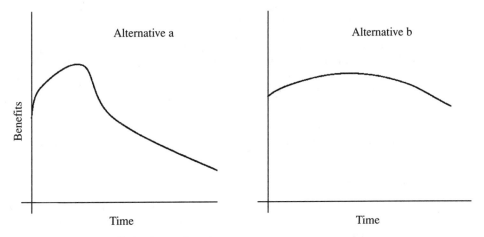

FIGURE 11.1 Time distribution for two alternatives.

should not be allowed at all in benefit-cost studies. Of course, this argument boils down to the assumption of a zero rate of discount. What follows in this section is a brief discussion of some of the arguments of various authors as to what the proper discount rate should be.

Some authors believe there is no "correct" discount rate that can be used for government investments. It is argued that since the private investment decision can be changed or sold on the market, it makes sense to discount it. However, with regard to government investments, society cannot change its benefit stream after the decision to invest is made. That is, society cannot sell part of the benefits from an investment on the market. Thus, there is no discount rate that is valid.

Other authors tend to think that a higher interest rate should be used for riskier investments to reflect the risk involved. Alternatives that realize payoffs in the more distant future are considered more risky. The argument is that unforeseen technological change may eliminate the need for the alternative exhibiting payoffs in the more distant future. A higher discount rate is recommended to reflect the riskiness of these more distant payoffs.

Others believe that the total opportunity cost of government borrowing is the correct discount rate, since debt policy can be used to achieve preferred time profiles of both taxes and borrowing. They argue that additional government borrowing displaces an equal amount of private investment. The total opportunity cost is the sum of the interest rate on long-term government bonds plus the loss of tax revenues associated with society's forgone investment income.

Some economists argue that rather than an interest rate being used to discount benefit and cost streams, the real growth rate of the economy should be used instead. The real growth rate should be used if estimates of future costs and benefits are made in constant dollars. If the estimates take changes in future price levels into consideration, the unadjusted growth rate should be used.

If the analyst has examined all the issues involved and still cannot find the appropriate discount rate, one alternative is to use an upper bound rate and lower bound rate to see if it makes any difference in the final analysis. This is a good use of sensitivity analysis in benefit-cost studies.

11.11   CRITERIA FOR CHOICE AMONG ALTERNATIVES

The next step in the analysis involves choosing a criterion or criteria on which to base a decision as to which alternative system or project is "best." This is one of the more difficult steps in an analysis.

Some authors continually search for the alternative that yields the greatest amount of benefit for the least amount of cost. There is no such alternative. Other authors warn against the "sole criterion fallacy." This is the fallacy that a single criterion can be selected to evaluate all the alternatives. The real world is never this simple. To try to evaluate all of the alternatives on the basis of one criterion is meaningless. This occurs because of the complexities involved in each system and because the systems may differ in many respects.

In most benefit-cost studies, the costs and benefits accrue over time. The prices of both the costs and benefits can change over time. The question that arises is: Should the analyst take these changes in price into account when evaluating the various alternatives? Most economists feel that the analyst should take into account changes in the relative prices of costs and benefits, but not changes in the general price level. All prices should be in constant dollars, which for convenience is usually the initial year of the analysis.

Some of the more commonly used criteria are benefit-cost ratios, net benefit, incremental benefit-cost ratios, and the internal rate of return on investment. Each of these methods is discussed in turn.

11.11.1   Benefit-Cost Ratios

One method used to evaluate the various alternatives is to form a ratio of the benefits and the costs accruing to each alternative project or system and select the alternative that exhibits the highest benefit-cost ratio. This method of selection has produced a great deal of controversy.

Probably the most common criticism of the use of benefit-cost ratios is the argument that the ratios ignore the relative magnitude of the various costs and benefits involved. Some economists believe that benefit-cost ratios can be used as long as the level of cost is approximately the same for the alternatives under consideration. Another defect of the benefit-cost ratio method is that the definition of benefits and costs can affect the outcome. In the highway field, maintenance costs, for example, can be treated as either positive costs or negative benefits. If the maintenance costs are put in the numerator of the ratio, the ranking of alternatives may differ from the case where these costs are placed in the denominator of the fraction. This is not a problem for the net benefit or incremental benefit-cost ratio method.

The primary advantage of using the benefit-cost ratio method for ranking alternatives is that it can deal with cases in which the benefits and the costs are not expressed in the same units. This advantage is unique to the benefit-cost ratio method when compared to all the other methods previously mentioned.

In many transportation benefit-cost analyses, the benefits and the costs are expressed on an annual basis. This is done through the use of the capital recovery factor mentioned earlier. An example of the ratio most commonly used in the highway field is given as:

$$\frac{R_0 - R_1}{S_1 + M_1 - S_0 - M_0} = \frac{R_0 - R_1}{(S_1 - S_0) - (M_0 - M_1)}$$

where S = investment costs on an annual basis
  M = maintenance costs on an annual basis
  R = road user costs on an annual basis

The subscripts 0, 1 refer to the existing and the proposed facilities, respectively. It should be noted that the benefits of a given highway improvement as expressed in the numerator of the ratio are given in terms of a decrease in road user costs. Most studies of highway improvements calculate benefits in this manner.

Thus, the measure of benefits in the highway field is usually limited to the reduction in road user costs that would result from a proposed facility. Some studies incorporate other costs and benefit categories as well, such as externalities associated with highway use.

As mentioned previously, the benefits and the costs that accrue because of highway improvement are usually expressed on an annual basis. However, benefits and costs usually do not accrue in uniform annual streams. It is necessary to annualize benefits and costs by taking the present worth of each stream first, and then distribute these benefits and costs uniformly over the life of the project.

11.11.2 Net Benefit

The net benefit method requires the analyst to subtract the costs from the benefits of each alternative; the alternative that exhibits the largest net benefit (a difference between cost and benefit) is selected. Looking at this in terms of production theory, total benefits can be considered the same as total revenue, total costs have the same meaning, and net benefit can be considered as total profit. Maximizing net benefit then corresponds to the firm maximizing total profit.

Use of benefit-cost ratios to evaluate the various alternatives does not always lead to the same alternative being chosen, as would use of net benefits of the various alternatives. For example:

Alternative	Benefits	Costs	Benefit-cost ratio	Net benefit
A	200	100	2	100
B	30	10	3	20

In this example, alternative A will be chosen if the analyst uses net benefit as the criterion of choice, but alternative B will be chosen if the benefit-cost ratio method is used. If a firm used the principle of the benefit-cost ratio as the criterion to decide at which level of output it should produce, it is unlikely that the firm would be maximizing profit. The one defect in the net benefit method is that it cannot handle situations in which the costs and benefits cannot be expressed in the same units.

11.11.3   Incremental Benefit-Cost Ratios

A third method frequently mentioned in the literature as a criterion of choice is the incremental benefit-cost ratio. A simplified example illustrates how this method works. Assume that a DOT must pick 1 alternative from among 10 mutually exclusive alternatives. There is no budget limitation, because the DOT has funds available to cover the cost of any one of the 10 alternatives being considered. For convenience of illustration, it is assumed that none of the alternatives is dominated by any other (that is, is overshadowed by reason of another being any more effective while costing the same or less). The various alternatives are ranked by increasing cost in Table 11.1.

The essence of the method is to examine the ratio:

$$\frac{(B_{k+1} - B_k)}{(C_{k+1} - C_k)}$$

If this ratio is greater than 1, alternative A_{k+1} is accepted and compared with alternative A_{k+2} in the same manner. If the ratio is less than 1, A_k is accepted and compared with A_{k+2}. These comparisons are continued until the last acceptable alternative is reached. This alternative is accepted as best. Table 11.2 illustrates the method.

TABLE 11.1   Example of Costs and Benefits of Alternative Projects

Alternative	Benefit, B	Cost, C	$B - C$
A1	10	2	8
A2	20	14	6
A3	50	25	25
A4	80	30	50
A5	90	45	45
A6	100	70	30
A7	140	75	65
A8	170	130	40
A9	220	160	60
A10	350	320	30

TABLE 11.2   Example of Incremental Method of Selection of Best Alternatives

Comparison	Incremental benefits	Incremental cost	Incremental benefits/cost	Decision in favor of
A1 vs. A0	10	2	5	A1
A2 vs. A1	10	12	5/6	A1
A3 vs. A1	40	23	40/23	A3
A4 vs. A3	30	5	6	A4
A5 vs. A4	10	15	2/3	A4
A6 vs. A4	20	40	1/2	A4
A7 vs. A4	60	45	4/3	A7
A8 vs. A7	30	55	6/11	A7
A9 vs. A7	80	85	16/17	A7
A10 vs. A7	210	245	6/7	A7

In this example, alternative $A7$ is considered best. In this situation, the incremental analysis leads to the same alternative being chosen as for the net benefit method. This can be readily seen, for, if

$$\frac{(B_j - B_i)}{(C_j - C_i)} > 1, \ (B_j - B_i) > (C_j - C_i) \quad \text{and} \quad (B_j - C_j) > (B_i - C_i)$$

The incremental benefit-cost ratio method for ranking alternatives suffers from the same defect as the net benefit methods. That is, it cannot handle situations where the benefits and the costs are expressed in different units.

11.11.4   Ratios and Net Benefits—A Comparison

It might be helpful to examine the relationships that exist among the benefit-cost ratio method, the net benefit method, and the incremental benefit-cost ratio method for ranking the alternative projects or programs. The benefits and the costs of six alternatives are shown in Figure 11.2, in which the origin represents the existing situation. The benefits that accrue to alternative $A2$ can be represented by the line segment $A2$–C_2. The costs accruing to this alternative are represented by the line segment $A0$–C_2.

The benefit-cost ratio, then, is $(A2–C_2)/(A0–C_2)$; but this is nothing more than the tangent of the angle $A2$–$A0$–C_2. As this angle increases, the benefit-cost ratio increases. Conversely, as the benefit-cost ratio increases, the angle increases. The tangent function is ever increasing in the interval $0°$ to $90°$. Thus, choosing the alternative that exhibits the largest benefit-cost ratio corresponds to picking the alternative whose ray from the origin is the highest. The $45°$ line indicates where the benefits are equal to the costs. All along this line the benefit-cost ratio is 1.0, as should be expected, because tan $45° = 1$. For all the alternatives above the $45°$ line the ratio is greater than 1.0; for those below this

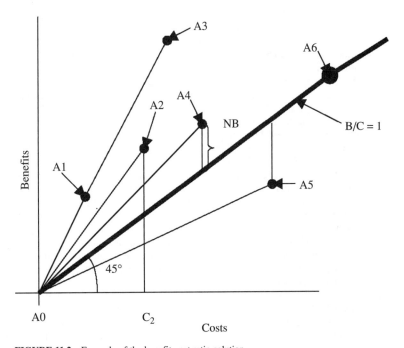

FIGURE 11.2   Example of the benefit-cost ratio solution.

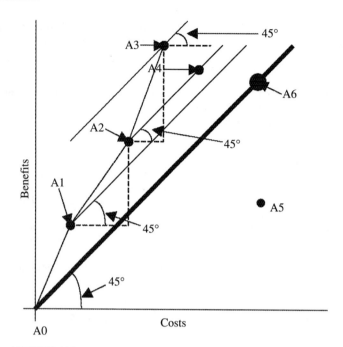

FIGURE 11.3   Example of the incremental benefit-cost ratio solution.

line, less that 1.0. The benefit-cost ratio solution would indicate indifference between alternatives $A1$ and $A3$, because they both lie on the highest ray from the origin.

Net benefit in this figure is represented as the vertical difference between the 45° line and the alternative. As can be seen clearly, all the other alternatives are dominated by alternative $A3$ since this alternative exhibits the greatest benefit-cost ratio and has the largest net benefit of all the alternatives. Alternative $A5$ has a negative net benefit and a benefit-cost ratio less that 1.0.

The same six alternatives are shown in Figure 11.3. The incremental benefit-cost ratio in this figure is represented as the line segment that joins two alternatives; thus, the incremental benefit-cost ratio in going from project $A1$ to project $A2$ is the line segment $A1$–$A2$. Now, the criterion of choice using the incremental benefit-cost ratio procedure is that if the ratio is greater than 1.0, the alternative exhibiting the greater cost is accepted over the other. For alternative $A1$ as compared to alternative $A2$, a ratio of 1.0 can be displayed as the 45° line originating at point $A1$. So, if the incremental benefit-cost ratio is greater than 1.0 it will be above the 45° line; if it is less than 1.0 it will be below the 45° line. But the increment in going from $A0$ to $A1$ is greater than 45°, so $A1$ is considered better than $A0$. The increment in going from $A1$ to $A2$ is also greater than 45°, so $A2$ is accepted over $A1$. This procedure is continued for all the alternatives. As can be seen from the figure, alternative $A3$ is accepted as best by this criterion because a line segment joining $A3$ to any other alternative will be below the 45° line originating at point $A3$. Notice also that alternatives $A5$ and $A6$ are below the 45° line originating at alternative $A1$. There is no need to compare them to any other alternative because $A1$ is better than either of these two and any other alternative that is better than $A1$ will be better than either of them.

Again, the same six alternatives are shown in Figure 11.4. As is evident from the diagram, the 45° line extending from each alternative merely projects the net benefit for each alternative and compares it to the net benefit of the other alternatives. That is, the vertical distance from the 45° line coming out of the origin is the net benefit that accrues when alternative $A1$ is employed. When, for example, $A1$ is compared to $A2$ by incremental procedure, the 45° lines become the references to decide if one

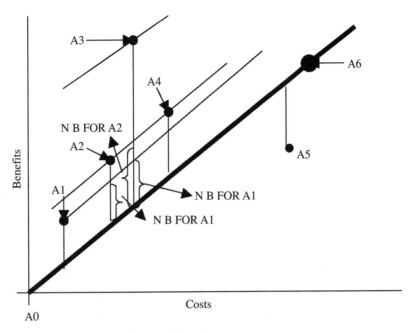

FIGURE 11.4 Example of the net benefit solution.

alternative is better than the other. If $A2$ were on the 45° line through $A1$, it would have the same net benefit as $A1$. Because $A2$ is better than $A1$ it lies on a higher 45° line than the corresponding parallel line through $A1$. This means that the net benefit for $A2$ is greater than that for $A1$. It should be clear from these three diagrams that the net benefit solution is equivalent to the incremental solution and the benefit-cost ratio solution need not necessarily be the same solution as the other two.

11.11.5 Rate of Return

The fourth criterion mentioned previously to evaluate the various alternatives is the rate of return method. This is that rate of interest that makes the annual benefits equal to the annualized costs.

Using an example from the highway field, annual benefits are given by $(R_0 - R_1) - (M_1 - M_0)$ and annual costs are given by $(I_1 - I_0) K(i, n)$. Thus, the rate of return is obtained by solving for i in:

$$(R_0 - R_1) - (M_1 - M_0) = (I_1 - I_0)K(i,n)$$

in which $K(i, n) =$ capital recovery factor,

$$K(i,n) = \frac{i(1+i)^n}{(1+i)^n - 1}$$

where $R =$ road user costs on an annual basis
$\quad\quad\quad M =$ maintenance costs on an annual basis
$\quad\quad\quad I =$ initial investment
$\quad\quad\quad n =$ the life of the project, in years
$\quad\quad\quad i =$ the rate of return

The alternative that produces the highest rate of return is accepted as best.

Some economists take issue with the use of the rate of return method. They believe that this method does not provide an adequate basis for comparing one alternative to another. They propose using the incremental rate of return method instead. This method is similar to the incremental benefit-cost ratio method discussed previously.

11.12 SUBOPTIMIZATION

Another problem that occurs frequently in benefit-cost studies is that it is sometimes difficult for the analyst to relate how well the various alternatives attain the stated objectives. Suppose a hypothetical family wishes to perform a benefit-cost analysis to ascertain the optimal combination of goods and services to buy so as to maximize happiness. Happiness, then, is the family's stated objective. Of course, it is very difficult, if not impossible, to decide which combination of goods and services will maximize happiness. It may be impossible even to quantify the degree to which happiness is attained through the use of one good or another. In other words, although a utility function for the family may exist, in practice it is impossible for the analyst to decide its exact form.

To get out of this difficulty, the family might try to formulate new objectives whose attainment could be more easily calculated and whose attainment may be an indication of the attainment of happiness. The family, in this case, is said to be "suboptimizing" or optimizing on a lower level. Some of these new objectives that the family may try to attain are the satisfaction of hunger and thirst, the attainment of more leisure time, and the attainment of shelter from the extremes of temperature. These new objectives are referred to as lower-level objectives, and happiness can be referred to as the higher-level objective.

Suboptimization occurs when objectives are subclassified and apportioned to a lower level. For example, if the objective of government is to promote social welfare, the government suboptimizes when it grants its DOT the jurisdiction over highways. The goal of the DOT may be to reduce the cost of transportation along the highways in its jurisdiction. The higher-level objective then is to promote social welfare; the lower-level objective is to reduce the cost of highway transportation. It would be difficult indeed for the transportation department to relate various courses of action to the higher-level objective. This lower-level objective is then used to relate the decision of the DOT to this higher-level objective. Whenever the analyst suboptimizes, it is important for him or her to understand and state how the attainment of the lower-level objectives is an indication that the higher-level objectives are attained.

11.12.1 A Highway Example

It is assumed throughout this discussion that one of the objectives of transportation improvements is the promotion of social welfare. It is further assumed that transportation improvement promotes social welfare either through the reduction of the amount of real input required to obtain a stated level of transportation services or through an increase in transportation services for a stated level of inputs. What is measured then is how the productive capacity of the economy has increased through transportation improvement.

Many transportation improvements result not only in an increase in the productive capacity of the economy, but also in a redistribution of income from one group in the economy to another. Many highway studies fail to take this redistribution of income into account. An example of counting a transfer of income as a benefit occurs in the valuation of time savings that accrue to commercial vehicles. One method to value these time savings is to assume that these savings will permit the same level of transportation services to take place with fewer inputs. One of the inputs that is decreased in the number of trucks engaged in these services. Thus, the commercial vehicle owners need a smaller fleet of trucks to do the same amount of work. Included as benefits would be not only the cost to the owners of the trucks they did not have to buy, but also the decrease in interest charges on the trucks. However, interest charges in no way constitute a using up of real goods and services for the economy as a whole. They are merely a transfer of income from the commercial vehicle owners to the lenders of money. Thus, the inclusion of interest charges as a benefit is not consistent with the previously stated objective and would be subotimization at a lower level. Of course, if the goal of the transportation department is limited to the reduction of the cost of operation of commercial vehicle owners, a reduction in interest charges is a valid benefit to include in an analysis. This, however, is an example of the lower-level objective being inconsistent with the higher-level goal.

11.13 ADVANTAGES AND LIMITATIONS OF BENEFIT-COST ANALYSIS

A variety of authors over the years have documented the advantages and limitations of benefit-cost analysis as a method of project appraisal. One of the big limitations is there is less confidence in the estimates of benefits than costs. Another limitation is incompleteness. It is argued that limitations on both time and money place limits on an analysis. Imperfect information also places a limit on what can be quantified. It is argued that benefit-cost analysis can never look at all relevant factors. There is always something left for the decision-maker.

Many authors believe that, in view of its limitations, benefit-cost analysis should be used only as an aid to the decision-maker. It is argued that the main role of the analysis is to sharpen the intuition and judgment of the decision-maker. It should not be used to make the decision. Even so, it is an important part of the decision process. If used as an aid to the decision-maker, it is probably the best aid that he or she has. If nothing else, the analysis can eliminate the really bad alternatives and give the decision-maker a shorter list to choose from.

11.14 QUANTIFICATION AND EVALUATION OF BENEFITS OF TRANSPORTATION PROJECTS

As stated earlier, a good benefit-cost analysis takes into consideration all the relevant costs and benefits. This involves the complete enumeration and specification of all the benefits that accrue to users as well as nonusers, benefits that are easily quantified and those that do not lend themselves readily to quantification. It includes the enumeration of market as well as extra-market consequences. This does not imply, of course, that all benefits must be quantified, nor does it imply that all benefits must have some value attached to them. What it does imply is that no prospective source of benefit should go unnoticed in the analysis.

In the highway field, the beneficiaries of highway improvement are usually categorized as users and nonusers. User benefits can occur as a decrease in the cost of motor vehicle operation, a decrease in time spent on the highway, a decrease in accident rates, or a decrease in the strain and discomfort of nonuniform driving. Nonuser benefits can accrue as an increase in land values, stimulation of economic growth along a new roadway, decrease in the cost of transportation-intensive goods, or increase in the efficiency of delivery of transportation-intensive goods.

Transit improvement benefits include decreases in user costs, including decreases in travel and waiting times, increased comfort and convenience, and reduced accidents and crime. Many of the beneficiaries of transit improvements may seldom if ever use transit. These are the users of the road network who benefit from reduced congestion resulting from increased transit use. A variety of external benefits accrue to transit improvement, such as increases in land values, economic growth, etc., similar to those of highway improvements. In addition, there may be reductions in auto-related emissions resulting from increased transit use. One problem with the evaluation of nonuser external benefits is that double counting of benefits may occur. The analyst must be careful to understand how benefits actually result.

Another aspect of benefit estimation concerns understanding the distribution of benefits that occur. This involves determining which groups, areas, municipalities, etc. benefit and which suffer a loss. One example concerns the provision of park-and-ride lots in suburban areas. A benefit-cost analysis may reveal that the social and regional benefits of such facilities far outweigh the costs. Increased transit use resulting from building of such facilities would result in decreased commuting costs for former auto commuters, reduced user costs for those that still commute by car, reduced air pollution, etc. However, there are a variety of cost and benefit impacts that accrue to the residents of the municipalities in which park-and-ride lots are built. These include increased congestion, pollution, and crime associated with increased traffic volumes and impacts on property values, tax base, and economic development in the vicinity of the park-and-ride lot. Unless municipalities are compensated for possible losses or the losses mitigated in some way, a great deal of opposition to such lots may result in these areas. An analysis of the distribution of benefits and costs may help to develop such compensation and mitigation measures.

Another complication in measuring the benefits of transit improvements is that there is an option value of transit that is not necessarily captured by the benefits associated with regular transit use. The argument is that some people benefit from the availability of transit in case the automobile cannot be used. For example, in bad weather conditions, or if the car breaks down or is not available, individuals benefit from the transit being there, even if they do not use it on a regular basis. Merely estimating the cost savings that these individuals experience when using transit underestimates the true benefits associated with this option value of transit. One way of attempting to estimate such option benefits is to value these using the Black and Scholes formula for valuing stock options. A discussion of this approach can be found in TCRP Report 78 (ECONorthwest and Parsons, Brinckerhoff, Quade & Douglas 2002).

11.15   ROAD USER BENEFITS

Since road user benefits accrue to both beneficiaries of road improvements and transit projects, it would be helpful to look at these in a little more depth. Much work has been done in attempting to quantify these benefits. The major work in this area is the AASHTO "Red Book" of 1977. This work is currently being updated by NCHRP Project 02-23, which should be released some time in 2003.

An important category of road user benefits is decreases in motor vehicle operating costs. These costs can be reduced by transportation improvements through a decrease in the consumption of fuel and oil, a reduction in tire wear, a decrease in maintenance and repairs, and a reduction in ownership costs. These operating costs differ for different types and conditions of roadway. Some of the more relevant factors that the analyst should take into consideration when measuring these operating costs are the number and arrangements of lanes, the type of roadway surface, the amount of grade along a road, the average running speed, the traffic volume, the alignment of the roadway, and whether the roadway is located in a rural or an urban area.

11.15.1   Fuel Consumption

One of the largest single components of vehicle operating cost is fuel consumption. Not all highway improvements will cause a decrease in this cost. Fuel consumption is affected by frequency of accelerations, grades, road roughness, horizontal curvature, operating speeds, and congestion, among other factors.

Generally, road-improvement projects that cause free-flowing average speeds to increase above 35 mph will cause fuel consumption to increase. This increase in speed will, however, produce time savings. Thus, there exists a trade-off between these two types of user costs. There exist many other trade-offs as well in transportation project appraisal. Thus, the analyst cannot really be certain that an improvement that reduces a few elements of cost will actually be the best improvement to make. It could be that other costs are increased so much that they far outweigh the savings produced by the given improvement. This example serves to emphasize that all the relevant costs and benefits should be taken into account in any project appraisal study. If only those elements of cost that are reduced are examined, the analyst may conclude that a given improvement is best or acceptable when in fact it is not.

11.15.2 Tire Wear, Oil Consumption, and Maintenance Costs

The effects of transportation improvements on tire wear, oil consumption, and vehicle maintenance costs have for the most part not been firmly established. The type of roadway surface, route shortening, and type of operation are the only factors that have been established as affecting one or more of these cost elements. Tire wear increases with surface roughness, wear being less for paved than for gravel surfaces. It also increases with increased stop and go driving. These same factors may also affect oil consumption and maintenance costs, although the effects are not well established.

11.15.3 Vehicle Ownership Costs

Vehicle ownership costs include garage rent, taxes, licenses, insurance, interest, and time depreciation. These costs are constant throughout the year regardless of vehicle use. If a transportation improvement, such as a light rail system, results in a reduction in the number of vehicles that are owned, then these costs will be affected.

Time depreciation is independent of vehicle use and is only a function of age. Another type of deprecation is that which is related to vehicle use. There is not much agreement in the literature as to all the factors that affect depreciation due to use, nor is there much agreement as to the percentage of total depreciation that should be attributed to time and use depreciation.

11.15.4 Time Savings

Another benefit that may accrue to road users is a reduction in travel time. Although it is no easy task in itself to measure the time savings in minutes or hours, there exists a much more conceptually difficult problem. This is the placing of some dollar value on the savings. Vehicles are usually divided into passenger cars and commercial vehicles. Separate values of time are calculated for each class of vehicle. Some authors also have tried to estimate the value of travel time to commuting motorists. The value of time varies with the purpose of use. Thus, the value of time for commuting motorists may differ from the value of time for leisure driving.

One problem with valuation of time savings concerns aggregation of such savings. Suppose a highway improvement results in a time savings of 1-minute per motorist. When the minute is then summed for all vehicles, 365 days a year, for the life of the project, it would seem that a very substantial savings would result. But what is the real value of a 1-minute savings per trip? For most vehicles, this savings has little or no value. However, if the 1-minute savings is then added to time savings from other transportation projects, a savings of a considerable block of time can result.

There are various ways that the analyst may go about placing a value on time savings. The revenue and the cost savings methods are two means frequently used to value the time savings of commercial vehicles. The willingness-to-pay method is a way used to value the times savings of private automobiles. Finally, there is the cost-of-time method, which is used not to value time savings but to ascertain the cost of providing such savings.

Revenue Method. The revenue method, also referred to as the net operating profit method, is a technique used to value the time savings of commercial vehicles. It is assumed that time savings will permit an increase in revenue-miles driven by the commercial vehicle, as the vehicle will be driven more miles in the same amount of time. For every hour of time saved by a transportation improvement project, the owner of the commercial vehicle is assumed to have 1 additional hour of revenue operation at no additional cost to those factors that vary with hours of operation.

Cost Savings Method. The cost savings method is also used for the evaluation of commercial vehicle time savings. While the revenue method assumes that time savings result in an increase in revenue-miles in the same amount of time, the cost savings method assumes that the time savings will permit the same amount of mileage to be driven as before, but with the use of fewer resources. Thus, according to the cost savings method, time savings will result in the use of fewer vehicles and drivers to do the same amount of work as before. The costs that are reduced are those costs that vary with hours of operation, or those associated with a reduction in the number of drivers and vehicles. These costs include interest, depreciation, and property tax on equipment and reduced wages and benefits.

Several of the costs that are assumed to be reduced are open to dispute. Interest charges are one of these. Transportation improvements are beneficial to society when they permit the expansion of the productive capacity of the economy through the use of fewer resources to obtain the same level of output or through the expansion of output with the use of the same level of resources. The cost to society in resources used to obtain a given level of output is not a money cost, but the cost in real goods and services whose consumption society must forgo to obtain this output. Money is only a means of measuring how many resources society must use to attain some stated level of output. Interest charges as a money cost do not use up any of society's resources. These are merely a transfer of income from highway users or highway-intensive goods users to bondholders. As such, reductions in these charges should not be counted as benefit to society.

The inclusion of property taxes in the cost savings method can be attacked on the same grounds as interest charges. That is, property taxes do not constitute a using up of real resources, and as such their reduction should not be calculated as a benefit of transportation improvement projects. These same arguments apply to the valuation of vehicle ownership costs mentioned earlier.

Willingness-to-Pay Method. There are several willingness-to-pay methods used to value time savings. All the techniques, however, involve the same basic methodology. The essence of the method requires the calculation of the opportunity cost to estimate how much of other goods and services, at market prices, motorists are willing to forgo to obtain 1 hour of time savings. This opportunity cost is then the estimated value of time savings. The classic approach is to find situations where motorists save time, but must pay a toll, over another route that takes longer but is toll-free.

Cost-of-Time Method. The cost-of-time method differs from the other methods because, instead of calculating the value of time savings itself, this method seeks to compare alternative transportation projects according to the cost of providing the savings. Because the various methods of valuing time savings can lead to different results, many analysts place little confidence in the estimated value of time savings. If little confidence can be placed in the estimate of the value of time, little confidence can be placed on the ranking of alternatives using this value. Instead, the cost-of-time method seeks to determine how the ranking of alternative projects will change for different assumed values of time.

11.15.5 Cost of Accidents

Another road user cost that can be reduced by transportation improvement projects is the cost of accidents. The analyst faces two rather difficult problems in attempting to use the cost of accidents in the analysis. The first problem involves the difficulty in obtaining statistically significant estimates of how much a given project will affect accident rates. The second problem involves attaching some dollar estimates to the cost of accidents. This latter problem is taken up in this chapter.

The usual method of placing a value on the cost of accidents involves enumeration of all costs associated with various types of accidents. The costs usually enumerated are the so-called direct costs of a given type of accident. These direct costs are then estimated for the various accidents that occur in the geographic area that the analyst studies during a given time period. Some average value is then formulated for the costs of each type of accident. Most studies classify accidents by severity. Thus, the usual estimates of the costs of accidents are presented as the cost of:

1. Property-damage-only accidents
2. Nonfatal injury accidents
3. Fatal accidents

Some approaches break down injury accidents into categories of severity.

The direct costs of accidents include:

- Damage to the vehicle itself
- Damage to property outside the vehicle
- Damage to other vehicles
- Cost of ambulance service
- Hospital and treatment services
- Funeral costs
- Value of work time lost
- Present value of the loss of future earnings by those fatally injured or impaired
- Other costs

Indirect costs are those costs incurred to deal with the accident problem as a whole. These include costs of police enforcement, driver licensing and education, overhead costs of automobile insurance, etc.

The direct and indirect costs of accidents can be lumped together and called the economic cost of accidents, since these involve the expenditure or loss of money. Noneconomic costs are those intangibles that do not involve money expenditure or loss. Intangibles include the pain, fear, and suffering of the victims due to death or personal injury.

Each of the elements of direct and indirect costs involves challenges of estimation. One approach is to use court damage awards as a proxy for these costs. This is an approach not without controversy since some analysts argue that these are merely income transfers that only indirectly relate to the true costs incurred. The court award of $7 million dollars, which was later reduced, for spilling hot coffee on a woman's lap is a case in point.

Estimation of expected lifetime earnings is also fraught with difficulties. The analyst must estimate the loss in potential earnings had the victim enjoyed a normal work life. Factors that can affect lifetime earnings include age, sex, employment status, level of education, etc. Thus, a male with a college degree will have a certain expected income, while a female without a degree will have another. An assumption must be made as to how many years the victim would have continued in active employment. Individuals who stay home to take care of children and retirees are difficult to deal with using this approach. Most studies assign arbitrary values to household and child-rearing services. Some studies assume the value of retiree lost earnings is zero, although this may lead to unacceptable conclusions, especially when maintenance costs are subtracted from income.

The intangible, noneconomic costs are even more difficult to ascertain. One approach is to carry the intangible costs through the analysis as a parameter, I. The analyst could then calculate all the economic costs of the project and arrive at a net benefit that includes the intangible cost. For example, suppose the net benefit of a given transportation improvement project is:

$$\text{Net benefit} = -\$100,000 + 0.1I$$

The analyst could then solve for the value of I that would make net benefit equal to zero. In this case, it would be:

$$I = \$1,000,000$$

That is, the intangible cost per accident must be at least $1 million to economically justify such a project.

This approach to dealing with intangibles can be used not only for accident costs but any intangible cost element in which the analyst does not have a good estimate of the possible costs or benefits.

11.15.6   External Benefits and Costs

External benefits and costs are those that accrue to the users of other facilities and nonusers. These are the externalities of transportation projects and should be taken into account in any benefit-cost analysis. These include air, water, and noise pollution effects, impacts on land values, economic development effects, decreases in the cost of transportation-intensive goods, and increases in the efficiency of delivery of transportation-intensive goods.

Some of these external benefits and costs are a using up of real resources or a real cost, such as increased levels of pollution, and should be taken into account. Some categories of external benefits and costs, however, may be either transferred user benefits or a transfer of benefits and costs from one group of nonusers to another. For example, if a new highway is built, land values in the vicinity of the new facility will rise. What is not considered in most analyses is that land values in another area may fall because of the new facility. This can occur to land along an older route that has traffic diverted from it to the new highway. It is important for the analyst to consider these impacts, to understand the distribution of benefits and costs of the improvement.

11.16   CONCLUSIONS

Transportation project appraisal using benefit-cost analysis is a powerful tool to prioritize transportation improvement projects in a systematic fashion. It attempts to substitute quantitative evaluations for the decisions based on intuition, judgment, and political considerations. However, such analyses cannot take all factors into account. There are many intangibles and unquantifiables that cannot be readily incorporated into an economic analysis. As such, benefit-cost analysis cannot be used to make decisions as to which projects to implement. The intuition and judgment of the decision-maker is still an important element in transportation project appraisal. Benefit-cost analysis, however, can shed light on important benefits and costs that can be quantified and valued. As such, it is an important aid to decision-making. It cannot ever replace the decision-maker. Its use in transportation project appraisal helps to ensure that wherever possible, such decisions are made in a businesslike basis.

REFERENCES

American Association of State Highway and Transportation Officials (AASHTO). 1977. *A Manual on User Benefit Analysis of Highway and Bus-Transit Improvements,* AASHTO, Washington, DC.

Barnum, H. N., J.-P. Tan, J. R. Anderson, J. A. Dixon, and P. Belli. 2001. *Economic Analysis of Investment Operations: Analytical Tools and Practical Applications.* Washington, DC: World Bank, February.

Cambridge Systematics, Inc, R. Cervero, and D. Aschauer. 1998. *Economic Impact Analysis of Transit Investments: Guidebook for Practitioners.* TCRP Report 35, National Research Council, Transportation Research Board, Washington, DC.

ECONorthwest and Parsons, Brinckerhoff, Quade & Douglas, Inc. 2002. *Estimating the Benefits and Costs of Public Transit Projects: A Guidebook for Practitioners.* TCRP Report 78, National Research Council, Transportation Research Board, Washington, DC.

Faiz, A., and R. S. Archondo-Callao. 1994. *Estimating Vehicle Operating Costs.* Washington, DC: World Bank, January.

Louis Berger & Associates. 1998. *Guidance for Estimating the Indirect Effects of Proposed Transportation Projects.* National Cooperative Highway Research Program Report 403, National Research Council, Transportation Research Board, Washington, DC.

Organization for Economic Co-operation and Development (OECD). 2001. *Assessing the Benefits of Transport.* OECD Code 752001091E1. http://www1.oecd.org/publications/e-book/7501091E.PDF Paris, April.

Prest, A. R., and R. Turvey. 1965. "Cost-Benefit Analysis: A Survey." *Economic Journal* 75:683–735.

Quade, E. S. 1965. *Cost-Effectiveness: An Introduction and Overview.* The RAND Corp., P-3134, May.

Texas Transportation Institute. 1993. *Microcomputer Evaluation of Highway User Benefits.* National Cooperative Highway Research Program Report 7–12, National Research Council, Transportation Research Board, Washington, DC.

Tsunokawa, K. 1997. *Roads and the Environment: A Handbook.* Washington, DC: World Bank, November.

van der Tak, H. G., and L. Squire. 1975. *Economic Analysis of Projects.* Washington, DC: World Bank, January.

INTERNET RESOURCES

- World Bank publications can be found at: http://publications.worldbank.org/ecommerce/
- OECD publications can be found at: http://www1.oecd.org/publications
- TRB publications and the status of NCHRP projects and TCRP projects can be found at: Transportation Research Board, http://www.nas.edu/trb
- Two software tools are available to evaluate the benefits and costs of transportation improvement projects. One is called the Sketch Planning Analysis Spreadsheet Model (SPASM) and the other is the Surface Transportation Efficiency Analysis Model (STEAM). Both were developed for the Federal Highway Administration. For further information:
  - Federal Highway Administration, *STEAM User Manual,* http://www.fhwa.dot.gov/steam/users guide.htm
  - Federal Highway Administration, *SPASM User's Guide,* http://www.fhwa.dot.gov/steam/spasm.htm
  - Federal Highway Administration, *Using SPASM for Transportation Decision-Making,* http://www.fhwa.dot.gov/steam/spasm.htm

CHAPTER 12
INNOVATIVE INFORMATION TECHNOLOGY APPLICATIONS IN PUBLIC TRANSPORTATION

John Collura
University of Massachusetts at Amherst
Amherst, Massachusetts

Carrie Ward
Capital District Transportation Authority
Albany, New York

12.1 INTRODUCTION

A major movement is taking place in the United States, Europe, and the Pacific Rim to improve public transportation facilities and services with the use of advanced information technologies (National Transit Institute 2001; Casey et al. 2000). These technologies include, for example, computer hardware and software, the Internet, satellite-based navigation and location systems, wire and wireless telecommunications, sensors, and advanced computational methods. Anticipated benefits associated with investments in these technologies are, for example, improvements in customer convenience, transit management and operations, and safety and security (Goeddel 2000), and activities are underway in the United States to promote the evaluation of the actual benefits and costs in operational field tests using these technologies (Casey and Collura 1992).

The primary aim of this chapter is to review the major areas in which transit operators in the United States have invested in information technology and to examine their experiences and lessons learned. These areas include public transportation management and operations, traveler information services, transit signal priority, and electronic payment and fare collection.

A special effort will be made in this chapter to discuss the way in which use of systems engineering concepts are employed in the design and deployment of systems consisting of one or more of these information technologies. Attention will also be given to describing the relationship of these concepts to the National Intelligent Transportation System (ITS) Architecture developed by the U.S. Department of Transportation. More detailed reviews of systems engineering concepts, the National ITS Architecture, and their potential application to enhance public transportation services are provided elsewhere (Gonzalez 2002; FHWA 1997).

While this chapter may be of interest to transportation engineers and transit planners, it may also be useful to other members of the transportation community, including transportation managers and policy-makers, for the purpose of enhancing their understanding of the intent, expectations, and potential merits of information technology investments in public transportation. Such an understanding facilitates discussions with technology and equipment vendors and systems integrators and enables public sector transportation professionals and policy-makers to make more informed choices in information technology investments.

12.2 *SYSTEMS ENGINEERING CONCEPTS*

Systems engineering has been generally defined as an "approach to building systems that enhances the quality of the end result" (Gonzalez 2002). Such systems that employ one or more of the information technologies mentioned above to improve transportation facilities and services have been called intelligent transportation systems (ITS). Intelligent transportation systems designed to enhance public transportation services have also been referred to as advanced public transportation systems (APTS).

Central to the application of the systems engineering approach in the design of intelligent transportation systems are the concepts of system objectives, system functional requirements, and system architecture.

A system objective is the intent that an intelligent transportation system is designed to achieve. System functional requirements are the "what's" that a system must perform in order to accomplish intended objectives. For example, a system objective might be to reduce incident response times to address more quickly the needs of disabled transit vehicles, and a functional requirement might be, for example, the capability to identify the location of the vehicle with the use of one or more of the location technologies mentioned above.

System architecture depicts the structure of a system design and may be of the physical or logical type. A physical architecture includes a depiction of the subsystems in the system, and the logical architecture represents the flow of data in the system. The graphic in Figure 12.1 represents

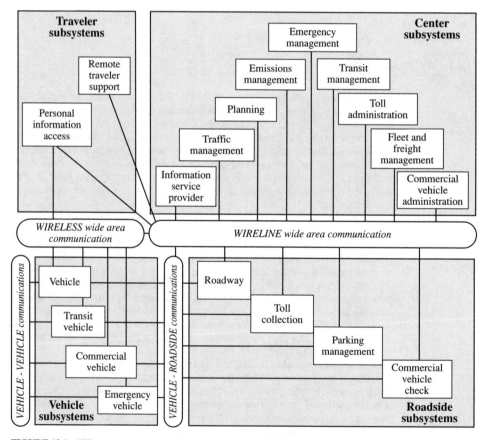

FIGURE 12.1 ITS Architecture subsystems and communication elements.

a very broad representation of the physical architecture and subsystems of an intelligent transportation system as defined in the National ITS Architecture, developed by the U.S. Department of Transportation.

Another concept employed in the National ITS Architecture is the ITS user service, which is a particular service provided to a user. The user may be a traveler, but could also be a public transportation dispatcher or vehicle operator. Table 12.1 provides a list of the seven general categories (or bundles) of ITS user services and the corresponding 31 individual ITS user services currently included in the National ITS Architecture.

The remainder of this chapter will review the four major application areas in which transit authorities in the United States have deployed intelligent transportation systems:

1. Public transportation management and operations (public transportation management, archived data function, ride matching and reservation, and public travel security)

2. Traveler information services (pretrip travel information and enroute transit information)

3. Transit signal priority (traffic control and public transportation management)

4. Electronic payments and fare collection (electronic payment services)

Noted in parentheses are the ITS user services that relate to these four application areas. These user services are highlighted in Table 12.1.

TABLE 12.1 ITS User Service Bundles Included in National ITS Architecture

Bundle	User services
1. Travel and transportation management	• Pretrip travel information • En route driver information • Route guidance • Ride matching and reservation • Traveler services information • Traffic control • Incident management • Travel demand management • Emissions testing and mitigation • Highway-rail intersection
2. Public transportation management	• Public transportation management • En route transit information • Personalized public transit • Public travel security
3. Electronic payment system	• Electronic payment services
4. Commercial vehicle operation	• Commercial vehicle electronic clearance • Automated roadside safety inspection • On-board safety monitoring • Commercial vehicle administrative processes • Hazardous material incident response • Commercial fleet management
5. Emergency management	• Emergency notification and personal security • Emergency vehicle management
6. Advanced vehicle safety systems	• Longitudinal collision avoidance • Lateral collision avoidance • Intersection collision avoidance • Vision enhancement for crash avoidance • Safety readiness • Precrash restraint deployment • Automated vehicle operation
7. Information management	• Archived data function

A special effort will be made to review these areas in terms of their systems, technologies, objectives, requirements, and associated ITS user services and architectures. In addition, selected ITS deployments will be reviewed in each of the four application areas and results and lessons learned will be presented.

12.3 PUBLIC TRANSPORTATION MANAGEMENT AND OPERATIONS

Public transportation management and operations involve a variety of day-to-day activities performed by the public transportation agency personnel. Two major activities include fleet management and passenger security (Casey et al. 1998). While the two activities are often viewed as inseparable and interrelated, fleet management tends to focus on the efficient and safe operations of vehicles and passenger security attempts to reduce risk and harm to transit patrons. To facilitate the conduct of public transportation management activities related to fleet management and passenger security, many public transit providers have made investments in public transportation management systems, including the use of the following systems and technologies:

- Advanced communication systems
- Automatic vehicle location systems
- In-vehicle diagnostic systems
- Transit operations application software
- Automatic passenger counters

The aim of this section is to review these systems and technologies as they relate to public transportation fleet management and passenger security and their intended system objectives, system architectures, and anticipated benefits. In addition, several system deployments in the United States will be examined and the deployment results and lessons learned will be discussed.

12.3.1 Technologies

In order to understand better the capabilities and limitations of the technologies mentioned above, it is important to consider their functionalities and associated benefits as presented below.

Advanced communication systems (ACS) may have the functionalities of transmitting data, voice, and video. A popular technology still being used to transmit voice between vehicle operators and dispatchers in the public transportation industry in the United States and abroad is the conventional two-way radio.

New technology being employed by some transit operators in the United States to transmit data and text is the radio frequency-based mobile data terminals and cellular digital packet data (CDPD).

Anticipated benefits of these new technologies include easing the strain on the transit operator's existing communication network and the ability to accommodate increased telecommunications needs for both data and voice purposes and perhaps video.

Automated vehicle location (AVL) systems are used to monitor vehicles and track real-time location of vehicles and to transmit this information to a central location (Spring, Collura, and Black 1997). This information has been used for a variety of purposes in U.S. industry, including paratransit scheduling and dispatching (Hardin, Mathias, and Pietrzyk 1996; Stone 1993); passenger information systems; and locating vehicles in case of an emergency (Casey et al. 1998).

To facilitate the integration of AVL data into a comprehensive database for transit planning and operations, the SAE J 1708 standard has been used by the transit industry in the United States. This location referencing message specification (LRMS) describes a set of standard interfaces for the transmission of location references among different components of an advanced public transportation system. LRMS interfaces define standard meanings for the content of location reference messages, and standard, public domain formats for the presentation of location references to application software.

AVL technologies employed in the transit industry in the United States include global positioning systems (GPS), signpost and odometer, ground-based radio navigation and location, and dead reckoning. Each technology has a different set of operational characteristics associated with advantages and disadvantages (Casey et al. 2000). A general trend taking place in the United States is that the transit industry is moving toward the use of GPS-based AVL. In some cases a combination of GPS and dead reckoning has been considered for use where GPS signals cannot be picked up by buses due to the so-called canyon effect resulting from tall buildings in high-density downtown areas.

Anticipated benefits associated with AVL are safety and service related. Safety benefits include more timely decisions in emergency situations, quicker response to vehicle mechanical problems, and increases in driver and passenger safety and security. Service-related benefits are improvements in dispatching efficiency, route and schedule planning, customer service, schedule adherence, and the collection of passenger information.

In-vehicle diagnostic (IVD) systems perform continuous measurement of vehicle components (e.g., oil pressure, engine temperature, status of electrical system, and tire pressure). In some IVD applications, out-of-tolerance conditions are sent to the dispatcher in real time without driver intervention. Examples of IVD functionalities include in cab diagnostic and fuel economy display, in-vehicle AVL linkage, and automatic data links to bus maintenance bays with the corresponding bus ID, odometer reading, and fuel level. Anticipated benefits of IVDs are quicker notification of mechanical problems and more rapid response to service diagnostics.

Transit operations software (TOS) is used in conjunction with AVC, AVL, and IVD technology and is employed by dispatchers and other transit staff in the delivery of fixed route and paratransit service. For example, in fixed-route bus service computer-aided dispatch (CAD), software is used to facilitate communication between the bus operator and the dispatch center, a function becoming more and more important due to the increase in data transmission that results from the use of AVL, AVC, IVD, and other advanced technologies. CAD applications reduce the amount of voice traffic. TOS applications also help in tracking the on-time status of each vehicle in a fleet, thus assisting operators, dispatchers, and customers by updating them about the location of vehicles. TOS may also be used to reduce bus bunching.

In rail transit, TOS applications may assist in integrating supervisory control and data acquisition systems (SCADA) with other control systems such as automatic train control (ATC), automatic vehicle identification (AVI), traffic signal loop detectors, and automated train dispatch. In addition, TOS systems in rail transit may aid in establishing vehicle location using signal block occupancy and/or AVI transmitters.

In paratransit operations, CAD software can be used for ride matching and assigning customers to demand responsive vehicles operating in a shared-ride, advanced reservation mode. Such CAD software is employed in combination with two-way data and voice communication systems, mapping software, and a global positioning system base station.

Anticipated benefits associated with the use of TOS applications and the other advanced technologies mentioned include improved scheduling and dispatching, more reliable service to the customer, improved efficiency of operations for the agency, increased productivity, and enhanced safety for drivers and passengers.

Automatic passenger counters (APCs) estimate the number of passengers boarding or alighting and thus provide data for service planning and operational purposes. Such data may be archived for long-term planning activities. APCs use two different types of technologies—infrared beams and treadle mats—and some APCs are integrated into AVL systems to the location of the passengers boarding and alighting. APC output is transferred in two ways: (1) in an off-line mode, in which data are stored on vehicle and downloaded at the end of the day at the transit center; and (2) in a real-time mode, in which a wireless communication system sends the data to a central computer at the transit center.

Anticipated benefits related to the use of APCs are reductions in data-collection costs, increase in the type and range of data available, decrease in the time to process collected data, and improved service planning.

12.3.2   Public Transportation Management and Operations System Objectives and Requirements

Primary objectives of public transportation management and operations systems designed to enhance fleet management and passenger security are:

1. To improve transit service effectiveness
2. To increase operating efficiency
3. To preserve passenger safety

Examples of system features directed at achieving such objectives include:

- A bus operator can activate a silent alarm discreetly and the dispatcher at the transit center can be alerted regarding a theft or robbery. With the use of a satellite-based automated location system (AVL), the dispatcher can determine vehicle location and then contact the appropriate authorities.

- With the aid of an AVL, a dispatcher knows when and where to send a replacement vehicle in the event another vehicle becomes disabled.

- With the help of in-vehicle diagnostic systems, the dispatcher may send a replacement vehicle before the vehicle becomes disabled.

- Computer-aided dispatch (CAD) software may be used to respond to real-time passenger requests for demand response door-to-door service or for fixed route/route deviation service.

- CAD software may also aid in the development of paratransit routes by assigning passengers and designing schedules, thus improving system efficiency by grouping riders and locating their origins and destinations.

Examples of system requirements to achieve the above objectives and features are wireless transmission of either data or voice between the vehicle operator and the dispatcher; direct communication between the dispatcher and emergency response officials; and tracking of vehicles in real time.

12.3.3   Relationship to National System Architecture

Advanced public transportation systems designed for fleet management and passenger security relate to the ITS User service of public transportation management, public travel security, ride matching and reservations, and archived data function, as defined in the National ITS Architecture. It is also useful to think of such systems in terms of their subsystems, as reflected in the National ITS Architecture's so-called interconnect diagram as shown in Figure 12.2. As can be observed, the shaded subsystems include the vehicle, a transit management center, and the transit traveler. On board the vehicle are technologies including the AVL receiver and other components, the silent alarm, and vehicle diagnostic equipment. At the transit center are also AVL components and transit and paratransit operations software and associated computers. A third subsystem may include the traveler interested in scheduling a paratransit trip. Connecting the transit center and the vehicle is some form of wireless communication system. Either wireless or wireline communication provides the link between the traveler and the transit center.

12.3.4   Deployments

Winston-Salem, North Carolina.   The Winston-Salem Transit Authority operates a 22-vehicle paratransit service called Trans-AID (Casey et al. 1998; Spring, Collura, and Black 1997). The service is designed to meet the special needs of elderly and handicapped individuals. The authority uses PC-based computer aided dispatch and scheduling software together with mobile data terminals

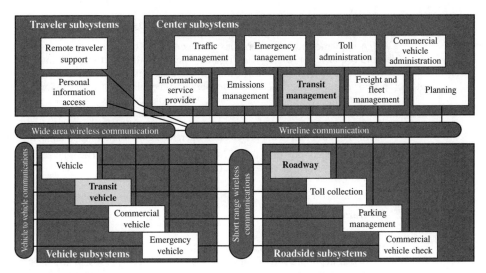

FIGURE 12.2   Architecture subsystem interconnect diagram: public transportation management and operations.

(MDTs) in the vehicles and a satellite-based AVL system. Results associated with the deployment of these technologies are as follows:

- Operating expenses decreased by 8.5 percent per vehicle mile and by 2.4 percent per passenger trip.
- Customer satisfaction increased by 17.5 percent.
- Client base increased by 100 percent.
- Passenger waiting time decreased by 50 percent.

Oakland, California.   From December 2004 to April 2006, a field operational test of the first transit-based smart parking system implemented in the United States was launched at the Rockridge Bay Area Rapid Transit (BART). District station in Oakland (Rodier, Shaheen, and Smirti, 2006; RITA/ITS Benefits website). During peak hours, parking at most suburban stations was at or near capacity. This smart parking system provided the ability to reserve one of fifty parking spaces pretrip by telephone, mobile phone, Internet, or PDA as well as en route by mobile phone. Space availability was displayed on roadside dynamic message signs on an adjacent commute corridor and the ability to reserve parking spaces at a transit station, either pretrip or en route, via. Reservations were initially free, although after 1 year a structure was introduced to charge participants $1 for en route, and $4.50 for pretrip reservations. Surveys of participants found:

- Sizable increases in BART mode share of between 5.5 and 4.0 percent for commutes.
- An average 5 percent decrease in commute time.
- A reduction in total vehicle-miles-traveled of 9.7 miles per month per participant.

Santa Clara County, California.   Santa Clara County, located south of the San Francisco area, operates a relatively large paratransit service called Outreach (Casey et al. 1998). The county uses computer-aided scheduling and dispatch software, a digital geographic database, and a satellite-based automated vehicle location system. Results attributed to this technology deployment include:

- An increase in shared rides from 38 to 55 percent
- A decrease in fleet size from 200 to 130 vehicles

Chattanooga, Tennessee.   The Chattanooga Area Regional Transportation Authority operates a bus network of 16 fixed-routes (Rephlo, Haas, Bauer, 2008; RITA/ITS Benefits website). In 2006, CARTA implemented new operations management software to support fixed-route scheduling and demand response scheduling and dispatch. The scheduling software can integrate information such as hourly rates, overtime rates, overtime policies, and driver break policies to produce runcuts (operator assignments developed from service schedules) that both minimize operating costs and ensure compliance with labor laws and CARTA policies.

Results attributed to the new scheduling software include:

- Savings of about 60 hours per week in operator labor, amounting to about $62,000 per year.

- Decreased runcutting time from 2 weeks to about 5 minutes. The time savings allows CARTA to run multiple scenarios and to select the optimum set of cuts.

12.4   TRAVELER INFORMATION

Traveler information systems provide travelers with information to assist in making travel decisions prior to their departure and/or during their trip (Casey et al. 1998). While some traveler information systems are designed to assist only transit patrons, other systems are more comprehensive and provide information on transit and other local transportation facilities and services including highways, parking garages, and airports.

12.4.1   Technologies

The technologies employed in the design of a traveler information system typically include computer software and hardware and some form of wire or wireless communication system. A traveler information system might be static in that scheduled bus arrival times, for example, are available on a website and accessed via a touch screen in the lobby of a hotel. On the other hand, the system could be dynamic in that it provides such information via a website that provides up-to-date information on transit services with the use of an automated vehicle location (AVL) system, tracks transit vehicles in real time, and provides actual bus arrival times on the touch screen, rather than merely scheduled times. Pretrip information sources include schedules, routes, geographic information, and estimated times of arrival and departure (ETA and ETD). Pretrip access mechanisms include wireline telephone, WWW, wireless devices, kiosks, and display boards. En route information sources include ETA, ETD, and delay information. En route access mechanisms include, for example, cell phones and other wireless devices, bus stop displays, terminal displays, and in-vehicle display boards.

12.4.2   Traveler Information System Objectives and Requirements

A primary objective of traveler information systems is to improve traveler convenience. For example, if individuals learn at work, perhaps via a Web-based dynamic traveler information system, that their afternoon commuter bus is leaving on schedule, then they are able to determine when they must leave the office; if the bus has been delayed, commuters may then decide to stay at work later or perhaps run an errand before getting to the bus station.

Another objective of a traveler information system may be to improve passenger safety and security. For example, if transit riders late in the evening know that the bus from the shopping mall or downtown is leaving on time, then they can plan accordingly. If they know the bus is running behind schedule, they are possibly able to avoid waiting an extended period of time late at night at a bus stop in an insecure location. It should also be noted that traveler information systems may also improve safety and security by facilitating communication between bus and train operators and dispatchers regarding problematic passengers on board transit vehicles and in terminal areas.

Another, more long-term objective often associated with the use of transit traveler information systems is the potential of improving overall ridership (and possibly revenue) by increasing the trip frequency of current riders and attracting new patrons. The underlying assumption is that over time, enhancements in transit service quality, convenience, safety, and security will allow transit to become more competitive as a result of the use of traveler information systems coupled with other transit improvements.

Requirements of static traveler information systems may simply include providing schedule and route information on a website and providing access to the website via wire and wireless communication systems. Dynamic traveler information systems will likely have additional requirements such as tracking locations of vehicles in real time; calculating estimated times of arrivals and departures (ETA and ETD) and comparing these times to scheduled times, and providing forms of communication from the information source (e.g., AVL system), to the information processors, to the website, and ultimately to the user interface and display.

12.4.3   Relationship to National ITS Architecture

Traveler information systems for transit incorporate aspects related primarily to two of the 30 ITS user services included in the National ITS Architecture. One is en route traveler information, which is provided during the trip either in the transit vehicle or at the transit station/stop; the other is pretrip traveler information made available prior to departure to assist in the selection of a mode, route, and departure time.

As depicted in Figures 12.3 and 12.4, several transportation subsystems (e.g., traffic and transit management centers, the vehicle) can be employed to provide information regarding transit and other modes (FHWA 1997). For example, the transit management center might communicate with transit

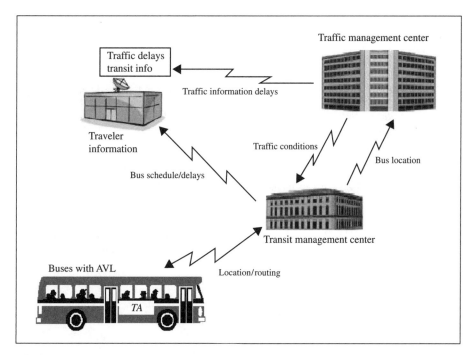

FIGURE 12.3   Traveler information systems.

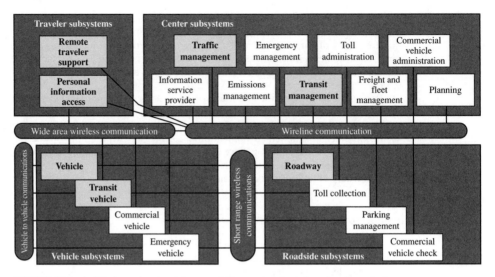

FIGURE 12.4 Architecture subsystem interconnect diagram: traveler information dissemination.

vehicles via a wireless communication system and, with the aid of an AVL system, provide travelers with actual vehicle locations and/or estimated vehicle arrival times, parking availability, and other information via the wireline telephone, cell phones, variable message signs, kiosks, and PCs with an Internet connection. In addition, a traffic management center can provide information on real-time roadway traffic conditions with the use of similar communication systems and devices.

12.4.4 Deployments

Seattle, Washington. A traveler information system for transit has been designed as an integral part of the Seattle Metropolitan Model Deployment Initiative (MMDI), a comprehensive, multimodal effort to implement a variety of ITS user services in the Seattle regional area (Collura, Chang, and Carter 2002). The Seattle MMDI is a partnership involving government, industry, and academia. The major participants are the U.S. DOT, Washington State DOT (WSDOT), King County Metro, the University of Washington (UW), and industry partners. As presented in Figure 12.5, a distinguishing

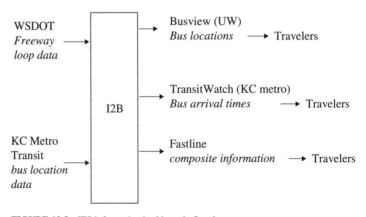

FIGURE 12.5 ITS information backbone in Seattle.

characteristic of the traveler information system in the Seattle MMDI is that it utilizes the Internet as a major communication medium between the transportation subsystems and the ITS information backbone (I2B).

The I2B, conceptually, is an information system design and specifies how participants (e.g., WSDOT, KC Metro, UW, private information service providers such as Fastline) organize information flows, consisting of gathering, processing, and dissemination steps, from the source to the actual users. From a functional standpoint, the I2B acts as a collector and distributor of information (i.e., it receives information from contributors and redistributes this information from processors, who in turn process the information, add value, and provide the information to travelers). The equipment components of the I2B consist of a series of geographically distributed computers located at the UW, WSDOT, and KC Metro. The computers host processes that interact and collaborate with each other via the Internet. The administration of the I2B is performed by UW, which also provides software tools for connecting to the I2B.

As shown in Figure 12.6, UW also operates Busview, a service that uses the World Wide Web (WWW) to display KC Metro bus locations graphically. Travelers, prior to their departure, can connect to the Busview website and click on a selected location of KC Metro's service. Icons will appear showing the route and bus numbers, direction of travel, and the time a bus passed that location.

UW and KC Metro also jointly operate TransitWatch, another transit-specific software application accessible on the Web, which displays estimated bus arrival times at key locations as presented in Figure 12.7.

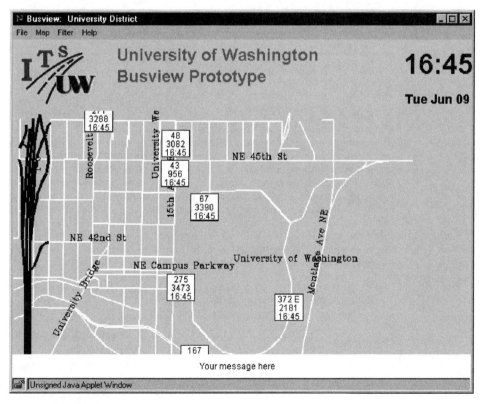

FIGURE 12.6 Busview screen.

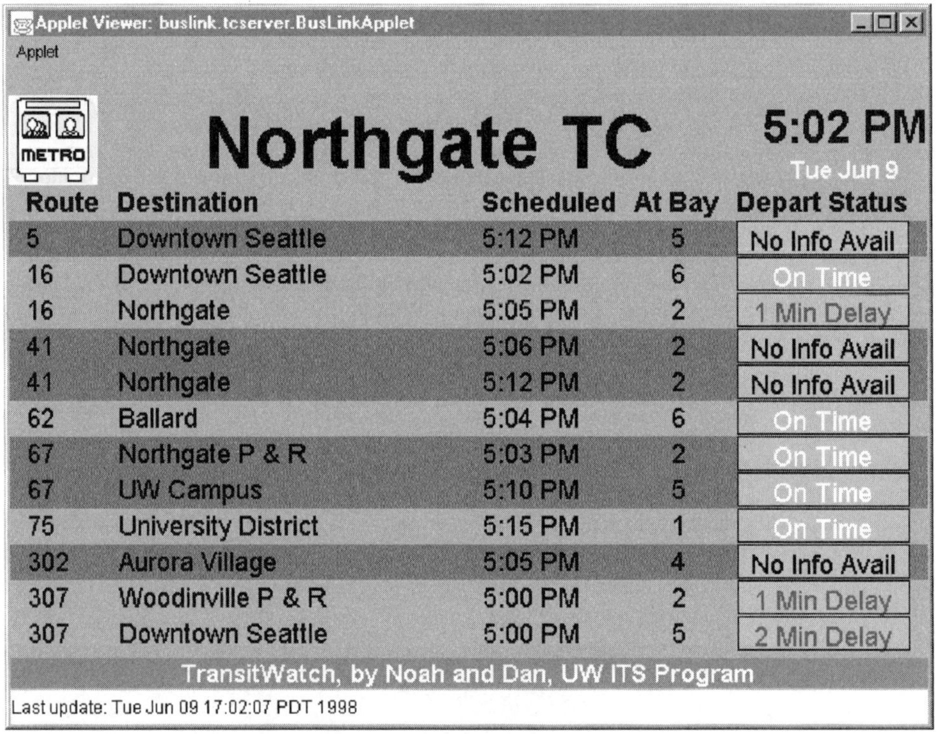

FIGURE 12.7   TransitWatch Buslink screen.

Bar Harbor, Maine.   Over 2.5 million people visited Acadia National Park in 2002, with increasing numbers expected annually (Zimmerman, Coleman, and Daigle, 2003; RITA/ITS/Benefits website). By providing more timely and accurate traveler information to visitors regarding the Island Explorer free shuttle bus service, researchers hoped to improve shuttle bus operations, reduce parking lot congestion and improve aesthetics and safety by decreasing the number of vehicles parked alongside roads. To disseminate accurate and timely information to visitors regarding on-site parking and bus arrival and departure times, real-time bus departure electronic message signs were installed at two locations, automated on-board next-stop annunciators were installed on each bus, and parking information was made available on the Park's website. In addition, a GPS system, on-board automatic passenger counters, and two-way voice communications were installed on buses. During the summer tourist season of 2002, postdeployment questionnaires with visitors and interviews with business managers revealed that:

- Over 80 percent of visitors indicated the real-time bus departure signs and on-board next-stop announcements made it easier for them to get around.
- 69 to 80 percent of visitors believed the ITS helped save them time.
- 44 percent of bus passengers who experienced real time parking information said it helped them decide to ride a bus.
- The Island Explorer bus system helped address congestion and parking management during the summer tourist season at Acadia, according to business managers.
- Drivers indicated ITS made their jobs easier and helped them cope with increasing traffic and ridership.

12.5 *TRANSIT SIGNAL PRIORITY*

Transit signal priority strategies provide buses and light rail vehicles with varying levels of preferential treatment at signalized intersections (ITS America 2002; Collura, Chang, and Gifford 2000). Priority may include an extension of the green interval and/or the truncation of a red interval. Preferential treatments are also provided to emergency vehicles (e.g., fire, ambulance, and/or police), usually in the form of an immediate green interval, often referred to as emergency vehicle signal preemption (McHale and Collura 2001). If such emergency preemption is provided, it is essential that its design and operation be considered in coordination with transit signal priority and that proper signal control transition strategies be included (Obenberger and Collura 2001).

This section deals primarily with the application of transit signal priority for buses on arterial streets. Transit priority system objectives, requirements, and architectures will be reviewed and the results and lessons learned in transit priority system deployments in the United States will be examined.

12.5.1 Technology

A relatively simple transit priority system, as depicted in Figure 12.8, might include a transmitter on board the vehicle that sends a signal to a receiver at the intersection via a short-range wireless communication link.

The receiver passes on the message to the signal controller equipment, requesting priority. When priority is requested and a green interval exists in the direction of transit vehicle travel, the green interval may be held and extended as needed to allow the transit vehicle to clear the intersection. Should a red interval exist, the transit vehicle may be required to wait; alternatively, the red interval may be truncated by shortening green intervals of other signal phases while still maintaining proper clearance times.

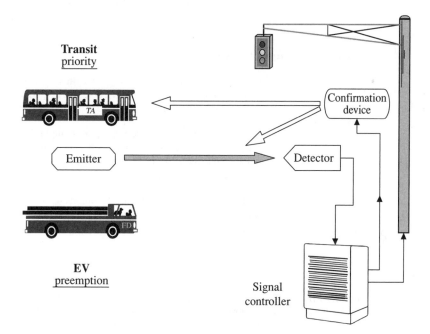

FIGURE 12.8 Transit signal priority systems.

Another, more technologically advanced transit priority system might include a transit management center that communicates via a wireless communication system with all fixed route buses to monitor their locations. Location data are then processed with the aid of an automated vehicle location (AVL) system and used by the transit management center to determine whether the bus is running behind schedule. If so, a message might then be sent to a central processing unit on board the bus to enable the bus then to request priority via a short-range wireless communications link as described above. A variation to this transit priority system might include determining bus lateness on board the vehicle. In addition, another variation might include the transit management center sending a message to the traffic management center for authorization. The traffic management center would accept and process the priority request via a centralized traffic signal control system.

12.5.2 Transit Signal Priority System Objectives and Requirements

An example of a transit priority system objective is to reduce vehicle running times for the purposes of increasing operational efficiency and maintaining scheduled times. Another objective might be to control vehicle flow when vehicles along the same route are following one another too closely. This phenomenon, referred to as bunching, is a common problem in some large urban areas along congested routes with short headways.

In order to meet one or more system objectives, transit priority systems are designed to meet certain requirements (Gifford, Pelletiere, and Collura 2001). For example, one requirement might be to detect buses prior to the intersection and to change signal phases in some prescribed manner (e.g., extending the green interval to allow a local bus to clear the intersection) for the purpose improving schedule adherence. Another requirement might be to grant priority selectively to certain buses to eliminate bus bunching and thus maintain proper headways. A third requirement might be to track actual vehicle location and time and compare them to scheduled time so that priority is granted only to buses running behind schedule by some predetermined threshold level. Other transit priority requirements might be to monitor traffic flow rates and/or to count passengers boarding and alighting so that bus priority can be granted depending on threshold levels of traffic congestion and/ or bus occupancy.

12.5.3 Relationship to National ITS Architecture

Transit priority relates to two of the 30+ ITS user services as defined in the National ITS Architecture. One is public transportation management, a user service supporting transit operational functions. The other is traffic control, a user service related to managing the movement of traffic, including transit and other vehicles along roadways.

The subsystems for a transit priority system may be more complicated than presented above in Figure 12.8. For example, a transit management center might be required to request priority authorization from a traffic management center. In this instance, the transit management and traffic management centers would likely communicate with one another via a wire line connection. The transit management center subsystem communicates via a wide-area wireless communication system with transit vehicles (the vehicle subsystem); the vehicle communicates with the AVL system; and the transmitter on the vehicle communicates with the detector (the roadside subsystem) via a dedicated short-range communication (DSRC) link.

12.5.4 Deployments

Charlotte, North Carolina. The City Department of Transportation deployed a transit priority strategy in the mid-1980s for express buses only along a 6-mile segment of a major arterial including some 14 signalized intersections (Casey et al. 2000). Priority requests include green extensions and red truncations in the inbound direction during the AM peak and in the outbound direction during the PM peak.

For safety purposes the priority phase ends after the bus clears the intersection. Also, to assist in reducing motorist confusion, no signal phases are skipped after the transit priority phase is completed. In addition, all nonpriority phases get a minimum green interval in the spirit of recognizing the importance of travel time as perceived by all travelers.

The physical system architecture of the priority system in place in Charlotte resembles the system in Figure 12.8. Essentially, an emitter mounted on the bus sends a strobe light signal to a detector at the intersection, which transmits the request to the signal controller. The emitter, detector, and controller equipment were provided by the 3M company.

Field studies in Charlotte indicate that bus travel times decrease by approximately 4 minutes and that there are no unacceptable cross-street delays. Anecdotal evidence suggests that other possible benefits include less wear and tear on braking systems, fewer rear-end collisions, decreases in idling time and localized emissions, reductions in driver stress, and improvements in traffic flow along the arterial.

The transit priority system in Charlotte is one of the earliest reported systems in the United States. A reason some cite for this early and relatively successful deployment relates to the institutional setting and decision-making structure in place in Charlotte, which includes the City Department of Transportation being responsible for both traffic and transit operations. Consequently, the centralized decision-making structure transportation authority made it easier to get the transit operator and the traffic engineering community to work together in a coordinated and cooperative effort. Such cooperation is essential in the planning and deployment of transit priority and other advanced public transit system applications.

Bremerton, Washington.   The transit priority system in Bremerton was deployed in the mid-1990s at 8 intersections on 40 buses (Casey et al. 2000; Williams, Haselkorn, and Alalusi 1993). The Bremerton system is very similar to the Charlotte system in terms of system objectives, requirements, and architecture. One difference is that phase-skipping is permitted in the Bremerton system.

Studies in Bremerton indicate that bus travel times were reduced by 10 percent, express bus travel times diminished by up to 16 percent, the recovery period back to signal coordination was about 19 seconds, and no significant changes in cross-street delays took place. Other anticipated benefits expected in Bremerton include reduced driver stress and improved morale and less stop-and-go thus leading potentially to an increase in brake life on the order of 20 percent.

Seattle, Washington.   The priority system in Seattle was deployed along two arterials, at 26 intersections, and on over 200 buses (Casey et al. 2000). While the system in Seattle is very similar to the systems in Bremerton and Charlotte in terms of system objectives, there are some differences in system requirements and architecture worth noting. For example, the Seattle system only provides a green extension and no red truncation. The system architecture in Seattle consists of the following subsystems: the in-vehicle subsystem includes a radio frequency (RF)-based tag; a roadside subsystem including pole-mounted antennas and readers which communicate with the vehicle via the RF signal and with the signal controller via a wireline connection; and center subsystems, including the King County transit center, which determine whether the priority request should be granted depending on time status and local traffic conditions established by the traffic management center staff.

Preliminary studies in Seattle indicate that signal-related stops encountered by buses in the AM peak in unsaturated flow conditions (i.e., level of service B) decreased by 50 percent, average stopped delay diminished by 57 percent, bus travel time dropped by 35 percent, and there were no significant increases in side street delays.

Snohomish County, Washington.   The Central Puget Sound Regional Transit Authority provides public transportation services to King, Pierce, and Snohomish counties (www.soundtransit.org). (Wang et al. 2007; RITA/ITS/Benefits website). Within this area, the South Snohomish Regional Transit Signal Priority (SS-RTSP) project was implemented at 17 intersections over two phases.

Transit vehicles were equipped with transporters to communicate with roadside antenna at intersections. Several measures of effectiveness were assessed to provide a comprehensive evaluation. To obtain accurate transit travel time data, in-vehicle GPS data loggers were used to record vehicle locations periodically. Traffic queue length data and signal cycle failure data were collected manually from video images, and the frequency of TSP calls were calculated from the logged transit vehicle requests. Using the field data, MOEs for average person delay and vehicle delays and stops were determined using a microsimulation traffic model. Evaluation results showed that the SS-RTSP system:

- Reduced transit corridor travel time by approximately 4.9 percent.
- Decreased delay by approximately 336,766 person-hours per year for peak-hour travel.
- Incurred insignificant negative impacts to local traffic on cross streets.

Minneapolis-St. Paul, Minnesota. The Minneapolis-St. Paul metropolitan transit agency has installed Global Positioning System (GPS) equipment in transit vehicles for the purpose of monitoring vehicle locations and schedules in order to provide more reliable transit services. (Davis and Liao 2007). Signal priority strategies implemented in various U.S. cities have mostly utilized sensors to detect buses at a fixed or at a preset distance away from the intersection. Signal priority is usually granted after a preprogrammed time offset after detection. The strategy developed in this research considered the bus's timeliness with respect to its schedule, location, and speed.

Helsinki, Finland. Helsinki City Transport provides all local public transport within Helskini, amounting to about 221 million journeys annually, on 500 buses, 90 trams, and 45 metro trains (HKL 2009; Lehtonen and Kulmala 2002). A pilot project in the City of Helsinki, Finland, provided signal priority to bus lines. Automated vehicle location (AVL) and computer assisted dispatch (CAD) systems were installed on Tram Line-4 and Bus Line-23. Real-time schedule information was also displayed at each tram and bus stop as part of the project.

- Delays at signals were reduced by 44 percent (1 min 13 sec) on the tramline, and 48 percent (3 min 18 sec) on the bus line. However, stop time for both modes increased slightly (3 to 5 sec) since drivers ahead of schedule increased their stop time to balance routing schedules.
- Total travel times decreased 1 percent (21 sec) on the tramline, and 11 percent (3 min 18 sec) on the bus line. The reduction was primarily the result of reduced signal delays.
- On-time arrival improved by 22 percent on the tramline, and 58 percent on the bus line.

In addition, simulations were used to evaluate fuel consumption and emissions for buses before and after system deployment. The calculations considered the average total length of delays, and the amount of time buses were in motion and standing still. The simulation model (VEMOSIM) indicated the following:

- Fuel consumption decreased by 3.6 percent
- Nitrogen oxides decreased by 4.9 percent
- Carbon monoxide decreased by 1.8 percent
- Hydrocarbons decreased by 1.2 percent
- Particulate matter decreased by 1 percent

12.6 ELECTRONIC PAYMENTS

Electronic payment systems (EPSs) in transportation may apply to toll roads, parking, and transit services (Collura and Plotnikov 2001). EPSs may also serve broad nontransportation functions, such as retail applications, telephone services, access systems, medical records, and social programs

(E-Squared Engineering 2000; Fleishman et al. 1998). EPSs may also be integrated with credit and debit cards in banking and other financial transactions.

EPSs in transportation are intended to address a variety of issues and problems that may be perceived by either the transit operators or the travelers (Dinning and Collura 1995). Issues and problems often seen by transit operators include costs and liability associated with coin and cash collection; the importance of accurate data collection and reporting; intermodal coordination; flexibility in fare policy implementation; and the need to reduce fare evasion and fraud. Issues and problems seen by transit riders include the need to have exact change; difficulties associated with intermodal transfer and multiple fares; and the desire for a single payment media accepted by various transit agencies, other transportation providers, and retail stores.

12.6.1 Technologies

Electronic payment systems being used on bus and rail transit employ a variety of technologies, such as card media, readers, vending and distribution equipment, computer hardware and software, wireless and wire telecommunication, and a clearinghouse (Hendy 1997).

Card media include magnetic stripe and microprocessor-based applications with varying levels of performance characteristics pertaining to storage capacity, processing speed, and security features. Microprocessor-based cards (also referred to as chip cards or smart cards) are both contact and contactless technologies and provide greater performance in terms of processing speed, storage capacity, and security, including accuracy, confidentiality, impersonality, data integrity, and repudiation. Finally, it should be noted that the card technology industry is currently considering the merits of a combination card that may employ two or more types of card media on one card.

Readers also vary in type. Some readers accommodate only magnetic stripe cards that are swiped manually through a slot in the reader. Other magnetic stripe cards are transported through the reader mechanically with a motorized unit, as is the case with the magnetic stripe cards currently used on some rail transit systems such as the Washington, DC, METRO and Bay Area Rapid Transit (BART) in San Francisco. A technology trend taking place in the United States and abroad is to deploy contactless smart cards which are read by readers using a wireless communication link. The expectation is that such microprocessor-based and wireless reader technologies will be less costly to maintain than the motorized readers using multiple moving parts. The basic function of the reader is to provide communication between the card and host computer in the electronic payment system. In addition, the reader may validate the card and provide a data processing function.

Vending and distribution technologies are used to assist the transit operator in the sale and distribution of card media to transit riders. Vending machines may be located at rail stations or major bus stop/terminal locations. The initial sale of cards and the process of placing additional value on cards may be carried out with the use of credit cards in person at major locations or on-line via the Internet.

Computer hardware and software and telecommunication systems are also critical in the deployment of the electronic payment system for a number of purposes, including data processing and distribution, database management and recordkeeping, security, and accounting activities such as billing and reconciliation. In the case of electronic payment systems on rail transit, computer systems usually operate in a centralized, on-line mode and data are transferred and processed with wire telecommunication on a dedicated or leased arrangement. For bus transit the card transaction and payment data are captured and stored on readers and transferred to the computer system in a distributive, off-line mode, typically at the end of each day. A thought among designers of electronic payment systems on bus transit is that a wireless transmission link [e.g., wireless fidelity (WIFI)] may someday facilitate the transfer of data from the bus (as it enters the maintenance yard) to the central computer. While the off-line mode has worked to date, it is believed by some transit operators that it may create a greater potential for data loss and fraud and significantly complicate testing, maintenance, and repair of readers on board the bus.

In closing the discussion on EPS technologies, it should be mentioned that some public transit operators and other transportation providers are considering partnering with banks, credit card companies, and other financial institutions in the deployment of electronic payment systems. The primary role of the financial institution is to provide a so-called clearinghouse function, thus possibly reducing the need for the transit operator to make investments in some of the EPS technology and associated activities described above. The clearinghouse function might include, for example, managing the central computer system; providing encryption-key and personal identification number (PIN) management; authenticating cards; validating transactions; and financial billing and reconciliation. It should also be noted that a third-party clearinghouse might be appropriate to operate an EPS in an open-system environment in which multiple transit operators and possibly other transportation providers (e.g., toll road agencies, parking authorities) are involved. An underlying aim of the inclusion of the clearinghouse is to maintain the integrity of the EPS and ensure its proper use by all the transportation operators and their users. Major questions surrounding the merits of a clearinghouse relate to the financial implications regarding costs and flow of revenues (Ghandforoush, Collura, and Plotnikov 2003; Lovering and Ashmore 2000).

12.6.2   Electronic Payment System Objectives and Requirements

Three commonly stated objectives of electronic payment systems in public transit are to enhance customer convenience and satisfaction, increase revenues, and reduce costs associated with payment collection and processing.

An example of an EPS requirement designed to achieve the above objective regarding customer convenience includes providing the ability to allow the transit customer to reload the value of the payment card with a credit card at a vending machine in the train station or via the Internet with a credit card. Requirements related to the other two objectives might include the use of a payment medium and reader to withstand certain temperature and precipitation levels, to satisfy certain minimum transaction processing speeds, to preserve customer confidentiality, and to ensure data security for fraud protection.

12.6.3   Relationship to National ITS Architecture

Electronic Payment Systems relate to the ITS user service included in the National ITS Architecture referred to as electronic payment services. Figure 12.9 suggests that the subsystems of a multimodal

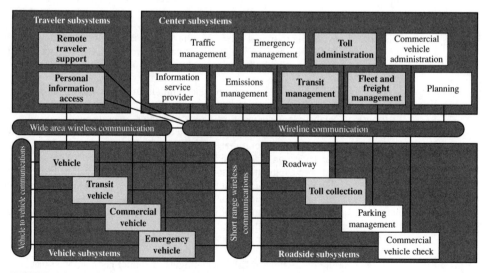

FIGURE 12.9   Architecture subsystem interconnect diagram: electronic payments.

EPS might include the toll administration, the transit management center, and the fleet and freight management center. The toll administration is involved in collecting payments on the toll roads from transit vehicles, commercial vehicle operators, and private vehicles. The transit management center in responsible for collecting the payments on board the buses and in the train stations and park-and-ride parking lots. The fleet and freight management center includes private commercial vehicle operators that have contractual payment agreements with the toll road agencies.

12.6.4 Deployments

Boston, Massachusetts. The Massachusetts Bay Transportation Authority instituted the CharlieCard in January 2007, and the system was entirely upgraded by the end of December. Passengers can purchase and add value to CharlieCards at in-station fare vending machines at all subway stops (MBTA website). Cards are embedded with a computer chip, and the appropriate fare is deducted by waving the card in front of a sensor at the farebox or fare gate. The MBTA encourages their use by charging customers not using the CharlieCard a slightly higher fare. Cards can also hold time-based passes.

- After 90 days, 60 percent of all boardings were by CharlieCard. After two years, the figure rose to 85 percent.
- 1 million CharlieCards were in use after two years.

Source: MBTA website, under Projects/Charlie. http://www.mbta.com/about_the_mbta/t_projects/

Washington, DC. The Washington Metropolitan Area Transportation Authority (WMATA) deployed a regionwide EPS project in 2001. While the primary objective of the EPS project, called SmarTrip, is to improve customer convenience, other objectives include accommodating seamless intermodal passenger transfers, reducing maintenance and operating costs, increasing reliability and security of the system, preventing fare fraud, and facilitating better data collection.

The SmarTrip electronic payment system consists of a number of subsystems. One subsystem is the transit management center, including the card-issuing administrator (WMATA treasury) and the central database (WMATA host computer). Another subsystem includes roadside components such as the Rail Station Monitoring and Control System (SMACS), the parking lot access system, and the SmarTrip vending machines located in rail stations. A third subsystem includes the traveler who carries the smart card and uses it as a means to pay the rail fare and enter and exit the boarding platform area.

The project involves the use of a contactless smart card to pay for WMATA METRO rail and park-and-ride services. The card costs $5 and can be recharged up to $180 in value. Any purchase over $20 earns the cardholder a 10 percent bonus in fare. A pin number uniquely assigned to each card keeps track of the transactions. The value of the card is reduced according to its usage. Fare cards can also be used as daily, weekly, or monthly pass.

The central database host computer system, where customer accounts are maintained, communicates with a network of computers located at rail stations and parking lots, which in turn controls SmarTrip card readers, express vending machines, faregates, and other system access devices. The network of computers at the stations also maintains system performance log and transaction history prior to passing it to the host.

The SmarTrip card reader facilitates contactless communication between the card and the central computer for transactions. When the card is brought within close proximity (about 3 inches) to the reader, the reader initiates data exchange with the card over a low-power radio frequency (RF). The card contains a plastic body with integrated chips and coil antenna inside. Approximately 150,000 passengers, or more than 50 percent of WMATA's daily ridership, uses the SmarTrip system. WMATA plans to put SmarTrip readers in its buses and to extend SmarTrip coverage to other transit systems in suburban Washington, including DASH (Alexandria, VA), Ride-On (Montgomery County, MD), Virginia Railway Express (VRE), and the MARC trains. Efforts are also underway to explore the implications of including a third-party clearinghouse for the reasons cited above in the technology section.

Puget Sound, Washington. Transit providers in the Puget Sound region include the following agencies: Sound Transit, King County Metro Transit, Pierce Transit, Community Transit, and Everett Transit. In 1999, they implemented a set of joint passes and in 2000 agreed to accept each other's transfers as payment of base fare. (King County Metro 2003; RITA/ITS/Benefits website) The Central Puget Sound Regional Fare Coordination (RFC) Project, implemented in 2003, featured smart card technology to support and link the fare collection systems of the major transit agencies operating in the Central Puget Sound region of Washington State (King County et al. 2003).

Even though little evaluation data is available to quantify the impacts of integrated transit policies and practices, transit agencies agree that service integration supports overall agency goals and benefits customers. A case study of the Puget Sound region of Washington State was selected for a case study examining various types of service integration policies related to infrastructure, schedule, information, fare payment, and special events/emergency services. The case study showed that riders have increasingly used multiple agencies for their travel in the Puget Sound region:

- The percent of Pierce Transit riders using the Puget Pass on other systems increased from 41 percent in 2001 to 60 percent in 2004. For King County Metro riders, this figure rose from 12 to 27 percent.

- The percentage of Pierce Transit riders using another transit system in the previous 30 days had risen from 19 percent in 1998 to 27 percent in 2004.

- Among King County Metro riders, the awareness that Puget Pass could be used on multiple systems increased from 61 percent in 1999 to 69 percent in 2000.

Chicago, Illinois. The major objectives of the Chicago Transit Authority's (CTA) Smart Card Customer Pilot Program, initiated in 2001, were to test the ease and acceptance of a smart card for transit customers and to explore economic and operational benefits of this technology (ITS 2000). Potential benefits of primary interest to the CTA related to the possible reduction in maintenance cost of fare-collection equipment and the durability and reliability of the new fare media. It was the nation's first operational, intermodal, multiagency smart card program. Smart cards can be used to pay fares on all CTA buses, at all CTA train stations, and on suburban buses operated by another transit agency.

Like the WMATA SmarTrip card, the smart card in Chicago is a permanent, rechargeable fare card. It is made of plastic and has an embedded computer chip that keeps track of the value of the card. The card can be charged in the vending machines at rail stations and other selected locations. The cost of the card is $5 and the card can be charged to a maximum value of $100. A bonus of $1 is earned for every additional $10. A pin number is assigned to each card and the transactions are kept track using the pin number. Users touch or wave their smart cards in close proximity to readers called Smart Card Touch pads mounted on rail turnstiles and bus fare boxes. The agency pays about $7 to $8 for a smart card versus 5 to 6 cents for a magnetic stripe card. In the long run, smart cards are more cost-effective because they can last for 5 or more years. If the card is lost or stolen, the CTA will issue a new card for a replacement fee of $5 with the value that was on the card at the time the customer reported the lost or stolen card. A cardholder is allowed to complete a trip even with insufficient value on the card. However, the negative balance will be adjusted during the next card-charging period.

The CTA is also considering partnerships with major financial institutions and is examining the potential of including multiple uses for the smart card, such as food, entertainment, and telecommunications purchases.

12.7 CHALLENGES AND FUTURE DIRECTIONS

While it is clear that some of the initial investments in information and communication technologies have benefited the public transit industry in the United States, there will be challenges along the way as more and more public transit providers consider the deployment of additional technologies in a widespread fashion.

One challenge will relate to securing the capital and operating funds associated with the investment of such technologies. Given that many transit providers operate under very tight fiscal constraints, there will be a continuing shortage of funds to invest in technologies such as new wire and wireless communication systems, computer hardware and software, automated vehicle-location systems, transit priority equipment, and electronic payment technologies.

A strategy transit providers should consider in light of the their fiscal situations is coordinating their plans to make these investments with other transit authorities and public agencies. For example, depending on the availability of capital and operating funds, a transit authority might collaborate with the local political jurisdiction (e.g., the county or city) to procure a radio system to meet the basic data and voice communication needs and provide the wireless communication link required for satellite-based AVL systems. The fixed cost of the radio system may require a smaller investment on the part of the transit authority since other agencies would share in the capital costs and the use of the system.

Other examples in which pooling financial resources might benefit the transit providers are in the areas of transit priority and electronic payment system planning and deployment. For example, transit providers might consider pooling funds with the fire and rescue community to design and invest in systems to deploy transit priority and emergency vehicle preemption strategies. Similarly, smaller suburban transit providers interested in a new electronic fare-collection system may benefit in collaborating with the larger metropolitan area transit provider, the net effect being both a reduction in the funds required from the suburban operators and the eventual establishment of a single, integrated fare-collection system region-wide.

In closing, it is important to stress that the deployment of information technologies is important to assist the transit agencies in the United States in providing their existing riders with continued improvements in transit service quality and safety. Such improvements are critical to the transit industry and, coupled with new and improved infrastructure investments, have the potential to aid the U.S. transit industry in maintaining stability in ridership and improving overall system efficiency and productivity.

REFERENCES

"CTA Begins Trial of Multi-Agency Smart Card." 2000. *ITS* 10(18).

Casey, R., and J. Collura. 1992. *Evaluation Guidelines for Advanced Public Transportation Systems.* Prepared for U.S. Department of Transportation, FTA, Washington, DC.

Casey, R., L. Labell, J. LoVecchio, R. Ow, J. Royal, J. Schwenk, L. Moniz, E. Carpenter, C. Schweiger, and B. Marks. 1998. *Advanced Public Transportation Systems: The State of the Art Update '98.* repared for U.S. Department of Transportation, FTA, Washington, DC.

Casey, R., L. Labell, L. Moniz, J. Royal, M. Sheehan, T. Sheehan, A. Brown, M. Foy, M. Zirker, C. Schweiger, B. Marks, B. Kaplan, and D. Parker. 2000. *Advanced Public Transportation Systems: The State of the Art Update 2000.* Prepared for U.S. Department of Transportation, FTA, Washington, DC.

Collura, J., J. Chang, and M. Carter. 2002. "Seattle ITS Information Backbone." *Transportation Research Record* 1753.

Collura, J., J. Chang, and J. Gifford. 2000. "Traffic Signal Priority Strategies for Transit: A Review of Selected Experiences in the United States." In *Proceedings of the World Congress on ITS,* Torino, Italy.

Collura, J., and V. Plotnikov. 2001. "Evaluating Electronic Payment Systems in Public Transit." In *Proceedings of the 11th ITS American Annual Meeting,* Miami, June.

Davis, G., and C-F Liao. 2007. "Bus Signal Priority Based on GPS and Wireless Communications." *ITS Institute Seminar Series,* University of Minnesota: September 25.

Dinning, M., and J. Collura. 1995. "Institutional Issues Concerning the Implementation of Integrated Electronic Payment Systems in Public Transit." In *Proceedings of the Second World Conference on ITS,* Yokohama, Japan.

———. 1996. "Evaluating Payment Systems in Public Transit." Prepared for the 6th ITSA Annual Meeting, Houston.

E-Squared Engineering. 2000. *Introduction to Electronic Payment Systems and Transportation.* Primer, ITA America, September.

Federal Highway Administration (FHWA). 1997. *ITS Deployment Guidance for Transit Systems Technical Edition.* U.S. Department of Transportation, FHWA, Washington, DC.

Fleishman, D., C. Schweiger, D. Lott, and G. Pierlott. 1998. *Multipurpose Transit Payment Media.* TCRP Report 32, National Research Council, Transportation Research Board, Washington, DC.

Ghandforoush, P., J. Collura, and V. Plotnikov. 2003. "Developing a Decision Support System for Evaluating an Investment in Fare Collection Systems in Transit." *Journal of Public Transportation* 6(2).

Gifford, J., D. Pelletiere, and J. Collura. 2001. "Stakeholder Requirements for Traffic Signal Preemption and Priority in the Washington D.C. Region." *Transportation Research Record* 1748.

Goeddel, D. 2000. *Benefits Assessment of Advanced Public Transportation System Technologies, Update 2000.* Prepared for U.S. Department of Transportation, FTA, Washington, DC, November.

Gonzalez, P. 2002. *Building Quality Intelligent Transportation Systems through Systems Engineering.* Prepared for U.S. Department of Transportation, FHWA-OP-02-046, Washington, DC, April.

Hardin, J., R. Mathias, and M. Pietrzyk. 1996. *Automatic Vehicle Location and Paratransit Productivity.* National Urban Transit Institute, September. (HKL) Helskinki City Transport, 2009. Official website. http://www.hel.fi/wps/portal/HKL.

Hendy, M. 1997. *Smart Card Security and Applications.* Norwood, MA: Artech House.

ITS America. 2002. *An Overview of Transit Signal Priority.* Washington, DC: ITS America.

King County Metro, Community Transit, Kitsap Transit, Sound Transit, Washington State Ferries, Pierce Transit and Everett Transit, 2003. "Interlocal Cooperation Agreement for Design, Implementation, Operation and Maintenance of the Regional Fare Coordination System." Seattle, WA.

Klodzinski, J., E. Gordin, and H. M. Al-Deek, 2007. "Evaluation of Impacts from Deployment of an Open Road Tolling Concept for a Mainline Toll Plaza." Paper Presented at the 86th Annual Meeting of the Transportation Research Board: January 21-25. Accessed on RITA/ITS/Benefits website, under Electronic Payment Systems: "In Florida, the addition of Open Road Tolling (ORT) to an existing Electronic Toll Collection (ETC) mainline toll plaza decreased crashes by an estimated 22 to 26 percent."

Lee, E.-B., and K. Changmo. 2006. Automated Work Zone Information System (AWIS) on Urban Freeway Rehabilitation: California Implementation. Paper presented at the 85th Annual Meeting of the Transportation Research Board. Washington, District of Columbia: 22-26 January. Accessed on RITA/ITS/Benefits website: "An automated work zone information system deployed on a California interstate greatly reduced traffic demand through the work zone resulting in a maximum average peak delay that was 50 percent lower than expected."

Lehtonen, M., and R. Kulmala, 2002. "The Benefits of a Pilot Implementation of Public Transport Signal Priorities and Real-Time Passenger Information." Paper presented at the 81st Annual Transportation Research Board Meeting. Washington, District of Columbia: January 13-17. Accessed on RITA/ITS/Benefits website, under Transit Management: "In Helsinki, Finland a transit signal priority system improved on-time arrival by 22 to 58 percent and real-time passenger information displays were regarded as useful by 66 to 95 percent of passengers."

Lovering, M., and D. Ashmore. 2000. "Developing the Business Case." *ITS International* (January/February).

MBTA website, under Projects/Charlie. http://www.mbta.com/about_the_mbta/t_projects/

McHale, G., and J. Collura. 2001. "Improving the Emergency Vehicle Signal Priority Methodology in the ITS Deployment Analysis System (IDAS)," In *Proceedings of the ITS Congress,* Sydney.

National Transit Institute. 2001. *Advanced Public Transportation Systems: Mobile Showcase One Day Workshop Notebook.* Brunswick, NJ: National Transit Institute.

Obenberger, J., and J. Collura. 2001. "Transition Strategies to Exit Preeemption Control: State of the Practice." *Transportation Research Record* 1748.

Rephlo, R. (SAIC), R. Haas (SAIC), and J. Bauer (SAIC). 2008. Chattanooga SmartBus Project Final Phase II Evaluation Report. US DOT-ITS Joint Program Office, Research and Innovative Technology Administration: June 10. http://www.itsdocs.fhwa.dot.gov/JPODOCS//REPTS_TE/14431.htm. Accessed on RITA/ITS/Benefits website: "In Chattanooga, Tennessee, fixed-route scheduling software improved operations by saving approximately 60 hours per week in operator labor, resulting in a savings of approximately $62,000 per year."

RITA/ITS/Benefits website: "In the Puget Sound region of Washington State, a fare payment integration system that used joint passes to allow base fares to be transferred between agencies increased the percentage of riders that made transfers."

Rodier, C., S. Shaheen, and M. Smirti. 2006. "Transit-Based Smart Parking in the U.S.: Behavioral Analysis of San Francisco Bay Area Field Test." Institute of Transportation Studies, University of California. Accessed on RITA/ITS/Benefits website, under 'Transit Management' "Outside San Francisco, a transit-based smart parking system contributed to an increase in transit mode share, a decrease in commute time and a reduction in total VMT."

Spring, C., J. Collura, and K. Black. 1997. "Evaluation of Automatic Vehicle Location Technologies for Paratransit in Small and Medium-Sized Urban Areas." *Journal of Public Transportation* 1(4).

Stone, J. 1993. "Paratransit Scheduling and Dispatching Systems: Overview and Selection Guidelines." In *Proceedings of IVHS America Third Annual Meeting,* April.

Wang, Y., et al, 2007. Comprehensive Evaluation on Transit Signal Priority System Impacts Using Field Observed Traffic Data. University of Washington: June 15. at http://www.transnow.org/publication/final-reports/documents/TNW2007-06.pdf. Accessed on RITA/ITS/Benefits website, under Transit Management: "In Snohomish County, Washington State, implementation of a transit signal priority system on two test corridors reduced average transit corridor travel time by 4.9 percent, and had insignificant negative impacts on local cross street traffic."

Williams, T., M. Haselkorn, and K. Alalusi. 1993. *Impact of Second Priority Signal Preemption on Kitsap Transit and Bremerton Travelers.* Seattle: University of Washington.

Zimmerman (Battelle), C. A., T. G. Coleman (Battelle), and J. Daigle, 2003. "An Evaluation of Acadia National Park ITS Field Operational Test: Final Report." Federal Highway Administration, U.S. DOT: June. Accessed on RITA/ITS/Benefits website under Traveler Information/Tourism and Events: "A survey of bus riders within Acadia National Park in Maine found that 44 percent who experienced real-time parking information agreed the information helped them decide to ride the bus."

CHAPTER 13
PARKING MANAGEMENT*

P. Buxton Williams
MPSA Partners, Oak Park, Illinois

Jon Ross
MPSA Partners, Chicago, Illinois

13.1 OVERVIEW

> Go to the village ahead of you, and at once you will find a donkey tied there, with her colt by her. Untie them and bring them to me.
>
> —Matthew 21:2, New International Version

Parking management—providing space where people can safely leave their form of transport (and link to other transportation) while they go about the business of working, shopping, and residing—has been a need of the common citizen for centuries. While the earliest documented reference to anything resembling a parking operation may be the excerpt above (it is unclear whether the village in question received any shekels from owners of donkeys who used the space), government has long faced the challenges posed by parking, and will continue to do so, no matter what vehicles and forms of transportation humans employ.

In modern times, government has discovered many ways to earn revenue for the use of parking spaces, a discovery that has, in turn, spawned a whole industry of hardware, software, management, and support services. Parking and related functions (such as transit, public works, and road/highway construction) are fundamental to the everyday activities of virtually every citizen. And, as the automobile is so central to the American way of life—and as the average American automobile becomes larger with SUVs dominating the landscape—parking as an industry and as a networked function of government faces new challenges.

To the average citizen, parking is like a trip to the dentist or buying an insurance policy. No one wants it, but when you need it, it had better be there for you. Parking is clearly taken for granted by those who use it. But parking, and the services and returns it can generate, are also undervalued and underutilized. Parking management in the bureaucracy of most American cities—including many of the largest cities—is isolated and marginalized, run by midlevel staff in police departments, municipal courts, and parking offices whose expertise is typically not shared with or leveraged across other agencies of government that could use their information to the economic and service benefit of the citizenry. As we will reaffirm later in this chapter, parking management has great potential to contribute to a host of key government activities, in areas from public works and transportation planning to economic development and tourism.

---

*Reprinted from the First Edition.

A look at industry statistics and economic indicators makes a solid case for elevating parking management out of its current stealth in most local governments. According to the International Parking Institute, the industry is responsible for more than $26 billion in U.S. economic activity—roughly evenly divided between the public and private sectors. More than a million Americans work in the field. Somewhere around 5 million parking meters are installed in the cities and towns of America. And, with well in excess of 100 million parking spaces in the United States—roughly two-thirds of them off-street—demand for parking is huge and growing.

Revenues from parking—even from parking fines alone—are significant. Even for cities with poor collection rates, there is a lot of money in parking—funds every city needs, regardless of the economic climate. To a megacity like New York, parking fines are worth hundreds of millions of dollars a year; to a top-50 city, revenues are typically in the tens of millions, and to a city of 75,000 people, parking revenues may fall under $1 million annually. But no matter the size of the city, optimizing parking management—in revenue generation and in terms of public safety and public service—is critical to the operation of local government.

For these reasons, this chapter will provide a highly strategic approach to parking management. It will discuss what a truly comprehensive parking management system includes and how it operates and will offer insight on how municipalities can develop and manage parking systems that control the use of on-and off-street parking spaces; optimize revenues, efficiencies, and customer service; and leverage the programs and information generated by a well-run parking management system into benefits across other levels of government.

13.2 SURVEY AFFIRMS MANY PARKING MANAGEMENT OPERATIONS ARE MARGINALIZED

To assess the state of the parking industry and its managers, and to confirm the authors' perspectives as delineated in this chapter, the authors conducted a survey. Surveys were sent to parking managers, traffic enforcement officers, and other city officials with responsibility for parking management in three categories of cities: the top 50 in population in the United States according to the 2000 Census, cities in the next tier of population (approximately 100,000–500,000), and cities in the 50,000–100,000 category. The questions spoke to issuance, enforcement, and collection practices; agencies/departments responsible for parking management and analytics; how parking revenues are generated and distributed; management and technology tools and practices (such as outsourcing); and quality assurance practices and metrics.

Surveys were completed on paper or via Internet, results were tabulated, and key findings were summarized below. Completed surveys were received from about 35 cities, distributed roughly evenly across the three categories of cities denoted above. While this number of responses is not sufficient to produce findings that are statistically significant, they do reveal and reaffirm many of the conclusions put forth by the authors.

13.2.1 Summary of Findings

- Police departments, courts, and public works are the primary or, in many cases, exclusive users and stewards of parking management systems across the United States. This practice generally mitigates against other departments using valuable parking and public safety data to enhance services to the general public and contribute to growth and new business initiatives.

- With few exceptions, revenues generated from parking management operations cover operating expenses and provide additional funds for other initiatives within the parking management system. This practice can stand in the way of directing resources toward other lucrative citation-related revenues (i.e., permitting, false alarm billings, and other fee/fine-based programs administered by municipal courts).

- Most parking management systems use total revenues and collection rates as the standard form of measurement of their efficiency. This suggests that these systems are not operating in a suitably sophisticated manner to provide sufficient analytics to gauge the impact of improvement initiatives.

- Parking management systems across the country are not yet taking full advantage of the Internet, credit cards, and other customer-driven alternatives as payment options. Comprehensive use of such measures is associated with improved collection rates, cash flow, and citizen/customer satisfaction.

- Municipalities are divided on outsourcing of parking management operations (i.e., ticket processing, collections, citation enforcement/issuance, and collections). However, most municipalities with a collection rate below 60 percent (below the minimal industry standard of 70 percent) do not outsource any of their parking management operations. This suggests that expanded outsourcing, when well managed, is likely to contribute to improved operations and efficiencies in parking systems.

- A major form of parking management outsourcing, automated ticket-writing, is on the rise. More cities today use handheld ticket-writers than they did 3–5 years ago. Consistent with this trend, more municipalities are automating ticket tracking and processing. This translates into improved efficiencies, faster collections, and fewer voidances of citations due to human error.

- Municipalities are reluctant to turn parking operations management over to a parking authority. These entities typically have more independence and flexibility than traditional city-run parking operations and can translate to more efficient management and generation of new revenue sources through economic and business development initiatives.

- Municipalities are increasingly recognizing the value of providing more accessible multimedia information to citizens about parking and related services. This practice contributes to citizen participation and satisfaction, and therefore improved operation and efficiencies.

13.3 *WHAT IS COMPREHENSIVE PARKING MANAGEMENT?*

In a perfect world, citizens would adhere to the rules and regulations, thereby eliminating the need for a parking management system. People would use parking meters for the allotted duration, pay the required amount, and remove their vehicles when their time was finished, thus freeing up the space for other motorists. And there would be no need to develop a system around the issuance of parking tickets, regulation of parking spaces, enforcement of parking laws, and provision of adjudication procedures.

But we do not live in such a world, so we need parking management systems to ensure that citizens have sufficient parking facilities, are accountable for parking violations, and have the opportunity to appeal any real or perceived miscarriage of justice.

In this real world, parking management means more than just providing spaces for parking cars. Truly comprehensive parking management involves a good deal of social engineering and planning because it must thoughtfully address a host of resource management issues: public safety, traffic flow and management, urban planning and design, land use/environmental management, and how to build, manage, and draw adequate revenue from parking operations and services. Fundamentally, a comprehensive parking management system serves the citizen in all these ways. This chapter discusses the major building blocks of comprehensive parking management relevant to a midsize village or a megacity. It discusses key facets of parking management: hardware/software functionality, enforcement measures, customer relations programs, and others. The chapter will also discuss methods for validating and verifying parking management system functions and measuring the impact and return on investment of parking management programs. Using samples and case studies derived from experience and a companion survey (see attachment), this chapter delineates best practices in parking management, including how to integrate these programs into economic development and growth initiatives.

13.4 COMPREHENSIVE PARKING MANAGEMENT—SYSTEMS AND OPERATIONS

Any discussion of parking management systems would be incomplete without making reference to the full spectrum of components these systems include. Parking management involves parking lots, garages and on-street parking; it includes parking meters, signs, booting and towing equipment, lift gates, loop detectors, smart cards, cellular phones, paving equipment, striping and chalking equipment, lighting, and construction; it uses parking tickets, permits, hang tags, envelopes, automatic ticket-writing devices, booths, and kiosks; it requires hardware and software for ticket processing, ticket dispensers, revenue control, and traffic controllers, to name a few. Needless to say, the parking industry is dominated by a wide variety of specialized vendors, manufacturers and service providers. In such a fragmented market, it is difficult to try to determine the origins of parking management as an industry or service without encouraging a serious debate among industry experts.

Generally speaking, a state-of-the-art parking management system should be proactive, responsive, accountable, and effective. It must be built upon a solid foundation of technology and well-conceived business processes. An exemplary system should include several key features:

- A ticket-writing process that minimizes or eliminates data entry errors due to human error such as poor handwriting (if possible, all tickets should be written using automated ticket-writing devices)
- Speedy processing of tickets via data entry or downloading from automated ticket-writing devices
- On-line interface with secretary of state/department of motor vehicles in order to obtain critical information about violators (if possible, in a real-time interface)
- Ability to generate notices (statements of outstanding violations) to violators
- Clearly defined enforcement measures

Additionally, the system must include the capability to:

- Track tickets from issuance through to final resolution.
- Identify violators with multiple vehicles separately from rental or fleet owners.
- Track and associate correspondence directly to the ticket or vehicle owner in question.
- Generate statistical reports that help to measure the effectiveness of the system.
- Provide citizens with multiple payment methods, including telephone and Internet options.
- Allow citizens to appeal decisions in person or electronically.
- Provide interactive Internet access to citizens who wish to make inquiries, complete applications for permits, provide change of address, or make a complaint or comment.
- Encourage payment through enforcement measures such as booting, towing, license suspension, and plate suspension.
- Interface with other city departments (i.e., public works agencies) to track damaged meters and signs and some state departments (such as motor vehicle administrations).
- Generate standard and ad hoc correspondence letters to citizens regarding inquiries, complaints, and general ticket status.
- Maintain multiple addresses in addition to a default address for violators.
- Furnish citizens with information about available permit spaces in city-run garages.

An oversimplified view of a parking management system can be reflected in a small spreadsheet showing the number of tickets issued, the violation committed, and whether payment has been received. This approach—one put in place in many smaller cities at the outset of their parking management programs, and still in operation in a good number of jurisdictions—enables a small village to determine at least the number of tickets being used, when to reorder tickets, and how much is collected from parking offenders.

Technological advances, double-digit population growth, and the need to become proactive to citizen needs and more responsive to citizen demands have driven local governments to initiate more robust parking management system solutions. Some of these solutions include document imaging, automated ticket-writing devices, wireless interfaces, booting and towing, workflow processes, interactive voice response (IVR) applications, e-commerce, webcasting, and use of the Internet for queries, appeals, and payments. Moreover, these solutions are flexible and scalable and allow processing of millions of tickets in nanoseconds.

Figures 13.1*a* and 13.1*b* use a citation flow diagram to show at a glance sample parking management system activities over a period of time. In the diagram, the horizontal lines with their respective labels represent specific stages in the life cycle of a citation. The vertical arrows represent the movement of a citation from one stage to the next over a period of time. For example, the first arrow from ticket issuance to data entry represents the change in status of the ticket from being issued to being entered in the system. The arrowhead indicates the direction of flow of the activity.

Figures 13.1*a* and 13.1*b* represent these two extremes of parking management system solutions. (Most systems are, of course, somewhere in the middle in terms of their complexity.) Figure 13.1*a* is a simple approach where the need for high-volume processing, fast-paced decision-making, and intricate data interfaces is not present. In this scenario, tickets are issued and sent for data entry. At times, tickets are paid before they are entered in the system. Some tickets are paid after data entry. Within a specified timeframe, violation notices are sent out for those tickets remaining unpaid after data entry. Later still, tickets remaining unpaid are sent to a collection agent. (Of course, tickets may be paid at any time during the process, as depicted with the arrows going from all other lines/ stages to the payment line/stage.) While this is a simplified process, it should be noted that there are municipalities that are not even at this level of processing and would need to mount a considerable effort to move up to this baseline.

Figure 13.1*b* provides a more sophisticated approach to parking management. This option is typically associated with larger cities that have a greater focus on parking as a function that enables economic development, improves public safety, and allows transportation engineers to gauge the impact of changes in traffic patterns. This type of system typically includes features such as ticket imaging, workflow process, interactive voice response (IVR), Internet interfaces, automated ticket-writing devices, e-commerce links, telephone inquiries, and citizen-friendly payment options. A system of this nature normally provides interactive access for citizens to make and track an appeal, request information, submit complaints or comments, and view transactions against their respective accounts. Also, a system with such a high level of functionality usually contains features and functions to allow easy management and control of city parking lots, garages, meters, and signs.

13.4.1 "Build-to-Suit" Parking Management Options

When trying to determine which parking management option or infrastructure is best for a given municipality, the public administrator has a variety of solution alternatives to consider.

Typically, cities opt for an in-house alternative, which typically is cheaper but can be plagued by operational limitations and their associated costs. Some available alternatives worth considering include:

- *Full-scale facilities management:* In this approach, the parking management operation is turned over to a private operator, who assumes responsibility for all parking management functions and personnel, including ticket-writing officers. Police officers continue to issue tickets but on a smaller scale since the bulk of the ticketing is covered by the private operator's employees. In some cases, the ticket-writing officers are managed by a third party to avoid conflict of interest.

- *Functional outsourcing:* This is a variation of the full-scale facilities management approach. Instead of turning all functions over to a single operator, the municipality determines the parking management functions best suited for outsourcing and turns those over to private operators or even internal departments (such as an information technology department).

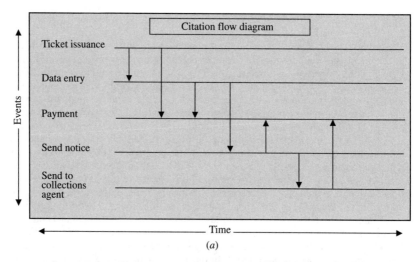

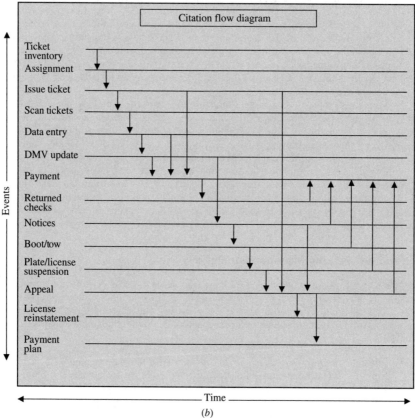

FIGURE 13.1   Citation flow diagram.

- *Software application purchase:* This option involves using technology and software to supplement an internal parking management operation and proves useful to many municipalities.

- *Private-public partnerships, revenue-sharing agreements, and privatization:* These options are becoming increasingly attractive to the public sector in the era of "reinventing government" programs, where privatization in particular has proven useful in delivering some infrastructure and public works functions. These three options have some significant application to parking and related areas, and because a private-public partnership usually includes a revenue-sharing agreement and a privatization element, they are necessarily linked for the purpose of this chapter. In a typical arrangement of this kind, an IT services company assumes the operation of the parking management system—possibly in a privatized or semiprivatized structure—for a fixed fee while the municipality handles all other functions. Although the private partner is responsible for the management of the technology solution, the citizenry hardly ever sees its staff, because municipal personnel handle all customer-facing activities. New revenues generated through greater efficiency of operation or from new programs may be split between the local government and the private contractor.

Of course, there are arguments for and against these approaches, depending on the municipality's perspectives and public/political sentiment. Full-scale facilities management is a boon to the municipality that stresses its core competencies of enhanced government-to-citizen services. All parking management issues are handled by the facilities management company, from a defective PC to a complaint about a parking meter malfunction. In a five-year agreement, for example (a standard term for a deal of this nature), the municipality has a fixed budget for this function, which means a consistent line item and a relief for many department heads. A properly negotiated contract will shield the municipality from the expenses of technology refreshment in order to keep pace with technology advancement, a responsibility of the facilities management contractor.

There are some potential downsides to this approach, and municipalities should perform careful due diligence before selecting a vendor for a facilities management arrangement. A local government should make sure the vendor is financially solid; if not, placing too much of the parking management operation in the hands of an unstable organization can be problematic. Also, cities should make sure they enforce sufficient checks and balances to prevent the contractor from abusing its position by harassing citizens with an inordinate amount of parking tickets. The municipality should also establish, through its contract, that timely and accurate analytical information will be provided. This information may be used to measure the performance of the contractor.

Functional outsourcing requires that the municipality conduct a thorough evaluation of its capabilities and expertise in order to determine which functions are best suited for outsourcing. With this approach, a municipality is able to nurture and develop its employees who will support the contractor responsible for the outsourced function. Also, this approach functions as a cost-containment measure because its cost is fixed over the contracted period. A negative impact of this approach is the possibility that a municipality may find itself overly dependent on a private contractor for key services. A typical example of functional outsourcing is the use of a company to manage meter installations, maintenance, and collections, with ticket processing provided in-house.

Similar to functional outsourcing, software application purchase requires an evaluation of the function for which the software application is required. Today, software application purchases with specific and refined procedures are commonplace in municipal purchasing departments. Following these procedures results in cost savings down the road. As in the other options addressed in this section, the right software application hinges on having an in-depth knowledge of the business process and its interfaces and impact on other departments. This approach puts responsibility for system and software availability onto the software contractor. Also, system upgrades and maintenance are left to the vendor, freeing municipal staff to focus on their duties and reducing software applications to just a tool. When considering this approach, cities need to account for additional training as the software vendor provides upgrades. Depending on the scope of the upgrade, training may require days as opposed to hours, and therefore demands coordination of schedules to limit service disruptions.

Private-public partnerships, revenue-sharing agreements, and privatization are innovative means of involving private-sector partners in efficiently and quickly finding cost-saving measures and alternate revenue streams. These options tend to move more quickly than other business models because the administrative and business processes are typically given legal authority to move faster and conduct business in a more enterprising way. (Parking authorities, for example, often operate in this fashion. Authorities are generally given more leeway in terms of taxing, land acquisition, and business transactions than traditional public-run governmental entities. The Miami Parking Authority, for example, is responsible for management of at least one mixed-use commercial building, in addition to public parking facilities throughout the city.)

In this approach, private partners receive additional payments based on a predefined percentage of savings and revenues they bring to the system. A plus is that contractors are paid for performance and therefore actively seek to increase revenues. For its part, the municipality typically has fewer burdens in terms of man-hours and hands-on activities and can take advantage of the economic benefits associated with privatized programs and public-private partnerships. City officials, though, should make sure not to give too much away in forging these kinds of business arrangements. Most business models of this type can be conducted so that local government maintains ownership of the land and infrastructure of its parking operation while turning over management of the system—and development of new programs, services, and revenue streams—to private partners.

13.4.2   System Planning

Planning for a parking management system is futile without first defining what and how much should be included. Depending on the size and complexity of the jurisdiction, a parking management system may be as simple as controlling and processing tickets issued for expired meters at local parking facilities. It may include controlling and processing tickets for expired meters, handicapped parking, and illegal parking; issuing residential permits; tracking citizen correspondence; conducting adjudication; performing collections activities; providing alternative payment methods; managing enforcement methods (booting, towing, and license/plate suspension; interfacing with the Internet; and providing alternative methods for citizens to obtain system information.

Whether the system is simple or complex, one critical area of concern is the municipality's responsibility to the citizen. The citizen, as the ultimate recipient of the service and its penalties, must be provided with every opportunity to provide input. Of course, the more complex the system becomes, the more need for closer management and control.

But handling a simple or complex parking management system is not necessarily a daunting task and, in fact, has proven to be quite enjoyable work, if you ask some parking directors. These leaders in the field are well informed about the value of the various industry associations, conventions, and publications, resources that are valuable to any parking professional and worth at least an initial review.

The most prominent association is the International Parking Institute (IPI; www.parking. org), which boasts members across the globe but holds its annual conventions primarily in North America. IPI publishes an annual directory of statistics and best practices about the industry (some are documented in this chapter). On a smaller scale is the Parking Industry Exhibition (PIE; www. parkingtoday.com), another group that brings together industry vendors and buyers. One other group, a network composed of industry vendors and service providers, is Expo1000 (www.expo1000.com/parking). All of these industry groups have regular conferences and seminars and publish reports and provide industry information for players throughout the parking field.

Other sources of useful information and networking include regional and state parking associations and municipal leagues, which offer memberships to parking industry leaders as well as other ranking municipal employees. Most hold an annual convention of some type. Many of their activities are tied in part to other government-related groups in public works, traffic engineering, planning/development, and other fields.

Most parking associations publish their own guides listing vendors and service providers, with *Parking Today* (published by PIE) and *The Parking Professional* (published by IPI) being the most reputable monthly industry publications. These provide interesting discussions on parking issues and general industry developments. While not specific to the parking industry, *Government Technology* magazine is also worth mentioning as a source for information on relevant industry issues. GT provides a broader spectrum of articles dealing with state and local governments and how they resolve social issues and enhance government-to-citizen relations.

13.4.3 Vendor Management—Tools of the Trade

An overview of parking management systems would not be complete without a discussion of vendor management, especially considering the large number of vendors who provide products and services to the industry. Municipal parking directors often find themselves unwittingly involved in partnerships with parking vendors of varying stripes. A successful parking management system requires willing participation from all vendors, along with consistent and reliable performance. To achieve this level of satisfactory performance, the parking director must pay more than lip service to vendor management. The parking director must nurture and develop a strong individual relationship with his or her vendors. He or she must include them in the initial stages of planning, seek their input on problems facing his or her department, know their product or service offerings, and be willing to trust their industry knowledge and expertise while still ensuring that they adhere to strict standards and milestones consistent with the jurisdiction and the contract.

13.5 FUNCTIONAL PARKING MANAGEMENT REQUIREMENTS

Parking management system solutions must take into account a wide variety of factors with direct impact on program success. These factors are numerous: placement and maintenance of signs and meters; coordination of on- and off-street parking facilities; partnering with private parking facility operators to help regulate traffic flow and enhance public safety; appropriate use of technology and supporting business practices; proper training of personnel to ensure proficiency in providing services to the general public; rigid audits and controls to ensure that checks and balances are performed as required to maintain a high degree of system transparency; and a continuous improvement program to keep pace with industry and technology advancements.

A comprehensive parking management system should begin with a solid infrastructure plan that details current operation components. This plan should identify areas of strengths and weaknesses, outstanding issues and concerns, departments that provide and receive system information, detailed documentation of the information in the system, the format and method of delivery of that information, and barriers to success. It should also project goals and objectives of the parking operation over a longer term (5 years minimally; 10 years optimally), including a general outline of initiatives that must be implemented to satisfy those goals and objectives. Also, clearly defined measurement criteria should be established in the plan so that the progress, improvements, and failures can be charted, evaluated, validated, verified, and acknowledged as lessons learned.

The process of developing the infrastructure plan gives the municipality a realistic view of the current program, a clear understanding of its services and how they affect the general public, its vendors, suppliers, and other municipalities or state departments with which it interfaces. Internally, this exercise reveals difficult issues and concerns among staff members and between departments. It uncovers artificial boundaries and needless activities that have lingered long after they have become obsolete due to procedural changes or system improvements. Upon completion of the infrastructure plan, the municipality will have a clear view of required steps to move forward. It may require major overhauling of the current system, including replacing applications, outsourcing some functions, business process reengineering, adding new technology, or turning over the entire operation to an outside contractor. Whatever the outcome, the municipality will have a thorough understanding and solid business process foundation in determining how to proceed in implementing its plan.

With an infrastructure plan that has the support and commitment of the core parking management officers and staff, what requirements must a comprehensive parking management solution now fulfill? Beginning with ticket control and issuance, it is imperative that the system be able to track all tickets from inventory through to resolution, whether tickets are written manually or by automated ticket-writing devices. Controls should be put in place to record and monitor the movement of tickets to and from the ticket inventory. Assignment of tickets to officers should be controlled to ensure that only the officer to whom the tickets were assigned can write those tickets. For manual tickets, this is accomplished by enforcing the requirement that the officer sign each ticket issued. (This is less of a problem for tickets written using automated ticket-writing devices, because the devices contain built-in features to facilitate ticket control and monitoring.) Tickets issued by the assigned officer should be entered into the system so that the officer data are retained as part of the profile, providing an internal mechanism to track the ticket and officer at every ticket-processing stage.

Once the ticket is issued and entered into the system, a robust ticket-processing application is needed to allow for proper tracking of tickets and correspondence with violators, aging of tickets, and generation of appropriate notices, fines and penalties. This application must also apply partial or full payments via Internet, telephone (by interactive voice response), mail-in, and in person; conduct on-line adjudication; provide interfaces to other departments and systems; and modify ticket-processing activities based upon external events and internal triggers. Also, this application must generate standard management, operations, and executive analysis reports as well as standard letters to inform citizens of the status of queries made against the system and to provide basic information. Additionally, the application must be able to generate ad hoc reports and customized letters.

Other features of the application should include the ability to:

• Obtain in-state and out-of-state license information.

• Furnish information about the parking management program to citizens via the Internet.

• Allow citizens to conduct inquiries, file complaints, and appeal against judgments via the Internet using ticket number and supporting data to ensure privacy and confidentiality.

• Manage information flow to departments such as public works, courts, city clerk, state department of motor vehicles, finance, and police.

• Accept payments for tickets issued but not yet entered into the system.

• Apply penalties for returned checks and halt, suspend or change processing of tickets in this category.

• Generate detailed transaction history by ticket, violator, plate number and account.

• Provide lists of violators whose vehicles are eligible for towing or booting.

• Facilitate suspension of license and denial of plate registration based upon a set level of outstanding fines.

• Monitor, control, and issue permits and renewals for city-owned garages via walk-in or Internet transactions.

• Assist the management of off- and on-street parking facilities.

• Provide adjudication officers with full view of ticket images and transaction history of each ticket.

• Generate court dockets and supporting correspondence to aid in the adjudication process.

• Manage maintenance of meters and signs.

An end-to-end view of the functional requirements of an automated parking management system for a mid- to large-sized city is depicted in Figure 13.2.

Typically, a system as depicted in Figure 13.2 would be implemented by a city with an annual ticket volume greater than 1½ million. Such a city would have the support infrastructure to facilitate the functions, features, and processes described without taking a large toll on other city departments. The automated workflow process enables the city to control ticket activities from inventory update through to data entry, payment, and on-line adjudication. Tickets and appropriate ticket images are moved to different work pools as ticket status and timing dictate. City personnel are able to move in and out of the different pools performing the necessary activities on the tickets in their pool. As this

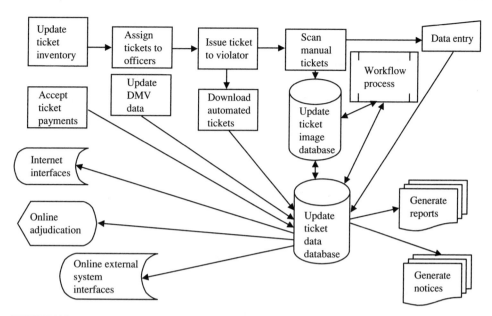

FIGURE 13.2   Functional requirements of an automated parking management system.

takes place, supervisors monitor the work pools and make corrections or reassign jobs to balance the workload and smooth out the process.

The ticket data entry process is necessary only for handwritten tickets. To improve ticket data processing accuracy and integrity, automated ticket-writing devices are recommended. Handwritten tickets should be kept to a bare minimum and eliminated whenever possible. To this end, a few cities have installed in-car devices for car patrol officers and handheld devices for foot patrol officers. This certainly increases the accuracy and effectiveness of the overall parking management system while reducing the need for ticket imaging.

The on-line adjudication process enables adjudicators to view all ticket activities, from inventory control up to and including the adjudication results. This puts appropriate information at the adjudicators'fingertips and expedites the process. Providing ticket information to users over the Internet gives citizens the opportunity to conduct interactive transactions without having to visit city hall or local parking management offices. Having online interfaces with external systems provides an added data security measure and in most cases reduces the need for redundant data input, increases data integrity, and expedites the overall process.

The reporting process in this system normally generates a large variety of analytical reports tailored to assisting managers and supervisors monitor the vital signs of the system to ensure increased productivity, improved data accuracy, and enhanced customer service. A few refined operational reports provide the necessary audit trails, system logs, and job statistics to ensure that the system achieves a high degree of transparency.

Of course, not all cities can afford to implement all the functions, features, and processes described in this approach. However, cities may use this example as a foundation for their plans and use the pieces that are feasible, then apply them to their specific situation. (This is beginning to take place at higher rates and in smaller cities. For example, a small Chicago suburb has eliminated handwritten tickets altogether by equipping all patrol cars with in-car ticket-writing devices and providing handheld ticket-writing devices to other officers.) Still, functions like interactive Internet processing, on-line adjudication, workflow process and online external system interfaces are deemed rather expensive measures for many cities to implement at this time.

Smaller municipalities with less need for automation, smaller ticket volume, and fewer interfaces will not have a need for such an extensive system. In fact, the smaller volume of tickets, and therefore revenue, precludes the use of expensive ticket-handling processes. It is important to remember that no city wishes to take on a parking management system that is going to lose money, no matter how noble the cause. Therefore, it is imperative that the municipality keep a tight rein on matters by sticking as closely as possible to its original infrastructure plan.

For most cities—those processing fewer than a million tickets annually—an efficient, citizen-responsive parking management system would maintain a number of core features:

- Ticket inventory
- Ticket assignment
- Ticket issuance procedures (for parking enforcement officers)
- Data entry of manual tickets
- Downloading of automated tickets
- DMV updates
- Payment processing
- Adjudication
- Internet interface
- Limited external system interfaces as warranted
- Small volume of analytical reports, audit-trail reports, system logs, and job statistics as necessary
- Notices
- Enforcement measures (i.e., booting, towing, plate denial, or license suspension)

Adherence to federal and state laws and regulations governing cities and municipalities is among the requirements that must be discussed for municipalities of any size. These requirements typically include regulations stipulating conditions that must be met in order to boot or tow a vehicle, suspend a driver's license, deny plate renewal registration, or escalate parking fines. Federal and state laws may also govern adjudication and determine what constitutes a proper mailing address and who must provide that address. Each municipality must conduct its own research to determine what federal and state laws and regulations apply to its particular parking management solution strategy, then enact appropriate laws to activate them.

13.6   GROWING THE PARKING MANAGEMENT SYSTEM INTO A COMPREHENSIVE CITATION MANAGEMENT SOLUTION

With a comprehensive infrastructure plan in place, a municipality can now begin to look at growing the parking management system into a centralized citation management system that incorporates processing activities for all citations and sets the stage for supporting economic development programs. The reasons why a city should integrate its parking and citation management programs are numerous and evident.

For one, IT integration of multiple, compatible government services and systems adds operational and bureaucratic efficiencies. There is less redundancy of effort and more integration of systems, leading to smoother operation and time and cost savings. Second, smart systems that unite back-office management of programs run by a common department (for example, a municipal court, which typically administers fine- or fee-based citations) streamline operation and enhance enterprise-wide systems management. Finally, it makes money—in raw revenues and by simplification and limiting duplication of effort.

The strategy for this approach should be established in the infrastructure plan so that the best suitable application and database designs are selected to accommodate this move. From the outset, a parking management system should be designed with flexibility, scalability, and a limited amount of proprietary software to allow ease of growth. Also, the system should be sufficiently robust to provide users with all the tools to perform operational responsibilities with an acceptable level of proficiency.

Certain functions are considered standard for any parking management operation and must be fully functional before moving into a centralized citation management system. These include enforcement measures such as:

- Ticket issuance, noticing, and warnings
- Booting and towing
- Driver's license and plate suspension
- Collection methods
- Adjudication procedures

Standard processing functions that must be accounted for in a basic parking management system being considered for expansion into a centralized citation management system include payment processing and interface with the department of motor vehicles and local police department. The parking management system should also accommodate changes resulting from federal and state regulations that must be taken into consideration. Again, the municipality may need to enact local laws to enforce these regulations. The municipality's legal counsel should be able to provide proper direction on this matter.

Once the groundwork has been laid to move forward with building a centralized citation management system, the municipality needs to evaluate the available programs in order to determine their suitability, the priority in which they will be added to the system, and how best to implement them. Basically all municipal programs that may generate a citation should be considered reasonable candidates. These include processing of:

- Permits
- False alarm violations
- City stickers
- Traffic violations
- Railroad at-grade crossing violations
- Railroad at-grade traffic interference violations
- Other city ordinances for which citations are issued

The City of Memphis and its Traffic Violations Bureau are a good example of how a parking management system can move into a comprehensive citation management program. The City of Memphis issues and manages 400,000-plus citations worth more than $12 million annually. To enhance this operation, it established a Traffic Violations Bureau (TVB) to manage and control all citations for traffic, parking, arrests, and other municipal ordinance violations. At the same time, the police department committed to eliminating handwritten tickets by placing automated ticket-writing devices in squad cars and using wireless technology to interface with the existing ticket processing system. The advanced nature of this initiative required that the existing ticket processing system be rewritten or replaced by a more comprehensive solution.

The City wanted to continue to maintain the operation and was not willing to enter into a transaction-based agreement with a ticket-processing vendor. Also, the City wanted to phase in the use of automated ticket-writing devices over time, starting with in-car units. Faced with a short implementation timeframe, limited resources, and a desire to implement a flexible and scaleable solution, the City embarked on a business process management solution, which included:

- Assessment of the current situation
- Identification, qualification, and selection of appropriate vendors

- Development of business requirements/objectives
- Review of entire business process and determination of implementation plan
- Scheduling and managing all phases of the request for proposal (RFP) process
- Preparation, development, and distribution of the RFP
- Review and selection of candidates for the bidding process
- Organization, definition, and facilitation of the bidders' conference
- Evaluation of bidder responses
- Development of recommendations for a final vendor

The City executed this strategy on a fast track over a 100-day period. The process was coordinated so that City operations were not unnecessarily interrupted or delayed. Extensive meetings were conducted with key stakeholders, including the City Court Clerk, senior officers of the City Court Clerk office, and supervisors and senior staff of the police department, in addition to the City's information technology outsource partner.

With this background and analysis in hand, the City then applied TVB's business requirements to development of the RFP and identified an initial field of vendors for consideration for bid. More than a dozen vendors were initially considered, but were quickly winnowed to a short list of candidates, who were forwarded to the RFP. From there, a preferred vendor was recommended and selected. The City plans to apply this business process approach to integrate all citation-related functions under one comprehensive citation management system.

13.6.1 Parking Management as a Foundation for Economic Development

As depicted in the hot-air balloon graphic in Figure 13.3, a comprehensive parking management system can serve as a foundation that enables the municipality to grow into a centralized citation management system and then use those roots to link with larger-scale planning and economic development programs. All of these can be incorporated into a properly developed parking management system without conducting a major overhaul of the program, assuming that the program was designed with sufficient modularity to allow existing components to remain unchanged. Therefore, adding permitting to a parking management system requires incorporating modules for a range of applications:

- Permit application, which should interface with the parking database to determine if there are outstanding parking fines and may interrogate other municipal databases to determine if the applicant has other outstanding debts with the municipality
- Permit tracking, which tracks permits issued and to whom and identifies renewals, which may be generated as part of the parking management system's noticing module
- Permit issuance, which prints permits for mailing to the approved applicant

Similarly, application modules are required for false alarm billing and city sticker registration/issuance. With careful planning these may be incorporated in the permit application module. The characteristics of false alarm billing citations—issuances applied to citizens or business whose home or office security systems improperly go off, unnecessarily requiring fire or police response—may be accommodated in the system by providing the necessary processing methods for citations that are marked as such. No special payment processing method is required for these. City stickers—annual fees levied by municipalities on vehicles whose owners reside in the municipality—are basically vehicle permits and may be treated the same way.

Traffic violations need very little explanation, as this is another citation with far more information than required for a parking ticket, and so it can be included in a centralized citation management system. Typically, only a separate data entry procedure is required for these violations because in most municipalities courts handle these violations. So the system must be able to generate appropriate court dockets and other court-specific documents to facilitate court processing. This does not

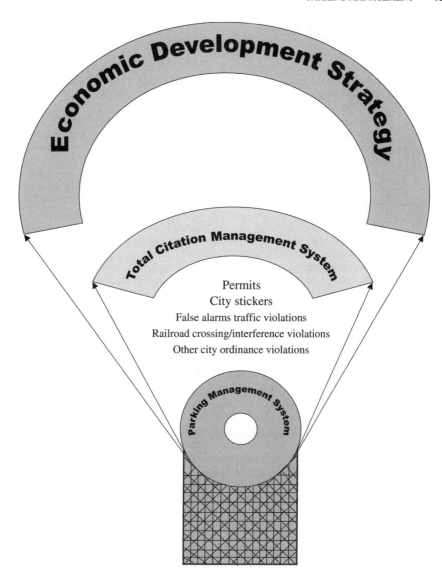

FIGURE 13.3 Parking management systems as a foundation for planning and economic development programs.

mean the system should take on processing functions better handled by a courts or justice management system; the citation management system will restrict itself to accepting the traffic violations, generating appropriate court documents, and accepting payments as directed by the courts.

Adding other municipal ordinances for which citations are issued to the centralized citation management program should not require major efforts. For example, adding noise, dumping, or building code violations would require some modification in the data entry module to accommodate these citations and, similar to traffic violations, would require a module to facilitate court processing. Depending on municipal laws, additional processing may be required. However, in all situations, in-depth analysis of the business processes must be conducted in order to ensure that the new system accommodates all business requirements.

In all situations described above, no special equipment is necessary to facilitate the addition of the business processes. That is the primary difference between these processes and the railroad violations—at-grade crossing violations and at-grade traffic interference violations.

At-grade crossing violations require installation of unique equipment to detect when a vehicle evades level crossing gates and must be able to capture the license plate of the vehicle as well as the date and time of the infraction. Also, the system must be able to reprint pictures of the infraction for mailing to the violator.

At-grade traffic interference violations require similar technology to capture the date and time of infraction as well as identify the railroad company whose train was in violation of the statutes. At this time, there are not many vendors in the marketplace that provide this technology, but there is sufficient competition to keep the prices affordable. (It should be noted that untapped revenues from at-grade traffic interference violations are potentially huge; one small city in the Midwest, for example, by instituting a relatively low-technology citation management system for at-grade crossings, generated $8 million in citation revenues, even after negotiating with several freight companies to simplify issuance/dispensation of tickets.) With the technology in place, the centralized citation management system needs to have a module that accepts the citations from the railroad violations system and process them according to federal, state, and local statutes.

Municipalities wishing to proceed with this action must first determine whether state statute allows them the right to charge railroad companies for this infraction and what is the procedure by which to do so. This process is not uniform across all states. In Illinois, for example, law allows only municipalities that reside in counties of a million or more in population to issue such citations against freight companies.

As each new type of violation is added to the centralized citation management system, expertise and proficiencies should also improve reducing the degree of difficulty for each. One benefit of this approach is a more robust and maintainable system that allows the municipality to streamline operations, reduce costs, and improve services. In the end, the totality of these initiatives allows municipalities to build the groundwork for broadening the reach of citation management into larger-scale planning and economic development plans.

13.6.2 Parking Management, Infrastructure Planning, and Economic Growth

So how is parking connected to the bigger world of planning and economic development? The answer is one word: money. State and local governments are in a period of nearly unmatched deficits and debt. According to the U.S. Department of Commerce, 1 in every 10 dollars budgeted by state and local government in 2002 was borrowed. During a flat economy, their increase in spending—on welfare, health, and education as well as on public works, public safety, and homeland security, all high priorities—outpaced their increase in revenues by nearly a third in 2002. It is the highest proportion of debt in state and local governments since the high-growth 1950s, when infrastructure fought to keep pace with the Baby Boom.

High federal deficits—forecast at $300+ billion well into the mid-00s—mean pressure on state and local government to provide services and balance budgets without raising taxes. Tax increases are never popular, especially in a down economy, which is why local governments raised taxes only 1.2 percent in 2002. Many states and cities face big deficits, even as they must provide key social and public services that will likely require them to spend more, not less. As debt increases, access to capital becomes more difficult. The most critical indicator of public finance—credit rating—has shown a downward trend. Six states (including California, New Jersey, Colorado, and Wisconsin) have seen their credit rating (as measured by Standard & Poor's) drop since 2001, and many major cities are in the same position.

So that means local governments must be particularly innovative in bringing new revenues into their coffers—not just to cut debt and balance budgets, but to be competitive by maintaining services and infrastructure that prevent citizens and businesses from going elsewhere. Parking and related programs are a major area of opportunity for government to capture new revenues, be more citizen-responsive, spur economic development, and generate maximum return on investment without raising taxes.

A look at citation management alone proves the point. A typical growth city or suburb of 50,000 that has a collection rate on parking tickets below the industry standard of 70 percent realizes net returns of $250,000 or more if improvements in its parking management system increase ticket collections by 15–20 percent. Add in enhanced efficiencies and better cash flow that result from improved business processes, and the city can realize another $100,000 or more. For a larger city, management of a comprehensive parking operation can be worth millions; for a major city, tens of millions—from parking operations alone.

Optimized parking management and infrastructure programs do not only generate revenues to pay for expanded public safety, emergency, and public works programs. Infrastructure development also connects government to citizens and the private sector by supporting economic activity and boosting quality of life. And a number of tools and programs associated with parking bridge into other key government functions—economic development, tourism/conventions, and public works, to name a few—that make the public sector more competitive and attractive to business and private investment. There are several practical examples of parking-related programs that are used to stimulate economic activity and quality of life.

Zoned/Managed Parking in Residential Neighborhoods.   Many communities, particularly those in densely populated urban areas with a healthy mix of private homes and commercial/retail businesses, find zoned or managed parking (through the use of permits or meters) a vital tactic to provide peace of mind to homeowners. Typically, this device is used to prevent commuters or nonresidents from using residential streets from parking their cars during the business day or overnight and to ensure parking for property owners in the neighborhood.

Evidence from many cities has shown that blocks or entire neighborhoods made up of private homes that employ zoned or managed parking realize many benefits. For one, they encourage use of public transportation and therefore minimize auto traffic and associated environmental effects. Second, property owners in zoned neighborhoods tend to keep their homes longer; they reinvest in their properties, which increases their value and in turn produces higher property tax revenues to cities. And, because (especially in residential communities that use meters) they hold the threat of a costly parking ticket, they essentially dictate that the overwhelming majority of people who park in these neighborhoods are shoppers—people who feed meters and retailers' cash registers. This, of course, contributes to cities' bottom lines in a number of ways.

Shared or Multiuse Parking Programs.   These programs strike a balance between use of parking and transit for commercial needs during the weekday and retail/entertainment demands at night and on weekends. They are designed to ensure the most efficient use of available parking and public transit facilities throughout the day and to meet the multiple and sometimes conflicting demands of residents, commuters and businesses. Parking facilities that accommodate 9 to 5 workers convert to service theater and restaurant patrons after hours. Permitting can be used to enable people to use parking for whichever purpose or purposes they require, and communication tools—the same technologies used in stock tickers or airline scheduling services—can provide real-time information to the businesses and citizens that require up-to-the minute parking availability. The current proliferation of in-car GPS applications makes this service even more attainable in the near future.

Many cities employ these programs to optimize parking/transit systems and traffic flow. Park-and-ride and kiss-and-ride programs of this nature are common, especially around public transportation stations and airports. Often, these programs are complemented by circulator services that minimize congestion in central business districts by encouraging perimeter parking, then use buses to shuttle drivers en masse.

The San Francisco/Oakland Bay Area Rapid Transit system employed a variation of this model in late 2002. BART allowed commuters to purchase decals that ensure rush hour parking at some of its busiest and parking-dense stations. Not only do the new revenues help plug budget gaps. The program also minimizes congestion at targeted lots and saves many commuters the time and hassle associated with inability to leave their cars at desired stations. BART officials also report that many commuters are happy to pay for such a service because it allows them to spend extra time at home with their families before they leave for work.

Shared and multiuse parking is more than a convenience and a smart business move. In many cases, it enables cities to acquire much-needed public and private funding for large-scale development and revitalization programs. In many cases, the inability to forge shared and multiuse parking operations prevents cities from pressing forward with vital development initiatives. Because many cities' development departments maintain strict rules and methodologies governing parking and traffic flow attached to any new building project, cities and developers must work together to develop flexible parking management systems. Otherwise, they—and the citizenry—often lose money, jobs, and other economic benefits.

Magnet Infrastructure Programs.   Some cities use a combination of infrastructure/land management, parking/transit measures, and tax incentives to harness and redeploy existing public resources to increase retail, commercial, and residential development. These magnet programs can be applied to an entire business district or single building, and many are built around a financing mechanism known as TIF (tax increment financing).

One of the few true economic development incentives left for local governments, TIFs pool property tax revenues into infrastructure development, often in blighted or largely vacant areas. TIFs provide for land cleanup, road and building repairs, and other infrastructure improvements so that the land and properties in these districts generate revenues, typically without raising taxes to the community. Through their special designation, TIFs enable development projects to happen at all, and faster than they might otherwise. At least 44 states maintain some form of TIF program for redevelopment and revitalization.

Because of their unique status, TIF districts integrate transportation and parking in innovative, revenue-generating ways. One large southwestern city, for example, has tied parking infrastructure (including a downtown shuttle, smart-card meters, a special parking infrastructure fund, and encouragement of alternative forms of transportation such as bicycles) to the central business district TIF. As a result, developers have thrust themselves into new retail and commercial building projects in the city's ongoing downtown development plan, and developers are now working with the city (with the parking management staff playing a central role) to expand building projects into the residential sector. The city is also exploring semiprivatization of its parking and transit systems to facilitate longer-term development and growth of the downtown area.

Privatization and Partial Privatization Programs.   These initiatives, discussed earlier in this chapter, can be extremely useful in expediting public development activities. Entities such as parking, transit, airport, and development authorities typically enjoy powers regarding land use, zoning, and taxing (including, in some cases, TIFs) that spur development and create new services and revenues. The new entity can be public or quasi-public and can be managed by the public or private sector (or in partnership between the two).

While some cities have chosen to go the route of full privatization—in which they turn over management responsibility for these operations to private contractors in exchange for a guaranteed amount of revenue, typically up front—public-sector leaders should be cautious of this approach. Revenue- or profit-sharing—even in combination with an up-front payment—may be a better option, simply because cities should enjoy the full benefitofnew revenues privatization can bring and not leave the lion's share of new revenues to private contractors. Nonetheless, privatization models can have significant benefits, as Figure 13.4 depicts.

Figure 13.4 represents an actual revenue-share privatization business model created for a small midwestern city. The city had no formal parking management operation or infrastructure, and required that a parking authority—owned by the city but managed by private partners—be created to this end. The plan created a governmental entity and parking authority appointed by the mayor and city council, with management and technical assistance provided by the private partner, which paid the city to turn over management responsibility, then split all profits with the city once that initial investment was recovered.

The parking management operation created by the new authority was also designed to form the foundation of the city's larger economic development plans. A regional hospital, the city's largest

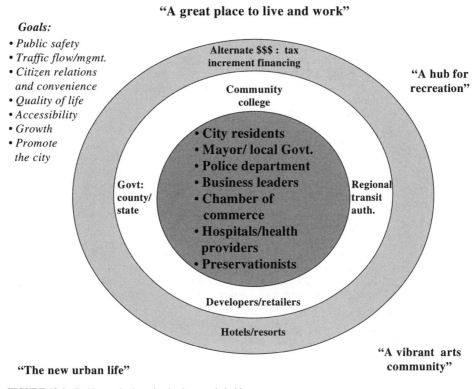

"A great place to live and work"

Goals:
- *Public safety*
- *Traffic flow/mgmt.*
- *Citizen relations and convenience*
- *Quality of life*
- *Accessibility*
- *Growth*
- *Promote the city*

Alternate $$$: tax increment financing

"A hub for recreation"

Community college

- **City residents**
- **Mayor/ local Govt.**
- **Police department**
- **Business leaders**
- **Chamber of commerce**
- **Hospitals/health providers**
- **Preservationists**

Govt: county/ state

Regional transit auth.

Developers/retailers

Hotels/resorts

"A vibrant arts community"

"The new urban life"

FIGURE 13.4 Parking authority privatization—stakeholder map.

employer, was expanding and required more parking and the land to build it. Four commuter train stations carry thousands of people, many of them who drive their cars to the city to take the train, to and from work. A local community college was also building a magnet campus in the downtown. Additionally, the city was continuing a long-standing movement from an industrial-based economy to a service-driven economy, which was to include development of riverfront land just outside downtown. Finally, city leaders wanted to highlight and heighten the city's health care and retail base by establishing a medical district and arts district in and around the downtown area.

Figure 13.4 depicts the strategic planning behind the parking authority and subsequent economic development. The inner circle of the bullseye shows the core stakeholders needed to spur the initial phases of the project. The next rings depict the next phases of partners/stakeholders and vehicles (such as financing) required to expand this infrastructure into broader-scale economic development. The phrases in quotes outside the bull's-eye suggest a few ways in which the city, citizens, and private sector could represent and market the plan within the city and to others outside of it, with an eye toward the goals listed in the upper left corner.

13.6.3   Bringing Parking Management's Role in Economic Development to the Table

While it is clear that parking management skills, technology, and business practices are potentially of infinite value, the function and discipline are still largely marginalized and fairly low on most cities' food chain. So how can those responsible for parking management and public-sector leaders better leverage the value parking management brings to the table and increase its return on investment? Several practical steps are required:

1. *Integrated management/decision-making:* Cities can do a much better job of aligning the people responsible for various parking-related functions—enforcement/ticketing officers, collections staff, courts officials, etc.—in overall planning and management. The integration of these staff, and simplified reporting relationships and lines of communication with the people to whom they report, streamline and therefore heighten the role of parking management in the bureaucracy.

2. *Better direct linkages between parking management and related infrastructure functions:* The economic, operational, and customer service benefits of parking management fall on deaf ears unless they are in front of decision-makers in other infrastructure-related public programs. Cities need to restructure to take full advantage. More direct roles, peer relationships, and bureaucratic linkages need to be made between the parking management function and those departments responsible for public works, transportation, housing, and public transit.

3. *Transform parking management skills, intelligence and business processes to larger government functions:* In a majority of cities today (including those surveyed in the material referenced in the accompanying sidebar), most of the information generated by the parking management function is rarely if ever shared with other departments related to infrastructure, much less those responsible for economic development and planning. The strategic value of parking and related infrastructure programs is too critical today for this trend to continue. Parking and infrastructure are central to revenue generation, public safety and citizen relations, and should be at the table in coordination of management of key agencies such as development authorities, planning/ development departments, tourism/convention centers, and related entities.

13.7 CUSTOMER RELATIONS

Customer relations is an area that unfortunately has been largely neglected by purveyors of parking management systems and municipal employees responsible for these systems. Typically, the first encounter between the municipality and a citizen occurs upon a parking infraction and continues through the process of further notification and, finally, payment of fines and fees. During that time, the citizen is not considered a customer to whom certain civil courtesy should be extended, but rather as the perpetrator of a crime and one who must be harassed into compliance—even if, as in most cities, parking infractions are limited civil, not criminal offenses. This attitude has led to the use of negative approaches geared toward forcing citizens into compliance with local laws and payment requests. Most people realize that finding a parking ticket on their vehicle, or finding their vehicle immobilized by a boot or towed due to overdue fines, is a disturbing and costly experience. Parking managers should understand this and better recognize the need to ease citizen discomfort.

There are a number of methods that may be employed to change this interaction with the general public. But first there needs to be a change in the way that parking managers view the people who commit parking infractions. They should not be viewed as intransigent, recalcitrant, or even scofflaws. By changing the way they view the general public, parking managers can develop proactive and responsive strategies such as innovative public awareness programs, alternative payment methods, and citizen inquiry facilities. The primary focus of these strategies is on encouraging a more amicable dialogue with the general public.

For example, a brief description (even pictorially) of how the infraction will be treated by the municipality and the various options available to the individual should accompany the parking citation notice. Additionally, as is done in Austin, Texas, a statement of overdue ticket should be included with each new ticket notice received by the individual. This approach provides citizens with knowledge of their total outstanding debts to the municipality. Also, it removes the excuse of pleading ignorance to the process, because details of the consequence of not paying are always provided with the notice. A by-product of this approach is the prevention of increased fines as a result of misapplied payments. For example, a citizen may have paid the fines but still be receiving a request for payment due to the municipality's poor bookkeeping.

Alternatively, municipalities may use Internet sites to provide details of how the parking management program works, updates to current practices, frequently asked questions, and guides to using parking services. Larger cities may look to provide regional service centers to disseminate information and accept in-person payments and inquiries from the general public, thus putting a friendlier face on the municipality in the eyes of the citizen.

Smaller municipalities may enter into agreements with small merchants to accept payments and hand out informative parking management brochures describing system updates and new procedures. Whatever the option or solution, the purpose of customer-service initiatives like these is to increase the public's willingness to comply with municipal laws through a greater understanding of policies, distribution of parking revenues, and public safety improvements. Other methods to improve public response and participation include:

- Improved maintenance of signs and meters

- Use of the Internet, TV, radio, newspapers, and frequent brochures to provide public information

- Warnings to motorists who are within weeks of being booted or towed or having their license suspended

- Special payment plan programs for motorists with very high outstanding balances (a variation of this is currently running in Memphis)

- Amnesty programs that allow citizens with high volumes of tickets to pay off citation debts without paying late fees or incurring additional civil penalties (Chicago instituted such a program in late 2002, with mixed results; the city did collect more than $10 million in unpaid parking tickets, however)

Of course, each municipality must devise its own public awareness and customer relations programs. Customer relations, though, should be an integral part of the parking management system designed to encourage public compliance and willingness to pay outstanding fines and penalties. Booting, towing, license suspension, and denial of plate registration renewal are sufficiently threatening measures in and of themselves and are almost always guaranteed to force compliance. Therefore, with these enforcement measures, municipalities would do well to encourage a friendlier interaction with citizens. Needless to say, criteria should be defined to determine the level of success of these programs so that educated decisions surrounding these programs can be made.

13.8 SYSTEM VALIDATION AND VERIFICATION METHODS

A comprehensive parking management solution must have validation and verification methods to promote and ensure accountability, efficiency, integrity, and accuracy. Today's parking management operations are singularly focused on maximizing enforcement measures and place little or no emphasis on monitoring manual and automated procedures for continuous improvements. As a result, systems tend to lack the functions and features necessary to identify faults and failures relating to accountability, efficiency, integrity, and accuracy.

The absence of these features can undermine confidence in the system and, over time, lead to general abuse of the system internally. This abuse manifests itself in increased public complaints, which can create animosity between government and citizens. To foster an environment of cooperation and harmony, the parking management system must be seen as unquestioned in terms of accountability, efficiency, integrity, and accuracy.

Citizens must be assured that when they follow the rules, they are dealt with fairly and equally. They should be confident that the municipality is operating the parking management system to provide sufficient availability of parking spaces (on-street, off-street, and in private and public lots) and to reduce operations costs by improving efficiencies, enhancing services to reduce complaints, and holding staff accountable.

The adage that "perception is reality" holds true for public services, and the general public must feel that government provides the highest level of service on a consistent basis. Anything less will give rise to apathy and general mistrust of the system. Therefore, it is imperative that system validation and verification methods be established as integral components of a comprehensive parking management solution. These are best suited to a continuous improvement strategy that solicits feedback from citizens, management, and operations staff. Mechanisms for obtaining feedback from the general public may include telephone surveys, focus groups, suggestion boxes in public locations, direct mail, and Internet questionnaires. Information obtained through these media should be carefully reviewed and evaluated to determine whether to apply the suggestions and how best to do so. Also, a mechanism should be in place to acknowledge appreciation for public participation.

13.9   MEASURING THE IMPACT AND EFFICIENCY OF THE PARKING MANAGEMENT SYSTEM

Many cities and municipalities are guilty of operating a parking management system that is separate and independent from other systems and departments and, in doing so, reducing or eliminating the opportunity to take advantage of successful business process innovations. In most cases these innovative measures generated cost savings and process efficiencies that would bring about exponential rewards to the parking management system. For example, a parking management system run by and for the police department is unlikely to share in the benefits gained from cost-saving and efficiency measures put in place by the city's IT department. Also, with very little sharing of information between departments, the limitations of the parking management system are not always visible. Notwithstanding, there are ways to ensure that the parking management system becomes a living, breathing organism and stays abreast of business process improvements and technology advancements.

There are some basic methods and procedures that are necessary to ensure the continuous relevance of any system. These include:

- A continuous improvement process
- System performance standards
- Business continuity strategy
- System backup and recovery procedures

Beyond these, there must be procedures and standards designed to measure the efficiency, impact, and progress of the program. Viewing the parking management program as a service to the general public, it stands to reason that regular public feedback should be sought to determine whether parking initiatives are having any effect. Also, it is important to have measures in place to ensure accountability of staff members, contractors, and managers. Too often the parking management function is treated as an isolated function or department with little or no interaction with other municipal departments, which in itself reduces the need for accountability and performance measurements. Measurements such as the revenues earned and revenues lost through negligence or inefficiencies are of equal importance to municipal governments. Other data, such as usage or lack of usage of parking facilities, may be of assistance to stakeholders in the planning, building, and licensing departments.

A properly run parking management operation should have procedures and standards that allow managers to gauge program progress. There should be standard response times for telephone and written inquiries with established procedures to measure staff responsible for these functions. The time for a citation to move through the system from initial issuance to final dispensation should be established and properly adhered to. Continuous improvement activities should be applied in order to ensure that these standards are met. Typically, it should take no more than 90 days for a ticket to move from initial issuance to final dispensation. As well, telephone inquiries should be answered within 24 hours and written responses should be answered within 40 hours. While these are guidelines and may be adjusted according to each environment, lack of standards like these will only lead to abuse of the system as well as mistrust of the system by the general public.

Applying continuous improvement activities to the system without obtaining feedback will not help to determine the success of the initiative. Municipalities may use surveys or focus groups to help determine whether an implemented improvement plan is gaining acceptance or needs to be changed. And like the continuous improvement plan, the surveys and focus group should be a regular feature for getting feedback internally and externally. As a matter of fact, economic indicators, measures, and tools should be incorporated in the feedback surveys and focus groups to help determine return on investment for each improvement initiative implemented. Like corporate America, government must establish acceptable payback periods for its investments in order to determine what programs are best for the general public in terms of social impact, citizen satisfaction, financial outlay, and acceptance by the broadest cross-section of the population.

13.10 BENCHMARKS AND INDUSTRY TRENDS

Industry benchmarks cover a broad cross-section of analytics, ranging from hourly wages to hourly meter rates, and cover segments such as airports, hospitals, municipalities and universities. They also cover a variety of geographic regions across the United States and Canada. Such a multidimensional array of analytical variables makes the term *industry standards* almost meaningless. Notwithstanding, there are some notable observations worth mentioning. Uppermost among these is the 70 percent collection rate noted elsewhere in this chapter, which is considered a basic benchmark for the industry.

Other observations that are best stated as trends rather than standards include statistics such as (IPI):

- 12 percent of municipalities do not use outside contractors for any parking management service function.
- 64.2 percent of municipalities rely on in-house support to administer their parking management program.
- Disabled parking violation incurs the highest fine of all parking violations.
- Parking revenues are higher than expenses, making most U.S. parking management operations profitable.
- 77.8 percent of municipalities treat parking violations as a civil offense.
- 85.7 percent of municipalities use handheld ticket-writing devices.
- 82 percent of municipalities assess penalties for overdue citations.
- 71.7 percent of municipalities issue citations via the police department.
- 92.6 percent of municipalities use towing as an enforcement measure. This number increases to 100 percent in Canada.
- 56 percent of municipalities use booting as an enforcement measure. This number is 0 in Canada.
- Most municipalities cite punishing scofflaws as the primary reason for booting and towing.

13.11 SUMMARY

Parking management is a much more complex and high-value area of operation than most cities currently give it credit for being. In most cases, cities treat parking systems as necessary evils rather than business and service operations. Greater strategic thinking and coordination are required to take better advantage of public safety, customer service, and economic benefits associated with parking management and related programs.

Through the combination of technology, financial management, business processes, and operations practices denoted in this chapter, public-sector leaders within multiple areas of government can build parking management systems that add service to the citizenry and dollars to the bottom line for many years to come.

REFERENCE

International Parking Institute (IPI). *Benchmarking the Parking Profession*. Fredericksburg, VA: IPI.

CHAPTER 14
TRUCKING OPERATIONS*

Amelia Regan
Computer Science and Civil and Environmental Engineering
University of California
Irvine, California

14.1 INTRODUCTION

14.1.1 The Importance of Trucking

Estimating the importance of the trucking industry is difficult, though all estimates point to its significance to our economy and the quality of our lives. In the latest data available from the U.S. Census Bureau, expenditures on transportation by truck in 2001 were over 318 billion dollars (U.S. Census 2003). Add in the costs of support facilities for road freight transportation and the number increases by 10–12 percent. Another way to look at it is that the nation's freight bill is about $1200 per person per year. The Census Bureau data note that trucks moved almost 87 billion miles in 2001—nearly 80 percent of them loaded. Further, the trucking industry employed nearly 3 billion employees in 2001. No matter which numbers we use, its clear that the trucking industry is vital to our economy.

14.2 A BRIEF HISTORY OF THE TRUCKING INDUSTRY IN THE UNITED STATES

In 1900, there were 8,000 automobiles in the United States. By 1920 there were 8,131,522 automobiles and 1,107,639 trucks. In the early years of trucking, companies competed with short-haul railroad operations and managed to be significantly more efficient due to the lack of regulation in trucking relative to the heavily regulated and high-fixed-cost rail operations (Herbst and Wu 1973). The trucking industry was heavily regulated from the mid-1930s, when the Motor Carrier Act was adopted, until 1980, when it was dismantled. In 1937, the first year in which interstate commerce commission (ICC) received reports from trucking companies, there were 54 trucking companies with incomes in excess of $1 million. By 1955, that number had increased more than fifteenfold (Taff 1956). The ICC, created in 1887, was the first regulatory commission of the U.S. government. It gained control of the trucking industry in 1935 and essentially ran the industry like a public utility until 1980. It lost most of its power then and was completely dismantled in 1995, when its remaining functions were transferred to the National Surface Transportation Board.

---

*Reprinted from the First Edition.

14.3 *CLASSIFICATION OF TRUCKING OPERATIONS*

Trucking companies were historically placed in the following classes for the purposes of regulation. Private carriers are those that are owned by the company for whom they provide exclusive service. Common carriers (first called motor common carriers to distinguish them from rail common carriers) are companies that move general commodities for any customer in need of service. Under regulation, common carriers were required to provide service to any customer who asked, at a fixed and publicly available price set by the ICC. Today's common carriers have short-term or long-term contracts with shippers or third-party logistics providers. Contract carriers are carriers that provide service under long-term contracts with shippers. Many of these have highly specialized operations. Under regulation, contract carriers could negotiate the rates charged with shippers directly. Those rates, and their operational costs, tended to be lower than those charged by the for-hire common carriers (Taff 1956). Exempt carriers were those that hauled loads specifically exempted from regulation and nothing else. The largest group of exempt carriers were those moving agricultural commodities. Some interesting court cases related to the exception of various commodities (seafood, frozen food, etc.) made it all the way to the U.S. Supreme Court.

Today, common and contract motor carriers of property are grouped into the following three classes: class I carriers having annual carrier operating revenues of $10 million or more; class II carriers having annual carrier operating revenues of at least $3 million but less than $10 million; and class III carriers having annual carrier operating revenues of less than $3 million. When carriers change classes, they are required to notify the U.S. Bureau of Transportation Statistics.

Companies are further classified by the type of services they provide. The first system of classification breaks the industry down into full truckload carriers, less-than-truckload (LTL) carriers, and package delivery services. LTL carriers are those that handle freight weighing less than 10,000 pounds, while truckload carriers haul heavier freight in full truckloads. Note that LTL carriers typically also provide truckload services. A further classification might specify what type of equipment is used by the carrier, including van (sometimes called dry van), flatbed, refrigerated (known as reefers), and tanker trucks, to mention a few.

Carriers are typically further classified as providing local, regional, or long-distance operations. The operations of these different companies and the working conditions for their drivers differ significantly. Drivers in local operations tend to spend much of their time loading and unloading and waiting their turn at dock facilities; in return for such inconveniences, they sleep at home every night. Long-distance drivers tend to stay on the road for 3 to 6 weeks at a time, sleeping in their cabs or occasionally in inexpensive motels. Long-haul drivers experience quite a bit of autonomy, despite the emergence of on-board computers that keep them in continuous contact with dispatchers and managers. Long-distance truck drivers are the cowboys of the 21st century.

14.4 *THE EVOLUTION AND IMPACTS OF REGULATION*

It would be impossible to do proper justice to the fascinating history of the regulation and subsequent deregulation of the U.S. trucking industry in this chapter, which is mainly concerned with contemporary issues in the industry. Nonetheless, we discuss some of the basic ideas here. Under regulation, the ICC, with the backing of the railroad industry, the International Brotherhood of Teamsters, and the newly formed American Trucking Associations, severely limited both entry into the industry and competition with respect to prices charged to shippers. Railroads felt they were subject to unfair competition from the trucking companies of the day because as common carriers, they were required to serve all customers. In addition, the railroads felt that the fact that trucking companies were not required to pay for the construction of the highway network made them unfairly competitive. Under the ICC, a carrier wishing to provide service along a route it had not previously served had to obtain a "certificate of public convenience and necessity" and would further have to provide evidence that no carriers currently servicing the route would be negatively impacted by increased competition. The rules by which the private, for-hire and exempt carriers operated and the way those rules changed over time is an endlessly entertaining and interesting subject. Rates under regulation were set by rate bureaus, which were collaborative groups of carriers exempt from federal antitrust laws. At the time of deregulation in 1980, there were 10 large rate bureaus and 55 small ones (Hirsh 1988). The rates set by these bureaus were typically not competitive at all. Labor unions were able to extract much higher wages than they would otherwise have received in a competitive market because carriers were able to pass their costs directly on to shippers. In many cases, companies could only secure rights for one-way service, and had to return empty after each loaded move. The postderegulation industry is significantly more efficient than its regulated counterpart. There are many measures of these increases in efficiency. One is that empty miles in the last few years have remained under 20 percent (U.S. Census Bureau 2003). This inefficiency of the regulated industry and the efficiency gains achieved postderegulation has been the subject of numerous studies. See, e.g., Bailey (1986) and Ying (1990). By 1983, prices for freight transportation by truck had dropped 25 percent, and they continued to drop by as much as 35 pecent. Interestingly, the main argument for the deregulation of the industry was the fact that it lacked the economies of scale necessary to require regulation. Economists argued that deregulation would lead the industry to break into smaller more competitive firms (Spady and Friedlaender 1978). However, deregulation has led to a dichotomous industry in which 70 percent of all companies control fewer than seven vehicles but major market segments are dominated by a few large carriers that own tens of thousands of vehicles and effectively control, through spot markets and subcontracting agreements, as many as hundreds of thousands of others. Technically, the economies of scale arguments might in fact be correct, but clearly all segments of the industry are subject to significant economies of density and many of economies of scope. High densities in the parcel and LTL segments allow carriers to make better use of their terminal networks, and high densities in the truckload segment allows carriers to minimize waiting times and empty moves.

The postderegulation entrants into the industry were mainly truckload carriers because of the relative ease of entry into that market, which does not require the investments in infrastructure required to run LTL operations. Since deregulation, virtually no carriers entering the LTL market survived while many truckload entrants were successful (Belzer 1995).

14.5 THE INDUSTRY AND UNIONS

Regulation made possible the rise of unionization in the industry, though there have always been a significant number of nonunion carriers. The height of the union movement in the trucking industry came in 1964 with the adoption of the International Brotherhood of Teamsters National Motor Freight Agreement. The Teamsters, led at the time by James R. Hoffa, achieved a long-time goal of effectively guaranteeing union wages nationwide. Unions had the largest impact on the intercity for-hire segment of the industry and were able to extract high wages in that segment in which companies were able to pass the higher costs directly off onto shippers because of industry-wide pricing agreements. Today, led by Hoffa's son, James P. Hoffa, the union is celebrating the hundredth anniversary of its creation in 1903. Union membership under regulation was around 60 percent of the industry, while by 1990, 10 years after deregulation, it had fallen to 25 percent (Hirsh 1988). According to the latest available data, in 2001 less than 20 percent of truck drivers were union members (Hirsh and MacPherson 2003).

14.6 TRUCKLOAD TRUCKING

The largest fraction of the trucking industry is the truckload sector, which moves full containers from shippers to their consignees. The largest long-distance truckload carriers use rail intermodal transportation whenever it is economically advantageous. Intermodal transportation is generally thought to be economically efficient for moves of more than 300 miles. The use of rail intermodal transportation became more popular during the mid- to late 1990s due to two simultaneous factors: a significant labor shortage in the over-the-road market and improvements in technologies that made intermodal operations more efficient. Long-haul truckload drivers, known as over-the-road drivers, typically spend 3 to 6 weeks on the road at one time. Many of these drivers drive sleeper cabs, and some work in teams of two known as doubles. Over-the-road drivers are typically paid by the loaded mile. A large portion of the local truckload market involves drayage operations—those operations in service to intermodal facilities. Dray operators pick up and deliver loads to railyards or ports and may move loads between such facilities. Dray operators typically move three to five loads per day and are alternatively paid per loaded distance or a flat fee for each load moved.

14.7   LESS-THAN-TRUCKLOAD

Less-than-truckload carriers typically haul loads that weigh less than 10,000 pounds. These carriers must operate consolidation terminals where multiple loads can be loaded onto vehicles en route to nearby destinations. Demands are picked up in a local area and delivered to a local terminal, where they are sorted and loaded into line-haul trailers. These line-haul trailers move loads to break-bulk transfer terminals, where the loads are unloaded and reloaded again. A load might be handled at several of these intermediate locations en route to its final destination. Figure 14.1, taken from

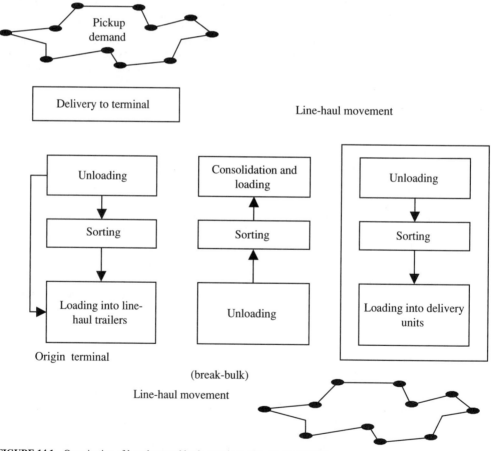

FIGURE 14.1   Organization of less-than-truckload operations. (*Source:* Roy 2003.)

Roy (2003), depicts the organization of these operations. Rates for LTL services are based on the class of product carried and the weight of the shipment. While the number of classes has varied over time, there are currently 18 different freight classifications. The tariffs charged for different classifications can be very different—base prices for the most expensive class (500) can be 10 times the price of goods in the least expensive class (50) of the same weight. Factors affecting the classification of goods include, at a minimum, the following: weight per cubic volume, value per pound, liability to loss damage or theft in transit, likelihood of injury to other freight, and risks due to hazard of carriage and expense of handling. The costs of LTL services are sufficiently high that for some shipments it is less expensive to move the goods as a truckload movement. For example, for many classes of goods, a 5000-pound LTL shipment can be moved less expensively as a full truckload movement; the breakeven point for LTL versus truckload moves depends upon the distance traveled, and the class of goods carried. The shorter the distance and higher the classification of the goods, the more likely a truckload move will be competitive in price with an LTL move. Similarly, charges related to the weight of a shipment are broken into categories such as less than 500 pounds, between 500 and 1000 pounds, between 1000 and 2000 pounds, between 2000 and 5000 pounds, and so on, but just as LTL freight might be better moved as a full truckload, sometimes it is more cost-effective to have a shipment classified in the next-higher weight category (these breakpoints are known as *break weights*). For certain items, the class rating structure based on weights is not a fair structure. Very light and bulky items are charged a rate based on the cubic size of the shipment rather than the weight. It should be clear that for companies with significant transportation needs, managing their freight transportation contracts can be quite complex. Most companies will have in-house transportation and logistics managers whose primary responsibility it is to manage their long and short-term contracts with trucking companies and to see that they pay the lowest fees possible. Much of their time is spent identifying ways to move LTL freight to truckload and move express freight to LTL.

14.8   EXPRESS SERVICES

There are two distinct types of express services: national and international package express services like United Parcel Service (UPS) and Federal Express (FedEx), and local express services, which handle both packages and irregular freight. The express package delivery companies got their start in the early 1900s with the emergence of the American Express Company, the United States Express Company, Wells Fargo Express Company, and Adams Express Company (Merritt 1908). These services were joined in 1912 by UPS, which, begun in 1908 as a courier service, began delivering goods from retailers to customers. In 1913, the United States Postal Service (USPS) extended its operations to include parcel post (Roper 1914). That move was not without its opponents (Merritt 1908) and remains controversial today. UPS gained the rights to expand its operations nationally in 1959, and from 1960 to 1970 its revenues grew from $4.9 million per year to $528 million; much of that growth at the expense of the USPS Parcel Post services (Nissen and Lago 1975). FedEx got its start in 1973 and was the first company to offer one-and two-day national express package delivery. FedEx's early operations benefited immensely from the deregulation of the air cargo industry in 1977 and relied heavily on a hub-and-spoke delivery system. Today UPS and FedEX run both ground and air operations and no longer rely as heavily on their hub-and-spoke systems. Their ground operations are built around local sorting facilities, which serve pickup and delivery operations and send partial or full truckloads to other sorting facilities across the country. These local operations are typically no more than 5 hours' driving time apart, so that the truck drivers can make a single round trip move and return to their home locations within one work shift.

Local express services are dominated by courier companies that provide 2-, 4-, and 8-hour express delivery services within small geographic regions. Every city in the United States is served by several of these, and large cities are served by many. The larger and more successful of these companies typically also provide third-party logistics services—that is, they balance out their irregular operations with contracts to make deliveries for retailers who provide same-day or next-day deliveries within a local area (e.g., express office supply stores).

14.9   PRIVATE FLEETS

A significant fraction of the trucking industry is made up of private fleets. These fleets are the transportation and logistics units of large manufacturing or distribution companies (large food and beverage manufacturers, large groceries, toy manufacturers, etc.). According to a relatively recent study, shippers spend more money transporting freight by private fleets than by for-hire carriers (Roth 1995). Private fleets were estimated in that study to be responsible for 54 percent of trucking industry receipts in 1995. While it is quite likely that the fraction has gone down during the past few years as the result of increased outsourcing, that fraction remains considerable. There have been many debates over the years about when companies should outsource their logistics services and when they should keep them in house. In addition, many major shippers maintain a private fleet for their mission-critical transportation and outsource the rest. One issue for private fleets is the difficulty of keeping empty miles low. For example, the 1995 study mentioned above points out that standard private tractor-trailer fleets were running empty 24 percent of the time, significantly more than for-hire fleets. Similar increases were found in the dry van and tank truck market segments.

14.10   THIRD-PARTY LOGISTICS

Since deregulation, third-party logistics providers (or 3PLs, as they are commonly called) have emerged as a new industry, providing a wide variety of logistical and management services. By combining contemporary information technology with traditional freight-handling systems, an increasing number of intermediary firms are already managing freight transportation for companies that choose not to handle their own shipping and receiving (for example, medium and large manufacturers who find their goods movements too complex to handle themselves). Many of these manufacturers deal with multiple trucking, ocean, rail, and air cargo providers. The complexities of booking and moving their freight are enormous. Freight transportation intermediaries provide a bridge between shippers and carriers, facilitating the flow of information and goods. Lately, these companies have been benefiting from the communications capabilities of the Internet.

Some 3PLs evolved from the prederegulation freight brokerages and shipping agents. The term *freight broker* referred to ICC-licensed truck brokers that acted as marketing agents and load matchers for smaller trucking companies, while shipping agents bought capacity from railroads and sold it to shippers. Until recently, most 3PL companies were affiliated with a parent transportation or warehousing company. 3PL providers are of two types: asset-based or non-asset-based. Today, both asset-based and non-asset-based providers are transforming themselves to respond to new market demands. For example, both asset-based and non-asset-based companies are integrating themselves more deeply into manufacturers' operations rather than focusing exclusively on warehousing and distribution activities.

Many 3PLs are non-asset-based, meaning they do not own transportation equipment or warehouse facilities. These tend to be either management- or knowledge-based consulting companies, focusing mainly on strategic or tactical activities rather than handling physical distribution themselves— although many work closely with asset-based carriers and warehouse managers.

Recently, new types of freight-transportation intermediaries and new business models have emerged. On-line logistics providers are attempting to use the power of the Internet and new software tools to interact efficiently and simply with shippers, carriers, and traditional third-party logistics providers. Some firms provide on-line freight marketplaces, enabling the purchase and sale of freight transportation capacity. These range from simple load-posting boards to sophisticated on-line exchanges. Some firms develop software tools to optimize freight operations or to simplify complex shipping problems. Others supply information on container ports or other intermodal facilities, or organize and aggregate buying power for various companies.

These new intermediaries supplant 3PLs by providing many of the services previously handled by traditional 3PLs. But they also offer opportunities for 3PLs to operate more effectively and provide better services.

These new on-line freight transportation intermediaries and infomediaries promise to transform the industries by enabling companies to move beyond traditional business paradigms, profiting from the synergies of information. Whether or not they will succeed in doing so remains to be seen. Intuitively, it would appear that the freight transportation industry, made up of many small firms and with many existing levels of intermediation, is ideally suited to benefit from the promise of the Internet. However, as an industry it has been very slow to change and to adopt new technologies, because of both a reluctance to do so and a lack of resources needed to make such investments.

Better information about congestion, queues at intermodal facilities, and border crossings and attractive purchasing agreements can increase equipment utilization and network efficiencies and reduce operating costs. These reduced costs may make these systems attractive to trucking companies that see an opportunity to leverage the technology investments of others.

14.11 *MAJOR INDUSTRY ISSUES*

14.11.1 Technology Adoption

All sectors of the industry are moving towards increased real-time operations management. For large companies, this move has been gradual and steady and has been taking place for the past 15 years or so. The largest companies in the industry, especially UPS and FedEx, have led the charge to adopt new technologies, including trip recorders, on-board computers, tracking and tracing technologies, advanced vehicle location systems, electronic data interchange (EDI) systems, and optimization-based routing and scheduling systems. The large truckload and less-than-truckload companies followed the large express package carriers and adopted a wide variety of technologies and optimization tools in the last decade. Their adoption was due in part to pressure from shippers who were familiar with the service guarantees provided by the express carriers. Smaller companies have been slower to adopt advanced technologies, both because their operations need them less and because the costs to small companies can be prohibitive. Technology adoption may well transform the way companies do business, but to date, with the exception of the express trucking industry, the impacts of adoption have been incremental—while the net effects may be large, the impacts of adoption of any specific technology have been relatively small.

14.11.2 Operational Costs

Keeping operational costs low is of grave importance in the trucking industry, where profit margins are very slim. While labor costs are an importance factor in profitability, insurance costs and fuel costs can significantly impact firms' success. Diesel prices rose approximately 20 percent in 2002 and have continued to rise in 2003.

14.11.3 Hours of Service Rules

The hours of service rules in place in the trucking industry through 2003 were developed in 1937 and adopted in 1938. Those rules limit drivers to no more than 60 hours of work in any consecutive 7-day period, and no more than 70 hours of work in any consecutive 8-day period. Drivers were further restricted to 10 hours of driving time in any 24-hour period and at least 8 hours off in the same period. An amendment put in place in 1962 replaced the limit of 10 hours of driving per 24-hour period with a requirement that drivers have at least 8 consecutive off-duty hours after 10 hours of driving. After this off-duty period, the driver could resume driving so that a driver could legally use an 18-hour cycle: 10 hours of driving, 8 hours off-duty, etc. A driver on this cycle would be driving 16 hours in some 24-hour periods. After several years of discussion and after holding extensive public hearings, in April 2003 the Federal Motor Carrier Safety Administration (FMCSA) changed the hours of service rules. Beginning in January 2004 the new rules allow drivers to drive for 11 hours after 10 hours off duty and are restricted to drive those 11 hours within a 14-hour period. These rules were intended to help drivers to get onto a 24-hour work clock. An exception to the 14-hour period is that local drivers (who sleep at home every night) can increase that period to 16 hours once per work week. The weekly limits mentioned above remain in place, with a new rule allowing any driver who remains off-duty for 34 hours to restart the week. For an extensive examination of the history of the rules, and the costs and benefits of amending them, see Belzer and Saltzman (2002) and Crum, Morrow, and Daecher (2002).

There have also been several studies on the impact of operational structures imposed by shippers on driver safety and the ability of companies to adhere to HOS regulations. For example, shippers set up pick-up and delivery schedules that do not take into account HOS regulations and the need for rest. In addition, just-in-time distribution policies require drivers to arrive at a specific time to load/unload, but then may delay loading and unloading. These and other practices may cause drivers to violate the HOS regulations, leading to an increase in fatigue related accidents (Fleger et al. 2002).

14.11.4   Inadequate Parking/Rest Facilities

The lack of sufficient parking and rest facilities for truck drivers is reaching crisis proportions. Truck drivers are often forced to drive in peak passenger commuting hours due to delivery schedules. If there were adequate rest facilities in urban areas, they might be able to reduce some of their peak period travel by arriving at these rest facilities early and resting and taking care of their paperwork until just before their deliveries are due.

14.11.5   Safety

Truck safety has vastly improved over the last few decades, and if new technologies are adopted it will continue to improve. Between 1981 and 2001, there was a 37 percent increase in the number of registered large trucks and a 91 percent increase in miles driven by these trucks. At the same time, there was an 8 percent reduction in the number of heavy duty trucks involved in fatal accidents and a 52 percent reduction in participation of trucks in those accidents, as a fraction of vehicles involve in such accidents (FMCSA 2003). Despite the gains due to better overall equipment, improved roadway design, and a better-trained workforce, problems will continue to emerge, particularly on freeways exceeding practical saturation of truck traffic. In the most congested areas, state and local government officials are seriously talking about the addition of truck-only lanes (possibly tollways and possibly partially automated). Western states have considered developing long-distance truck-only toll lanes that would allow for longer combination vehicles to travel at relatively high speeds. Weigh-in-motion (WIM) systems, which are increasingly replacing manual weigh stations, can be used to measure the axle loads of a vehicle traveling at highway speed. WIM equipment provides information on wheel and axle loads, gross vehicle mass, axle spacing, vehicle speed, distribution of traffic movement through the day, headway between vehicles, freight movement, and traffic volumes and mix. These data can be used for a range of applications, including identifying overloaded vehicles, collecting weight-dependent tolls, verifying optimal load distribution, and providing inputs into pavement management systems. WIM systems usually use axle sensors to detect and classify vehicles and allow for complete electronic screening of compliance and safe carrier status. They can also be used to measure weight, height, and speed of a truck traveling at highway speed, and use this information to warn driver of unsafe conditions. Other technologies that will improve safety are various warning systems being built into new vehicles. These will surely see more use as time goes by.

14.11.6   Changes in Contracting

The development of electronic commerce has impacted the trucking industry just as it has impacted other industries. Its primary impact has been in the area of contracting. In addition to the development of on-line spot markets, where shippers and obtain quotes and drivers can find available loads, the industry is seeing the development of collaborative marketplaces and the use of combinatorial auctions. The collaborative marketplaces take many forms—some develop networks of shippers who manage their transportation needs as though they were a single, large company—others allow carriers to collaborate and for small companies to subcontract loads from the large carriers or from groups of large carriers. Other collaborative models are likely to emerge in the near future (Song and Regan 2003b).

Combinatorial auctions are those in which the auctioneer places a set of heterogeneous items out to bid simultaneously and in which bidders can submit multiple bids for combinations of these items. Bids can be structured so that bidders can express their desire for a bundle of inseparable items (known as atomic bids), a collection of bids with additive values (known as OR bids), or a collection of atomic bids which are mutually exclusive (known as XOR bids). Shippers are increasingly using such auctions, in order to award contracts to trucking companies. The items put out to bid are lanes with associated volumes (for example an average of five loads per week on the lane from Los Angeles to Dallas). In order to bid effectively in such auctions, the carriers have to solve a number of very difficult problems. Large carriers can spend several days constructing their bids for major shippers, and will still not develop optimal bids. Recently, several researchers have begun to examine the optimization problems inherent in both the bid construction and winner determination in combinatorial auctions (Song and Regan 2003a).

14.12 CONCLUSION

The trucking industry is vibrant and active. The demands for movements of freight keep rising and, due to increasingly global manufacturing and distribution systems, they will continue to rise. Like all other industries, it is being impacted by emerging technologies and innovations in operating procedures. The past two decades have seen enormous productivity increases that are not likely to slow in the years ahead.

REFERENCES

Bailey, E. E. 1986. "Price and Productivity Change Following Deregulation: The US Experience." *Economic Journal* 96(3):1–17.

Belzer, M. H. 1995. "Collective Bargaining after Deregulation: Do the Teamsters Still Count?" *Industrial and Labor Relations Review* 48(4):636–55.

Belzer, M. H., and G. M. Saltzman. 2002. "The Case for Strengthened Motor Carrier Hours of Service Regulations." *Transportation Journal* 41(4):51–71.

Crum, M. R., P. C. Morrow, and C. W. Daecher. 2002. *Motor Carrier Scheduling Practices and Their Influence on Driver Fatigue*, Federal Motor Carrier Safety Administration Report FMCSA-RT-03-005.

Federal Motor Carrier Safety Administration (FMCSA). 2003. *Large Truck Crash Facts 2001*. FMCSAR1-02-011, U.S. Department of Transportation, FMCSA, Washington, DC.

Fleger, S. A., N. Owens, L. Rice, and K. J Chen. 2002. "Assessment of Non-Carrier Encouraged Violations of Motor Carrier Regulations: Possible Solutions." In *Proceedings of the International Truck and Bus Safety Research and Policy Symposium,* 501–10.

Herbst, A. F., and J. S. K. Wu. 1973. "Some Evidence of Subsidization: The U.S. Trucking Industry, 1900–1920." *Journal of Economic History* 33(2):417–33.

Hirsh, B. 1988. "Trucking Regulation, Unionization and Labor Earnings, 1973–1985." *Journal of Human Resources* 23(3):296–319.

Hirsh, B., and D. Macpherson. 2003. "Union Membership and Coverage Database from the Current Population Survey." *Industrial and Labor Relations Review* 56(2):349–54.

Merritt, A. N. 1908. "Shall the Scope of Government Be Enlarged So as to Include the Express Business?" *Journal of Political Economy* 16(7):417–35.

Nissen, D. H., and M. Lago. 1975. "Price Elasticity of the Demand for Parcel Post Mail." *Journal of Industrial Economics* 23(4):281–99.

Roper, D. C. 1914. "Fundamental Principles of Parcel Post Administration." *Journal of Political Economy* 22(4):522–35.

Roy, J. 2003. *The Impact of New Supply Chain Management Practices on the Decision Tools Required by the Trucking Industry in Applications of Supply Chain Research and eCommerce in Industry.* Dordrecht: Kluwer, in press.

Song, J., and A. C. Regan. 2003a. "Approximation Algorithms for the Bid Valuation and Structuring Problem in Combinatorial Auctions for the Procurement of Freight Transportation Contracts." University of California, Irvine Working Paper.

———. 2003b. *An Auction Based Collaborative Carrier Network, Transportation Research E,* under review.

Spady, R. H., and A. F. Friedlaender. 1978. "Hedonic Cost Functions for the Regulated Trucking Industry." *Bell Journal of Economics* 9(1):159–79.

Taff, C. A. 1956. "The Competition of Long-Distance Motor Trucking: Farm and Industrial Productions and Supplies (in The Changing Patterns of Competition in Transportation and Other Public Utility Lines)." *American Economic Review* 46(2):508–20.

U.S. Census Bureau. 2003. *2001 Service Annual Survey: Truck Transportation, Messenger Services and Warehousing.*

Ying, J. S. 1990. "The Inefficiency of Regulating a Competitive Industry: Productivity Gains in Trucking Following Reform." *Review of Economics and Statistics* 72(2):191–201.

CHAPTER 15

THE ECONOMICS OF RAILROAD OPERATIONS: RESURGENCE OF A DECLINING INDUSTRY

Wesley W. Wilson[*] and Mark L. Burton[†]

I5.1 INTRODUCTION

The railroad industry is one of the oldest in the United States, dating from the inception of the Baltimore & Ohio in 1830 (Pegrum 1968; Wilner 1997). Very early in their history, railroads were relatively small firms that provided specialized services to local markets over small networks (Wilner 1997). At the time, rail transportation represented a tremendous advance over other available modes. Networks quickly connected, the mode became dominant and, in fact, spurred other industrial and agricultural developments. By the end of the 19th century, America's railroads stood as the centerpiece in a bourgeoning industrial economy.[‡]

As the railroads grew and firms consolidated through the latter half of the 1800s, abuses of market power, growing concentration, and rent-seeking led to passage of the Interstate Commerce Act in 1887. This legislation provided for the economic regulation of railroads. This statute was followed by a number of other regulatory initiatives that further tightened the federal government's control over the industry's economic decision making.[§]

In the first half of the 20th century, motor carriage and commercial aviation emerged as powerful competitors. By the end of World War II, railroad transportation had lost its dominance. The

[*]Dr. Wilson is professor of economics at the University of Oregon and an affiliated researcher at the Upper Great Plains Transportation Institute. He was formerly president of the Transportation and Public Utilities Group of the American Association of Economists, and president of the Agricultural Chapter of the Transportation Research Forum. Contact: Department of Economics, University of Oregon, Eugene, OR 97405-1285 and wwilson@oregon.uoregon.edu.

[†]Dr. Burton is an associate professor of economics at Marshall University. He is director of the Center for Business and Economic Research and Assistant Director for Program Development and Management at Marshall's Rahall Transportation Institute. He is a member of the Transportation Research Board's Agricultural Transportation Task Force and Committee on Inland Navigation. Contact: Center for Business and Economic Research, Marshall University, Huntington, WV 25755, burtonm@marshall.edu.

The authors gratefully acknowledge the editorial and research assistance of Chris Clark in preparation of this book chapter.

[‡]The history and development of the railroad industry is available in many places. Locklin (1972), Pegrum (1968), and Wilner (1997) are examples of excellent reviews of the evolution of the railroad market.

[§]Between 1887 and 1920, there was a steady increase in the control exercised by the federal government and, in fact, the government ceased control of the railroads during the First World War, but returned them to private hands at that conflict's conclusion. The magnitude of regulatory oversight reached its zenith with Transportation Act of 1920. Beginning with that act, the railroad industry entered a 50-year period in which it had little or no control over its economic destiny.

postwar decades saw a further decline in rail usage and in the industry's financial well-being. By the mid-1970s, it was clear that the highly regulated environment in which the railroads had operated was no longer consistent with a functioning and financially viable industry.[*] This realization was quickly followed by series of federal laws, i.e., the Regional Rail Reorganization Act (3-R Act) passed in 1973, the Railroad Revitalization and Regulatory Reform Act (4-R Act) passed in 1976, and the Staggers Rail Act of 1980. Each of these legislative actions were designed to help a threatened industry through a series of actions that created Conrail, provided subsidies to the industry, and most importantly provided for a substantial reduction of the economic regulations under which railroads operated for nearly 100 years.

In the wake of deregulation, the rail sector has reemerged as vital component of the nation's overall transportation system. Deregulation both induced and facilitated a number of cost-reducing activities.[†] Today, the industry is characterized by relatively few, very large firms. These firms provide services over large and complex networks, with substantially less fuel and labor usage than in the past.[‡] Railroads actively participate in two types of markets—the general movement of high-volume, low-value bulk commodities and the transport of higher-valued goods in geographically limited, high volume long-distance corridors. While U.S. carriers face measurable competition in most rail-served markets, there are circumstances in which large, bulk commodity shippers have very limited transportation alternatives. These "captive" shippers contend that postderegulation rail industry gains have been achieved at their expense. Consequently, further modifications to the regulatory environment are constantly under consideration.

Perhaps because of its pivotal economic role, the railroad industry has remained controversial. Historians have both lauded and lamented the role the railroads play and have played in economic development.[§] The industry has been chastised for noncompetitive behavior and inefficient production and is a frequently used as an example to illustrate the rationale for economic regulation. It has also been used to demonstrate the inefficiencies of regulation and the savings to society from deregulation.[¶] Regardless, however, of one's vantage, few would argue that the railroad industry is unimportant as a means of transporting freight. To the contrary, the industry moves billions of tons of commerce each year and is the principal mode used to transport a number of bulk commodities such as coal and grain. Moreover, the railroads have emerged as an important component in the nation's system of intermodal transportation, a system that is vital to participation in international markets.

Due to their private ownership of infrastructure, spatial production technology, and historical importance, the economics of railroading tend to be somewhat more complicated than the economics other transport industries. The purpose of the current chapter is to carefully describe the railroad industry, its operations, the markets it serves, the current regulatory environment, and the specific economic conditions that motivate firm decision-making. Accordingly, Section 15.2 outlines railroad operating practices; Section 15.3 describes rail-served transport markets; Section 15.4 provides a history of rail industry regulation; and Section 15.5 reveals railroad pricing motivations and constraints through a series of economic models.

---

[*]See Friedlaender and Spady (1981), Gallamore (1999), Keeler (1983), Wilner (1997), Winston and Grimm (2000) and a variety of others.

[†]See Barnekov and Kleit (1990), Boyer (1977, 1981, 1987), Burton (1993), MacDonald (1987, 1989, 1998), MacDonald and Cavalluzzo (1996), Wilson (1994, 1996, 1997), Winston and Grimm (2000).

[‡]Bitzan (1999), Bitzan and Keeler (forthcoming), Ivaldi and McCullough (2001), and Wilson and Bitzan (2003).

[§]See Gillen and Waters (1996) who introduce a special volume of the *Logistics and Transportation Review* dedicated to transport infrastructure and economic development. In addition, the reader is referred to Cain (1997), Nelson (1951), Owen (1959). See also excellent discussions in Harper (1978), Locklin (1972), Pegrum (1968), and Winston (1985).

[¶]See the classic books by Meyer et al. (1959) and Friedlaender and Spady (1981) along with more recent research by Barnekov and Kleit (1990), Boyer (1977, 1981), Burton (1993), MacDonald (1987, 1989), MacDonald and Cavalluzzo (1996), McFarland (1989), Wilson (1994; 1996; 1997), Winston (1985; 1993) and Winston et al. (1990) and a litany of others underscoring the effects of regulation and the savings of partial deregulation. Winston (1993), in particular, provides a concise synopsis the distortions created by regulation.

15.2 RAILROAD OPERATIONS

US Class I railroads are typically organized around three primary operating departments. Engineering departments are responsible for the construction and maintenance of track, signals, and other right-of-way structures. Mechanical departments oversee locomotives and freight cars. Transportation departments are responsible for both terminal and line-haul train operations.[*]

Common to all transport modes, railroads operate over networks comprised of links (line-haul trackage) and nodes (terminals, yards, and junctions). The configuration and quality of any specific network element depends on local geography, topography and the nature of the traffic that uses that element. There is also considerable variation in railroad vehicles. While diesel locomotives represent the standard source of power, freight cars vary widely in size and form, depending on the nature of the commodities they are designed to accommodate.[†] Generally, U.S. railroads own the network over which they operate. However, in some relatively isolated settings, carriers will share facilities or operate over the tracks of a connecting railroad.[‡] Similarly, U.S. carriers typically own fleets of locomotives and freight cars. However, it is not at all unusual for a carrier to use equipment owned by another railroad or, in the case of freight cars, equipment that is owned by shippers.[§]

Railroad traffic ranges from single car shipments of relatively high-valued goods to multiple car shipments of lower-valued bulk commodities. Typically, smaller movements are retrieved from shippers by switch crews who move the freight to classification yards where it is consolidated or "blocked" with other shipments destined for the same general location.[¶] This process is repeated in reverse when the shipment is near its destination. Larger shipments, however, often move directly between shipper and receiver. These operations are referred to as "unit train" movements. By eliminating a substantial portion of terminal costs, unit trains can achieve substantially lower per ton-mile costs.[**] Much of the growth in railroading discussed in Section 15.3 resulted from growth in multiple car and unit train traffic stimulated by partial deregulation, the accompanying rate flexibility, and relaxed merger guidelines, which have dramatically increased the size of railroad networks.

In addition to carload and unit train service, "intermodal" traffic has emerged over the past two decades as an important source of railroad activity.[††] Intermodal service combines the line-haul movement of shipping containers or truck trailers by rail with traditional truck service at the origin and destination of the movement. Typically, non-railroad-owned drayage firms retrieve containers or trailers from the shipper and deliver them to railroad intermodal facilities in advance of a pre-scribed cutoff time. The trailers and containers are then loaded by crane onto specially designed railroad equipment for the line-haul move.[‡‡] The process is then reversed as the intermodal shipment

---

[*]Our focus is on the latter of these three operating departments. See Armstrong (1998) for a more complete discussion of all three areas.

[†]During the 20th century, the single most important change in locomotion was the switch from steam to diesel. Recently, more subtle developments, including AC traction and computerization of engine control systems, have produced significant advances in locomotive power and fuel efficiency.

[‡]In some cases shared facilities or trackage rights over a connecting carrier are the result of cooperative agreements between the railroads. In other cases, these outcomes have been imposed by the Interstate Commerce Commission or Surface Transportation Board as a condition for granting merger approval. The payment for the use of rail lines has not been well developed in the economics literature. Obviously, the granting of traffic rights provides for the possibility of greater competition on a route, but the effectiveness of the competition depends critically on the level of payment by the tenant carrier.

[§]Of course, if the railroad does not own the equipment there is a corresponding decrease in the rate as their costs are lower.

[¶]For a more in depth discussion of yard operations see Armstrong (1998).

[**]There are a number of ways to measure railroad outputs. One standard measure is the movement of one ton of freight one mile. This measure is referred to as a ton-mile.

[††]Trailer-on-flat-car (TOFC) services have been routinely available for nearly a half century. However, the rapid growth in intermodal traffic dates to the mid-1980s. Projections suggest that in 2003, the movement of containers and trailers will replace coal as the single larges source of railroad traffic.

[‡‡]In the case of trailer-on-flat-car (TOFC) movements, the most modern equipment consists of articulated "spine" cars that minimize the ratio of equipment weight to lading. Currently, the most efficient form of container-on-flat-car COFC movement involves stacking the containers two-high in articulated "well" cars. This process is referred to as "double-stacking."

nears its final destination. Most intermodal trains are operated as unit trains without being switched en route. Currently, Class I railroads operate approximately 200 intermodal terminals located in 42 states.[*]

Regardless of whether trains are assembled in classification yards, by switch crews, or dispatched directly as unit trains from an on-line shipper or intermodal facility, the transportation process between terminals is likely to be the same. Most line-haul route segments are divided into blocks of varying lengths. Generally, a centrally located train dispatcher controls a number of route segments and the blocks that comprise them. Trains progress from one location toward another as they are given permission to occupy successive blocks. This permission is conferred to the train crew through a variety of methods, including but not limited to, written train orders, electronic lighted signals, and voice-issued track warrants. Regardless of which method is used, the dispatcher attempts to move each train from one location to another as quickly as possible, while maintaining the requisite separation between all trains (and any track maintenance crews) that are present on the portion of the railroad he or she controls. In the earliest days of railroading, all train movements were controlled by strict timetables. However, throughout most of the 20th century, freight trains were run on irregular schedules as "extras" whenever a sufficient amount of traffic was gathered to form a train.[†] Within the past decade, however, the emergence of time-sensitive intermodal traffic, combined with labor demands, has caused a partial return to closely scheduled freight service in limited settings.

The capacity of a particular line-haul route segment depends on a variety of factors. First, this capacity is significantly affected by the number of mainline tracks. Route segments with two or more main tracks are vastly more efficient because they allow trains moving in opposing directions to meet and pass at any location along the route segment. Similarly, faster trains are able to easily overtake slower trains moving in the same direction. The same sort (but not degree) of operating flexibility is afforded by the presence of passing sidings, so that longer, more frequent sidings imply greater route capacity. Unfortunately, multiple main tracks and longer, frequent sidings impose significantly higher construction and maintenance costs. Consequently, they are used only on those locations where high traffic density justifies the additional expense.

Route link capacity and operating ease are also a function of the signaling system in place. "Dark" track segments with no signals, where dispatchers must control operations through track warrants or train orders, require a greater degree of train separation to ensure safety than do route segments with automatic block or dispatcher-controlled electronic signals.

Finally, the capacity of a specific route segment can be influenced by track alignment and grade. The ability to affect the efficient flow of trains is degraded as the variability of train speeds increases. To the extent that relatively sharp curves or steep grades reduce train speeds at specific locations along the route segment, the overall capacity of that segment is reduced.

From an operating perspective, labor and fuel usage constitute the two most important cost items. This is true with respect to both line-haul and terminal operations. Labor costs accounted for 27 percent and fuel accounted for 43 percent of total transportation expenditures in 2008.[‡] However, the railroad industry has made tremendous strides in reducing both labor and fuel usage.[§] In 1980, the industry produced 932 billion ton-miles of freight transportation services with a work force of approximately 550,000. In 2005, Class I railroads produced 1.57 trillion ton-miles of service with only 162,000

---

[*]The number and location of intermodal facilities was determined by the authors based on railroad promotional materials. It should be noted that many of the terminals are small and exist to serve individual customers.

[†]Train sizes vary considerably based on the nature of the commodity or commodities, the nature of the service, and the configuration of the trackage over which the train is to pass. Trains providing local service may be only a few hundred feet in length. Fast moving intermodal trains are generally in the neighborhood of 7000 feet long, whereas heavy unit coal trains may extend to a length of 10,000 feet or more.

[‡]See *Surface Transportation Board, R-1 Reports*, 2008. Readers should note that this discussion focuses on operating costs to the exclusion of capital costs. Also 2008 saw extreme petroleum price volatility that likely exaggerated fuel's share of total transportation costs.

[§]See Davis and Wilson (1999) and Bitzan and Keeler (forthcoming).

employees.* This represents a fivefold increase in output per worker over the span of two decades. Similarly, the average fuel efficiency of rail transport has improved dramatically. The average number of ton-miles of transport services per gallon of fuel consumed has increased from approximately 250 to nearly 400 over the past decade.†

Unlike most freight modes, railroad traffic is routinely interchanged between different railroads. Although, railroad mergers have reduced the frequency of interchange, in 2005, roughly 21 percent of all rail traffic was handled by two or more carriers.‡ At most locations, interchange is handled directly between connecting carriers. However, in some larger locations, traffic is exchanged through the services of a terminal railroad that serves multiple carriers. From an operational standpoint, the existence of interchange means that a carrier's ability to efficiently utilize its own fleet of rail cars is very often dependent on the extent to which connecting railroads return interchanged cars in a timely fashion.§

Like most transport modes, U.S. rail carriers are continuously evaluating emerging technologies in search of ways to improve their operations. The railroads' search for improvement is generally driven by three factors. First, competition between carriers and from alternative modes, forces a continual search for ways to reduce operating costs. Second, from a shipper's perspective, the predictability of transit times is increasingly essential. Thus, rail carriers routinely search for cost-effective ways to improve service reliability. Finally, because railroad accidents can be extraordinarily costly, the nation's railroads continually search technologies that reduce the likelihood of such accidents. These motivations have led to the adoption of numerous new technologies over the past decade. Examples include, but are not limited to new computer and electronic technologies that improve both the performance and fuel efficiency of locomotives, the use of global positioning systems (GPS) in train dispatching, the development of car identification systems that improve shipment tracking abilities, the development of new way-side detection devices that help eliminate the need for cabooses, and the use of remote-control locomotives in yard switching activities.¶

15.3 THE DEFINITION AND EVOLUTION OF RAILROAD MARKETS

In defining markets for analysis, economists do not have a generally accepted prescribed methodology. Various rules have been used in practice but, by and large, the definition of markets is at best a "murky" science. Stigler and Sherman (1986) begin by stating: "The role of the market is to facilitate the making of exchanges between buyers and sellers" (p. 55). They define markets as "...that set of suppliers and demanders whose trading establishes the price of a good" (p. 55). Historically and generally, there are two dimensions of focus in defining markets—the identification of products and the geographic locales of suppliers and demanders. However, neither the product nor geographic limits to a market are well defined. Indeed, as noted by Slade (1986), "[b]ecause markets so frequently overlap, market definition is rarely an easy task" (p. 291).

Despite the difficulties encountered in practice, however, market definitions are necessary for economic analysis and for policy implementation. In economic analysis, market definitions are often arbitrarily chosen with occasional specification tests used to identify whether other products

---

*See www.aar.org. Davis and Wilson (1999) develop and estimate a model of railroad employment. As they discuss, industry employment has fallen dramatically during a time period in which industry output has been increasing. Labor has become therefore more productive. In their model, they find that innovations such as increasing use of unit trains, longer lengths of haul, and mergers have played significant roles in reducing the quantity of labor used by firms. Bitzan and Keeler (forthcoming) also examine productivity, focusing on labor saving devices e.g., the elimination of the caboose.

†See "Available Navigation, Fuel Consumption and Pollution, Abatement: The Missouri River Basin," U.S. Army Corps of Engineers, Omaha, Nebraska, July, 1998.

‡This value was developed through the use of the Surface Transportation Board's 2001 Carload Waybill Sample.

§Initially, concerns regarding the timely return of rolling stock made many U.S. carriers hesitant to interchange traffic with Mexican carries. However, reforms in the operation of Mexico's railroads have largely eliminated such issues.

¶In the fall of 2008, Congress passed legislation mandating that all routes hosting passenger rail service and heavily used freight-only routes be equipped with Positive Train Control (PTC), a system that will override human control inputs when track occupancy or speed authorities are violated. The legislation calls for full implementation by 2015.

or locations not in the defined market affect results of the analysis. In antitrust policy, market definition is often a major issue, particularly in monopoly and merger cases. Given a court determined market definition, a variety of summary measures, e.g., concentration ratios, Herfindahl indices, etc., are then used to evaluate the structure of the market to infer some notion of market power. A narrowly defined market may give overstated degrees of market power, while broadly defined markets may yield underestimated degrees of market power.[*]

Given both the necessity and the importance of market definitions, a variety of approaches have evolved. Theoretically, substitution and complementary relationships among products interconnect markets together. These substitution or complementary relationships may be present for either demanders or suppliers, but the determination of the limits of market interconnections is not an easy task. Uri and Rifkin (1985) discuss the range of practices used. These include a consideration of how "reasonable" is the interchangeability of products, to an examination of prices and their movements. But, as Slade (1986) states, "In practice, market definition frequently involves rules of thumb and may rest on complex legal distinctions that bear little relationship to economic principles" (p. 292). Beginning in the early 1980s, the definition of a market used by the Department of Justice in antitrust proceedings changed to "a product or group of products and a geographic area in which it is sold such that a hypothetical, profit-maximizing firm, not subject to rate regulation, that was the only present and future sell of those products in that area would impose a 'small but significant and nontransitory' increase in price above prevailing or likely future levels."[†] Such a definition, as noted by Scheffman and Spiller (1985) and by Spiller and Huang (1986), relies on counterfactual rather than observed behavior.

With network technologies, the complications of market definitions are more difficult. Connections of markets may emanate from the demand side of the market, e.g., wheat from North Dakota and wheat from Kansas may be substitute movements by a receiver in Portland. Connections may also emanate from the supply-side, e.g., if economies of density and/or scope exist.[‡] That is, the pricing of a commodity on a link in the network is affected by pricing of commodities either on the same link or, if economies of scope exist, by pricing on other links in the network.[§] Such interdependencies require recognition that railroad firms are multiproduct firms that operate over spatially separated nodes in a network. This recognition is very much consistent with Winston (1985) who points out that the output of a transportation firm "...is the movement of a commodity or passenger from a specific origin to a specific destination over a particular time period" (p. 60).

Following the discussion above and Winston (1985), railroads operate in many markets, hauling of different commodities between lots of different origin-destination pairs. In aggregate, however, railroads dominated U.S. freight transportation during the latter half of the 19th century and through the first few decades of the 20th century. In the early part of the 20th century, new modes developed and the dominance of the 19th century eroded. Today, although railroads continue to play an important role in the nation's overall transportation system, the role they play is much narrower. As reported in the 1997 Commodity Flow Survey, railroads provided 38.4 percent of all tonmiles moved, pointing to their continued importance. However, they provided only 4.6 percent of transportation services expressed in FOB value of shipments,[¶] suggesting their service is larger for lower valued commodities. Further, while they hauled 38.4 percent of ton-miles, this figure corresponds to only 14 percent of all tonnage moved. Together, these statistics suggest they haul relatively low valued commodities over relatively longer distances.

In the remainder of this section, we describe the specific markets in which railroads operate. Our discussion initially focuses on identifying the products hauled and the extent of operations in each

---

[*]We note that the Department of Justice uses Herfindahl indices as one factor in considering whether to contest a merger between firms in a market. The Herfindahl index is simply the sum of squared firm market shares.

[†]The *Journal of Economic Perspectives* (1987) has a series of papers on the issues related to the definition of markets. See Schmalensee (1987), Fisher (1987), and White (1987) papers for relatively detailed discussions of market definitions. In addition, Uri and Rifkin (1985) and Westbrook and Buckley (1990) provide excellent discussions related to the rail market.

[‡]Economies of density means that as output over a given link increase, per unit costs of providing the service fall. Economies of scope means that a single multiproduct firm can produce two or more outputs more cheaply than can separate specialized firms who together product the small level of outputs.

[§]Berry (1992) and DeVaney and Walls (1999) point out these and related interdependencies over a network.

[¶]FOB means free on board. It does not include the logistics costs of movements.

product category. We then describe the supply-side of the market. In this regard, we describe the network over which railroads operate and how it has changed over time. We then turn to describing the number of firms, concentration levels, and discuss how changes in regulation, firm behavior, and network characteristics have transcended into changes in operating characteristics and costs.

15.3.1 Products Hauled

Railroad markets defined in terms of origins-destinations and commodities suffer from data availability and sheer magnitudes of the number of markets and potential markets served. Data at this level of disaggregation are extremely difficult to access and use in characterizing markets. However, as part of the reporting requirement to the Surface Transportation Board (formerly, the Interstate Commerce Commission), railroads do provide aggregate measures of the products they haul. These data are summarized, by Standard Transportation Commodity Code at the two digit level (STCC-2 code), in Table 15.1. In this regard, we provide, the commodities hauled (at the two digit level),

TABLE 15.1 Commodity Codes, Tonnage, Revenues, and Rates 1983 and 2005 (All Monetary Values Are in Nominal Dollars)

STCC	Description	1983 tons	1983 revenues	1983 revenues per ton	2005 tons	2005 revenues	2005 revenues per ton
1	Farm Products	130.98	1,805.05	13.78	156.72	3,902.72	24.90
9	Fresh Fish or other Marine Products	0.02	0.65	28.05	0.12	6.41	52.53
10	Metallic Ores	66.56	326.98	4.91	70.56	608.32	8.62
11	Coal	426.92	4,013.19	9.40	687.54	9,197.89	13.38
13	Crude Petroleum, Not Gas or Gasoline	1.40	16.62	11.83	0.79	16.59	20.96
14	Nonmetallic Minerals	55.44	434.18	7.83	148.97	1,414.91	9.50
19	Ordnance or Accessories	0.27	9.93	37.15	0.18	21.51	119.41
20	Food or Kindred Products	67.94	1,302.71	19.17	110.29	3,810.33	34.55
22	Textile Mill Products	0.32	11.34	35.37	0.33	26.96	81.84
23	Apparel	0.11	5.33	50.63	2.16	180.74	83.85
24	Lumber or Wood Products	48.60	813.75	16.74	72.78	3,094.08	42.51
25	Furniture or Fixtures	0.45	29.43	65.60	0.95	101.72	107.07
26	Pulp, Paper, or Allied Products	26.86	526.58	19.60	51.20	2,628.10	51.33
27	Printed Matter	0.19	5.47	29.52	0.48	36.55	76.61
28	Chemicals or Allied Products	82.33	1,462.53	17.76	170.52	6,182.13	36.25
29	Petroleum or Coal Products	32.00	498.40	15.57	58.16	1,694.59	29.14
30	Rubber or Misc. Plastic Products	1.30	48.20	37.18	1.82	162.59	89.10
32	Clay, Concrete, Glass, or Stone	33.69	449.59	13.34	61.31	1,704.04	27.79
33	Primary Metal Products	26.73	414.75	15.51	65.90	2,060.39	31.27
34	Fabricated Metal Products	0.75	26.99	36.01	0.83	68.47	82.74
35	Machinery	0.83	36.47	43.86	1.03	102.28	98.85
36	Electrical Machinery or Equipment	1.09	58.18	53.61	1.97	235.18	119.18
37	Transportation Equipment	21.13	1,033.00	48.87	55.09	5,577.68	101.25
38	Instruments or Photographic Goods	0.05	1.95	39.25	0.12	11.80	102.37
39	Misc. Products or Manufacturing	0.11	6.18	55.76	0.64	58.47	91.15
40	Waste or Scrap Metal	20.74	264.78	12.76	47.42	1,226.98	25.88
41	Misc Freight Shipments	1.16	47.36	41.00	2.10	234.47	111.63
42	Containers, Shipping, Returned Empty	0.92	25.98	28.32	12.82	1,004.23	78.33
44	Freight Forwarder Traffic	1.29	50.40	39.04	3.24	230.03	71.00
45	Shipper Assoc. or Similar Traffic	7.89	429.83	54.45	0.24	17.64	73.76
46	Misc. Mixed Shipments	30.14	1,209.04	40.10	133.45	7,871.02	58.98
48	Waste Hazardous or Substances				1.04	61.07	58.70
	ALL COMMODITIES	1089.01	15384.04	14.12	1,920.77	53,549.88	27.95

Source: 1983, 2005 Carload Waybill Sample.

along with *originated* tonnages,[*] revenues, and the revenue per ton received from each product category.

In total, the industry hauled approximately 1.1 billion tons in 1983. By 2005, the amount hauled by the industry was about 1.9 billion, an increase of about 73 percent since in a little more than 10 years. These tons reflect movements of 35 different commodity (STCC-2) codes and encompass a wide array of different types of movements. The diversity of products hauled by railroads is large, and includes coal, farm products, chemicals, machinery, freight forwarder traffic, etc. Despite the degree of diversity in the commodities that railroads haul, coal movements have dominated and continue to dominate railroad tonnage. Specifically, coal is far and away the leading commodity (in terms of total ton-miles) in both 1983 and 2005. Coal accounted for about 39 percent of all railroad traffic in 1983 and about 31 percent in 2005. When examining the contribution of coal to total revenues, however, the percentage is much lower. That is, in 1983 only 26 percent of railroad revenues were from coal ($5.693 billion), and that figure despite the increase in market share, fell to 14.6 percent in 2005. After coal, there is something of a drop off in terms of percentage of tons originated and the revenue received from the various other commodities. In 2005, the next three largest commodities hauled (intermodal shipments, chemicals, and farm products) accounted for a total of 33 percent of ton-miles and about 36 percent of total revenues, so that together, the top four commodity groups account for roughly two-thirds of rail traffic and revenues.

Different commodities are and have been priced differently. These commodities have very different demand, cost, and competitive characteristics and, as a result, it is not surprising that rates vary across commodities. We summarize average rates per ton by two-digit Standard Transportation Commodity Code (STCC) in Table 15.1. Overall, railroads earned about $14.12 per ton in 1983, which increased to $27.95 per ton in 2005 (weighted average by tonnage and expressed in nominal terms). In real terms (2005 base), the railroads earned $25.35 per ton in 1983, compared to $27.95 in 2005—a modest increase in real per-ton rates during a period where average shipment distances increased by more than 28 percent. The leading commodity hauled (coal) had an average 2005 rate per ton of about $13.38. This is well below the overall average rate per ton for all tonnage moved (which was $27.95). The rates per ton for the other major commodities hauled by the railroad (intermodal, chemicals, and farm products) are $58.98, $36.25, and $24.90, respectively. For other commodities, the rates tend to be much higher. For example, the average rates for transportation equipment (37), instruments or photographic goods (38), and freight forwarder (44) traffic are each in excess of $50 per ton. However, the combined traffic share for these relatively high rated commodities (measured in tons) is only 3.7 percent. In contrast, the combined market share of the leading five commodities (in tonnage), which yield lower rates is, 71.4 percent.

15.3.2 Shipment Characteristics

From Table 15.1 it is clear that railroads tend to haul a lot of different commodities, but specialize in commodities that do not command high rates. There are several possible explanations. Some of these explanations relate to the characteristics of the commodities (and their associated demand and supply characteristics), and some relate to characteristics of how the service is provided. The characteristics we describe include the value of the commodity shipped, shipment sizes, density (weight) of the product, distances shipped, and the number of railroads involved in the movement.

Demand and supply characteristics of the commodity shipped have an important and obvious influence on the rate that can be charged. Table 15.2 provides representative commodity values per ton.[†] These data reinforce a number of points we have made throughout this chapter. In particular, railroads tend to haul low-value bulk commodities, which tend to be shipped over long distances

---

[*]Railroads often interline with other railroads to serve origin-destination pairs. The tonnage figures we report are the sum of tonnages originated on the railroad lines regardless of whether the service terminates on the associated railroad's lines or another railroad lines.

[†]These data were calculated using data from the 1993 Commodity Flow Survey.

TABLE 15.2   Rail Shipment Characteristics by Commodity—2005

STCC	Description	Cars	Tons per car	Shipment distance	Number of RR's	F.O.B. value
1	Farm products	56	101	1,014	1.19	223.74
10	Metallic ores	108	91	257	1.18	235.58
11	Coal	108	112	719	1.16	20.75
14	Nonmetallic minerals	34	100	274	1.11	11.58
20	Food and kindred prod.	6	87	1,029	1.22	996.65
24	Lumber and wood prod.	1	88	1,181	1.58	190.94
26	Pulp and paper prod.	1	71	1,047	1.83	897.66
28	Chemicals	6	93	895	1.49	977.08
29	Petrol and coal prod.	27	87	767	1.42	190.61
32	Clay, concrete, glass, or stone	6	98	619	1.42	114.28
33	Primary metal prod.	7	89	699	1.38	858.12
37	Transportation equip.	2	32	939	1.14	446.89
40	Waste or scrap	3	82	553	1.27	139.49
46	Misc. mixed shipments	1	19	1,506	1.07	N.A.

Sources: 1983, 2005 Carload Waybill Sample and 1993 Commodity Flow Survey (value data).

with large shipment sizes. Indeed, coal (11) and nonmetallic minerals (14), commodities which ranked as number 1 and 5 in terms of tonnages in 2005, each have FOB values considerable less than the other commodity classes. Higher-valued goods, such as transportation equipment (37), tend to flow by other modes. These higher valued commodities have higher inventory costs and, as a result, motor carriers, which offer faster transit times, tend to dominate the movements of these commodities. This was directly predicted by the work of Baumol and Vinod (1970) and other related research discussed in the surveys by Oum, Waters, and Yong (1992) and Winston (1985). These studies point to the notion that rate is just one element of the cost of alternative modes. Service characteristics do have a strong and important effect on the decision of shippers regarding which mode to use.*

In addition to demand considerations, the specifics of how service is provided also has an enormous influence on costs and through costs on rates. In this regard, there are a number of differences across commodity groups. For example, coal movements tend to involve larger shipment sizes that allow railroads to use unit trains and multiple cars in meeting the demands. Such operations are much more efficient than single car movements and, as a result, point to lower costs and rates. Coal and nonmetallic mineral movements have the largest average shipment size, at more than 108 cars. Other major commodity groups such as farm products and metallic ores also tend to be multiple car shipments. The remaining commodities are dominated by single car movements, and we note that the average rates per ton-mile is larger than are the rates for coal, farm products, metallic ores and nonmetallic minerals.

Bulk commodities also tend to have larger weights per car. Certainly, this likely represents differences in the type of car used but also reflected is the density of the product. For example, on a per ton basis, feathers are far more expensive to ship than lead is on a per-ton basis. Thus, one would expect that the rate attached to feathers is much higher. Indeed, the rates per ton-mile are lower for higher car weights.

Overall, railroads ship commodities long distances. In 12 of 14 cases, rail shipment distances are in excess of 500 miles. In the 1997 Commodity Flow Statistics, the average shipment distance for all shipments (all modes) is 472 miles. For comparison purposes, for-hire truck, private truck, and shallow draft water are 485, 53, and 177 respectively. The provision of a service involves both variable

---

*There are scores of demand studies which reinforce the notion that service characteristics play an important role in shipper demand decisions. See surveys by Winston (1985) and Oum, Waters, and Yong (1992), and Small and Winston (1999). Also, see Inaba and Wallace (1989) for an excellent example of such research.

costs as well as "quasi" fixed costs. Quasi fixed costs are costs that are incurred with the provision of service but do not vary with the extent of the service provided. It is well documented that as average lengths of haul increase, rates fall owing to the reduced per unit cost of providing the service as mileage increases. It is also well documented e.g., Locklin (1972) that greater shipment distances generally favor rail movements versus truck since rail per unit costs fall with distance faster than do truck per unit costs. Again, across commodities, there are considerable differences. Coal travels an average of 719 miles, intermodal shipments travel the furthest (1506 miles), while other primary products including farm products (01), nonmetallic minerals (14), and chemicals (28) travel 1014, 274, and 895 miles, respectively.

Another element, provided in Table 15.2, is the number of railroads involved in a movement. Specifically, railroads often work with other railroads to complete a movement from one location to another. An individual rail network may not be able to service a given origin-destination pair. But through "interlining" railroads can extend the number of origin-destination pairs that can be serviced. A value of one means that one railroad both originates and terminates the movement on its own lines. While there is no discernable pattern across commodities, we do note that there is a tendency for less interchange for the bulkiest of commodities and slightly more interchanged for higher-valued, less bulky commodity groups.

In summary, the traditional bulk commodities (coal, grain, minerals, and ore) still dominate rail tonnage. However, the rapid growth in higher valued intermodal traffic and the steady flow of more highly rated chemical shipments are essential elements in providing the railroad industry with a stable and adequate revenue stream. While these outcomes certainly reflect demand and commodity characteristics, costs are also a major determinant of pricing. Larger overall shipments, shipments where cars are more heavily loaded, longer shipment distances, and fewer interchanges all imply lower railroad rates and higher probability that shippers will choose rail carriage over available alternatives.[*] As we report in the following sections, many of the changes occurring in the last twenty years reflect innovations, pricing, and consolidations that have occurred largely as a result of partial deregulation and that have yielded tremendous efficiency gains to railroads.

15.3.3   Network Characteristics and Firm Size

When the Stagger's Rail Act was passed in 1980, railroads provided services over 180,000 miles of road. Over time, this figure has fallen to about 120,000 route miles (Figure 15.1 depicts this information from 1983 through 2007). Two primary factors explain the reduction in network size. Through the 1970s and 80s, there has been on-going pressure of the railroads to abandon unprofitable low-density lines (branch lines). Under partial deregulation, some of the lines have indeed been abandoned, while others have been sold to form new short-line and regional railroads.[†] Overall, the size of the rail network has fallen. Corresponding to the fall in network size, however, there has been an increase in the ton-miles (the unit of output traditionally used in railroad economics). Indeed, industry ton-miles increased from about 800 billion in 1983 to about 1.5 trillion in 2007 (Figure 15.1) with a corresponding increase in network utilization (Figure 15.2).

Within firms, the changes are dramatic. Since partial deregulation, there has been an unprecedented consolidation of firm outputs and network through merger activity. The number of Class I carriers has fallen dramatically from 28 in 1983 to just 7 in 2007 (Figure 15.3).[‡] Associated with the consolidation of firms has been a tremendous increase in the size of firm networks and in output levels. While the size of the industry network has fallen, the average size of firm networks has increased

---

[*]The primary source of competition for these commodities is generally either commercial navigation or local commodity usage, where truck transport is feasible such as local grain processing, local livestock operations, or local electricity generation.

[†]Allen (1990), Klindworth (1983), and Tye (1990) describe some of these changes and effects.

[‡]The decreases in firm numbers is due to the declassification of firms as Class I railroads. From 1983 to 1997, six railroads (BLE, BM, DH, DMIR, FEC, PLE) were declassified. In addition, to these declassifications, there were 12 mergers which accounted for remaining disappearance of firms. See Bitzan (1999) and Wilson and Bitzan (2003) for greater detail.

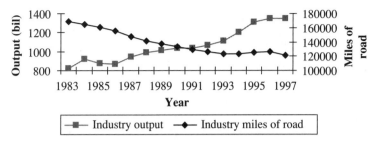

FIGURE 15.1 Industry output and network size.

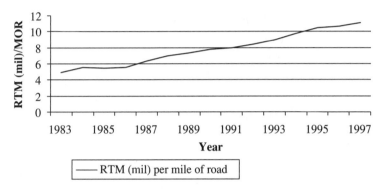

FIGURE 15.2 RTM (mil) per mile of road.

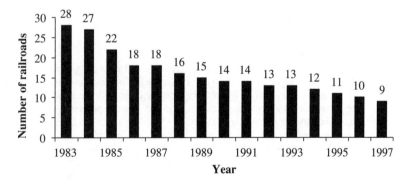

FIGURE 15.3 Number of rail roads.

from about 6000 miles per firm in 1983 to over 17,000 miles per firm in 2007 (Figure 15.4). Changes in terms of output are even more dramatic. In 1983, Class I railroads operated an average of about 29 billion revenue ton-miles (Figure 15.4). By 2007, this figure had increased by more than a factor of eight to over 253 billion revenue ton-miles (Figure 15.4). These figures together suggest that network utilization has been tremendously impacted by partial deregulation and continues to grow. The increases in outputs, networks, and utilization are very important in terms of realizing greater economies of density.

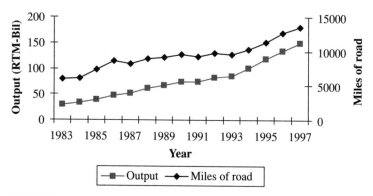

FIGURE 15.4   Average firm size.

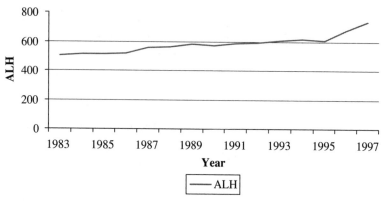

FIGURE 15.5   Average length of haul.

In addition to growing outputs, networks, and network utilization, there are a number of other sources of change in the industry. Specifically, economists have long examined the structure of costs in the railroad industry.[*] In so doing, they have provided a number of measures of operating and/or network characteristics e.g., average length of haul, percentage of unit train traffic, percentage of interlined traffic, etc. Many of these have been impacted and have changed dramatically over the last two decades. By and large the primary variables are similar to those described earlier in Table 15.2.

As discussed earlier, longer lengths of haul are generally thought to be efficiency enhancing. As we show in Figure 15.5, average length of haul has been increasing consistently over the span of the data. In 1983, average length of haul across railroads was about 360 miles, but increased just over 800 miles by 2007 (tonnage-weighted mean across railroads). A change in average length of haul from 360 to 800 miles is more than a 100 percent increase and points to dramatic reductions in costs.

Another significant change in railroad operations is the growing emphasis on unit train traffic. Unit train traffic follows a much different and much more efficient operation if the volumes are large enough, as we discussed in Section 15.2. Such traffic occurs in large volumes between a specific shipper and receiver who have made investments, e.g., high-speed loaders that allow large shipments.

---

[*]See, for example, see Brown et al. (1979), Keeler (1974), Braetigam and Daughety (1982), Caves et al. (1980, 1981, 1985), Tolliver (1984), Vellturo et al. (1992), Berndt et al. (1993), Friedlaender et al. (1993), Wilson (1997), Bitzan (1999), and Ivaldi and McCollough (2001).

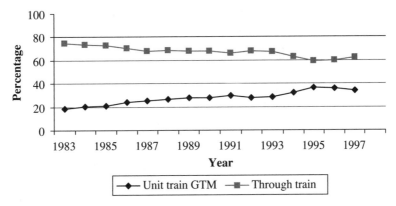

FIGURE 15.6   Type of train traffic.

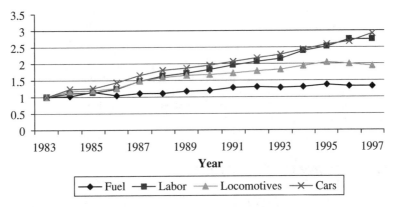

FIGURE 15.7   Productivity measures.

In Figure 15.6, we compare the use of unit train versus through train operations[*] and show that there was increased use of unit trains through the 1980s and 1990s, while the more traditional use of through trains fell. This pattern continues today.[†]

The increase in firm sizes, the changes in networks, the consolidation of firms, and the changes in traffic patterns have made firm inputs much more productive. In Figure 15.7, we have calculated the weighted (by revenue ton-mile) outputs per inputs. The inputs we use are labor, fuel, locomotives, and rail cars.[‡] In all cases, we normalized the variables by 1983 values. In evaluating these figures it is clear that, across the board, the primary inputs used by railroads have yielded efficiency gains. The production of both labor and equipment are the largest. The other input discussed here—fuel (network and route miles were discussed earlier)—has also increased, just not as dramatically. Even so output per gallon of fuel has increased by about 30 percent.

---

[*]Way train movements are the residual.

[†]Notably, continued growth in unit train usage is largely attributable to the breath-taking growth in intermodal shipments that are generally moved under the unit train model, albeit with significantly different equipment.

[‡]We do note that the percentage of both types of equipment that are owned by the firms has fallen through time. In 1983, railroads owned 76 percent of the locomotives and 82 percent of the rail cars used in moving commodities (weighted averages by revenue ton-miles). By 1997, each these figures had fallen to 67 percent, reflecting a now long-standing trend to the use of leased equipment.

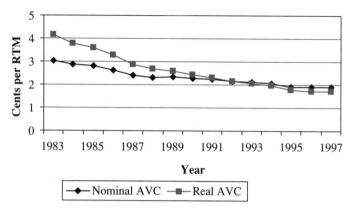

FIGURE 15.8 Average variable costs.

Obviously, when output per unit input is increasing *for all inputs,* cost per unit produced is falling. Indeed, econometric research, e.g., Wilson (1997) supports the fact that railroad costs have fallen and fallen measurably since partial deregulation. In documenting the outcome of the efficiency enhancements, we simply plot average variable costs over time (Figure 15.8). Variable costs in this regard include fuel, labor, equipment, and materials and supplies. To express the values in real terms, we used the GDP price deflator with a base year of 1992.[*] By any measure, the decreases in per unit cost have been large. In nominal terms, per unit costs fell from 3.03 cents per RTM in 1983 to 1.92 cents per RTM in 1997—a 36 percent reduction in per unit costs. In real terms, the change is much larger. In real terms, per unit costs fell from 4.14 cents per RTM in 1983 to 1.72 cents in 1997—a 58 percent reduction in per unit costs.[†]

15.4 CHANGES IN MARKET STRUCTURE

In addition to demand and cost (network) characteristics, the structure of the marketplace theoretically may have a large effect on explaining market outcomes. As noted earlier, there have been a number of mergers over the last two decades. By any standard, this represents a major consolidation of market power. Industrial organization economists use a variety of measures to gauge the level of market power. Historically and most commonly, both Concentration Ratios and Herfindahl indices are used.[‡] The employment of such measures rests heavily on the definition of the market, and as we discussed earlier, network markets are best characterized in terms of the flows of a commodity between origin-destination pairs. We present these measures in this section, simply as a description of the consolidation of output among firms operating in the United States.

Figures 15.9 and 15.10 provide a four firm concentration ratio (CR-4) through 2007 and the Herfindahl (H) ratio in the US from 1983–1997.[§] As these figures indicate, the consolidation of firms through merger activities has led to an astounding increase in concentration. In 1983, the CR-4 was

---

[*]They do not include way and structures. We also include an opportunity cost in the measurement of capital items. For a complete discussion of variable definitions, see Bitzan (1999) and Wilson and Bitzan (2003).

[†]Data limitations make it difficult to extend this series beyond 1997. However, anecdotal information suggests that the cost impacts of deregulation were largely exhausted by the mid-1990's and that average costs have flattened or even increased slightly over the intervening years.

[‡]These measures have been subject to some debate, and recent advances in industrial organization have led to more direct measures of market power. Nonetheless, these two measures, CR and H continue to be routinely used to describe the structure of the marketplace.

[§]The CR-4 is the sum of the largest four firm's market shares. The H is the sum of the squares of market shares for the firms in the market.

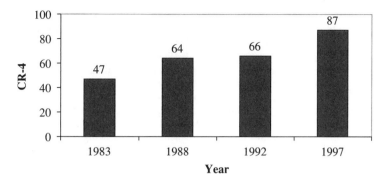

FIGURE 15.9   Four-firm concentration ratio (CR-4).

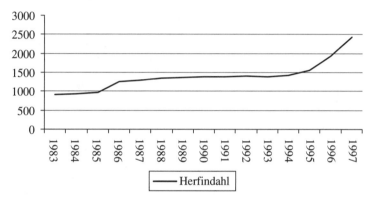

FIGURE 15.10   Herfindahl index.

only 47 percent, but by 1997 it had risen to 87 percent and was only slightly lower (86 percent) in 2007. Hence, four firms provide 86 percent of the Class I ton-miles produced.[*] By any standard, this level of concentration is large. The Herfindahl index yields a comparable conclusion. But, in examining the data, we note that most of the mergers in the latter period fall within Department of Justices' anticompetitive guidelines. That is, the postmerger Herfindahl index is in excess of 1800 and the effects of the merger are to increase the H by more than 100 points. Specifically, the UP-SP merger in 1995 and the BN-ATSF merger in 1996 would have fallen under DOJ merger guidelines. These measures point to increased market power in the railroad industry. While both costs and revenues per unit have been falling, the increase and growth in concentration point to potential inefficiencies in the market.

15.5   RAILROAD MARKETS AND THE FUTURE

This section has attempted to summarize, albeit briefly, railroad markets. In this regard, we found that railroads operate in many different commodity and geographic markets. However, the traffic is dominated by commodities that are low in value and can be shipped in large volumes and intermodal

---

[*]Currently Class II railroads and short lines (Class III) provide roughly 20 percent of all rail freight ton-miles. However, from a competitive standpoint, the role of these small railroads is akin to the competitive fringe in a dominant firm model.

shipments. Rates for bulk shipments tend to be relatively low, and rates for intermodal movements are, at least, truck competitive. This pattern of shipments is one that has been longstanding and will likely remain so. But, since deregulation, there have been a number of changes in railroading that point to considerable new efficiencies. These include a sizable reduction in the size of the industry network, an increase in firm size, and improvements in operating characteristics such as output per route mile, average length of haul, and increases in shipment size, that explain tremendous efficiency gains. Some of these efficiencies have been gained as a result of pricing practices (e.g., unit train rates), but some are due to mergers and innovations geared to labor- and fuel-saving devices. In total, these effects explain the fact that per unit costs have fallen, and have fallen dramatically over the last 25 years.

Weighing against these cost reductions is a growing level of concentration in the U.S. railroad industry. Our data pertain to the whole of the United States as the market (all U.S. Class I firms). However, all of the firms have limited geographic scope and the preponderance of shippers are captive shippers, i.e., served by one railroad. As a result, the overall level of concentration is likely understated here, and the extent of market power is likely larger than one might infer from these figures. Indeed, at the highest level of disaggregation—commodity movements between specific origin-destination pairs—the rigor of intramodal, intermodal, geographic, and product competition is essential for competitive outcomes.

The railroads have actively promoted larger shipment sizes through both pricing and operating practices. Bulk commodity shippers are given more favorable rates when shipments exceed specific size thresholds. Moreover, the railroads have worked to consolidate geographically dispersed shipping locations into more centralized, high-volume, origins, and destinations. The result is that grain shipment size increased by 69 percent between 1995 and 2001, while coal shipment size increased by 410 over the same period (Waybill Statistics).

While railroads have generally conceded high-valued, low-volume traffic to motor carriers, they have worked hard to retain high-valued traffic where shipment volumes are sufficient to allow unit train operations. Thus, we observe a continued strength in automobile, chemical and intermodal traffic. Higher-valued traffic will probably never dominate overall tonnage, but it continues to grow in importance as a source of revenue.[*]

Looking to the future, it is likely that both trends will continue. Other available transport modes do not have the capacity (trucking) or the route flexibility (commercial navigation) to compete effectively for most of the bulk commodity traffic that currently moves by rail. At the same time, general and significant constraints on the potential growth of truck traffic of any kind mean that the railroads have the potential to recapture an increasing share of higher-valued shipments when and where shipment volumes are relatively large and/or shipment distances are relatively long.[†] Whether or not railroads are successful in doing so depends heavily on whether they can provide a level of service that, at least, approaches that offered by motor carriers.

15.6 REGULATION AND PARTIAL DEREGULATION[‡]

Regulation of railroads, as noted by Keeler (1983), was rooted in British common law dating to the 16th century. From common law emanated common carrier obligations. That is, modes predating railroads were often granted operating certificates, and this practice was extended toward railroads. Keeler notes "…the notion behind common carriage was that the government would grant the carrier certain powers and privileges, generally conferring an exclusive right to make a profit from

---

[*]Indeed, in 2007, the year prior to the economic collapse of 2008, intermodal shipments surpassed coal as a revenue generator for Norfolk Southern.

[†]The nation's highway system is largely complete. Moreover, the ability to further expand this system is measurably constrained by both environmental concerns and fiscal realities. Thus, the availability of additional motor carrier capacity depends heavily on our ability to better use the network that is currently in place.

[‡]The material covered in this section can be found in a number of sources. See, for example, Hilton (1969), Keeler (1983), Locklin (1972), Pegrum (1968), Winston et al. 1990, and Winston (1993).

transportation. In return for these privileges, the carrier was expected to assume certain obligations." (p. 20) Such obligations were that carriers may not refuse service, must provide service at a reasonable price, must provide for all equally and are responsible for safe delivery.[*] Owing to difficulties in implementing such regulations through the court system, due to growing and more complex railroad networks,[†] numerous states passed laws to regulate railroads during the 1870s and 1880s.[‡]

Passage of the Interstate Commerce Act of 1887 was the first piece of federal regulation to regulate railroads.[§] Beginning with this legislation and various amendments, economic regulation by the Interstate Commerce Commission grew to include maximum rate regulation (Hepburn Act of 1906), minimum rate regulation (Transportation Act of 1920), entry and exit (Transportation Act of 1920), and promotion of mergers (Transportation Act of 1920).[¶]

Throughout the 1920s and 30s, other modes began to develop and the railroads' dominance of transportation began to erode. During this time period, other modes came under regulation. The Motor Carrier Act of 1935 and the Transportation Act of 1940 brought much of the motor carrier industry and some of the barge industry under ICC jurisdiction. By the end of World War II, virtually every aspect of railroad operations was tightly controlled by the Interstate Commerce Commission (ICC). In particular, the ICC had jurisdiction over rail rates, line abandonment, service discontinuities, rail mergers, car flows, and interchange rules.

The control, however, was subject to considerable criticism. Value-of-service pricing resulted in commodities captive to railroads paying higher prices than commodities where the railroads faced competition regardless of cost similarities. Small shippers tended to pay the same price as large shippers despite large cost differences in the provision of service. Innovations such as the Big John Hopper car by the Southern Railway and unit/multiple car movements were discouraged to protect other modes of competition and shippers. Rail lines could not be abandoned even though they were no longer profitable and mergers were given much scrutiny, with proceedings lasting for years before a decision was reached.

By the early 1970s, it became clear to policy-makers that the then current set of regulations and the associated regulatory practice was no longer serving its purpose. New innovations were not being adopted, market shares continued to fall, and there were a series of high-profile railroad bankruptcies.[**] In response, the federal government began the process of relaxing economic regulation of the railroad industry with passage of the Railroad Revitalization and Regulatory Reform (4R) Act of 1976. This legislation provided subsidies to ailing railroads and began the process of regulatory change. The regulatory changes included some rate flexibility for railroads, easing of merger restrictions, and the relaxation of some abandonment rules. However, a general assessment of that legislation and, in particular, its application by the ICC suggests that it did not offer the sort of meaningful regulatory relief needed to rescue the ailing industry.[††] Consequently, Congress acted again in 1980 to institute greater regulatory relief through the passage of the Staggers Rail Act.

The Staggers Act effectively ended the regulation of most railroad rates. Continuing to work toward the objective of the 4-R Act, a regulatory screening process was put in place. Specifically, the reasonableness of the rate could only be considered if the movement in question was found to be

---

[*]Daggett (1928) and Keeler (1983). In addition to common law principles, such obligations were manifested in corporate charters.

[†]Such difficulties included differences in court decisions, differences in the conflicts in corporate charters, rate wars on competitive routes and extremely high rates with competition was not present. Finally, in 1886, a supreme court ruling, Wabash, St. Louis, and Pacific v. Illinois, 118 U.S. 557, 1886, found that states were not "empowered" to regulate interstate commerce (Keeler 1983).

[‡]In addition to Keeler (1983), Locklin (1972) and Pegrum (1968) provide excellent overviews of the history of railroad regulation and its roots.

[§]Since passage there have been a number of legislative bills enacted. Most notably, these include the Transportation Acts of 1920 and 1940, the 3-R Act in 1973, the 4-R Act in 1976, the Stagger's Rail Act of 1980, and the ICC Termination Act of 1995.

[¶]This statue allowed the ICC to force mergers to "save" railroad service from weak railroads.

[**]Such bankruptcies included the Penn-Central, the Erie-Lackawanna, Lehigh Valley, the Reading, Central of New Jersey, the Katy, and latter the Rock Island and the Milwaukee Road. The Regional Reorganization Act of 1973 created the US Railway Association; an agency created to deal with the problems of railroads in the Northeast. One solution was to plan for public ownership.

[††]See Winston et al. (1990).

"market dominant."[*] The decision of whether a particular movement was the result of market dominance, involved two stages. First, if the revenue-to-variable cost ratio (R/VC) was less than some given threshold, the rate could not be considered market dominant. If the R/VC ratio was in excess of the given threshold, then there was a qualitative evaluation performed to determine whether or not competitive pressures (i.e., intramodal, intermodal, product or geographic competitive elements) were present. If so, then the market was not considered market dominant and therefore not subject to reasonableness proceedings. As noted in Friedlander and Spady (1981), the presence of competition was defined relatively liberally, except for coal markets. As a result, railroads had considerable latitude in the establishment of rates. The new law also allowed rail carriers to establish confidential contracts with shippers.[†] This provision effectively ended any requirement that rates be nondiscriminatory, thus opening the door to a number of revenue-enhancing and cost reducing innovations. Most notable among these practices was the use of multiple car shipments and unit trains. Finally, Staggers also radically reduced the regulatory hurdles formerly associated with approval of mergers and branch-line abandonment, and as we documented above, the result has been an unprecedented consolidation of railroad operations into larger firms.

Because of the railroads' ailing financial condition, there was a general expectation that deregulated railroad rates would increase. Indeed, initially after passage, rates did rise for a number of commodities, but it was not long before the rates for most commodities (in real terms) fell below pre-partial deregulation time periods (Wilson 1994). Under partial deregulation, the cost-cutting activities were implemented, and as a result, real operating costs per unit output fell, and fell dramatically.[‡]

The Staggers Act also severely limited the activities of the rate bureaus through which railroads had historically gathered to recommend rates to the ICC. This provision, coupled with the ability to engage in confidential contracts, opened the door to rate competition between railroads and, often, where direct rail-to-rail competition did not exist, the competitive pressure exercised by motor carriage and commercial navigation assumed a disciplinary role.

The end result of all of these measures was a dramatic decrease in average operating costs that were rapidly transformed into lower rail rates. In addition, there were also tremendous improvements in the quality of service. During the first years after deregulation transit times fell by 14 percent and the variability of transit times fell by 36 percent.[§] With pared-down route systems and marginally greater profits, the railroads began to invest more heavily in their remaining facilities. These investments helped sustain the continuing reduction in operating costs.

Given the number of changes in the railroad operating environment along with motor carrier deregulation, further change was implemented when President Clinton signed into law The ICC Termination Act of 1995. This statute replaced the ICC with a much smaller Surface Transportation Board (STB). Currently, it is the STB that is responsible for adjudicating shipper-based rate cases brought under the Staggers Act; it is the STB that now evaluates proposed mergers; and it is the STB that is more generally responsible for assuring there is adequate competition in rail-served markets.

Given the market outcomes since 1980, it is hard to imagine any dissatisfaction with the deregulation process. However, there were other results of the Staggers Act that have drawn a less favorable review. At the time of the statute's passage (1980) there were approximately 40 Class I railroads

---

[*]See Eaton and Center (1985) and Wilson (1996) for a complete discussion of the rules, and Wilson (1996) for a discussion of railroad pricing behavior under such rules.

[†]See Hanson et al. (1989, 1990) and Barnekov and Kleit (1990) for further discussion.

[‡]Richard C. Grayson, former president and CEO of Burlington Northern once remarked, "We started with costs; we didn't know a thing about pricing, but we knew how to cut costs." Many of these activities are documented in Wilson (1997). MacDonald (1998) also provides a excellent synopsis of cost-cutting measures. Of course, industry consolidation has had an effect as well. Wilson and Bitzan (2003) document some of these effects. They find that the consolidation of output among firms in the industry account for a 20 percent reduction in industry costs over the time period 1983–1997. Finally, less viable but important innovations include the replacement of cabooses with way-side detectors and End-Of-Train (EOT) devices, the elimination of the fireman's position from locomotive service, increased mechanization of maintenance of way activities, and an across-the-board reduction in labor force. For a full description of the increase in railroad productivity see Bitzan and Keeler (forthcoming) and Wilson (1996) and Davis and Wilson (1999, 2003).

[§]See Grimm and Smith (1986).

operating in the United States. As noted above, this number has now fallen to only 7. Wilson and Bitzan (2003) and others, have indicated that Staggers streamlined the merger approval process and the rail carriers were quick to seize this new opportunity. The result of the sustained merger activity is that most rail shippers have very limited choices among rail options. When shipment and commodity characteristics are such that alternative modes are good substitutes and when those alternative modes are available, the paucity of rail competition is less confounding. However, for those shippers who must use rail, the reduction in the degree of rail-to-rail competition has lead to increasing "captivity" i.e., shippers without meaningful competitive alternatives.

At the same time the ICC was being replaced with the STB, the railroads undertook another round of significant mergers.[*] Not only did this further reduce the degree of direct rail-to-rail competition, the implementation of these mergers routinely leads to severe service disruptions. While the STB generally approved proposed mergers, in many cases, it also granted trackage rights to alternative carriers in order to partially mitigate the loss of direct rail competition. Moreover, when merger implementation led to service quality degradation, the STB imposed strict monitoring systems to ensure a timely service recovery.

15.7 MODELS OF RAILROAD DECISION-MAKING

With partial deregulation, the business of railroading has become more complicated. In the course of the business of railroading, firms make an enormous number of decisions. We categorize these decisions into network design, input, and output/pricing decisions. Network design decisions involve the size, the configuration, and the maintenance of the network. Input decisions are those related to the employment of economic resources, such as labor, fuel, equipment, etc., to produce some given set of outputs, i.e., movements between origin-destination pairs. Output and pricing decisions include decisions of whether or not to serve a market (participation decision) and how much to supply. In this section, we develop a series of relatively simple models to characterize the pricing and output decisions over a network. The network size, configuration and maintenance decisions have received some attention in the economics literature and the input decisions have often been addressed jointly with cost considerations.[†]

15.8 CAPTIVE SHIPPERS AND RAILROAD PRICING

We first describe the pricing of services to a single shipper, who ships a single commodity between two locations vis a vis Winston (1985) and Section 15.3 of this chapter. This simple model illustrates how competition enters into railroad decision-making. In the ensuing sections, we generalize the model to capture flows over a more complicated network involving multiple commodities under conditions of decreasing average costs (economies of density) and ultimately render a model that allows for intramodal competition from a separate network. We conclude this section of the novelty of this pricing model from others that have appeared in the literature.

We begin by noting that there is an array of choices confronting an individual shipper.[‡] If the producer of the good is the decision maker, the decisions include how much to ship, where to ship,

---

[*]In the west, the Union Pacific acquired the Southern Pacific and Burlington Northern merged with the Santa Fe. In the east, rivals CSX and Norfolk Southern jointly acquired, and then divided Conrail. In the central US, Illinois Central, a north-south carrier, was acquired by the Canadian National.

[†]In contrast to the cost function literature (enumerated in footnote 34), there are a series of relatively recent papers e.g., MacDonald and Cavalluzzo (1996), Davis and Wilson (1999, 2003), and Bitzan and Keeler (forthcoming) which examine labor decisions and the influence of partial deregulation on both employment and costs.

[‡]In this section, the model follows from Wilson (1996). The ensuing sections represent modest extensions of this model. The models are fully consistent with discrete choice models of shipper demands. For surveys of this literature see Winston (1985), Oum, Waters, and Yong (1992) and Small and Winston (1999). Specific models that are useful for understanding our model are Daughety and Inaba (1978), Inaba and Wallace (1989).

and by what mode to ship. If the receiver is the transportation decision maker, choices may, again, include how much to ship and by what mode, but also from where to receive the product. In pricing services to a single shipper, these choices may or may not limit the pricing power of a railroad. In fact, such principles are present in the market dominance standards under partial deregulation to represent the adequacy of competition in limiting rates.[*]

We assume that shippers are price-takers in all markets and have a well defined production technology for each of the choices, with appropriate regularity conditions. Shippers are assumed to be profit-maximizing, given the prices of inputs and outputs and the technology that maps inputs into outputs. Given these conditions, shippers maximize profits by choosing appropriate levels of inputs and outputs for each of the choices described above. Using these profit-maximizing input and output levels, the maximum profits associated with *each choice* can be expressed as a function of the associated input and output prices i.e., $\pi = \pi(P, w, m)$, where P, w, m represent the output price (the price the shipper receives for the product shipped), the input prices (the price of labor, capital, etc.), and the price of transportation by mode m, respectively.[†] The shipper has a number of alternatives indexed by $i = 1, 2, ..., I$. The actual choice observed is the choice i that yields the maximal profits from the set of profit functions representing options to the shipper.

In pricing services to an individual shipper, railroads must remain the preferred mode of the shipper to provide the service. The railroad then chooses a price and output combination that maximizes profits subject to the constraint that it yields shippers a profit level that is greater than or equal to the shippers next best alternative. Formally, the railroad's problem is stated as:

$$\text{Max } \pi = R(r)r - C(R(r)) \qquad s.t. \ \pi_r^S(r) \geq \pi_m^S \qquad (15.1)$$

where r is the rail rate, $R(r; P, w)$ is the shipper's demand for railroad service,[‡] $C(.)$ is the railroad's cost function, and $\pi_r^S(r) \geq \pi_r^S$ reflects the constraint that railroad service to shipper (S), by the railroad offering rate (r), yields profits at least as great as profits by the next best alternative mode (m). The first-order conditions for this problem can be written as:[§]

$$\frac{r - mc}{r} = \frac{\lambda - 1}{\varepsilon} \qquad (15.2)$$

This is the Lerner Index of market power (the percentage markup of price over marginal cost). In this expression, λ is the multiplier attached to the participation constraint and ε is the shipper's elasticity of demand. The basic point of this derivation is that the rate charged in a particular railroad market may or may not be constrained by alternative choices of the shipper. Specifically, λ may take a value from zero to one. If zero, the monopoly price obtains with the result that the railroad charges a monopoly price and is not constrained by competitive alternatives. This is the extreme case of a captive shipper, who, as discussed earlier, has lost more and more alternatives over time. If λ is in

---

[*]Essentially partial deregulation introduced a screening mechanism into regulatory design. Before a rate could be considered for its reasonableness, a ruling that the movement was market dominant was necessary. Market dominance criteria included a revenue to variable cost threshold and a qualitative evaluation of whether intramodal, intermodal, geographic and product competition was not significant in limiting railroad rates. Recently, the geographic and product competition standards were removed as criteria [Bitzan and Tolliver (1998)]. Also, see Eaton and Center (1985) and Wilson (1996).

[†]While transportation is an input to the shipper, we represent it separately here because it is the focal point of the ensuing discussion. We further note that the price of transportation is not the rate but the rate and the associated handling charges, inventory costs etc.

[‡]We note that under the assumptions of the model the railroad demand function(s) can be derived from the associated shipper profit functions as $\dfrac{\partial \pi_i(P, w, r)}{\partial r} = -R(P, W, r)$.

[§]This condition can also be written as $\dfrac{r}{mc} = \dfrac{1}{\left(1 - \dfrac{(\lambda - 1)}{\varepsilon}\right)}$ which under the assumption that $mc \approx vc$ is the foundation for

the r/vc threshold used in market dominance proceedings.

the range from zero to one, the railroad price in a market is limited by the alternative choice of the shipper, which may reflect alternative terminal markets, alternative modes, alternative products, or some combination thereof. The closer these alternatives are to the movement in question, the more constrained is the railroad rate. In the limiting case of $\lambda = 1$, the railroad prices at marginal costs.

In the development of these profit-maximizing rates, we note that marginal costs play an important role. The lower marginal costs are, the lower is rate due to standard economic theory. Events through the 1980s and 1990s that have led to consolidation, innovations afforded by pricing flexibility (e.g., contract rates, unit trains, shuttle trains, etc.) and greater freedom to rationalize the network by abandoning or selling railroad lines have given railroads lower costs, and greater levels of market dominance.

We should finally discuss conditions of economies of density. At the individual shipper level in this model, economies of density simply means that the cost of serving a large shipper are lower than the cost of serving a small shipper. To the extent that alternative modes, e.g., truck do not operate under economies, railroads would tend to offer large shippers lower rates and the level of dominance of railroads over competing alternatives (modes) is larger.

15.9 MULTIPLE SHIPPERS AND NETWORK EXTERNALITIES

In this section, we add a second shipper to the model. While the generalization to multiple shippers seems innocuous, it complicates the model except under very specific circumstances that are not generally thought to hold in railroad economics. This second shipper may ship the same commodity or a different commodity. In presenting the model, there are two specific changes. First, the railroad has a participation constraint for each shipper. That is, the shipper may produce a different commodity or may have differing characteristics, e.g., size of firm, loading facilities, etc., which may affect the payoffs attached to using rail vis a vis other discrete alternatives. Second, in providing the services to two shippers, the railroad becomes a multioutput firm. Depending on the nature of the technology the multiple outputs may or may not be aggregated into a single output term. A few different possibilities are described below.

The development of shipper alternatives and choices was outlined in the previous subsection. With appropriate indexing of the two shippers (1 and 2), the railroad's profit-maximization problem becomes (given they choose to serve both shippers):

$$\text{Max } \pi = R_1(r_1)r_1 + R_2(r_2)r_2 - C(R_1(r_1),\ R_2(r_2)) \qquad s.t.\ \pi_r^{S_1}(r_1) \geq \pi_{m_1}^{S_1}$$
$$\pi_r^{S_2}(r_2) \geq \pi_{m_2}^{S_2} \tag{15.3}$$

The associated first order conditions can again be rearranged to form the Lerner index. These expressions are given by:

$$\frac{\partial L}{\partial r_1} = R_1'(r_1)[r_1 - C_{R_1}(R_1(r_1),\ R_2(r_2))] + R_1 - \lambda R_1 = 0$$

$$\frac{\partial L}{\partial r_2} = R_2'(r_2)[r_2 - C_{R_2}(R_1(r_1),\ R_2(r_2))] + R_2 - \lambda R_2 = 0 \tag{15.4}$$

$$\frac{\partial L}{\partial \lambda_1} = \pi_r^{S_1}(r_1) \geq \pi_{m_1}^{S_1}$$

$$\frac{\partial L}{\partial \lambda_2} = \pi_r^{S_2}(r_2) \geq \pi_{m_2}^{S_2}$$

Similar to the previous treatment, these expressions can be rearranged to define Lerner indices for both shippers, and can be solved to yield profit-maximizing price levels. A key point and the purpose of writing out the full set of first order conditions is to note the interdependence of the rates paid by each shipper. This dependence enters into the expression through the cost function. There are a number of cases to consider.

First, if the cost function exhibits constant marginal costs (i.e., $C = c_1 R_1 + c_2 R_2$) then the prices do not marginally depend on one another, and the optimization is separable.[*] In this particular case, there is no interdependence of the rates paid by each shipper. Over such cases are probably not observed in railroad economics.

Second, the cost function applies to outputs over the link i.e. $C = C(R_1 + R_2)$ but costs are not constant. In this case, the interdependence of shipper 1 and 2 flows remains. Under conditions of decreasing marginal costs (i.e., $C'' < 0$), the more that is shipped on a route the lower are the marginal costs of service for each shipper. Thus, the larger is the shipper, the lower are the rates to the other shipper. In addition, the greater are the pressures of competitive alternatives in one market (the larger is λ), the lower is the rate in that market. This added competition induces the associated shipper to ship more by rail (due to the lower rates), in turn making this shipper larger. But, the effect spills over to the other shipper. Since there are greater flows over the network (since the volume of the second shipper has increased), the costs of serving the other shipper (now shipper 1) are lower. Thus, again rates are lower, inducing even larger flows. To our knowledge, such interdependencies have not been considered in the literature.

The final case is the most general case which relates to the general first-order conditions. In this case, if outputs are complementary (that is, if added operations to one shipper reduces marginal costs to the other shipper) the end result is the same as described above under the decreasing marginal costs case. If on the other hand, outputs are substitutes, then the results are opposite of those described above. That is, increased operations for one shipper increases costs for the other shipper (e.g., congestion), and rates for the latter would increase. Added competitive pressures (increases in λ) on one shipper reduce rates to that shipper, thus increasing volumes of that shipper. Hence, the increased volume in turn increases the cost of providing service to the other shipper, increasing rates.

15.10   MULTIPLE SHIPPERS AND MULTIPLE RAILROADS

In this section, we extend the model in Section 15.6 to reflect intramodal competition between different originations. Specifically, we assume there is one shipper (or groups of identical shippers) serviced by two railroads with different networks. The terminal point may be common with different cost characteristics across railroads, or, alternatively, the terminal points may also be different without loss of generality. In this context we examine the oligopolistic behavior of railroads competing for service to this shipper (or group of shippers).

In the model developed to this point, railroads choose the rate. The results obtained in the simple model apply directly. That is, the shipper chooses the railroad/terminal location that offers maximal profits. If all else is the same, the railroad offering the lowest rate wins the traffic. If there are costs differences across railroads, the low cost railroad will offer the lowest rate, that being the rate at which the other railroad no longer wishes to provide the service. Such a framework is rich for explaining developments over the last several years. For example, when a merger offers a railroad greater network connectivity (more direct service at lower costs), it may have the effect of diverting traffic flows from another railroad and reducing rates to the shipper(s). Further, the granting of traffic rights to another railroad affords that railroad the opportunity to become the preferred alternative. Again, this results in a diversion of traffic from existing railroads. In cases in which traffic is not diverted, there are still important implications and price effects from this model. In particular, if the merger or the presence of traffic rights does indeed positively affect the nonparticipating carrier's

---

[*]Consider, for example, a case in which two shippers are served on entirely different segments of the route network.

ability to serve a given shipper, but does not improve it enough to make it the low cost firm, the railroad providing service may still need to lower price to remain the preferred mode.

In this chapter, the pricing of services to shippers has been developed in terms of discrete choices of a shipper. In this context, we have demonstrated that there is a trigger price at which the railroad may no longer provide the service i.e., if the railroad prices a shipper too high it will lose the service priced. It may be that the shipper receives or sends the product from or to another location, ships by a different mode, or alternatively, in the case of a receiver, the product may be priced out of the market.

In a more richly developed model, a shipper may use two or more railroads, ship to more locations, etc. In this regard, we simply note that such a shipment plan remains a discrete choice from a menu of choices. We assume that all other options do not bind railroad rates (i.e., $\lambda = 0$) for all other options. A consequence of this framework is that in equilibrium, the shipper is indifferent between the two railroad services.

To illustrate various oligopoly models, we assume there is a demand for transportation service to a location. This demand is provided by two different railroads with rates r_1 and r_2. Both railroads provide service to the shipper, although again, the demand may be the serviced from two different locations and the railroads may have different networks and/or cost conditions. There are a large number of plausible assumptions of rivalry between firms.[*] One example of this is the product differentiated demand model (i.e., the railroads provide service but they are differentiated in some regard).[†] In this model the demand functions for each railroad are now interdependent. That is, the shipper(s) use both railroads but the level of service from each is dependent on the railroads' choice variables (r_1 and r_2). Demands can then be written as: $R_1 = R_1(r_1, r_2)$ and $R_2 = R_2(r_2, r)$ where the demands are decreasing in the first term and increasing in the second term (the railroads provide substitutes).[‡] Within this framework, each railroad makes profit-maximizing decisions, wherein the first order conditions can be written in a general form as:[§]

$$\frac{\partial \pi_1}{\partial r_1} = \left(R_{r_1} + R_{r_2} \frac{dr_2}{dr_1} \right)(r_1 - MC_1) + R_1 = 0$$

$$\frac{\partial \pi_2}{\partial r_2} = \left(R_{r_2} + R_{r_1} \frac{dr_1}{dr_2} \right)(r_2 - MC_2) + R_2 = 0 \tag{15.5}$$

In the literature, the terms $\frac{dr_1}{dr_2}$ and $\frac{dr_2}{dr_1}$ (from the above equation) are typically termed a "conjectural variation," which can be interpreted as a firm's belief of how the other firm will react to its rate change (Martin 2002, p. 45). For our purposes, we simply include these terms as a way to index the nature of rivalry between firms. In the standard Nash framework, each firm takes the other firm's choice variable as given (the conjectural variation terms are zero) and these two equations are solved jointly for Nash equilibrium prices. In such a model, as the level of product differentiation rises, firms garner more market power.

There are a number of ways to proceed through the remainder of the oligopoly models. While not "vogue," a conjectural variation framework probably provides the best presentation for our purposes. In this regard, we follow Martin (2002) [who follows Bowley (1924) and Hicks (1935)], and treat the

---

[*]Because our emphasis is on rates, we limit ourselves to assumptions of rivalry pertaining to prices. Thus, the homogeneous Cournot model does not apply.

[†]This can be horizontal product differentiation, i.e., at the same price and service attributes different shippers may chose different railroads or vertical product differentiation, i.e., one railroad provides a better service than the other so that at the same price, shippers would prefer the high quality of service firm.

[‡]Burton and Wilson (2007) describe a network situation in which an origin-destination pair is served by a sequence of different firms (rail-barge) in a network. Such a situation can allow for complementary effects in the demand model, i.e., the second term would also be negative.

[§]Again, we note that all other possible combinations do not limit rates. The lagrangian multipliers are equal to zero for all other choices that the shipper(s) can make.

conjectures as constants. If the conjectures are positive, then the railroads expect that as they increase their own price, their rival (the other railroad), will increase its price. The consequence of a price change, then, is an increase (decrease) in a firm's price, which induces a decrease (increase) in output originating from its own price increase (decrease). But, since the other firm's price also increases (decreases), there is a corresponding increase (decrease) in output. In effect, the greater is the positive conjecture, the more intense the competition, and the lower are the prices and the markup.

Such a framework is often used to model origin-destination movements in railroad economics. For example, MacDonald (1987) used a markup model similar in spirit to the model described above.[*] In particular, in his paper, he modeled confidential waybill data pertaining to Corn, Soybean, and Wheat movements. His empirical model is based on a model wherein the "[f]irm-specific elasticity of demand depends on the market elasticity of demand for the commodity, M_k, the extent of competition at that location, C_i, and the nature of rivalry among sellers, which is represented by the conjectural variation term...." (p. 153). While such a framework is not identical to the model developed above, it is based on the same principles. In his model, he includes a variety of different measures to account for interfirm rivalry. These measures allow for intramodal (i.e., between railroads) and intermodal (i.e., between a railroad and barge modes) rivalry. The specific measures used were the distance a shipper was located to water and the inverse of the Herfindahl index. He found that both miles to water and the inverse Herfindahl had important effects on rates, and that the effect of these measures differed across commodities. Specifically, his results showed that miles to water increase rates. In his base models, a one percent increase in miles to water increases rates by 0.0864, 0.0856, and 0.2576 percent for corn, soybeans, and wheat, respectively. Conversely, he found that a one percent change in the inverse Herfindahl index leads to a 0.28, 0.19, and 0.11 percent reduction in rates. These results provide substantial evidence for the role of both intra and inter modal competition effects, even at the individual shipper level. In the context of the models presented in this chapter there is much. For example, the attractiveness of an alternative (i.e., water) falls as that shipper is located further from water. In the context of the first subsection, the railroad becomes more market dominant as this distance increases.[†] Further, the attractiveness and rivalry between railroads developed in this subsection relates to the inverse Herfindahl measure. In particular, as the number of railroads increase, accompanied by a decrease in market shares in a region (i.e., competition in a region increases) there is a corresponding decrease in railroad rates. In the context of the model presented here, the notion is that the conjecture term is positive and increasing in the Herfindahl index.

15.11   SUMMARY AND CONCLUSIONS

The post-deregulation productivity gains achieved by US railroads effectively reversed a trend that would have otherwise signaled their elimination as an important transport mode. Fortunately, 30 years after the implementation of the Staggers Rail Act, a healthy railroad industry is well positioned to respond to the capacity constraints that challenge other transport modes. However, whether or not U.S. freight railroads play an increasingly important role within the overall transportation landscape depends on many factors, some of which are well beyond the railroads' control.

First, policy-makers must realize that the productivity gains that restored the vitality of the railroad industry were the direct result of competition. As Section 15.5 demonstrates, rail carriers have every incentive to exercise market power when that power exists. This simultaneously reduces the incentive to engage in innovative cost-reducing activities. In those settings where competing transport modes cannot supply adequate competitive pressure, policy-makers must stand ready to ensure that there is sufficient rail-to-rail competition. This may require regulatory mechanisms that are not currently in evidence.

---

[*]Wilson (1994) used a similar type of model as well. In his case, he used a markup model to examine railroad rates over a wide range of STCC level 2 codes to estimate the effects of partial deregulation across commodities.

[†]There are other empirical studies that link rail rates to the availability of commercial navigation. See, for example, Burton (1996).

Shippers (even of bulk commodities) continue to increase the level of service they demand from transportation providers. Thus, if the currently observable rail renaissance is to continue, it will be necessary for US railroads to continually improve the level of service they offer. This, in turn, will require the railroads to invest in improving the quality of both vehicles and infrastructure. Currently, there is a trend toward the pooling of private and public funds to support rail infrastructure projects that benefit both rail customers and other affected constituencies. Such partnerships may well be necessary to achieve the required level of future investment.

The explosion in intermodal traffic is directly traceable to the tremendous growth in international trade. If this trade growth declines, so will the importance of increased intermodal capacity. It is likely, however, that the ability to increase the role of international economic activity depends on the ability to ensure secure container shipments. Thus, there is little the rail industry can do to affect such an outcome.

REFERENCES

Abdelwahab, W. M. (1992). "Modeling the Demand for Freight Transport." *Journal of Transport Demand and Policy,* 26(1):49–70.

Abdelwahab, W. M. (1998). "Elasticities of Mode Choice Probabilities and Market Elasticities of Demand: Evidence from a Simultaneous Mode Choice/Shipment–size Freight Transport Model." *Transportation Research Part E,* 34(4):257–266.

Allen, R. A. (1990). "Railroad Line Sales: Their Uncertain Legal Status after Pittsburgh & Lake Erie." *Transportation Practitioners Journal,* 57(3):255–280.

Armstrong, J. H. (1998). *The Railroad, What It Is, What It Does,* 4th ed. Omaha: Simmons–Boardman Books, Inc.

Barnekov, C. C., and Kleit, N. (1990). "The Efficiency Effects of Railroad Deregulation in the United States." *International Journal of Transport Economics,* 17:21–36.

Berndt, E. R., Friedlaender, A. F. , Chiang, J. S. W, Showalter, M., and Vellturo, C. A. (1993). "Cost Effects of Mergers and Deregulation in the U.S. Railroad Industry." *Journal of Productivity Analysis,* 4(1–2):127–44.

Berry, S. (1992). "Estimation of a Model of Entry in the Airline Industry." *Econometrica,* 60(4):889–917.

Bitzan, J. D. (1999). "The Structure of Railroad Costs and the Benefits/Costs of Mergers." *Research in Transportation Economics* 5:1–52.

Bitzan, J. D., and Keeler, T. E. (forthcoming). "Productivity Growth and Some of Its Determinants in the Deregulated U.S. Railroad Industry." *Southern Economic Journal.*

Bitzan, J. D., and Tolliver, D. D. (1998). "Market Dominance Determination and the Use of Product and Geographic Competition," A Verified Statement Before the Surface Transportation Board.

Bowley, A. L. (1924). *The Mathematical Groundwork of Economics.* Oxford: Oxford University Press.

Boyer, K. D. (1977). "Minimum Rate Regulation, Modal Split Sensitivities, and the Railroad Problem." *Journal of Political Economy,* 85:493–512.

Boyer, K. D. (1981). "Equalizing Discrimination and Cartel Pricing in Transport Rate Regulation." *Journal of Political Economy,* 89:270–286.

Boyer, K. D. (1987). "The Cost of Price Regulation: Lessons from Railroad Deregulation." *Rand Journal of Economics,* 18:408–416.

Braeutigan, R. R., and Daughety, A. F. (1982). "The Estimation of a Hybrid Cost Function for a Railroad Firm." *Review of Economics and Statistics,* 64(3):394–404.

Brown, R. S., Christensen, L. R., and Caves, D. W. (1979). "Modeling the Structure of Cost and Production for Multiproduct Firms." *Southern Economic Journal,* 46:256–273.

Burton, M. L. (1993). "Railroad Deregulation, Carrier Behavior, and Shipper Response: A Disaggregated Analysis." *Journal of Regulatory Economics,* 5:417–34.

Burton, M. L. (1996). "Rail Rates and the Availability of Barge Transportation: The Missouri River Basin." U.S. Army Corps of Engineers, Omaha, Nebraska.

Cain, L. P. (1997). "Historical Perspective on Infrastructure and U.S. Economic Development. *Regional Science and Urban Economics,* 27(2):117–138.

Caves, D. W., Christensen, L. R., and Swanson, J. A. (1980). "Productivity in the U.S. Railroads, 1955–74." *Bell Journal of Economics,* 11:166–81.

Caves, D. W., Christensen, L. R., and Swanson, J. A. (1981). "Productivity Growth, Scale Economies, and Capacity Utilization in the U.S. Railroads, 1955–1974."

Caves, D. W., Christensen, L. R., Trethaway, M. W., and Windle, R. J. (1985). "Network Effects and the Measurement of Returns to Scale and Density for U.S. Railroads." In *Analytical Studies in Transport Economics*, edited by A. F. Daugherty. New York: Cambridge University Press.

Daggett, S. (1928). *Principles of Inland Transportation.* New York: Harper Press.

Daughety, A. F., and F. S. Inaba. (1978). "Empirical Aspects of Service-Differentiated Transport Demand." Proceedings of the Workshop on Motor Carrier Economic Regulation. National Academy of Sciences, 329–355.

Davis, D. E., and Wilson, W. W. (1999). "Deregulation, Mergers, and Employment in the Railroad Industry." *Journal Regulatory Economics,* 15:5–22.

Davis, D. E. and Wilson, W. W. (2003). "Wages in Rail Markets: Deregulation, Mergers, and Changing Networks Characteristics." *Southern Economic Journal,* 69(4):865–885.

DeVany, A. S., and Walls, D. W. (1999). "Price Dynamics in a Network of Decentralized Power Markets." *Journal of Regulatory Economics,* 15(2):123–140.

Eaton, J. A., and Center, J. A. (1985). "A Tale of Two Markets: The ICC's Use of Product and Geographic Competition in the Assessment of Rail Market Dominance." *Transportation Practitioners' Journal,* 53:16–35.

Fisher, F. M. (1987a). "Horizontal Mergers: Triage and Treatment." *Journal of Economic Perspectives,* 1(2):13–40.

Fisher, F. M. (1987b). "One the Misuse of the Profits–Sales Ratio to Infer Monopoly Power." *Rand Journal of Economics,* 18(3):384–96.

Friedlaender, A. F., Berndt, E. R., Chaing, J. S. W., Showalter, M., and Vellturo, C. A. (1993). "Rail Costs and Capital Adjustments in a Quasi–regulated Environment." *Journal of Transport Economics and Policy,* 27(2):131–52.

Friedlaender, A. F., and Spady, R. H. 1980. "A Derived Demand Function for Freight Transportation." *The Review of Economics and Statistics,* 62(3):432–41.

Freidlaender, A. F., and Spady, R. H. 1981. *Freight Transportation Regulation: Equity, Efficiency, and Competition in the Rail and Truck Industries.* Cambridge, MA: MIT Press.

Gallamore, R. E. (1999). "Regulation and Innovation: Lessons from the American Railroad Industry." *Essays in Transportation Economics and Policy*, Chapter 15:493–529. Washington D.C.: Brookings Institution Press.

Gillen, D., and Waters, W.G. (1996). "Transportation Infrastructure and Economic Development: A Review of the Recent Literature." *Transportation Research: Part E: Logistics and Transportation Review,* 32(1):39–62.

Grimm, C. M., and Smith, K. G. (1987). "The Impact of Rail Regulatory Reform on Rates, Service Quality, and Management Performance: A Shipper Perspective." *The Logistics and Transportation Review,* 22:57–68.

Hanson, S. D., Baumel, C. P., Schnell, D. (1989). "Impact of Railroad Contracts on Grain Bids to Farmers." *American Journal of Agricultural Economics,* 71(3):638–46.

Hanson, S. D., Baumhover, S. B., Baumel, C. P. (1990). "Characteristics of Grain Elevators that Contract with Railroads." *American Journal of Agricultural Economics,* 72(4):1041–46.

Harper, D. V. (1978). *Transportation in America.* Englewood Cliffs, NJ: Prentice–Hall, Inc.

Hicks, J. R. (1935). "Annual Survey of Economic Theory: The Theory of Monopoly." *Econometrica,* 3(1):1–20.

Hilton, G. (1969). *The Transportation Act of 1958.* Bloomington, IN: Indiana University Press.

Inaba, F. S., and Wallace, N. E. (1989). "Spatial Competition and the Demand for Freight Transportation." *The Review of Economics and Statistics,* 71(3):614–625.

Ivaldi, M., and McCullough, G. J. (2001). "Density and Integration on Class I U.S. Freight Railroads." *Journal of Regulatory Economics,* 19(2):161–182.

Keeler, T. E. (1974). "Railroad Costs, Returns to Scale and Excess Capacity." *Review of Economics and Statistics,* 56(2):201–208.

Keeler, T. E. (1983). *Railroads, Freight, and Public Policy.* Washington, D.C.: The Brookings Institute.

Klindworth, K. A. (1983). "Impact of Staggers Rail Act on the Branchline Abandonment Process." *Proceedings of the Twenty Fourth Annual Meetings of the Transportation Research Forum,* 24(1):451–460.

Koo, K. W., Tolliver, D. D., and Bitzan, J. D. (1993). "Railroad Pricing in Captive Markets: An Empirical Study of North Dakota Grain Rates." *Logistics and Transportation Review,* 29(2):123–137.

Locklin, P. D. (1972). *Economics of Transportation,* Irwin Series in Economics. Homewood, IL: Richard D. Irwin Inc.

MacDonald, J. M. (1987). "Competition and Rail Rates for the Shipment of Corn, Soybeans, and Wheat." *Rand Journal of Economics,* 18(1):151–163.

MacDonald, J. M. (1989). "Railroad Deregulation, Innovation, and Competition: Effects of the Staggers Act on Grain Transportation." *Journal of Law and Economics,* 32:63–96.

MacDonald, J. M. (1998). "Railroad Deregulation, Innovation, and Competition: Effects of the Staggers Act on Grain Transportation." *The Foundations of Regulatory Economics,* 3:274–306.

MacDonald, J. M. and Cavalluzzo, L. C. (1996). "Railroad Deregulation: Pricing Reforms, Shipper Responses, and the Effects on Labor." *Industrial Labor Relations Review,* 50:80–91.

Martin, S. (2002). *Advanced Industrial Economics,* 2nd ed. Malden, MA: Blackwell Publishers, Ltd.

McFarland, H. (1989). "The Effects of United States Railroad Deregulation on Shippers, Labor, and Capital." *Journal of Regulatory Economics,* 1:259–70.

Meyer, J. R., Peck, M. J., Stenason, J. R., and Zwick, C. (1959). *The Economics of Competition in the Transportation Industries.* Cambridge, MA: Harvard University Press.

Nelson, J. C. (1951). "Changes in National Transportation Policy: Highway Development, the Railroads, and National Transport Policy." *The American Economic Review,* 41(2):495–505.

Oum, T. H. (1979). "A Cross Sectional Study of Freight Transport Demand and Rail–Truck Competition in Canada." *Bell Journal of Economics,* 10(2):463–482.

Oum, T. H. 1979. "Derived Demand for Freight Transport and Inter–Modal Competition in Canada." *Journal of Transport Economics and Policy,* 13(2):149–168.

Oum, T. H., Waters II, W. G., and Yong, J. S. (1992). "Concepts of Price Elasticities of Transport Demand and Recent Empirical Estimates." *Journal of Transport Economics and Policy,* May:139–154.

Owen, W. (1959). "Special Problems Facing Underdeveloped Countries: Transportation and Economic Development." *The American Economic Review,* 49(2):179–187.

Pegrum, D. F. (1968). *Transportation: Economics and Public Policy,* Irwin Series in Economics. Homewood, Illinois: Richard D. Irwin Inc.

Scheffman, D. T., and Spiller, P. T. (1985). "Geographic Market Definition under the DOJ Guidelines." Discussion Paper, *U.S. Federal Trade Commission, Bureau of Economics.*

Schmalensee, R. (1987). "Horizontal Merger Policy: Problems and Changes." *Economic Perspectives,* 1(2):41–54.

Slade, M. E. (1986). "Exogeneity Tests of Market Boundaries Applied to Petroleum Products." *The Journal of Industrial Economics,* 34(3):291–303.

Small, K., and Winston, C. (1999). "The Demand for Transportation: Models and Applications." *Essays in Transportation Economics and Policy: A Handbook in Honor of John R. Meyer.* Washington, D.C.: The Brookings Institution Press, pp. 11–55.

Spiller, P. T., and Haung, C. J. (1986). "On the Extent of the Market: Wholesale Gasoline in the Northeastern United States." *Journal of Industrial Economics,* 35:131–146.

Stigler, G. J., and Sherwin, R. A. (1986). "The Extent of the Market." *The Journal of Law and Economics,* 28:555–585.

Surface Transportation Board. (2001). R–1 Reports, Industry Composite.

Tolliver, D. D. (1984). "Economies in Density in Railroad Cost Finding: Applications to Rail Form A." *Logistics and Transportation Review,* 20(1):3–24.

Tye, W. B. (1990). "Regulatory Financial Tests for Rail Abandonment Decisions." *Transportation Practitioners Journal,* 57(4):385–403.

Uri, N. D., and Rifkin, E. J. (1985). "Geographic Markets, Causality, and Railroad Deregulation." *The Review of Economics and Statistics,* 67(3):422–428.

U.S. Army Corps of Engineers. (1998). "Available Navigation, Fuel Consumption and Pollution, Abatement: The Missouri River Basin." Omaha, Nebraska.

U.S. Bureau of Transportation Statistics, Department of Transportation. 2001. *National Transportation Statistics.*

U.S. Department of Commerce, Economics and Statistics Administration. *1997 Economic Census–Transportation.*

U.S. Department of Transportation, Bureau of Transportation Statistics. *1993 Commodity Flow Survey.*

U.S. Department of Transportation, Bureau of Transportation Statistics. *1997 Commodity Flow Survey.*

Vellturo, C. A., et al. (1992). "Deregulation, Mergers, and Cost Savings in the Class I U.S. Railroads, 1974–86." *Journal of Economics and Management Strategy,* 1(2):339–69.

Westbrook, M. D., and Buckley, P. A. (1990). "Flexible Functional Forms and Regularity: Assessing the Competitive Relationship between Truck and Rail Transportation." *The Review of Economics and Statistics,* 72(4):623–630.

White, L. J. (1987). "Antitrust and Merger Policy: A Review and Critique." *Journal of Economic Perspectives,* 1(2):13–22.

Wilner, F. N. (1997). *Railroad Mergers: History, Analysis, Insight.* Omaha, NE:Simmons–Boardman Books, Inc.

Wilson, W. W. (1994). "Market–Specific Effects of Rail Deregulation." *Journal of Industrial Economics,* 42:1–22.

Wilson, W. W. (1996). "Legislated Market Dominance." *Research in Transportation Economics,* 4(1):33–48.

Wilson, W. W. (1997). "Cost Savings and Productivity Gains in the Railroad Industry." *Journal of Regulatory Economics,* 11:21–40.

Wilson, W. W., and Bitzan, B. (2003). "Industry Costs and Consolidation: Efficiency Gains and Mergers in the Railroad Industry" mimeo.

Wilson, W. W., and Burton. M. (2003). "Network Pricing and Vertical Exclusion in Railroad Markets. mimeo.

Winston, C. (1981). "A Disaggregate Model of the Demand for Intercity Freight Transportation." *Econometrica,* 49(4):981–1006.

Winston, C. (1985). "Conceptual Developments in the Economics of Transportation: An Interpretive Survey." *Journal of Economic Literature,* 23(1):57–94.

Winston, C. (1993). "Economic Deregulation: Days of Reckoning for Microeconomists." *Journal of Economic Literature,* 31(9):1263–90.

Winston, C., Corsi, T. M., Grimm, C. M., and Evans, C. A. (1990). *The Economic Effects of Surface Freight Deregulation.* Washington, D.C.: The Brookings Institute.

Winston, C., and Grimm, C. (2000.) "Competition in the Deregulated Railroad Industry: Sources, Effects, and Policy Issues." *Deregulation of Network Industries: What's Next?* Washington, D.C.: The Brookings Institution Press, pp. 41–71.

CHAPTER 16
AIRLINE MANAGEMENT AND OPERATIONS*

Saad Laraqui
Business Administration Department
Embry-Riddle Aeronautical University
Daytona Beach, Florida

16.1 EXECUTIVE SUMMARY

The overall structure of firm-level management infrastructures in the airline industry is important to understand when analyzing airline operations management. More integrated relationships between strategic, financial, marketing, and other functions of management and operations and its management are increasingly evident in increasing cooperation among airlines to include cross-cultural integration among different regions and markets. Integrating mechanisms include a wide variety of strategic approaches, such as alliances, bilateral agreements, code-sharing, and other techniques. Integrating management and operations has been found to be a basis for maintaining a competitive advantage and a robust financial portfolio. Driven by adverse conditions in the competitive environment, airlines must persistently seek new approaches to create synergy between the management approach and the operations that they manage.

Collaboration on a global level is the best way to achieve a more cohesive relationship within an airline's operational structure. In this chapter, several variables will be analyzed relative to the successful combination of management structures and airline operations. Anecdotal evidence presented here suggests that there is a relationship between management style and function and operational utility and performance. No longer operating independently of each other, airlines seek to be part of larger networks and integrate their operations in the global environment. Therefore, airlines must consider not only the effects of competitors within the same region, but also of those that operate in other regions. The airline environment is a truly global enterprise, where interactions among airlines in different regions are crucial for survival.

16.2 INTRODUCTION

Management infrastructures and dynamics have a pronounced effect on the successful execution and flow of airline operations. Airlines fail regularly even in times of unprecedented growth, suggesting that failure is not entirely driven by external conditions such as pressures on operating economics due to high fuel prices. Tides of financial success are not simply reflections of favorable forecasts and stout economic indicators. Success in the airline industry appears to be better correlated with the

---

*Reprinted from the First Edition.

functions of management, particularly management's ability to rationalize and synchronize strategies and maintain good working relationships with labor to ensure success of the organization's business model.

Several startup airlines have demonstrated an understanding of the importance of acquainting all levels of management with actual operations, building a solid link with the operating employees. Southwest, JetBlue, and a host of other dynamic airlines have clearly expressed this relationship as fundamental to survival. But the benefits of linking management strategy with airline operations are not exclusive to newer, smaller airlines. This strategy to diminish the seemingly enormous fissures that exist between upper management and those involved in everyday operations can work for any airline. Certainly, the financial turnarounds of SAS in the 1980s under Jan Carlzon and of Continental Airlines in the late 1990s under Gordon Bethune are testimony to this.

Concurrent with this trend is the evolution of cooperative alliances, which have enabled airlines to rationalize their route structures and adjust their market strategies to compete with other newly formed global alliances and aggressively capitalized startups. While it is not evident at the present time due to union contract issues between carriers in an alliance, in time the trend toward globalization may cause greater integration of flight crews and maintenance personnel in the actual day-to-day operations of alliance-wide flight schedules. Already Continental Airlines and Northwest Airlines are coordinating the scheduling of Northwest's Asian hub at Narita, where a fleet of Continental B-737-700s and Northwest A-320s are based to feed the Northwest-operated B-747 long-haul fleet. In the recovery of the airline industry after the September 11, 2001, terrorist attacks, the level of coordination is likely to push levels that could not have been contemplated by managers or union leaders in the 1970s–90s.

Prior to the Airline Deregulation Act of 1978, airlines in the United States faced severe restrictions on their ability to expand under the Civil Aeronautics Board (CAB) policies of the day. Managements trying to improve the efficiency of their route networks or to add new service had to petition the CAB under carefully promulgated rules of administrative procedure. Even getting authorization to introduce a new fare type took up management time. In its attempt to manage a 12 percent return on equity to airline shareholders, following strict public utility doctrine, the CAB carefully controlled the growth of each carrier and the overall level of capacity being added, as well as fares, subsidy levels to small communities, and even the levels of flights offered between major cities pairs. Following the economic havoc caused by the 1973–74 energy crisis, several of the local service carriers and the newly emerging cargo carriers like Federal Express began to lobby for deregulation. Starting with modest liberalization during the Ford administration on fare and route regulation, most notably on dropping service to subsidized small communities where jet economics no longer worked at fuel prices 10 times higher than in 1972, the momentum built under the Carter administration to deregulate completely. On October 1, 1978, the Kennedy-Cannon Airline Deregulation Act became law and the CAB began to phase itself out of existence. Airlines operating in the environment that evolved after the 1978 Deregulation Act faced new challenges that stifled many airlines and their management.

The Carter administration embarked on a mission to reduce regulatory controls and ensure that they were kept to a minimum. One of his election pledges was to support the interests of the consumers: "In air transport, as in other industries, this meant less regulation and more choice" (Doganis 2001, 23). In a short time, there was pressure from the newly freed airlines to extend the domestic freedoms to the international sphere. To achieve this, negotiation of bilateral air service agreements was necessary. The aim was to promote the following (Doganis 2001, 24):

- Greater opportunities for innovative and competitive pricing
- Elimination of restrictions on capacity, frequency, and route operating rights
- Elimination of discrimination and unfair competitive practices faced by U.S. airlines abroad
- Flexibility for multiple designation of U.S. airlines
- Authorization of more U.S. cities as international gateways
- Liberalization of rules regarding charter flights

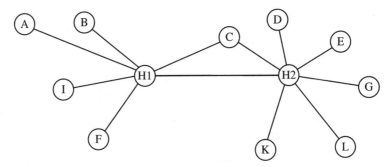

FIGURE 16.1 Hub-and-spoke airport structure. (*Source:* Button 2002, 30.)

In time, in light of the rapid expansion of the numbers of passengers transported in the United States from 250 million in 1978 to nearly 600 million in 2000, other governments around the world began to abandon the public utility theory of airline regulation and adopted the U.S. approach toward airline deregulation.

Today, most countries have liberalized domestic and international air operations with much the same result in terms of expansion of airline fleets and passengers transported as fares generally come down and frequency over an expanded city pair matrix increases. One unexpected benefit to many governments has been the substantial increase in revenues from landing fees, air navigation charges, gate rentals, and concession participations. The upward slope of revenue growth has permitted the self-funding expansion of infrastructure such as new runway construction and passenger terminal additions. At many airports around the world, the terminal has become a shopping mall rather than a sterile place where passengers are processed to and from flights. This is partly due to the structure changes after the deregulation in 1978. With the hub-and-spoke structure, passengers have longer waits between their connections, and this encourages terminals to generate more nonoperational revenue. Figure 16.1 shows the current airport structure.

Unfortunately, the industry is not immune to geopolitical problems and economic recession. However, with the long-term trend toward globalization, air transportation will continue to play a vital role in air commerce. The thin margins of a global commodity business will require nimble management working closely with operating employees under their own control and those who are nominally under control of the alliance partners as the networks become more finely meshed.

In the next section, an historical view of airline operations will be presented with a focus on Europe and the United States. Notable successes and failures will be compared and contrasted. Afterward, the new and continuously changing airline business environment will be discussed, suggesting that structural changes need to be met with a strategic paradigm that integrates strategic, financial, and operations management into a cohesive whole.

16.3 OPERATING ENVIRONMENT AMONG AIRLINES IN DIFFERENT REGIONS

16.3.1 The Growth Initiative

The operating environment of an airline should dictate the type of organization that it is or will become. Levels of management and functions such as planning and marketing and the structure of operating departments such as flight, maintenance, and station operations should reflect the dictates of the operating environment. Although elements of the airline business are constantly changing, airlines must develop and implement specific strategic plans in order to survive, focusing on their most important asset: customers.

While international service introduces additional complexities to which domestic operators are not fully exposed, the global nature of air commerce affects the manner in which all management strategies are articulated. In addition to the challenges of filling seats and running airplanes on time, managers with international exposure have to deal with foreign regulations, labor practices, and a varying mix of cultures, customs, and political meddling. As the industry globalizes, many of the daily challenges will resolve themselves. Outside of the United States, the ICAO and IATA approaches to operational standardization have facilitated the process of acquainting operating personnel and line managers with airline procedures, supplemented by customer service standards developed by the alliances. Of course, the level of difficulty in achieving a harmonized approach varies between regions. Some regions present less hostile environments than others. Several examples will be discussed, illustrating the challenge of operating across dissimilar environments that can present difficulties to airline management.

16.3.2 Defragmenting Europe

Prior to the beginning of the 1990s, the European airline industry was heavily regulated, with the goal of protecting the large national carriers that had evolved, country by country, since the end of World War II. With the creation of the European Union, the dropping of the bilateral restrictions between member states led to progressive liberalization of capacity controls and the procedures to obtain new routes. New domestic and intra-EU markets were opened, and a number of European carriers capitalized on the opportunities for new growth by reconfiguring their route systems and improving their marketing strategies. Some of the changes to the route network and fleet mix affected operations and caused new labor issues to emerge. Moreover, pressure from the United States on "open skies," and the great success of the U.S.-Netherlands open skies agreement and the near merger of Northwest Airlines and KLM, opened management eyes to the opportunities of adopting U.S.-style hub-and-spoke operations, feeding short-haul domestic, intra-EU into long-haul transatlantic and other global destinations. Figure 16.2 illustrates this type of network structure.

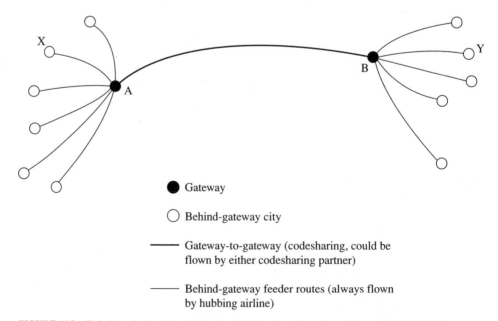

● Gateway

○ Behind-gateway city

—— Gateway-to-gateway (codesharing, could be flown by either codesharing partner)

—— Behind-gateway feeder routes (always flown by hubbing airline)

FIGURE 16.2   Global long-haul and short-haul network hub-and-spoke structure. (*Source:* Spitz 1998, 493.)

TABLE 16.1   Revenue and Traffic Effects of Code-Sharing Alliances

Alliance	Annualized impacts	
	Revenue ($ millions)	Traffic (000 pax)
Northwest/KLM	$125–175 (NW)	200 (NW)
	$100 (KL)	150 (KL)
BA/USAir	$100 (BA)	150
	$ 20 (US)	
United/Lufthansa	n/a	219
Delta/Virgin Atlantic	$100	250

Source: Spitz 1998, 493.

Table 16.1 summarizes revenue and traffic effects of major code sharing alliances in 1998 values. As can be seen from the table, Northwest/KLM benefits from their alliance a lot more than others do.

The EU operating environment changed as a result of the integration of several different operating environments, prompting a more unified and allied approach than was at first thought possible due to linguistic differences, labor practice, and operating regulation. A key element of success was the adoption of the Joint Airworthiness Authority (JAA) between the member states, which served to consolidate a mix of national operating regulations into a cohesive suite of rules for operating EU aircraft throughout Europe and the world. The adoption of certification rules for large and small carriers, as well as aircraft maintenance organizations, suddenly made it easier than ever in the past to start, grow, or restructure an airline in Europe.

The opening of new markets created challenges for both major carriers and new low cost carriers (LCCs) that entered the market. The new LCCs implemented strategies never seen in the European airline industry, designed to compete for passengers long transported by the largely state-owned carriers, which were also under great pressure from the EU to privatize. The LCCs developed a "lean and mean" management organization that had never been seen before in Europe, where the state-owned carriers had become top-heavy in a strictly regulated environment that approved most requests for increases in fares to match requests for higher wages paid to growing workforces. These new operators took advantage of growing price and cost pressures, using lower fares as a way of luring passengers away from established, well-known carriers (Daly 1995). Efficiency was the key to their success, and a new game plan was fashioned to accommodate the need to develop their cost structures against long-established charges for landings, handling, and navigation by the European civil aviation authorities.

Airlines in different regions have shared a similar evolution irrespective of their location. The period of heavy regulation is followed by a gradual liberalization, which in most cases leads to complete deregulation of the operating environment. This change in the regulatory environment triggers a number of startup airlines. Most startups usually begin flying early in the newly deregulated environment. Soon the economic realities of the thin-margined airline business serve to eliminate the poorly capitalized or poorly conceptualized carriers. In the bigger countries, the market becomes saturated and a round of consolidation occurs, eliminating many of the startups. Those that are well capitalized or capture enough market share to continue to grow usually survive, particularly if their management is nimble.

The Italian airline market serves as an example. Italy's industry historically consisted of mainly one major carrier, Alitalia, which emerged as a result of consolidation of a number of carries started in the post-World War II era. This government-owned entity controlled the domestic market and maintained hegemony in all international markets to and from Italy. Deregulation of the Italian marketplace opened up new opportunities, and new operators started to appear. The postderegulation Italian industry shared a pattern similar to that of the U.S. industry, where price wars between the new entrants and Alitalia eliminated all opportunities for profit and, more importantly, returns to shareholders who had invested in the startups. With much lower fares between the larger cities, the volume of Italian domestic traffic has greatly increased over prederegulation levels. Alitalia has fought off the new entrants very aggressively by developing an extensive code-share network with the Italian regional operators. Italian labor law makes it hard for a startup to have a major labor advantage, and the airport ground-handling monopolies make it difficult for a new operator to gain competitive advantage through lower boarding costs. Under the pressure of higher fuel prices since the summer of 2000, the Italian industry has consolidated into two large and two smaller groups. Some of the independent carriers have shut down or shifted their focus to increased charter activity. The shake-out among the jet operators has created some opportunities for regional turboprop operators, who soon become affiliated with Alitalia, which has been successful in maintaining market dominance, albeit at the price of profitability of its domestic route system. With its international network under severe economic pressure, Alitalia is struggling to right-size itself and find a long-term alliance partner. The EU rules make it very difficult for the Italian government to orchestrate an economic bailout of the carrier as it has done in the past.

The first challenge to European airline managements was to cope with the need of the industry to undergo changes of ownership from state governments to privatized entities. The second major challenge was to restructure operational procedures, oftentimes requiring renegotiating longstanding labor practices with unions. The third challenge was the need to integrate multicountry operating environments into successful operational units, a challenge the U.S. managers did not have to face when the United States began deregulating. It took some time for approaches to the legal, economic, and political problems that arose to be developed. Moreover, since the formation of the EU, airlines have found themselves in the midst of an arduous process to privatize carriers that now operate in a highly competitive, unrelenting air transport market.

Closely related to deregulation is privatization. The issue of privatization has deeply affected the airline environment in Europe. After partial and then full deregulation, European airlines were able to test their strategic visibility through privatization and, consequently, through carrier alliances, some involving cross-ownership or other financial participation. Privatization in the EU occurred in two phases: (1) the privatization of British Airways in 1987 via a public stock flotation, and (2) the phased reduction of state ownership of KLM Royal Dutch Airlines, Lufthansa German Airlines, and ultimately most of the other European state-owned airlines.

Privatization was facilitated by a strong European economy and booming stock markets and a period of low fuel prices, ensuring most of the carriers good earnings as they made their hub-and-spoke networks work around fortress hubs in the major cities of their original origin. The early success of some airlines like British Airways with its London hub and KLM with its hub at Amsterdam and business relationship with Northwest, which gave it near complete access to the U.S. and Asian market, created a climate of great expectations in the stock markets of Europe. Tides of economic success usually facilitate expansion, whereas uncertainty about the economic climate forces many airlines to contract. Many other newly liberated European airlines adopted a "me too" strategy, trying to improve earnings as best they could to effect an offering of stock to the investing public. Some carriers were more successful than others at coping with all of the challenges of operating profitably to satisfy shareholders and other stakeholders, knowing that under the new EU rules they could not go back to the national government for cash or a fare increase when they failed to meet budget projections.

Olympic Airways: Finding Interface between Management and Operations. Like some other ailing airlines in Europe at the time, Greece's national flag carrier, Olympic Airways, faced symptoms of what was referred to as the "distressed state airline syndrome," common during the aftermath of the Gulf War in the early 1990s when the world-wide industry was nearly put out of business by the steep increase in fuel prices and subsequent economic recession (Doganis 2001). Once controlled by Aristotle Onassis, Olympic had become a ward of the state during the period of unprofitabilty that commenced after the 1973–74 energy crisis, and the changes in the traffic makeup of its long-haul network to Australia, Canada, and the United States. Olympic's politically appointed executives had to deal with the indecisiveness of its indifferent operating managers, and its board of directors consistently failed to face the challenges to the structural integrity of the organization. In fact, during the 1-year period between February 1995 and March 1996, Olympic's board of directors was changed three times (Doganis 2001). The focus was more on entertaining political objectives than on the airline's fiscal and operational strategy. The problems that the airline frequently faced, extensively impaired its ability to successfully operate, were:

- It was largely undercapitalized, with huge debts.
- It had serious financial difficulties securing credit.

- It was frequently overpoliticized.
- It had overweight, poorly planned labor networks.

After years of dire financial trauma, executives and closely thereafter the government began to probe the extent of the problem. After months of investigation, it was concluded that yield and operating expenses were aspects of the business that were simply ignored. Unusually low utilization was damaging Olympic's ability to maintain its competitive edge against other airlines in Greece and Europe. Specifically, the economic perspective of operating the airline was visibly disregarded. According to a senior economist for the MITRE Corporation, "[E]conomic analysis provides more precise estimates of the causal relationship among airline operating expenses, yield, and demand" (Homan 1999, 506).

Economic analysis requires management organizations to locate problem areas within the operational structure and focus their efforts on reducing unnecessary utilization of assets and finding ways to redeploy them. Its own aircraft heavily burdened Olympic Airways. For example, the airline used an Airbus A300-600R, an aircraft with a capacity of 260 passengers, on early morning routes from Athens to the northern Greek city of Thessaloniki. The route's main occupants were Athens newspapers ferried for distribution in Thessaloniki, the second city of Greece. Rather than managing the aircraft and the route for profit, Olympic used it on this route for four years simply because the marketing department had once decided that most peak time service between Athens and Thessalonki would be flown in a wide-body, irrespective of experienced demand patterns. New management eventually realized the effect this scheduling practice had on the value-generating capacity of the aircraft: "Aircraft cost plays an important role in operating economics . . . and this dictates the profitability of individual transactions" (Vella 1999, 385).

The process of transforming the airline came at a time when airlines worldwide were facing their most pronounced financial crisis ever. Although new Airbus A-340s were delivered to alleviate economic losses being incurred by operating aging Boeing 747s on many long-haul routes, the airline failed to regain its competitive position and serve its entire customer base. During this period, the Greek government decided to allow the startup of new carriers to serve the domestic market in response to demands by regional politicians for an alternative to moribund Olympic. Ultimately management discovered a valuable lesson in its evolution, namely that the concept of "the one-size-fits-all airline is dead" (Gertzen 2002). How Olympic learned that valuable lesson, and how revenue management is impacting airlines' thinking process, is shown in Table 16.2.

As a consequence, Olympic continues to restructure its European operations and is struggling to salvage the few parts of its long-haul operations that are economically viable, given the fundamental facts that fewer Greeks are emigrating to countries like Australia and that third-generation immigrants do not return to the motherland in the same numbers as did the first and second generations.

TABLE 16.2   How Revenue Management Is Impacting Airline Thinking

Conventional thinking	Revenue management thinking
Airlines control the price consumers will pay for their products and services.	Consumers determine price through buying behaviors; successful airlines respond to changing consumer signals.
Consumers pay for products and services.	Consumers pay for the opportunity to buy at the price and conditions acceptable at a particular moment in time.
Airlines segment consumers by standard demographic and psychographic factors (age, sex, income, and so forth).	Airlines segment consumers by variations in buying behiaviors based on price sensitivity and time sensitivity.
Prices are set on a cost-plus profit margin basis.	Airlines rework costs to accommodate consumer-driven prices and still achieve profits.
Discounts are determined by salespeople to close sales and meet volume objectives.	Revenue management objectives determine discounts in terms of revenue maximization goals.

Source: Cross 1998, 307.

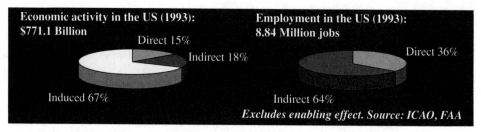

FIGURE 16.3   Aviation industry's impact on U.S. economy. (*Source: ICAO.*)

16.3.3   The North American Precedent: Operational Optimization

The airline industry in the United States is one of the largest industries in terms of sales, employment, capital requirements, and importance to the overall economy. Figure 16.3 shows the impact that the aviation industry has on the U.S. economy.

North American airlines have long been exposed to challenges of integrating a strategic approach to the business by developing route networks across a broad swath of geography while meeting the demands of public private and institutional shareholders who financed the industry almost from infancy, labor, and government policies. While the scheduled business was initially more developed in pre-war Europe, the United States and Canada were pioneers in approaching the airline industry from the perspective of dynamic management and intelligent operations. Management dynamics is the cornerstone of American and Canadian business, and the degree to which airlines develop their management organizations has a significant impact on the way an airline operates. The management structures at the major carriers like American Airlines and Air Canada evolved in a manner reflecting the regulatory requirements of the time, while dealing with the problems of running a continent-wide airline operation in a period when telecommunications was primitive. This pioneering class of carriers had to invent solutions ranging from a standardized approach to air traffic control (initially without significant government involvement) to demanding that manufacturers develop specific airliners such as the Douglas DC-3 and later the DC-8 as well as the Boeing 707 to meet growing requirements of their customer base. In the early 1970s, competitive pressures forced Southwest Airlines to develop management tactics that got everyone involved in the overall operation of the aircraft from station to station. For instance, flight attendants assisted in the boarding of passengers in the gate area, which was unheard of at such well-established competitors as Braniff International. Braniff tried to fend off the incursion of pesky Southwest into the Texas marketplace by selling below its cost, anticipating falsely that the unheard-of ticket prices were also below Southwest's internal costs, which were further refined to deal with the lowered ticket yields. The success of this approach provided the benchmark for many of the world's airlines, which have tried to emulate the unusually strong bond among the people at Southwest.

Airline management is responsible for ensuring that the industry does not collapse on technicalities such as safety of flight, gross management incompetence, and clumsy business decisions. Over the years, many of the surviving U.S. airlines have taken the initiative and implemented several strategies aimed at bridging the gap between management and operations. For example, American Airlines pioneered yield management and focused its strategy on developing a solid relationship with the customer. Subsequently, the airline established what is referred to as customer relations management (CRM). This conceptual approach to operations management is described as "how one can maximize or create the right sort of experience every time" (Ott 2000, 52).

The ultimate goal of U.S. airlines has been to optimize operations and integrate them with the goals of upper-level management. The objective of this approach, as in the case of the American Airlines initiative, was to create an interface of cooperation among organizational structures. In turn, this "should be tailored to match specific measurable corporate objectives of the company based on a well-defined mission statement" (Asi 2002). Another approach U.S. airlines have advanced is compensating travelers for inconveniences: "on a specific consumer level, airlines must use information on a customer's operational history to set the stage for ways an airline could compensate for negative experiences . . . knowing a customer's wants and needs in greater detail will help develop 'the right product for the consumer'" (Ott 2000, 51).

People Express: Pioneering Leadership Development. People Express began operations in April of 1981 and quickly emerged as a model airline with an innovative management team and an aggressive pricing and frequency strategy. According to Donald Burr, founder and CEO of People Express, the most predominant reason for establishing the new company "was to try and develop a better way for people to work together . . . that's where the name People Express came from" (People Express 1990). Having cut his teeth with Frank Lorenzo, in the turnaround of Texas International Airlines in the early 1970s, Mr. Burr called the traditional style of management a "deadening grind and a lack of vision" (People Express 1990).

From its inception, the airline focused on combining the strategies of management with the interests of its customers. Profit was no longer the core focus; instead, serving people and customers were precursors to profit, and that became the focus of management (People Express 1990). People Express channeled its energies into finding a crossover point between the organization's design and the successful execution of that design. By assembling "management teams," People Express effectively improved the performance of its operations. The airline was one of the first to pioneer leadership development in its long-term strategy. However, its pioneering labor approach and its rapid expansion strategy failed to produce earnings, and in 1986 Burr surrendered control of the People Express experiment to Continental Airlines, then run by Frank Lorenzo, who went on to integrate the People Express Newark hub successfully into the dominance of the New York market. In the aftermath of the Gulf War, Mr. Lorenzo's hard-nosed approach with the labor unions in wrestling cost savings forced Continental into bankruptcy for the second time in a little over a decade, which provided an opportunity for new management to focus the Continental strategy narrowly around three hubs at Cleveland, Houston, and Newark, and set about to simplify and rejuvenate the fleet.

16.3.4 Operations Management Among Airlines in an Alliance

Overview. The emergence of worldwide airline alliances over the past decade has instigated a gradual reevaluation of airline operations among participating carriers. The need to become a more cohesive marketing organization providing a seamless experience to alliance customers has prompted airline management to consider new ways of integrating joint scheduling, gate operations, and reservations data. According to industry sources, half of the world's jet fleet is controlled by only 17 airlines, and almost half of world passenger traffic flies on one of the four largest alliances: Star Alliance, Wings, OneWorld, or SkyTeam (Barry 1998). The fifth worldwide airline alliance, which was recently restructured, is Qualiflyer. Table 16.3 summarizes these five alliances and their member airlines.

The environment in which an alliance operates should be clarified. A *strategic alliance* is one where present or potential competitors commingle their assets in order to pursue a single or joint set of business objectives. Such commingled assets may include terminal facilities, maintenance bases, aircraft, staff, traffic rights, or even capital resources. The Northwest/KLM alliance is an example of a successful crossborder, crosscultural strategic partnership, originating in KLM's financial participation in the 1988 leveraged buyout (LBO) of Northwest Airlines in a hostile takeover by an LBO group. *Marketing alliances*, on the other hand, are not strategic alliances, in that partners use their assets independently of one another and pursue their individual objectives. Alliances of this sort are quite common: code-sharing agreements, joint frequent-flyer programs, and other agreements are widespread, but the degree of cooperation is limited by the degree to which each airline offers its assets to another. Table 16.4 illustrates different phases of airline alliances common between international carriers.

The Star Alliance spearheaded by Air Canada, Lufthansa, and United Airlines is perhaps the best example of a strong marketing alliance that has expanded to encompass a global network of more than a dozen carrriers (Doganis 2001).

TABLE 16.3   Five Worldwide Airline Alliances and Their Member Airlines

Alliance	Members	
	Before 2001	After 2001
SkyTeam	Delta Air France Alitalia Korean Air Aero Mexico TSA-Czech Airlines	SAME
Qualiflyer	Air Littoral Austrian Airlines AOM French Airlines LOT Polish Airlines Air Europe Sabena Crossair Swissair PGA Portugalia Airlines Turkish Airlines Volare Airlines TAP Air Portugal	Air Littoral Air Liberte LOT Polish Airlines Swiss Air Lines SN Brussels Airlines TAP Air Portugal
Wings	Northwest Airlines KLM Continental	SAME
Star	United Airlines Tyrolean Airways Thai Airways International Singapore Airlines Scandinavian Airlines Varig Brazilian Airlines Asiana Airlines Austrian Airlines Air New Zealand Lauda Air Mexicana Airlines LOT Polish Airlines Lufthansa German Airlines Spainair All Nippon Airways BMI British Midland Air Canada	SAME
OneWorld	American Airlines British Airways Air Lingus Cathay Pacific Finnair Iberia LanChile Qantas	SAME

Source: Oster 2001, 22.

TABLE 16.4   Phases of Airline Alliances to Full Merger

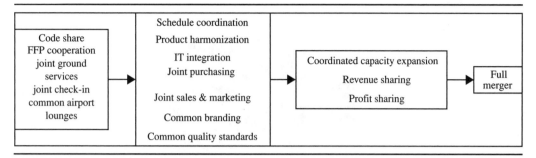

Table 16.5 summarizes the features of alliances and the advantages of being in an alliance in various ways.

Why Alliances Fail.   Despite the success of many global alliances, several have failed. The lack of clear objectives and a changing economic marketplace, compounded by conflicting management objectives, have been the major causes of alliance disruption and disintegration.

In order to avoid failure airlines should consider the process described in Figure 16.4 before they choose their global partners. This will ensure the alliance to last for long time periods.

No greater example of such a failure exists than the collapse of the Swiss-based Qualiflyer Group, in which Belgium's SABENA, France's Air Liberte, and Switzerland's SWISSAIR all collapsed within a very short time of each other. To avoid such operational and financial malfunctions, management's role is to define, implement, and measure its short-and long-term strategic goals.

A difficulty in assessing the viability of integrating airlines in alliances lies in the complexity of maintaining streamlined operations in the regulatory environment. For example, the integration of airlines in Europe and North America has been characterized as complicated and convoluted. A number of factors have conspired to stifle many attempts of international airlines to cooperate effectively and, to a lesser extent, merge their operational units into one. While in 1987–89 Northwest and KLM successfully lobbied the U.S. government for the most integrated approach from a financial, marketing, and operational standpoint, the U.S. administration, favorably disposed to unions, placed limitations on KLM's role. North-west needed more financial involvement from KLM in the period after the Gulf War and before the public markets allowed Northwest to go public in 1993, and the

TABLE 16.5   Alliance Feature

Network extension	Operational craft reduction	Competition reduction
Complementary code share	Shared airport lounge and	Parallel code sharing
Schedule coordination	gate facilities	Joint pricing and yield management
Special prorate	Ground handling	Revenue pooling
Through baggage	Catering	Joint flight
Single check-in	Maintenance	Joint venture
Shared or proximate gates	Joint purchasing	Multiple listing of code-shared flights
Marketing alliance	Joint insurance purchasing	
Frequent flyer	Joint marketing	
Block seat	Management contract	
	Wet-lease	

Source: Oster 2001, 24.

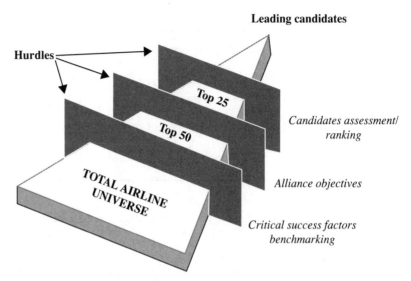

FIGURE 16.4 The rigorous process for selecting global partners. (*Source:* Baur 1998, 537.)

U.S. government allowed KLM to take up to a 49 percent equity position in Northwest, provided KLM did not violate the 25 percent voting stock limitation on foreign ownership. Conversely, when American Airlines and British Airways attempted to unify transatlantic operations in the United States and Britain, the principal competitors, United Airlines and Virgin Atlantic Airways, mounted a lobbying effort aimed at triggering a change in the U.S.-U.K. bilateral agreement. Slots at London Heathrow airport were reallocated, which ultimately led to the abandonment of most features of the service agreement, except for the OneWorld alliance. Had the deal overcome regulatory hurdles, the alliance would have given the two airlines 60 percent control of the transatlantic market (Barry 1998). Moreover, the longer-term consequences would have fundamentally challenged foreign ownership limitations placed on air carriers by nationalistic governments, as well as how flight crews are flowed through an integrated global route network. In short, it was too much too soon for politicians on both sides of the Atlantic.

Converging Strategies: Successful Allied Cooperation. The success of many alliances is attributed to the ways that operational utility and flexibility are managed. As evidenced by the increasing number of alliances, the operational environment and management strategy play significant roles in their successful implementation. Again, one of the most longstanding major alliances is Northwest and KLM. Beginning with KLM's participation in the 1998 leveraged buyout of Northwest, the alliance quickly adopted code-sharing as a means of flowing traffic, and hence money, between themselves. By late 1992, the U.S. Department of Transportation granted them antitrust immunity. This allowed unobstructed entry and capacity privileges between the United States and the Netherlands, especially considering the open skies agreement between the two governments (Wang and Evans 2002). With the broadened agreement, they were able successfully to implement a strategy that would facilitate a mutual opportunity for expansion and market access. While flowing some Northwest airplanes into Amsterdam from cities such as Boston and Detroit, Northwest was able to use the KLM fleet to gradually establish a Northwest-branded hub at Amsterdam. Meanwhile, KLM mimicked the Northwest flight offerings to extend its reach into the United States beyond the principal Northwest hubs. As the results proved positive, the two companies developed a joint marketing strategy to extend the network and reach farther into Europe. This was aided by the U.S. and Dutch governments by establishing less restrictive bilateral agreements with the newly emerging CIS states and Eastern Europe.

16.4   THE CHANGING AIRLINE BUSINESS ENVIRONMENT

During the past decade, full-service airlines have responded to several challenges:

- New entrants and low-cost competition
- Increased pressures on costs
- Simplified product and process function with strategies that average down labor cost

With the wave of attempts to average down labor cost, managers hope to include changes to labor contracts. This approach to defend and protect market share has not been effective, despite the laying off of numerous employees and outsourcing maintenance and airport operations (Collings 2002).

In the United States, established labor law makes it difficult to achieve significant changes to labor contracts and work practice, except where bankruptcy restructurings force a renegotiation. In Europe, managers have had greater success in affecting changes to the restructuring of work, but have found less success in changing compensation rates, in many cases established by national government decrees. Further, in Europe, national carriers have long contracted out ground handling at locations outside their countries and have used third-party maintenance contractors. The average U.S. major, in contrast, has more integrated station and maintenance operations, making it harder to change work practices by contracting to third parties.

Where the approach of replacing airline workers with third-party contractors has been attempted, airline managers expect or imagine a company where employees understand what it takes for the airline to make money. Improvements from a streamlined airline would improve cash flow so that greater investment can be made in customer service, attracting prospective clients and turning them into loyal customers. Improved customer service would include features that passengers want at prices they are willing to pay. Unfortunately, to date such a scenario or understanding is unrealistic or rare in the full-service airlines. Few managers can state clearly or view the business model of their airlines, much less how their companies make money or protect market share: "This is clearly visible with respect to the competition with new entrants, comprised of operators such as Jet Blue, and low-cost carriers like EasyJet" (Collings 2002).

However, there are indications that this is changing in the United States. Carriers have tried to reinvent themselves in the wake of the September 11 terrorist attacks, made worse by the subsequent recession triggered by instability in energy markets. Of the U.S. majors, Continental has had the greatest success with this approach. Begun in the turmoil of the acquisition of Eastern Air Lines by the then-parent of Continental Air Lines, Texas Air Corp., in the mid-1980s, the new management headed by Gordon Bethune, installed after the Gulf War, was able to build on the approaches of the previous managements to improve the product. The fleet was streamlined from more than 10 aircraft types to 3, most of which were relatively new, reducing maintenance costs and improving reliability. Hubs were carefully analyzed route by route, and in some cases service was abandoned in order to ease self-induced air traffic control delays, improving the ability of the carrier to run on time. This was a major factor in the refusal of prospective customers to fly the carrier. Above all, employees at the operating level were empowered to solve customer service issues as they arose. The internal slogan of "Work Hard, Fly Right" was extended to become the marketing theme pitched to Continental customers, who have responded enthusiastically to the changes at the carrier, greatly improving earnings to shareholders and workers.

Full-service airlines have begun to change their approaches in the implementation and adoption of more realistic business models where the core logic lies in creating value. Some approaches have included:

- A set of values that the airline offers to its stockholders and employees
- An operating process to deliver these values
- Arrangement as a coherent system (point to point, hub-and-spoke, international)
- A system that relies on building assets, capabilities, and relationships

By carrying out these steps, airlines have begun to realize the goal of creating value. For example, "[Managers at] American airlines [have] finally announced [that] they recognize that they have to change their business model, and cannot focus on maximizing revenues; they must focus on maximizing efficiency" (Collings 2002).

Managers of successful airlines find that no single business model is guaranteed to produce financially superior results; however, successful models do share certain characteristics. Successful airlines such as JetBlue offer unique value, sometimes in the form of a new idea where the carrier started up with new equipment instead of used aircraft. This has been coupled with adopting a distinctive approach to customer service at a time when major carriers were cutting back on inflight service. Often, a good formula is a combination of product and service, features that offer more value or lower price for the same benefit, or deliver more benefit for the same price.

Southwest and JetBlue have radically different approaches to creating value, even though they exploit the same low-fare market segment. Southwest offers a simple, reliable product favored by business clients and leisure passengers who like the frequency of its short-haul services. JetBlue has developed a lower-frequency, longer-haul product aimed more narrowly at the leisure market, providing the low-fare filler for the U.S. majors in coach.

A winning business model can be difficult to imitate. This is often achieved by establishing a key differentiator, such as customer attention or superb execution. True differentiation creates a barrier to entry that protects the main profit stream. Successful business models are grounded in reality and are based on accurate assumptions about customer behavior, financial structures, and day-to-day operations. As they have become globe-circling giant network carriers, it appears that many full-service airlines lack understanding of where they make money, why customers prefer their offerings, and which customers only drain resources.

Since airlines compete for customers and resources, a business model must highlight what is distinctive about the firm, how it wins customers, attracts investors, and earns profits. Effective business models are rich and detailed, and components reinforce each other. It is important to remember, however, that regulatory and economic changes, such as increased fuel prices, can make a once-effective business model less capable of generating earnings to the carrier, without which it slowly dies.

16.4.1 Airline Pricing—A Historical Perspective

A key objective of any company that produces and sells a commodity product is to gain as much control as possible over the supply of that product. The organization can thus avoid selling at commodity price levels, which are always just marginally above the production costs of the lowest-cost producer (Greenslet 2002). Historically, airline strategy has followed this fundamental rule, resulting in attempts to control the supply of seats in markets. This has led to the creation of primary hubs for all the major carriers, allowing them to offer customers a vast network of routes and markets. This tactic, when combined with the enticement of frequent flyer programs, has, however, led to the narrowing of options for many consumers who are forced to select the airlines that serve their hub cities.

Generally, network airlines have learned that there is no earnings leverage in competing aggressively against each other at major hubs. This has led to the emergence of patterns of regional dominance through "fortress" hubs, where a single carrier may have as much as an 80 percent share of the hub airport. Over the years, airlines have been able to employ this strategy most effectively in maintaining a yield premium servicing the business passenger, who historically has proved willing to pay a higher fare for the convenience of frequency. Business travel has always been the foundation of airline revenue generation. A United Airlines study in the 1990s stated that 9 percent of its passengers accounted for 43 percent of total revenue (Greenslet 2002, 1). Since the availability of frequent and reliable service has stimulated business travel over many years, business travelers have developed very short travel planning horizons, even on international trips in a multinational work environment. One consequence is that the airlines have been better able to segment passengers by fare type and offer an array of fares attractive to many different passenger segments on the same aircraft.

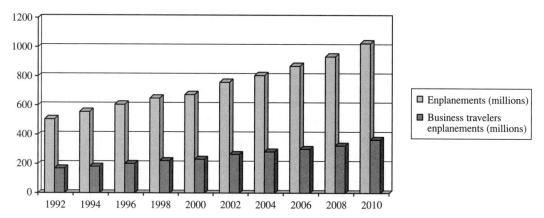

FIGURE 16.5   Enplanements and business travelers enplanement growth rates between 1992 to 2010. (*Source:* Baur 1998, 537.)

Generally, the shorter the booking horizon, the higher the fare. With short travel horizons, business travelers have become accustomed to paying higher fares. Additionally, restrictions on discount fares that required a Saturday night stay made them unattractive to business travelers. The full fare in the 1980s was about twice the average discount fare, and this disparity increased threefold during the economic boom of the late 1990s. While perhaps only 5 percent of the inventory of the major airlines sold at the highest fare levels as the price segmentation became more and more refined, many of the major corporations contested this pricing disparity by negotiating company discounts. Moreover, the carriers competed aggressively for identifiable major corporate travel accounts. This allowed the biggest customers greater access to a wide range of hidden discounts. But with the trend to smaller firms and self-employed people in the changing information age economy, a growing segment of customers traveling at the last minute were forced to pay high fares.

At the same time, more and more conferences and conventions, which account for as much as 33 percent of all annual business travel, began to be scheduled in such a manner as to allow access to the discounted fares. Airlines attempted to manipulate their yield-management systems in such a way as to preserve revenue streams. This had the effect of pushing some short-notice coach and business class fares in key markets up at a very aggressive rate in the late 1990s, just as the information technology boom, which had fueled traffic growth, was ending.

Figure 16.5 shows the number of enplanements and business traveler enplanements between years 1992 and 2010. From this graph it can be seen that the growth rate for business travelers is slightly less than the total enplanement growth rate.

16.4.2   Impact of the Global Economic Downturn on Operations Management

The business traveler has often complained about the disparity that exists between full fare prices and the many discount prices that are offered to entice the leisure traveler. Until December 2000, due to a strong global economy, the aviation industry was able to extract a yield premium in exchange for schedule flexibility to the business passenger.

However, as soon as the information technology sector began to decline, particularly as the e-commerce/telecom bubble collapsed in late 2000, airline revenue from full-fare travel fell drastically from a $10 billion annual rate to about $7 billion. It appears that in 2001 this may have declined by as much as 30 percent more, to $3.7 billion (Greenslet 2002). The primary reasons for this drastic impact on the bottom line of airline financial statements are that:

1. The business traveler is no longer a captive customer of the large full-service airlines. He or she is aware of the price and service options that are now available and easily accessible due to the

World Wide Web. Moreover, travelers are willing to change lifestyles sufficiently to travel in periods when low fares are available, as small and large companies are more open to compensatory flextime in exchange for holding a weekend meeting or traveling through the weekend.

2. In the wake of September 11, major carriers reduced many of the restrictions on discount fares in attempts to build market share and regain customer loyalty. Initially, airlines that attempted to reinstate these restrictions, such as US Airways, appeared to have induced passengers to switch to flying on other carriers. More recently, carriers have developed an approach to increased revenues through surcharges at airports rather than at the time of booking.

In the United States, established network carriers have had to face a challenge not faced by any generation of airline managers since the tenfold increase in fuel prices from 1972 to 1975. U.S. managers are under intense pressure to develop innovative solutions to restore earnings. First, they must rebuild their traffic bases and encourage customers to fly again, as opposed to driving or taking other modes of transportation. Managers at U.S. major airlines are beginning to realize that they may not be able to restore the most important element of the traditional business model—the business traveler. A permanent loss of this revenue stream may prove detrimental to the financial health of airlines, since it will not allow them to offset the increases to their cost structures resulting from schedule cutbacks.

In the United States, in the period from September 11, 2001, until this writing, yield per passenger mile has decreased by about 1 cent while costs have increased by about 1 cent. With load factors trending slightly down, this is an unbearable condition in an industry that considers a 5 percent net operating margin to be exceptional. This implies that a structural change in the prevailing full-service airline business model will be ultimately necessary in order to ensure success in the long-term: "Mr. Schenk, president of the New York-based firm Airline Associates, said if the airline business wanted to be profitable again it had to ditch many of the strategies which had brought rich reward in the past" (Collings 2002). Even the best-designed business model cannot last forever. Business models must change to keep pace with shifting customer needs, markets, and competitive threats.

Second, full-service airlines may have to modify their existing business models by expanding geographically into other markets in order to add customers, adjust prices, and extend products and services. All these are incremental changes that can boost the returns from an existing model.

Another business approach being adopted is to renew the distinctiveness of an existing business model. This focuses on revitalizing value to counteract the forces that encourage competition based solely on price. Decades after the creation of Southwest, many full-service airlines believe that their future success lies in cutting costs.

Some successful airlines replicate their business model in new domains by taking new products into new markets. For instance, EasyJet, the successful U.K. low-cost carrier, has extended the underlying pricing concept into rental cars and Internet cafés. Airlines adopting their existing advantages to new models will grow by building around unique skills, assets, and capabilities.

Operations managers will have to clarify their roles in improving their airlines' focus. They need to understand the organization's business model and how it contributes to the overall company strategy. Flight operations must understand what makes the airline distinctive and how the airline makes money. Likewise, management needs to establish a team-wide framework to allow the carrier to be a more nimble competitor. The business model becomes a series of strategic building blocks, expanded in strategic range by experimentation with new blocks, then mixed and matched to create profitable new combinations.

16.5   AN ERA OF CHANGE IN THE AIRLINE INDUSTRY

Air transport is regarded as an initiator of economic development and has traditionally experienced greater growth than most economic sectors (ICAO 2002). The economic stimuli of airlines, airports, and their direct affiliates contributes more than 4.5 percent of world output and accounts for 15.4 million jobs. The ramifications of the recent global economic downturn that has been magnified by terrorist attacks serve to remind us of the crucial role played by civil aviation.

In the weeks following September 11, 2001, airlines, manufacturers, and analysts attempted to envision the long-term implications of the terrorist attacks (Morrison 2002). Their analysis yielded the following two worst-case scenarios:

1. The world economy would go into a meltdown prompted by a collapse in consumer spending, rising oil prices, and plummeting stock market.

2. Passengers, worried about airline security and safety, would refuse to fly again.

It has two years since those ghastly attacks took place, and both of the above scenarios are still hanging. The U.S. economy has largely weathered the recession and is trying hard to begin a recovery phase, prompting European economies to try hard to gain lost ground and regain investor confidence. The military actions in Afghanistan, the threat of war in the Middle East, and the implementation of security measures on board aircraft and at airports have not helped the economic situation. These steps have won back some customer confidence, and there has been resurgence in passenger numbers, both domestic and international. Airlines, however, face a deeper crisis resulting from the fact that there are too many airlines chasing too few customers. This problem is particularly acute in Europe and the United States, where there is a dire need for consolidation. The economic boom of the late 1990s had masked the problems of inefficient operations and incompetent management at most major carriers. These inadequacies were revealed once the economic boom collapsed and airlines were forced to compete for lucrative business customers.

There presently exist many barriers to structural reform in the airline industry that must be overcome before real change occurs and a healthier industry results (Morrison 2002). First and foremost, airlines must switch their focus from maintaining market share to building more profitable operations. The obsession of major carriers with size is essential to maintaining a hub-and-spoke network and supporting giant infrastructures. However, airlines need to identify and retain their core operations while outsourcing all other functions to external contractors.

Second, the outdated regulatory framework that plagues airlines globally needs to be replaced by transparency and trust among member states. Ownership restrictions that ban foreign nationals from owning majority stakes in domestic airlines are antiquated and do not reflect changes that have been implemented in various other industries such as auto manufacturing. Bilateral agreements between the United States and individual European countries restrict competition and promote inefficiency. As former International Air Transport Association (IATA) director-general Pierre Jeanniot insisted, the airline sector needs to become like a normal industry.

The third area that requires immediate change is pricing strategy. As discussed, airlines have realized the folly of complex business models and are now striving to achieve simplicity in their pricing strategies. A risk that full-service airlines face is the evolution of air travel as a commodity. By offering customers a unique experience, airlines can compete on value and profit rather than price and revenue.

16.6   RISK MANAGEMENT AT AIRLINES

Mercer Management Consulting recently analyzed aviation industry risks for the 10-year period from April 1991 to April 2001 (Zea 2002). The primary risks facing the industry can be broadly classified into the following four categories:

1. *Strategic risks* are defined by business design choices and their interaction with external factors. Challenges from new competitors, shifts in customer preferences and industry consolidation are examples of strategic risks.

Southwest Airlines has mitigated these risks successfully by building a business model that is simple and operationally cost-effective (Zea 2002). Its use of secondary airports insulates from direct competitive pressures and reduces aircraft turnaround times. Profit sharing with employees and introducing "fun" in the workplace has been an integral part of Southwest's corporate culture and its recipe for success.

Lufthansa's diversification into nonflying businesses began in 1994 with four companies being created in engineering, cargo, services, and systems (Zea 2002). This decision was made to reduce strategic risk and the volatility of its earnings base from the passenger airline business. Thus, Lufthansa is now placed in a better position than its competitors, due to reduced earnings volatility while simultaneously maintaining its core airline segment.

2. *Financial risks* involve the management of capital and cash. External environmental factors that affect the variability and predictability of cash flow such as general economic conditions or currency exchange rates are also taken into consideration.

Ryanair, Europe's low-cost carrier, pursues an aggressive cash-management strategy. It made history in January 2002 by placing a record-breaking order with Boeing for 100 737-800s along with 50 options. Ryanair thus overcame the airline crisis by adding capacity rather than retrenching and downsizing company personnel. Chief Executive Mike O'Leary's perspective on this issue was, "When many of our high-fare competitors were grounding aircraft and canceling flights, we did the opposite and put a million tickets on sale at £9 ($13) each—our forward bookings went through the roof" (Campbell and Kingsley-Jones 2002, 35).

Fuel hedging is also a common way to manage the financial risk of volatile price changes. Cathay Pacific relies on long-term fuel hedging contracts and has saved approximately $80 million over the last five years. In 2000, 15 percent of Cathay's operating profits were directly attributable to savings from fuel hedging.

3. *Operational risks* arise from the tactical aspects of running airline operations. Examples include crew scheduling, accounting and information systems, and e-commerce systems. Although airlines address major risks such as business interruption quite effectively, they often fail to recognize subtle issues such as managing relations with government agencies. These regulatory issues cost the industry millions in operational inefficiencies and legal actions.

Operations managers need to reshape their units by incorporating familiar concepts such as process reengineering, contingency planning, and improved communication at all levels of the organization. The challenge facing airlines today is that they need to look at risks holistically, and evaluate each potential response, by gauging the impact on shareholder value (Zea 2002).

4. *Hazard risks* include risks arising from unforeseen events such as terrorist activities, war, and environmental hazards. Ironically, it seems that hazard risks were the least likely to result in loss of value to the aviation industry (Zea 2002).

Airlines have turned toward the world of technology to help deal effectively with increased security measures at airports and optimize internal operations. Self-service kiosks are now available at many ticket counters, and airlines strongly encourage passengers to use them to help speed up the boarding process (Gilden 2002). This is done through publicity campaigns, having ticket agents at check-in counters who advise travelers to use the automated service, and awarding frequent flier bonus mileage. Some carriers have gone a step further and now allow customers to check in at home using the Internet, a boon for those who are traveling without bags or plan to use the curbside baggage check. James Lam, former chief risk officer at General Electric, aptly summed up the issue: "Leaders recognize that over the long-term, the only alternative to risk management is crisis management" (Zea 2002, 3).

16.7 CONCLUSION

The convergence of management and airline operations is a symptom of the trend to integrate strategy and everyday performance. The strengthening of this relationship ensures a synthesis among management and business processes and creates an interrelated working environment. The changing face of international aviation management and the volatile environment of today make it essential that these processes be implemented. Although the crisis of an ever-widening financial burden threatens the economic conditions and fiscal viability of the airlines, management organizations are well positioned to offset the consequences.

As the tide of consolidation and increasing cooperation among airlines develops, the future of operations management and the industry as a whole will coevolve. Consequently, passengers will find an increasing role in the future development of the air transport paradigm as new measures to ensure safety and security are employed. The realization of these measures has already been seen and effectively employed by airlines and airports around the world. However, while management's initiative to accelerate the implementation of strategies has been slowed by waves of unpleasant economic indicators, by no means has it been halted. In time, the recovery of the industry is inevitable, although the environment in which these business processes are carried out will be considerably changed.

REFERENCES

Asi, M. 2002. "Institution of 'Corporate Culture' in Jordanian Air Transport." *Middle East News Online* (May 19), http://www.middleeastwire.com.

Barry, J. 1998. "EU Fears Alliances Foster High Fares." *The International Herald Tribune* (September 7): 17.

Baur, U. 1998. "Winning Strategies in a Changing Global Airline Environment." In *Handbook of Airline Marketing,* ed. G. F. Butler and M. R. Keller. New York: McGraw Hill, 533–44.

Button, K. 2002. "Airline Network Economics." In *Handbook of Airline Economics,* ed. D. Jenkins. New York: McGraw Hill, 27–34.

Campbell, A., and M. Kingsley-Jones. 2002. "Rebel Skies." *Flight International* (April 9): 30–39.

Chrystal, P., and S. LeBlanc. 2001. "Alliances: Beyond Marketing Facing Reality." In *Handbook of Airline Strategy,* ed. G. F. Butler and M. R. Keller. New York: McGraw Hill, 269–94.

Collings, R. 2002. "U.S. Air Industry 'Must Change.'" BBC News Online (August 19). Accessed October 13, 2002, from the World Wide Web: http://news.bbc.co.uk/1/low/business/2202442.stm.

Cross, R. G. 1998. "Trends in Airline Revenue Management." In *Handbook of Airline Marketing,* ed. G. F. Butler and M. R. Keller. New York: McGraw Hill, 303–18.

Daly, K., ed. 1995. "Networkers of the Future." *Air Transport Intelligence* (*ATI*) *Online* (May).

———. 2002. "Consolidation Time in Italy." *Air Transport Intelligence* (*ATI*) *Online* (August).

Doganis, R. 2001. *The Airline Business in the 21st Century.* New York: Routledge.

Gertzen, J. 2002. "Airline Industry Faces Fear of Failing: Inevitable Shakeout Expected to Lead to Higher Fares, Fewer Flights." *Business Newsbank* (September 8).

Gilden, J. 2002. "The World of Travel Keeps on Changing." *Orlando Sentinel* (May 12). Accessed September 26, 2002, from the World Wide Web: http://www.orlandosentinel.com/travel/la-sourcebook-security.story.

Greenslet, E. 2002. Executive Summary. *The Airline Monitor.*

Homan, A. C. 1999. "Charges in Airline Operating Expenses: Effects on Demand and Airline Profits." In *Handbook of Airline Finance,* ed. G. F. Butler and M. R. Keller. New York: McGraw Hill, 503–10.

ICAO Secretariat. 2002. "11 September's Negative Impact on Air Transport Is Unparalleled in History." *ICAO Journal* 57(2):6–8.

Laney, E., Jr. 2002. "The Evolution of Corporate Travel Management: Reacting to the Stresses & Strains of Airline Economics." In *Handbook of Airline Economics,* ed. D. Jenkins. New York: McGraw Hill, 477–86.

Morrison, M. 2002. "Time to Worry." *Flight International* (April 23): 3.

Oster, C. V., and J. S. Strong. 2001. "Competition and Antitrust Policy." In *Handbook of Airline Strategy,* ed. G. F. Butler and M. R. Keller. New York: McGraw Hill, 3–34.

O'Toole, K. 2002. "Keeping It Simple." *Airline Business* (September).

Ott, J. 2000. "Customer Service Drive Yields More Coach Space." *Aviation Week and Space Technology* (February 28): 50–51.

People Express Airlines: Rise and Decline. 1990. Harvard University Business School Case Study.

Spitz, W. H. 1998. "International Code Sharing." In *Handbook of Airline Marketing,* ed. G. F. Butler and M. R. Keller. New York: McGraw Hill, 489–502.

Vella, S. L. 1999. "Aircraft Asset Value Management." In *Handbook of Airline Finance*, ed. G. F. Butler and M. R. Keller. New York: McGraw Hill, 405–22.

Wang, Z. H., and M. Evans. 2002. "The Impact of Market Liberalization on the Formation of Airline Alliances." *Journal of Air Transportation* 7(2):26–30.

Zea, M. 2002. "Is Airline Risk Manageable?" *Airline Business* (April).

CHAPTER 17
THE MARINE TRANSPORTATION SYSTEM*

James J. Corbett
Department of Marine Studies
University of Delaware
Newark, Delaware

17.1   INTRODUCTION

This chapter describes the role of the marine transportation system (MTS) within a context of more familiar transportation modes, such as automobile, trucking, and rail. Within this context, the role and complexity of the MTS can be better understood in overview. The waterway network may not have rail track or asphalt defining its limits and extent, but navigable waterways are more like highways than an uncharted horizon. Specific features unique to the MTS will be summarized, including fleet characteristics for vessels engaged in cargo transportation, fishing, or other service. Like any broad system of different vehicles and destinations, the MTS cannot be summarized by only one or two images of ships. Diversity in vessel technologies, national fleet ownership and registration, and concentration of trade activity all affect the way the MTS is managed by industry and regulated by government. This diversity is driven by trends in the transportation system toward increasing global trade in an era where governments and businesses are requiring greater transparency in the MTS to foster commerce and satisfy security and environmental objectives.

17.2   OVERVIEW OF MARINE TRANSPORTATION

The marine transportation system (MTS) as a primary mode of trade transportation is global in nature and is older than any other modern form of transportation except perhaps the dirt road. Before locomotives crossed the continents, shipping enabled long-distance exploration and trade. Before roads linked metropolitan centers on a global scale, major cities arose near coastlines to take advantage of marine transportation. And before people knew the earth was not flat, sailors used ships to trade with distant cultures as civilization (both western and eastern) used the MTS to explore and expand their horizons.

Today, the MTS is key to global and multimodal transportation of energy resources, raw materials and bulk goods, manufactured products, and even people (through tourism and ferry transit). While land and air transportation is most familiar to the traveling public, the MTS is nearly invisible to the average person, who nonetheless benefits from its service. In its broadest sense, the marine transportation system is a network of specialized vessels, the ports they visit, and transportation infrastructure from factories to terminals to distribution centers to markets. In this regard, the marine transportation system depends upon the land-based modes for point-to-point movement of goods and people.

---

*Reprinted from the First Edition.

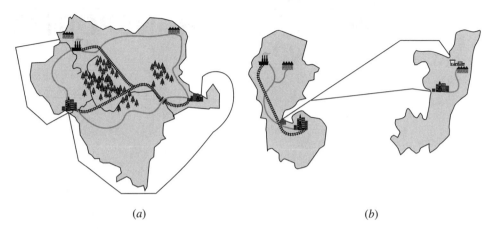

(a) (b)

FIGURE 17.1 Marine Transportation System illustration as (a) substitute for and (b) complement to land modes.

Maritime transportation is a necessary complement to and occasional substitute for other modes of freight transportation (see Figure 17.1). For many commodities and trade routes, there is no direct substitute for waterborne commerce. (Air transportation has replaced most ocean liner passenger transportation, but it carries only a small fraction of the highest-value and lightest cargoes.) On other routes, such as some coastwise or short-sea shipping or within inland river systems, marine transportation may compete with roads and rail, depending upon cost, time, and infrastructure constraints. Other important marine transportation activities include passenger transportation (ferries and cruise ships), national defense (naval vessels), fishing and resource extraction, and navigational services (vessel-assist tugs, harbor maintenance vessels, etc.).

17.2.1 Role of Marine Transportation in Global Trade and Regional Transportation

The role of marine transportation in commerce and society today is often overlooked because of the public visibility of roads and rail. However, marine transportation in the 21st century is an integral, if sometimes less publicly visible, part of the global economy. In the United States, more than 95 percent of imports and exports are carried by ships (EC 2001; MARAD 2000). On a worldwide basis, some 35,000 oceangoing vessels move cargo more than 13 billion tonne-km, annually. International trade by water modes connects international trade with all modes of transportation in nearly every nation. In the United States, the Maritime Administration (MARAD) reports that waterborne commerce within the United States moves cargo more than 2.1 billion tonne-km annually (see Figures 17.2 and 17.3). In the European Union, marine transportation moves more than 70 percent (by volume) of all cargo traded with the rest of the world; in the United States, more than 95 percent of imports and exports are carried by ships (EC 2001; MARAD 2000).

The MTS provides two basic functions in the multimodal transportation system: first, the MTS is the only mode capable of efficiently transporting large volumes of high-density cargo between continents and therefore provides a necessary complement to other modes. This includes most of our energy resources (coal and oil), raw materials (ore, aggregate sands and stone), food products (grains, fruit, vegetables, and meat), and manufactured goods (machinery, shoes, apparel, furniture). (Air transportation provides higher-speed service, but it generally moves only small volumes of low-density, high-value cargoes like people, electronics, and last-minute surge deliveries of apparel or other goods.) Second, the coastal waterway and inland river network enables the MTS to substitute in many areas for many land-based modes such as automobile, heavy-duty truck, and rail. This includes passenger ferry service as part of commuter networks in New York, Seattle, and San Francisco, inland-river barges providing connections between rail and highway networks, and coastwise ships moving cargo between major trade centers.

FIGURE 17.2   United States waterway network. (*Source:* NDC Publications and U.S. Waterway Data CD, U.S. Army Corps of Engineers, 2001, latest version available at http://www.iwr.usace.army.mil/hdc/.)

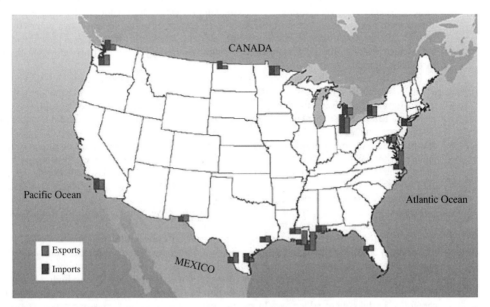

FIGURE 17.3   Top 20 international freight gateways by cargo tonnage in the United States (1998). (*Source:* U.S. Department of Transportation, Federal Highway Administration, Freight Analysis Framework, 2002.)

Marine transportation is connected with the other modes of transportation so intrinsically that some studies include the road and rail networks as part of the system to deliver marine cargoes. The U.S. Department of Transportation reported to Congress in 1999 that the MTS included "the network of railroads, roadways, and pipelines" (Slater and MTS 1999). In terms of the work performed—measured in tonne-km or ton-miles—the marine transportation system rivals the other modes. The relative share of cargo moved by water compared with truck and rail modes varies from year to year, but U.S. waterborne commerce represents between 22 and 24 percent of the total tonne-km of cargo movements in recent years. Truck and rail modes in the United States each account for about 25 to 29 percent of annual cargo tonne-km. In Europe, waterborne commerce between nations (called shortsea shipping) accounts for nearly 41 percent of the goods transport market, compared to 44 percent for trucking and only 8 percent for rail and 4 percent for inland rivers (EC 2001).

One of the reasons that marine transportation moves so many tonne-km of cargo is cost efficiency. Choice of mode is a function of cost (shipping is usually cheaper), time (shipping is often slower), and other quality-of-service factors. Only pipelines move goods at lower average cost than shipping, because of their fixed capital and ability to move fluids (liquids and gases) with very low energy requirements. Bulk shipment of liquid and dry goods such as oil and grain costs about two to three times as much per tonne-km as pipelines, averaging between 2.5 and 3.5 cents per tonne-km, according to U.S. national transportation statistics (BTS 1996, 1998, 1999). These average costs are generally less than but similar to rail, whereas marine transportation of bulk goods can cost much less than trucking. Except for high-value containerized goods that are shipped intermodally, the cost per tonne-km for trucking can be more than an order of magnitude greater than the cost for marine transportation.

Of course, intermodal shipping of market-ready items can be more time-sensitive than cost-sensitive, which makes the price (freight-rate) to ship a container by sea more similar to the price to ship a container by road or rail. These differences in the demand for cargo transportation that is low-cost, timely, and reliable are the primary determinants in modal choice. Another factor includes the volume of material required, where just-in-time (JIT) inventories may favor single container deliveries or even partial loads that trucks are best suited to move. However, when the distances are large and demand is high, water transportation is often the mode of choice. Examples include primary commodity flows where the market for goods is either dominated by a few closely located firms or other receivers with large demand for cargo (like crude oil destined for major port refineries in Texas, New York, and California) or consumer-ready commodities where the market is large and can be centrally supplied via ports on navigable waterways (such as East Coast markets for juice or fruit via the Port of Wilmington, Delaware, or for home furnishings via the Port of New York/New Jersey).

For most transportation engineers, planners, and policy-makers, it is this intermodalism that demands attention. Where a transportation network can handle cargo in similar packets, the ability to transfer cargo between modes benefits from greater capacity, greater security, and greater speed. These characteristics enabled a relatively well-defined transportation infrastructure to accommodate much greater numbers of vehicles moving freight. In the next section, containerized vessel designs that facilitated this intermodal uniformity will be discussed. Containerization of ship cargoes, in its infancy in the 1950s, revolutionized ocean trade and now accounts for over 70 percent of nonbulk ocean cargo. This shift to containerization accompanied the boom in global trade during the later part of the 20th century that is associated with increased wealth and consumption and also with increased road congestion and environmental problems.

Regionally, passenger ferry service and local freight movements by barges or small vessels have become increasingly important to the marine transportation system. As automobile transportation and the construction of bridges and highways enabled greater personal mobility, many regions shifted away from ferry transportation. Similarly, as trucking became more efficient and subsidized highways made freight movement more cost-competitive, coastwise shipping declined. Today, road congestion in many regions exceeds infrastructure capacity, and adding highway lanes is not always a good choice. Building new bridges into New York City, for example, and adding parking garages if bridges were added would require exorbitant costs and consume precious real estate. However, passenger ferry service across the waterfront business centers in New York, San Francisco, and other major cities provides a much more viable and attractive solution to congested passenger transportation. As a result, commuter ferry service has emerged in the United States as one of the fastest-growing domestic markets for marine transportation. Similar efforts are underway in Europe to increase overall transportation capacity through better integration of marine transportation with road and rail. Transportation planners and policy-makers have begun advocating shortsea shipping, inland-river barge service, and passenger ferry transportation as a necessary part of a successful transportation network.

17.2.2 Complex Marine Transportation System

The MTS is a complex element of global trade. For many, the most visible elements of the MTS are the vessels. Complexity of marine transportation may be seen most easily by examining the complex mix of ships and boats that comprise the system's vehicles. Ships, barges, ferries, towboats, and tugboats are among the most recognized and romanticized parts of the MTS, perhaps second only to the water itself. Unlike automobiles, where many thousands of each model are built and sold each year and a popular car design may remain largely the same for years, each ship design is unique to only one or a few vessels.

Because of different service requirements, ship hull designs, power plants, cargo loading and discharge equipment, and other features vary greatly. Inland-river and coastal towboats range in power from 0.5 megawatts (MW) to 4 MW (600 Hp to ~5000 Hp), and can push from 1 to more than 30 barges in a tow. Passenger ferries often have capacity to carry 150 to 300 passengers and range in power from 2 MW to 4 MW (~2600 Hp to ~5000 Hp), depending upon speed requirements; some ferries, like the Staten Island ferries in New York, have capacity for 6,000 passengers. Roll-on, roll-off ships (or RoRos, so called because vehicles and other rolling cargo are driven onto the vessel rather than loaded by crane) can carry between 200 and 600 automobiles and have installed power between 15 and 25 MW (~20,000 and ~33,000 Hp). Tankers and dry bulk carriers can carry more than 250,000 tons of cargo, with installed power often in the range of 25 to 35 MW (~33,000 to ~46,000 Hp). And while container ships are not often as large as tankers, they have much larger power plants to accommodate greater vessel speeds. Average container ships carrying between 1,750 and 2,000 standard shipping containers have installed power of 20 to 25 MW (~26,000 to ~33,000 Hp); larger containerships with capacity for more than 4,000 containers can have installed power of 35 to 45 MW (46,000 to 60,000 Hp), and the largest container ships can carry more than 6,000 standard containers with engines rated as high as 65 MW (86,000 Hp).

Generally speaking, the MTS is concerned with cargo or passenger transport vessels—ships that move cargo or passengers from one place to another in trade. These vessels account for almost 60 percent of the internationally registered fleet and are analogous to on-road vehicles because they generally navigate well-defined (if unmarked) trade routes similar to a highway network. Other vessels are primarily engaged in extraction of resources (e.g., fishing, oil or other minerals) or primarily engaged as support vessels (vessel-assist tugs, supply vessels). Fishing vessels are the largest category of nontransport vessels and account for more than one-quarter of the total fleet. Fishing vessels and other nontransport ships are more analogous to nonroad vehicles, in that they do not generally operate along the waterway network of trade routes. Rather, they sail to fishing regions and operate within that region, often at low power, to extract the ocean resources.

Cargoes, vessels, ports, and terminals also vary greatly. The primary common goals at ports and terminals are to achieve rapid vessel loading and unloading and to maximize the rate of cargo transport through the port facilities. Many ports are designed to facilitate container movements between ship, truck, and rail. Recently, transportation planners and engineers have been working to speed this port transfer with on-dock rail, larger cranes, and improved queuing schedules for trucks. Some ports are megahubs that attract much of the world trade (by value and volume), where other ports are networked ports that handle much of the niche cargoes or serve as feeders for megaports. And each of the world's ports is unique by location (proximity to market), waterway configuration, and ecosystem, which makes general characterizations difficult.

17.2.3   Cargo Networks: Shortsea Shipping and Multimodal Networks

In terms of the marine transportation system, the network of ports, waterways, and associated rail and highway provides the linkages between commodity supplies and their markets. In addition to global trade, regional transportation—both domestic and international—relies upon water modes for movement of bulk goods. In fact, regional trade of containerized and general cargoes by water is gaining attention, especially in Europe. This is termed shortsea shipping.

In general, shortsea shipping applies to certain maritime transport services that do not involve an ocean crossing. This includes shipping along coastlines or inland rivers within the North American and European continents, although politically the term applies more to Europe because so much of European trade among nations involves shipment of this type. As defined politically by the European Commission (EC 1999), shortsea shipping "means the movement of cargo and passengers by sea between ports situated in geographical Europe or between those ports and ports situated in non-European countries having a coastline on the enclosed seas bordering Europe." However, the United States has begun considering its coastwise and inland shipping as shortsea shipping as well, and there is no fundamental difference since shortsea shipping can refer to both domestic and international cargoes.

One of the primary advantages of shortsea shipping is that these networks may enable reverse cargo flows (backhauls) and higher average cargo capacities. In this sense, shortsea shipping may improve the economics of certain feeder services. This may be more complementary than competitive in most transportation logistics, because the average distance of a metric ton transported in the 1990s has been 100 kilometers for road, 270 kilometers for inland waterways, 300 kilometers for rail, and 1,385 kilometers for shortsea shipping (EC 1999). In other words, shortsea shipping may be most economic when moving cargo longer distances, complementing transport efficiencies of shorter routes by other modes.

Similar statistics exist for shipments in the United States, although less containerized cargo is carried by water modes because cross-country highways are favored over longer coastwise routes. However, the north-south coastal (and perhaps inland river) corridors could relieve more highway congestion in the Western, Eastern, and (perhaps) Central United States. Recent studies sponsored by the U.S. Department of Transportation (Federal Highway Administration) have produced a Freight Analysis Framework that evaluates how cargoes are moved in the United States. This important work may be motivated by the need to understand highway traffic and congestion, but it reveals important information about the maritime transportation system as well. Importantly, there is a strong cross-country link to East Coast states; this suggests that trucking is an important link for cargoes between the major coastlines and that Pacific Rim shipping routes carrying cargo destined for New York and other states may use California as the port of call. (These flows appear very similar whether one picks New York or California, showing symmetry in the Freight Analysis Framework.)

However, heavy freight flows along north-south highway networks illustrate the potential for waterborne commerce to complement current truck transportation routes along the major highways between Seattle and Los Angeles in the west and between Maine and Florida in the east. The potential for the marine transportation system to assist in the movement of cargoes on shortsea routes will depend upon three things: policy action to make sustainable shipping a priority and remove barrier legislation; business practices to demonstrate that coastwise and shortsea shipping can be reliable and administratively efficient; and transportation planning that emphasizes and develops the multimodal network.

17.3   VESSELS

Vessels are the most numerous and active link in the marine transportation system. Whether one considers international oceangoing trade, coastwise shipping, or inland river and harbor activity, ships are the means by which transportation between nodes is accomplished. This section discusses the international fleet, and United States domestic fleets of vessels. By way of these examples, it can be seen that vessels in domestic service are not simply smaller versions of international fleets.

TABLE 17.1   Profile of World Oceangoing Fleet[a]

Ship type	Number of ships in world fleet	Percent of world fleet
Transport ships		
Container vessels	2,662	3.0%
General cargo vessels	23,739	26.8%
Tankers	9,098	10.3%
Bulk/combined carriers	8,353	9.4%
Miscellaneous passenger	8,370	9.4%
Non-transport ships		26.4%
Fishing vessels	23,371	10.5%
Tugboats	9,348	4.2%
Other (research, supply)	3,719	
Total	88,660	100.0%

[a]Vessels greater than 100 gross registered tonnes.

17.3.1   International Fleet Profile (Fleet Characteristics by Vessel Type, Nation of Registry, Geographic Operation)

Despite the variation among vessels, general types of internationally registered ships can be categorized according to their service and design characteristics. A profile of the world oceangoing fleet is presented in Table 17.1; vessels larger than 100 gross registered tons (grt) are generally considered capable of ocean or coastal commercial transit.

Container ships are specialized cargo vessels that carry standard-sized boxes stacked together in cargo holds and on deck. These vessels represented a revolutionary innovation in shipping because they were designed to enable cargo to be packaged into a standard container suitable for intermodal transport by ship, truck, and rail (even some aircraft are equipped to carry containers). The standard unit of measure for container ships is the 20-foot-equivalentunit (TEU), which represents a container that is 20 feet long, 8 feet wide, and 8.5 feet high. Today, most shipping containers are twice as long and are referred to as 40-foot-equivalentunits (FEUs). However, container ships are uniformly referred to by the number of TEUs they can carry when fully loaded. Container ships account for only about 3 percent of the world fleet, but over the decades since their introduction the rate of growth for these vessels has been about twice as great as for other vessel types. This is because of their versatility in carrying relatively high-value cargoes, from refrigerated fruits and meats to furniture and shoes, and because the container is so well suited to the other modes of freight transportation. Container ships are typically faster than other vessel types, with average cruise speeds greater than 18 nautical miles per hour (knots); recently built container ships often have speeds greater than 24 knots. These higher speeds enable container ships to provide scheduled reliable service, whereas much of the world fleet sails between ports without published schedules, following the cargoes and markets. Also, because of economies of scale and trade growth, recent container ships have been very large. The fleet average container capacity is around 2,000 TEUs per vessel, but recently built container ships have exceeded 6,000 TEU capacity and new container ship designs exceed 10,000 TEU. While these vessels are getting larger and faster, there may be an upper bound to these economies of scale—perhaps not in vessel design, but because very large container ships may be larger than most world ports can accept without significant investment in channel deepening or terminal equipment.

General cargo ships include a large variety of vessel types, but all are capable of carrying cargoes in various package configurations. These vessels were the predecessors to modern container ships and still constitute the majority of cargo vessels in the world fleet. However, the average age for these vessels is more than 10 years old, their average speeds are slower (~12 knots), and their cargo capacities are typically smaller than those of container ships. Noncontainerized nonbulk cargoes, referred to break-bulk, typically require more labor to load and discharge because the configurations for cargo packaging are not standard. Unlike the container ships, which rely on terminal cranes that can offload dozens of TEUs or FEUs per hour, general cargo ships may have cargo-handling equipment on deck or rely on smaller terminal equipment. Slower cargo handling tends to limit the economic value of larger capacities because transport vessels are most productive when time in port is minimized. Their size, speed, and capacity make general cargo ships better suited to regional and coastwise transport.

For the purposes of this chapter, certain specialized vessels are included in the general cargo ship category although they serve more specialized cargo markets. This includes vehicle carriers, refrigerated cargo ships, and others. It is worth noting that some of these vessels are more like container ships than general cargo ships in terms of design standardization, speed, and specialized port handling. However, the markets for their cargoes is smaller and they account for only about 1 percent of all general cargo ships (less than 0.4 percent of the world fleet).

Liquid and dry bulk carriers are similar in terms of their service, sizes, and speeds. Liquid bulk carriers, or tankers, are familiar to many because of their service in moving energy resources, particularly crude oil. Cargoes carried by these vessels include mostly crude oil and petroleum products, but tankers also are used to carry bulk juices, wines, and beer. (It is much less expensive to move juice, wine, or beer in bulk and bottle it near the consumer market than to move these products in bottles, cans, or crates.) Dry bulk vessels carry cargoes such as grain and other agricultural products, ores and minerals, and coal. Advantages of moving liquid and dry cargoes in bulk are that loading and discharge can be faster and less labor-intensive and that unrefined cargoes can be transported from ports near their point of harvest or extraction to other ports that are nearer to refineries, factories, or markets. These cargoes can then be processed, packaged, or used to manufacture products for shipment to consumers. On average, tankers are smaller than bulk carriers, although the larger tankers have more than twice the capacity of the largest bulk carriers; this is primarily because many smaller tankers serve regional or feeder markets. Average speeds for bulk carriers are around 13–15 knots, with liquid tankers typically a bit slower. The average fleet ages for both liquid and dry bulk ships are roughly the same, and older than container or general cargo vessels.

Passenger vessels in the international fleet represent both ferry transport and cruise excursion vessels. By way of analogy, ferry transport vessels are more like scheduled bus service on highways, while cruise excursion vessels are more like tour buses. Both move passengers along the waterway network, but cruise ships carry the same passengers both to and from destinations during a given voyage. Ferries tend to serve markets where passengers in one location want to go to another location for work, tourism, or other purposes. Passenger vessels include many specialized designs, but two general characteristics dominate. Some passenger vessels are conventional monohull designs that operate at slower speeds (~13 knots on average), while others are twin-hull (catamaran) designs that operate at higher speeds (greater than 22 knots on average). Certain routes have enough demand for fast service to justify the significantly higher power and fuel requirements of faster vessels; twin hulls, or catamarans, can achieve these speeds efficiently with adequate stability and passenger comfort in typical sea conditions.

17.3.2 U.S. Domestic Cargo Fleet Profile

The U.S. domestic cargo fleet can be seen simply as a subset of vessels in the world fleet, but this would be misleading. As in other major nations with active commercial coastlines and inland river networks, many ships in the U.S. cargo fleet are smaller than internationally registered vessels. Moreover, nonpropelled vessels such as barges are counted in U.S. inventories that would not be reported in global inventories. Tables 17.2 and 17.3 present the summary of cargo vessels in the U.S. fleet, by those of 1,000 gross tons and greater and by those less than 1,000 gross tons, respectively. Passenger vessels and military craft are not included.

Of the 462 self-propelled vessels in Table 17.2, those 127 in foreign trade are most comparable to the world fleet profile for cargo vessels, although some of the 164 vessels in domestic trade may also be oceangoing. The government ships are typically operated in dedicated civilian service to

TABLE 17.2 U.S.-Flag Cargo Carrying Fleet (Vessels of 1,000 Gross Tons and Over) by Area of Operation[a]

	Total		Liquid carriers		Dry bulk carriers		Containerships		Other freighters[b]	
	No.	Tons	No.	Tons	No.	Tons	No.	Tons	No.	Tons
Grand total	**3,869**	**30,495**	**2,196**	**15,714**	**759**	**5,889**	**123**	**3,108**	**791**	**5,784**
Foreign trade	**268**	**5,319**	**53**	**946**	**116**	**1,115**	**61**	**2,510**	**38**	**748**
Oceanborne	263	5,214	51	927	113	1,029	61	2,510	38	748
Great Lakes	5	105	2	19	3	86	0	0	0	0
Domestic trade	**3,430**	**21,921**	**2,116**	**13,887**	**643**	**4,774**	**57**	**512**	**614**	**2,748**
Coastal	1,344	13,299	567	8,770	355	2,146	57	512	365	1,871
Inland waterway	2,000	6,454	1,542	5,075	227	576	0	0	231	803
Great Lakes	86	2,168	7	42	61	2,052	0	0	18	74
Government	**171**	**3,255**	**27**	**881**	**0**	**0**	**5**	**86**	**139**	**2,288**
Total self-propelled	**462**	**14,914**	**114**	**6,230**	**69**	**2,600**	**90**	**2,898**	**189**	**3,186**
Foreign trade	**127**	**4,588**	**17**	**771**	**12**	**579**	**61**	**2,510**	**37**	**728**
Oceanborne	127	4,588	17	771	12	579	61	2,510	37	728
Great Lakes	0	0	0	0	0	0	0	0	0	0
Domestic trade	**164**	**7,071**	**70**	**4,578**	**57**	**2,021**	**24**	**302**	**13**	**170**
Coastal	105	5,063	68	4,559	2	71	24	302	11	131
Inland waterway	0	0	0	0	0	0	0	0	0	0
Great Lakes	59	2,008	2	19	55	1,950	0	0	2	39
Government	**171**	**3,255**	**27**	**881**	**0**	**0**	**5**	**86**	**139**	**2,288**
Total non-self-propelled[c]	**3,407**	**15,581**	**2,082**	**9,484**	**690**	**3,289**	**33**	**210**	**602**	**2,598**
Foreign trade	**141**	**731**	**36**	**175**	**104**	**536**	**0**	**0**	**1**	**20**
Oceanborne	136	626	34	156	101	450	0	0	1	20
Great Lakes	5	105	2	19	3	86	0	0	0	0
Domestic trade	**3,266**	**14,850**	**2,046**	**9,309**	**586**	**2,753**	**33**	**210**	**601**	**2,578**
Coastal	1,239	8,236	499	4,211	353	2,075	33	210	354	1,740
Inland waterway	2,000	6,454	1,542	5,075	227	576	0	0	231	803
Great Lakes	27	160	5	23	6	102	0	0	16	35

Source: U.S. Maritime Administration (MARAD), current July 2002. Adapted by MARAD from Corps of Engineers, Lloyd's Maritime Information Service, U.S. Coast Guard and Customs Service Data. Self-propelled vessels ≥1,000 gross tons; excludes one domestic coastal passenger vessels of 3,988 dwt and eleven other passenger vessels of 96,474 dwt.

[a]Carrying capacity expressed in thousands of metric tons.

[b]Includes general cargo, Ro-Ro, multipurpose, LASH vessels, and deck barges; excludes offshore supply vessels.

[c]Integrated tug barges of 1,000 grt and greater are contained in non-self-propelled categories as follows: foreign trade—1 dry bulk (36,686 tons), 1 other freighter (20,000 tons); domestic coastal—11 liquid (449,370 tons), 3 dry bulk (111,000 tons); U.S./Translakes—Great Lakes 2 liquid (18,955 tons) and 3 dry bulk (85,514 tons).

meet military sealift demands and do not compete directly for general commercial cargoes. The self-propelled vessels in Table 17.3 are typically not oceangoing vessels, as shown by their service area description as coastal, inland waterway, or Great Lakes.

It is interesting to note that the U.S. inventory of ships includes so many non-self-propelled vessels, mostly barges on inland rivers. This is partly an artifact of the inventory methodology, but the information can be important to transportation planners and engineers, particularly in terms of designing waterway locks and dams or when considering port terminal designs.

Not shown in these tables are some 1,330 commercial passenger vessels (mostly ferries carrying less than 150 passengers each), more than 5,400 tugs and towboats that assist ships in U.S. ports or push barge tows along rivers and coastlines, and some 1,600 workboats, including crewboats, supply, and utility vessels. And another significant element of the domestic MTS is recreational boating. These small vessels are rarely registered internationally and do not typically transit oceans, operating most of the time in domestic waters on a weekend or seasonal basis.

TABLE 17.3 U.S.-Flag Cargo Carrying Fleet (Vessels less than 1,000 Gross Tons) by Area of Operation[a]

	Total		Liquid carriers		Dry bulk carriers		Containerships		Other freighters	
	No.	Tons	No.	Tons	No.	Tons	No.	Tons	No.	Tons
Grand Total	**32,229**	**46,381**	**2,214**	**3,965**	**23,010**	**36,438**	**4**	**2**	**7,001**	**5,976**
Foreign trade	**109**	**50**	**3**	**1**	**106**	**49**	**0**	**0**	**0**	**0**
Oceanborne	109	50	3	1	106	49	0	0	0	0
Great Lakes	0	0	0	0	0	0	0	0	0	0
Domestic trade	**32,120**	**46,331**	**2,211**	**3,964**	**22,904**	**36,389**	**4**	**2**	**7,001**	**5,976**
Coastal	3,930	3,562	241	982	573	741	1	1	3,115	1,838
Inland waterway	27,890	42,394	1,961	2,975	22,211	35,478	3	1	3,715	3,940
Great Lakes	300	375	9	7	120	170	0	0	171	198
Total self-propelled	**384**	**948**	**77**	**797**	**4**	**2**	**0**	**0**	**303**	**149**
Domestic trade	**384**	**948**	**77**	**797**	**4**	**2**	**0**	**0**	**303**	**149**
Coastal	256	902	71	795	0	0	0	0	185	107
Inland waterway	109	27	2	0	0	0	0	0	107	27
Great Lakes	19	19	4	2	4	2	0	0	11	15
Total non-self-propelled	**31,845**	**45,433**	**2,137**	**3,168**	**23,006**	**36,436**	**4**	**2**	**6,698**	**5,827**
Foreign trade	**109**	**50**	**3**	**1**	**106**	**49**	**0**	**0**	**0**	**0**
Oceanborne	109	50	3	1	106	49	0	0	0	0
Great Lakes	0	0	0	0	0	0	0	0	0	0
Domestic trade	**31,736**	**45,383**	**2,134**	**3,167**	**22,900**	**36,387**	**4**	**2**	**6,698**	**5,827**
Coastal	3,674	2,660	170	187	573	741	1	1	2,930	1,731
Inland Waterway	27,781	42,367	1,959	2,975	22,211	35,478	3	1	3,608	3,913
Great Lakes	281	356	5	5	116	168	0	0	160	183

Source: U.S. Maritime Administration (MARAD), current July 2002. Adapted by MARAD from Corps of Engineers, Lloyd's Maritime Information Service, U.S. Coast Guard and Customs Service Data.
[a]Carrying capacity expressed in thousands of metric tons.

17.4 PORTS

Perhaps the most visible element of the marine transportation system, ports are the MTS nodes where cargo and passengers transition from vessel to on-road vehicle or rail. These ports range from megahubs of world trade to fishing or recreational centers of activity. Trends indicate that port activity overall will increase dramatically as cargo volumes double or triple over the next decades and as more of the world's population uses coastal waters for commerce, fishing, and recreation. These trends create conflicts over waterfront real estate and environmental resources and provide strong incentive for new technology and modernization. To ensure that marine transportation continues to meet our needs, planners must recognize, understand, and project the intermodal nature of the MTS and its inherent interfaces and complexities.

17.4.1 World Port Ranking

Table 17.4 presents a summary of the top 40 world ports, ranked according to the volumes of cargoes they handle annually, of total cargoes and containerized cargoes handled annually, in tons and TEUs, respectively. As shown, less than half (17) of the top 40 world ports appear in the top 40 for both tonnage and container TEUs. This is important for transportation engineering because it means that the transportation infrastructure is very different when considering the needs for large ports handling bulk cargoes versus containerized cargoes. Moreover, the value of cargo (per ton)

TABLE 17.4   World Port Ranking[a]

	Total cargo volume 2000 Metric tons (000s), except where noted					Container traffic 2000 Twenty foot equivalent units (TEUs)		
Rank	Port	Country	Unit[b]	Tons	Rank	Port	Country	TEUs
1	Singapore	Singapore	FT	325,591	1	Hong Kong	China	18,098,000
2	Rotterdam	Netherlands	MT	319,969	2	Singapore	Singapore	17,090,000
3	South Louisiana	U.S.A.	MT	197,680	3	Pusan	South Korea	7,540,000
4	Shanghai	China	MT	186,287	4	Kaohsiung	Taiwan	7,426,000
5	Hong Kong	China	MT	174,642	5	Rotterdam	Netherlands	6,274,000
6	Houston	U.S.A.	MT	173,770	6	Shanghai	China	5,613,000
7	Chiba	Japan	FT	169,043	7	Los Angeles	U.S.A.	4,879,000
8	Nagoya	Japan	FT	153,370	8	Long Beach	U.S.A.	4,601,000
9	Ulsan	South Korea	RT	151,067	9	Hamburg	Germany	4,248,000
10	Kwangyang	South Korea	RT	139,476	10	Antwerp	Belgium	4,082,000
11	Antwerp	Belgium	MT	130,531	11	Tanjung Priok	Indonesia	3,369,000
12	NewYork/New Jersey	U.S.A.	MT	125,885	12	Port Kelang	Malaysia	3,207,000
13	Inchon	South Korea	RT	120,398	13	Dubai	U.A.E.	3,059,000
14	Pusan	South Korea	RT	117,229	14	New York/New Jersey	U.S.A.	3,050,000
15	Yokohama	Japan	FT	116,994	15	Tokyo	Japan	2,899,000
16	Kaohsiung	Taiwan	RT	115,287	16	Felixstowe	U.K.	2,793,000
17	Guangzhou	China	MT	101,521	17	Bremer Ports	Germany	2,712,000
18	Quinhuangdao	China	MT	97,430	18	Gioia Tauro	Italy	2,653,000
19	Ningbo	China	MT	96,601	19	San Juan	U.S.A.	2,334,000
20	Marseilles	France	MT	94,097	20	Yokohama	Japan	2,317,000
21	Osaka	Japan	FT	92,948	21	Manila	Philippines	2,289,000
22	Richards Bay	South Africa	HT	91,519	22	Kobe	Japan	2,266,000
23	Kitakyushu	Japan	FT	87,346	23	Yantian	China	2,140,000
24	Qingdao	China	MT	86,360	24	Qingdao	China	2,120,000
25	Hamburg	Germany	MT	85,863	25	Laem Chabang	Thailand	2,105,000
26	Dalian	China	MT	85,053	26	Algeciras	Spain	2,009,000
27	Kobe	Japan	FT	84,640	27	Keelung	Taiwan	1,955,000
28	Tokyo	Japan	FT	84,257	28	Nagoya	Japan	1,905,000
29	New Orleans	U.S.A.	MT	82,400	29	Oakland	U.S.A.	1,777,000
30	Dampier	Australia	MT	81,446	30	Colombo	Sri Lanka	1,733,000
31	Vancouver	Canada	MT	76,646	31	Tianjin	China	1,708,000
32	Corpus Christi	U.S.A.	MT	75,461	32	Charleston	U.S.A.	1,629,000
33	Beaumont	U.S.A.	MT	75,032	33	Genoa	Italy	1,501,000
34	Newcastle	Australia	MT	73,871	34	Seattle	U.S.A.	1,488,000
35	Tubarao	Brazil	MT	73,182	35	Le Havre	France	1,486,000
36	Tianjin	China	MT	72,980	36	Tacoma	U.S.A.	1,376,000
37	Port Hedland	Australia	MT	72,914	37	Barcelona	Spain	1,364,000
38	Hay Point	Australia	MT	69,379	38	Cristobal	Panama	1,354,000
39	Le Havre	France	MT	67,492	39	Hampton Roads	U.S.A.	1,347,000
40	Port Kelang	Malaysia	FT	65,227	40	Melbourne	Australia	1,328,000

[a]Primary source: American Association of Port Authorities, www.aapa-ports.org; AAPA Sources: *Shipping Statistics Yearbook 2001*, 359; U.S. Army Corps of Engineers, *Waterborne Commerce of the United States CY 2000; AAPA Advisory*, May 21, 2001; various port authority Internet sites.
[b]Abbreviations: MT = metric ton; HT = harbor ton; FT = freight ton; RT = revenue ton.
Note: The cargo rankings based on tonnage should be interpreted with caution since these measures are not directly comparable and cannot be converted to a single, standardized unit.

will be very different and may influence the economic importance of different ports more than the volume of cargo transported. This observation will be repeated at the national level for most large trading nations.

17.4.2 U.S. Port Ranking

There are over 300 ports in the United States, but most of the cargo and ship trafficis handled by fewer than 100 ports. Table 17.5 summarizes statistics for the top 30 ports in the United States, cargo and containerized cargo, respectively. As with the world port profile, less than half (10) of the top 30 U.S. ports appear in the top 30 for both tonnage and container TEUs. Illustrations of principal U.S. ports are shown in Figures 17.4 and 17.5.

TABLE 17.5 United States Port Ranking[a]

	Total cargo volume 2000 Short tons (000s)				Container traffic 2002 Twenty foot equivalent units (TEUs)	
Rank	Port name	Total	Domestic	Foreign	Rank	Port name	Total
1	South Louisiana, Port of	217,757	119,141	98,615	1	Los Angeles	4,060,000
2	Houston	191,419	62,617	128,802	2	Long Beach	3,184,000
3	New York	138,670	72,273	66,397	3	New York	2,627,000
4	New Orleans	90,768	38,316	52,452	4	Charleston	1,197,000
5	Corpus Christi	83,125	23,989	59,136	5	Savannah	1,014,000
6	Beaumont	82,653	16,043	66,609	6	Norfolk	982,000
7	Huntington	76,868	76,868	0	7	Oakland	979,000
8	Long Beach	70,150	17,400	52,750	8	Houston	851,000
9	Baton Rouge	65,631	42,505	23,126	9	Seattle	850,000
10	Texas City	61,586	20,330	41,256	10	Tacoma	769,000
11	Port of Plaquemine	59,910	38,864	21,046	11	Miami	752,000
12	Lake Charles	55,518	20,476	35,042	12	Port Everglades	370,000
13	Mobile Harbor	54,157	24,232	29,925	13	Baltimore	302,000
14	Pittsburgh	53,923	53,923	0	14	New Orleans	216,000
15	Los Angeles	48,192	6,065	42,127	15	Portland	185,000
16	Valdez	48,081	46,409	1,672	16	San Juan PR Harbor	159,000
17	Tampa Bay	46,460	31,662	14,798	17	Palm Beach	142,000
18	Philadelphia	43,855	14,066	29,788	18	Wilmington, DE	133,000
19	Norfolk	42,377	10,505	31,872	19	Gulfport	132,000
20	Duluth-Superior, MN & WI	41,678	28,165	13,512	20	Philadelphia	115,000
21	Baltimore	40,832	14,535	26,297	21	Jacksonville	114,000
22	Portland, OR	34,334	16,357	17,977	22	Boston	80,000
23	St Louis	33,338	33,338	0	23	Wilmington, NC	71,000
24	Freeport	30,985	5,599	25,386			
24	Chester	59,000					
25	Portland, ME	29,330	2,337	26,993	25	Newport News	57,000
26	Pascagoula	28,710	10,454	18,256	26	Freeport	54,000
27	Paulsboro	26,874	9,186	17,688	27	Port Bienville	41,000
28	Seattle	24,159	8,718	15,441	28	Richmond	36,000
29	Chicago Harbor	23,929	20,063	3,866	29	Honolulu	32,000
30	Marcus Hook	22,584	8,872	13,712	30	Ponce	29,000

[a]Primary source: U.S. Army Corps of Engineers Navigational Data Center (for tonnage data) and American Association of Port Authorities, www.aapa-ports.org (for containerized data).

Note: The cargo rankings by tonnage are for a different year than the rankings by TEU and should be considered representative of relative port ranking, which may vary from one year to another.

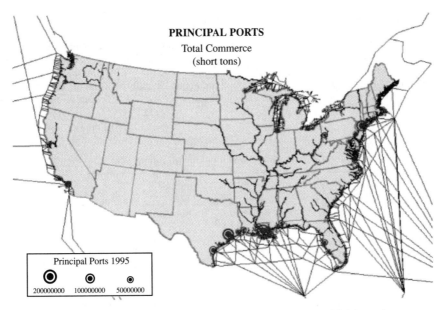

FIGURE 17.4 Principal ports in the United States waterway network, identified by total cargo tons. (*Source:* NDC Publications and U.S. Waterway Data CD, U.S. Army Corps of Engineers, 2001, latest version available at http:// www.iwr.usace.army.mil/ndc/.)

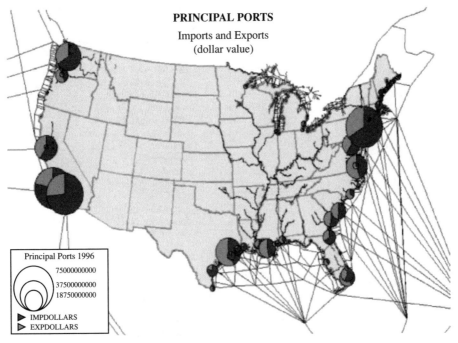

FIGURE 17.5 Principal ports in the United States waterway network, identified by cargo value. (*Source:* NDC Publications and U.S. Waterway Data CD, U.S. Army Corps of Engineers, 2001, latest version available at http:// www.iwr.usace.army.mil/ndc/.)

17.4.3   Major Fishing Ports

Major fishing ports look quite different than the major cargo ports. Unless supported by factory or mother ships or fitted with freezing equipment, fishing boats need to get to port and send their cargo on to the market quickly. Table 17.6 presents the top 40 fishing ports, both by volume and value of fish landed. Note that these ports are generally different from the top cargo ports. Most of these ports are smaller and more clearly dedicated to fishing activities and related industry (including tourism).

TABLE 17.6   Top 40 Fishing Ports, by Commercial Landings in Dollars and Pounds 1999

Rank	Port	Pounds (millions)	Port	Dollars (millions)
1	Dutch Harbor-Unalaska, AK	699.8	New Bedford, MA	146.3
2	Cameron, LA	414.5	Dutch Harbor-Unalaska, AK	124.9
3	Empire-Venice, LA	396.2	Kodiak, AK	94.7
4	Reedville, VA	366.8	Brownsville-Port Isabel, TX	88.6
5	Intracoastal City, LA	321.7	Dulac-Chauvin, LA	68.1
6	Kodiak, AK	289.6	Empire-Venice, LA	61.6
7	Los Angeles, CA	254.7	Honolulu, HI	56.0
8	Pascagoula-Moss Point, MS	199.9	Hampton Roads Area, VA	52.8
9	Port Hueneme-Oxnard-Ventura, CA	162.2	Key West, FL	50.6
10	Astoria, OR	130.1	Port Arthur, TX	49.3
11	Newport, OR	102.3	Bayou La Batre, AL	48.9
12	Sitka, AK	95.5	Cameron, LA	47.6
13	New Bedford, MA	89.0	Portland, ME	45.4
14	Beaufort-Morehead City, NC	68.4	Golden Meadow-Leeville, LA	44.9
15	Naknek-King Salmon, AK	63.1	Sitka, AK	44.6
16	Portland, ME	62.8	Palacios, TX	41.8
17	Cape May-Wildwood, NJ	59.9	Point Judith, RI	41.4
18	Point Judith, RI	59.3	Galveston, TX	40.8
19	Ketchikan, AK	57.3	Los Angeles, CA	38.1
20	Moss Landing, CA	50.5	Naknek-King Salmon, AK	37.3
21	Atlantic City, NJ	50.4	Seward, AK	35.8
22	Dulac-Chauvin, LA	48.2	Gulfport-Biloxi, MS	35.5
23	Gloucester, MA	40.1	Homer, AK	30.9
24	Point Pleasant, NJ	38.2	Delcambre, LA	30.8
25	Westport, WA	37.5	Gloucester, MA	30.0
26	Wanchese-Stumpy Point, NC	33.3	Reedville, VA	29.5
27	Petersburg, AK	32.3	Cape May-Wildwood, NJ	28.6
28	Brownsville-Port Isabel, TX	29.2	Astoria, OR	28.0
29	Seward, AK	28.6	Grand Isle, LA	26.4
30	Rockland, ME	28.3	Atlantic City, NJ	26.1
31	Honolulu, HI	27.0	Newport, OR	24.6
32	Golden Meadow-Leeville, LA	26.9	Intracoastal City, LA	24.5
33	Bayou La Batre, AL	23.0	Wanchese-Stumpy Point, NC	24.0
34	Hampton Roads Area, VA	22.7	Freeport, TX	22.8
35	Morgan City-Berwick, LA	20.2	Tampa Bay-St. Petersburg, FL	20.4
36	Ilwaco-Chinook, WA	19.8	Port Hueneme-Oxnard-Ventura, CA	20.2
37	Coos Bay-Charleston, OR	19.2	Delacroix-Yscloskey, LA	20.1
38	Grand Isle, LA	18.2	Ketchikan, AK	20.0
39	Homer, AK	18.1	Petersburg, AK	19.4
40	Bellingham, WA	18.0	Stonington, ME	18.0

Source: Fisheries Statistics & Economics Division of the National Marine Fisheries Service; www.st.nmfs.gov/st1/commercial/

17.4.4 Port Infrastructure Development

While these data are important indicators of vessel traffic and other modal traffic to major ports, transportation planners must also consider the number of vehicles (ships, trucks, trains) in addition to the cargoes transported. In this regard, port calls are an important direct measure of marine transportation infrastructure demand. Table 17.7 presents a summary of the top 25 world ports by the number of ship visits, or port calls; this table also provides estimates of the observed capacity of these ports in terms of the total deadweight tonnage of the ships that called on each port. This may qualitatively indicate relative demand for intermodal infrastructure to carry cargoes into or out of the port facilities.

TABLE 17.7 Top 25 World Port Calls and Observed Capacity by Vessel Type 2000[a]

Port	Tanker[b] Calls	Tanker[b] Capacity	Dry bulk Calls	Dry bulk Capacity	Containership Calls	Containership Capacity	Other general cargo[c] Calls	Other general cargo[c] Capacity	Total Calls	Total Capacity
Singapore	5,351	436,844	4,581	242,709	11,286	354,686	3,232	63,656	24,450	1,097,895
Hong Kong	637	26,774	1,040	34,262	12,462	412,264	1,360	24,240	15,499	497,540
Kaohsiung	773	48,032	1,387	69,756	5,808	199,284	692	12,504	8,660	329,576
Pusan	150	4,555	1,181	51,191	5,217	164,795	1,009	18,681	7,557	239,222
Rotterdam	2,112	121,957	900	73,730	2,528	110,192	1,579	33,243	7,119	339,122
Antwerp	990	34,071	921	41,747	2,111	76,312	2,183	44,795	6,205	196,924
Yokohama	505	36,129	530	17,725	3,298	103,399	1,663	30,212	5,996	187,465
Keelung	256	10,350	491	13,545	4,344	94,522	542	8,704	5,633	127,121
Port Kelang	425	10,480	522	18,797	3,950	109,883	668	12,035	5,565	151,195
Los Angeles & Long Beach	911	66,045	783	37,568	2,955	124,281	677	15,057	5,326	242,951
Nagoya	265	28,669	814	51,991	2,699	91,331	1,374	24,596	5,152	196,587
Houston	2,988	134,809	748	28,342	614	19,799	779	24,881	5,129	207,831
New Orleans	1,371	81,956	2,676	119,270	388	10,853	655	21,957	5,090	234,036
Kobe	301	9,012	381	14,049	3,325	116,447	660	11,149	4,667	150,657
New York & New Jersey	1,271	65,965	301	10,099	2,172	87,463	861	23,104	4,605	186,631
Taichung	668	25,561	1,228	54,158	1,998	33,604	513	9,276	4,407	122,599
Laem Chabang	207	15,027	495	20,057	2,600	49,820	442	7,860	3,744	92,764
San Francisco & Oakland	787	50,653	626	22,619	1,936	82,958	226	6,841	3,575	163,071
Santos	637	17,342	727	31,262	1,547	42,749	637	14,336	3,548	105,688
Hamburg	440	14,349	565	32,753	1,745	74,067	764	16,210	3,514	137,379
Tokyo	1	260	222	7,692	2,987	102,198	238	4,547	3,448	114,697
Durban	442	23,604	809	27,354	1,043	29,088	1,115	21,028	3,409	101,074
Shanghai	180	6,208	782	44,157	1,763	47,449	582	10,718	3,307	108,532
Le Havre	699	53,308	104	6,681	2,013	82,329	433	9,768	3,249	152,086
Osaka	95	5,244	478	17,638	2,030	57,659	475	9,424	3,078	89,966
Top 25 world ports	22,462	1,327,204	23,292	1,089,151	82,819	2,677,433	23,359	478,822	151,932	5,572,609
Total world port	138,296	8,751,934	126,246	5,917,050	180,766	5,406,073	115,127	2,344,277	560,435	22,419,335
Top25 as percent of world total	16.2%	15.2%	18.4%	18.4%	45.8%	49.5%	20.3%	20.4%	27.1%	24.9%

[a]Primary source: U.S. Maritime Administration, office of statistics, 3/8/2002; MARAD source: Lloyd's Maritime Information Services, Vessel Movements. *Observed capacity* is estimated by multiplying the number of vessel calls by vessel deadweight tonnage capacity; it does not represent total capacity for a given port. Excludes calls by vessels under 10,000 dwt tons.

[b]Includes petroleum, chemical, and gas carriers.

[c]Includes Roll-on/Roll-off (Ro/Ro), Ro/Ro container, vehicle carriers, general cargo, partial containership, refrigerated, barge carrier, livestock carrier, and combination carriers.

From this table, it is clear that frequent calls to port are typically associated with general cargo and containerized cargoes, more so that with bulk cargoes. Containerized cargo vessels calling on the 25 ports shown in Table 17.4 account for some 46 percent of all container ship port calls worldwide. Moreover, about half of the ports receiving the most port calls are also listed as ports with the greatest cargo volumes for both total and containerized cargoes. These ports may qualify as true megaports, which are primary hubs for all types of marine transportation. Another interesting note is that there are more ports in Table 17.4 that also appear as major container ports. Lastly, 4 of the busiest 25 ports in terms of port calls are not listed among the top world ports by total cargo volume or by container volume. This suggests that on a global basis, at least, transportation planning and engineering needs to consider the total system when developing new infrastructure for maritime ports.

17.5   TRENDS, OPPORTUNITIES, AND CHALLENGES

The marine transportation system is distinct in many ways from land-based modes, but it has much in common with truck, rail, and even air transportation systems. In particular, the MTS must respond to increasing trade over the longer term, meet emerging and stricter environmental regulation and safety requirements, and adopt innovative technologies. Responding to these trends presents opportunities and challenges to all modes of transportation, but for the marine transportation system the environmental challenges are relatively new and unfamiliar (compared to trucking, for example, which has been more heavily regulated than shipping, especially in the United States and Europe). And unlike other modes, where vehicle life is measured in years and/or where periodic overhauls involve reengining with newer designs, ships last for decades without major design changes to hull or power train.

17.5.1   Increasing Trade Volumes

Global trade has increased dramatically over the past century, and projections continue to forecast a doubling of trade volume for major nations, and as much as a tripling in cargo volumes to major world ports. Carriage of these increasing trade volumes will by necessity include water transportation. Three possible ways exist for the marine transportation system to accommodate these forecasted cargoes. First, the existing fleet might simply carry more through better logistics and more fully loaded ships to maximize efficiency. Second, additional ships may be added to carry the additional cargoes resulting in an increase in ship traffic overall. Third, the fleet might transition to larger vessels without increasing the overall number of ships in service in order to provide greater capacity within the same network of routes. There are advantages and challenges to the MTS under each alternative. In practice, each of these alternatives may be applied in some combination.

On certain routes and for certain cargoes, it may be possible to carry more cargo with the current fleet. However, there is an upper bound to this alternative. Most shipping routes are practically unidirectional, meaning that most of the cargo flows in one direction. For example, oil is primarily imported into the United States, and grain is primarily exported. Bulk carriers typically transit empty (or under ballast) in one direction and return with a load of cargo; this results in a practical upper bound of 50 percent capacity overall on routes for bulk carriers (e.g., averaging 100 percent one way and 0 percent the other, or more realistically, averaging 80 to 90 percent one way and 10 to 20 percent on the return). For container and general cargoes, some routes are able to secure a backhaul; however, there is limited ability for ships to realize this idealized goal. Where a strong backhaul market can be established, the backhaul freight rates usually are much lower and may not directly cover the costs of the voyage; here, as for bulk carriers, the freight rate for inbound cargoes must cover most of the cost of the round trip voyage. For inland-river and shortsea shipping, backhauls are more practical because vessel capacities are smaller. Even so, backhauls on inland rivers are relatively few and are commodity specific. In fact, the industry appears to behave as though there is an upper bound. Most ships carry loads that average 50 to 65 percent capacity or less. When cargo capacities exceed 70 percent, it can be an indication that too few ships are available for the route (Abrams 1997); under these conditions, ships are added to the route.

Adding ships to increase route capacity is easier than it might be for onroad vehicles in the current transportation system. This is primarily because ocean routes are unconstrained highways. For most route distances, few modifications will be needed to enable more vessel traffic. However, in and near port regions this may not hold. The number of berths at terminals and cargo handling rates provide at least two important limits on how many vessels a port can accommodate in a given period. As discussed above, no consistent metric to evaluate port capacity exists, but clearly the capacity of the marine transportation system will be limited if ships have to wait for a berth at a port.

17.5.2 Environmental and Safety Constraints

Environmental impacts from shipping have been the focus of increasing attention over the past decades, primarily focused on waterborne discharges and spills of oil, chemical, and sewage pollution. A number of international treaties and national laws have been adopted, along with industry best practices, to prevent these pollution releases through accident or substandard operation. However, the shipping industry is being seen as an important hybrid between land-based transportation and large stationary power systems. On the one hand, these are non-point-source vehicles operating on a transportation network of waterways; on the other hand, these vehicles operate ship systems that compare to small power-generating plants or factories. As the industry itself and policy-makers at state, federal, and international levels recognize these facts, environmental performance is being measured and regulated. Recent efforts to mitigate environmental impacts from shipping, including invasive species in ballast water, toxic hull coatings, and air emissions, are relatively new for the industry and will take decades to address.

Environmental regulation of the shipping industry typically lags behind regulation of similar industrial processes (engines, etc.) on land. At the international level, the International Maritime Organization has effectively developed a set of treaties addressing much of these ship pollution issues. Collectively called the International Convention for the Prevention of Pollution from Ships (MARPOL 73/78), these regulations apply global standards to environmental protection practices aboard ship. However, national and state jurisdictions can apply stricter regulations to certain vessel operations and to the ports that receive these ships. For example, the U.S. Environmental Protection Agency recently regulated shipping under authority mandated by the Clean Air Act. Another example includes efforts by port states in Sweden and Norway and by individual ports in California to apply market-based or voluntary regulations to reduce air pollution from ships entering their ports.

In terms of ship safety, increasing regulations are also seen as a result of international security concerns, not to mention efforts to standardize best practices for crew training and safe vessel navigation. Specific efforts include automated ship identification systems, more transparent cargo tracking processes, and additional shipboard and corporate responsibilities for ship security to reduce the risk of theft, terrorism, or smuggling.

These increasing regulations are imposing changes in both ship operation and ship equipment. For many vessels this will simply mean retrofitting the minimum necessary equipment to comply with regulations. However, as seen in shoreside stationary systems, there may be a point at which retrofits may become more costly than replacement. At this point, emerging regulations may speed up the normal process of fleet technology and modernization.

17.5.3 Technology and Modernization

As growth in trade and stricter regulations continue over the coming decades, vessel technology will adapt through the two processes of expansion and modernization. Technology and modernization affect both ports terminals and vessels. Internationally, the trend toward larger vessels—as large as 10,000 TEUs—is advancing the concepts of megaports and net-worked ports. Additionally, the trend continues in certain ship types for increased speed. More fundamental than these leading trends in certain segments of the MTS are crosscutting efforts. This modernization affects different industries in the marine transportation system very differently, depending on the rate of technology capitalization and expected working life.

In perspective, efforts to innovate in marine systems are not at all new. The efficient use of energy in marine transportation has been important since before Greek and Roman times. Shipping itself has been an important human activity throughout history, particularly where prosperity depended primarily on commerce with colonies and interregional trade with other nations (Mokyr 1990). From ancient times when ships sailed mostly in sight of land using wind-driven sails or hand-driven oars to today's oceangoing fleet of steel ships powered by the world's largest internal combustion engines, the history of shipping includes many examples of technical innovations designed to increase ship performance while conserving three factors: energy, capital, and labor. Whether the primary energy for ship propulsion was wind and sail or engines and petroleum, all major innovations in marine transportation throughout history involve balancing these factors.

However, fleets typically have built up during brief periods (typically during and following major wars) and then aged without much replacement. Some nations have instituted planned replacement policies for over-age ships, but most rely on industry to modernize when vessel economics indicate. Other nations have implemented unsuccessful policies that protect domestic shipbuilding and/or other policies that remove incentives to replace older ships. For these nations' fleets, shipboard technologies tend to be older than the state-of-the-art.

In this regard, the U.S. fleet has modernized more slowly than the world fleet. Over the past decades, the U.S. fleet has not built new ships at the same rate as the world fleet. Considering all vessels greater than 100 gross tons, the U.S. fleet construction rate has been less than 0.5 percent annually (MARAD 1996), while the average rate of construction for the world fleet of ships has been about 2 percent (LMIS 2002; Lloyd's Register). Under current market and policy conditions (e.g., U.S. versus foreign labor rates for merchant vessel construction and operation), there are limited economic incentives to modernize the U.S. fleet at the pace of the world fleet. This results in an average fleet age for the U.S. (23 years old) that is equal to or greater than the average age that most ships are scrapped (UNCTAD 1995).

Many ship operators, especially those using ships purchased on the second-hand market or in nations like the United States, currently face the alternatives of life extension for these vessels (either with or without technology and efficiency improvements) or new construction. This choice will continue to conflict with efforts to improve fleets for greater trade, stricter environmental demands, and safer operation. Ports and transportation systems that connect the MTS to the general transportation network will need to plan for the emerging MTS at the same time as they accommodate less-preferred technologies that will be replaced at some point in the future.

REFERENCES

Abrams, A. 1997. "Ship Cargo Capacity Tightens in 2nd Quarter." *Journal of Commerce.*

Bureau of Transportation Statistics (BTS). 1996. *1993 Commodity Flow Survey*. U.S. Department of Transportation, BTS, Washington, DC.

———. 1998. *National Transportation Statistics*. U.S. Department of Transportation, BTS, Washington, DC.

———. 1999. *1997 Commodity Flow Survey*. U.S. Department of Transportation, BTS, Washington, DC.

Corbett, J. J., and P. S. Fischbeck. 1997. "Emissions from Ships." *Science* 278:823–24.

———. 2000. "Emissions from Waterborne Commerce in United States Continental and Inland Water." *Environmental Science and Technology* 34(15):3254–60.

———. 2001. "Commercial Marine Emissions and Life-Cycle Analysis of Retrofit Controls in a Changing Science and Policy Environment." In *Marine Environmental Engineering Technology Symposium (MEETS) 2001*. Arlington, VA: ASNE/SNAME.

Corbett, J. J., P. S. Fischbeck, and S. N. Pandis. 1999. "Global Nitrogen and Sulfur Emissions Inventories for Oceangoing Ships." *Journal of Geophysical Research* 104:3457–70.

CHAPTER 18
GLOBAL LOGISTICS AND THE MARITIME TRANSPORT SYSTEM

John Mangan
School of Marine Science and Technology
Newcastle University
United Kingdom

18.1   THE ECONOMIC IMPORTANCE OF MARITIME TRANSPORT

Maritime transport plays a critical role in the global economy moving large volumes of freight between (and within) countries. It has existed as a mode of transport for thousands of years and has developed in line with the evolution of international trade inherent in shaping the modern world. Today, many of the world's economies are becoming increasingly interrelated as a result of increasing trade and the growing trend toward globalization of production. Over the past half-century, most countries have seen an increase in exports as a share of GDP, with the vast bulk of these exports transported by sea. A number of trends affecting the maritime sector have been central to efficiency and productivity gains. These include better, faster, and larger vessels, and improvements in cargo handling at ports.

Relative to the other transport modes, maritime transport has the following cost and operating characteristics [Mangan, Lalwani, and Butcher 2008, p. 134]:

- Fixed cost is on the medium side including vessels, handling equipment, and terminals.

- Variable cost is low because of the economies of scale that can be enjoyed from carrying large volumes of freight, this is the main advantage of the water mode, together with its capability to uplift large volumes of freight. Like air, it cannot offer direct consignor to consignee connectivity, and vessels are sometimes limited in terms of what ports they can use. It is also quite a slow mode.

Maritime freight is typically classified as: liquid bulk (the most significant subcategory here is oil), dry bulk (such as coal and some agricultural products), unitized freight [which comprises both lift-on/lift-off containers (i.e., Lo-Lo) and roll-on/roll-off units (i.e., Ro-Ro)], and other general freight. The key attribute of maritime transport when compared to the other transport modes is its ability to move large volumes of freight over long distances relatively cheaply. The value to volume ratio of maritime freight is usually lower than is the case with freight carried by other modes, although as speed and handling efficiency increases, the maritime mode (in particular using containerization) is becoming more attractive to shippers. Another attribute of the mode is that it is perceived to have a lower impact upon the environment than is the case with other transport modes. The United Nations Conference on Trade and Development (UNCTAD 2009, p. 7) estimates that total maritime trade during 2008 was 8.17 billion tons comprised of oil (34 percent); iron ore, grain, coal, bauxite/alumina and phosphate (26 percent); and other dry cargo (40 percent). Major loading areas were located in developing regions (60 percent), followed by developed economies (34 percent) and countries with transition economies (6 percent). Asia continued to dominate the picture, with a share of 40 percent of total goods loaded, followed in descending order by the Americas, Europe, Africa, and Oceania (*ibid*, p. 7). Total freight movements vary according to region, commodity and freight origin/destination. In the European Union (EU), for example, it is estimated that the ports sector handles more than 90 percent of the Union's trade with third countries and approximately 30 percent of intra-EU trade, as well as over 200 million passengers every year.

The maritime transport chain comprises a number of sets of actors:

- The consignors of freight, generally referred to as "shippers"
- Ports and terminal operators within ports
- Maritime transport providers

Each of these sets of actors is discussed in turn in this chapter. While these are the principal actors in the maritime transport chain, it is important to point out that there are many other sets of actors also, however there isn't room in this chapter to review all of these. In shipping there is for example a large and complex support industry in finance, insurance and related areas. The ownership structure in shipping is also generally complex with a variety of different actors (owners, brokers, operators, charterers, and so forth).

Before however any of the preceding topics are discussed, it is important to first place maritime transport within the wider context in which it plays a role, namely the end to end supply chain for moving freight between consignors and consignees.

18.2   *MARITIME TRANSPORT, SUPPLY CHAIN MANAGEMENT, AND LOGISTICS*

It is now generally accepted that supply chains, and not individual firms or products, are the basis of much marketplace competition (Christopher 1992). Christopher defines the supply chain as the network of organizations that are involved, through upstream and downstream linkages, in the different processes and activities that produce value in the form of products and services in the hands of the ultimate consumer. There is a growing appreciation of the impact that best practice logistics and supply chain management (SCM) can have on firm success. In the intervening years since Christopher first put forward his proposition in 1992, many successful organizations have learned how to use their supply chains both to differentiate their offerings from those of the competition and to compete with other actors in the market. Much has been written on and learned from the experiences of these organizations, many of which are now household names [see, for example, Ferdows Lewis, and Machuca (2004) on the Spanish clothing retailer Zara, and Lee (2004) on the computer manufacturer Dell]. Today, many organizations are aware of the fundamental importance of using effective and efficient logistics and SCM practices in order to drive down costs and concomitantly add value to their marketplace propositions.

At this juncture, it is important to first clarify one important question: Are logistics and SCM one and the same? Both terms are often used interchangeably. Much work has, in fact, been done on this issue of distinguishing and understanding both terms; this chapter adopts the "unionist perspective," that is that logistics is a part of SCM, and that SCM is a wider, intercompany, boundary-spanning concept than is the case with logistics.[*]

Supply chains, and their concomitant flows and activities, are thus at the heart of much of global commerce today. To quote Gattorna (2006, p. 1), "supply chains are the business." Indeed, many companies are coming to realize that not only are logistics and SCM a key part of any business

---

[*]For a discussion on the different perspectives on logistics see, for example, Mangan, Lalwani, and Butcher (2008, p. 11–12); recent empirical studies support the "unionist perspective"—see, for example, Larson, Poist, and Halldorsson (2007).

strategy, they are realizing that in some instances a good logistics and SCM strategy can actually drive overall firm strategy.[*] Detailing the various factors that have led to the emergence of logistics and SCM as key areas of activity today is a topic outside of the remit of this chapter, but suffice to say such disparate factors include, inter alia, a reducing transport intensity of freight, falling product prices, deregulation of transport, an increased emphasis on the role and costs of inventory, and productivity improvements in areas such as asset tracking. Undoubtedly, however, one of the most influential of all drivers has been the emergence and growth of maritime containerization, which has allowed cheap mass transportation over long distances (we will consider this topic in more detail in Sec. 18.5, which deals with maritime transport providers). And intertwined with all of these developments has been the emergence and increased influence of the phenomenon of globalization. Indeed somewhat of a "which came first: the chicken or the egg?" scenario is evident: Has globalization facilitated the evolution of logistics and SCM, or were logistics and SCM key to the evolution of globalization? Levinson (2010, p. 1) noted that "the container didn't cause globalisation, but globalisation could hardly have occurred without the decline in transport costs that containerization made possible." Dicken (2007, p. 410), in his seminal book *Global Shift*, notes that:

> [T]he *circulation processes* that *connect* together all the different components of the production network are absolutely fundamental. They have become central to the abilities of transnational companies to operate their complex, geographically dispersed operations. The logistics industries themselves are huge, worth around $4 billion per year.

In today's interconnected and globalized world, highly developed logistics systems thus facilitate global supply chain activity. And many such supply chains are geographically dispersed, complex and involve flows of large volumes of freight, often being moved on maritime transport. Consolidation in many industries (witness, for example, the widespread merger and acquisition activity within the automobile and pharmaceutical industries) leads to an even greater focus by the merged companies on integrated SCM. Transport services (links in supply chains) and transport infrastructure (nodes in supply chains) are key elements in these efficient logistics systems. Maritime transport (comprising ports as nodes and shipping services as links) is the dominant mode for international freight movements and is thus crucial to international trade and a vital component of many supply chains. When difficulties arise in the maritime chain, the results can be quite dramatic, as was the case for example in the second half of 2004 when congestion at the port of Long Beach in the United States, which had its origins in labor disputes, led to ships remaining idle and anchored at sea for up to four days with a considerable knock-on effect on consignors and consignees (Marshall 2005). Later in this chapter, we will look in particular at the various roles which ports can play within wider supply chains and consider the emergence of a concept known as "port-centric logistics."

18.3   CONSIGNORS AND MARITIME TRANSPORT

The preceding section highlighted the important role of maritime transport in the context of the wider supply chain. Other transport modes, too, are also used to provide transport services within many supply chains. The relative use of maritime transport thus depends upon which modes and services are used, and who makes the routing decisions. Reflecting wider logistics and supply chain management issues, there has been recognition within the literature concerning the evolution of the demand for maritime transport. Panayides (2006), for example, notes that "the demand for maritime transport nowadays cannot be solely considered to be a derived demand emanating from the need for products, but rather as an *integrated demand* emanating from the need to minimise costs, improve reliability, add value, and a series of other dimensions and characteristics pertaining to the transportation of goods from the point of production to the point of consumption." This *integrated demand* thus makes use of the concept of the generalized costs of transport (see, for example, Mangan, Lalwani, and

---

[*] See, for example, Fabbe-Costes and Colin (2001).

Butcher 2008, p. 70), which typically seeks to capture all of the costs associated with a product's movement (transport costs, insurance and risk-related costs, opportunity cost of the capital tied up in the freight, etc.). The routing decision thus is broadened from a selection between a number of competing services to a wider consideration of a multiplicity of factors. The interaction and relative importance of these various factors has been reported in many academic papers, while the hierarchy first put forward by Davies and Gunton (1983) and illustrated in Figure 18.1, remains apposite in that it highlights the relative importance of various typical choice criteria.

The *locus* of the mode choice and routing decisions is also an important consideration and we cannot assume that this is always the consignor, in fact many consignors will delegate that decision to a logistics service provider who will choose the appropriate transport services. Many consignors choose a single dedicated logistics service provider who can provide a seamless end-to-end transport solution for the consignor; the end result of this is that the consignor will often not know which routes and services their freight utilizes. This is not to suggest that consignors are not or should not be concerned with the specific details of mode and route choice; research by Saldanha et al. (2009), for example, has shown how appropriate ocean carrier selection can help reduce total end-to-end logistics costs. We will also see in Section 18.6 how the carbon footprint of the transport elements of a supply chain can be reduced by use of appropriate modes and routings.

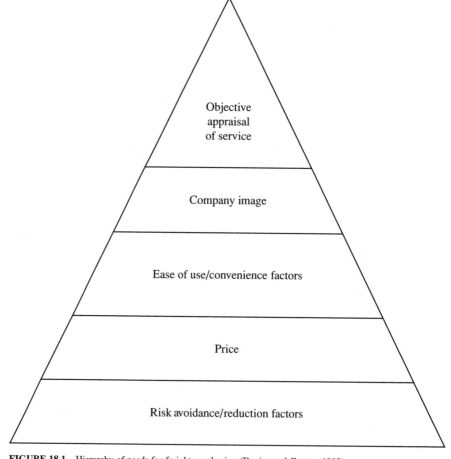

FIGURE 18.1   Hierarchy of needs for freight purchasing (Davies and Gunton 1983).

18.4 *PORTS AND TERMINAL OPERATORS*

According to the World Bank (2001) there are more than 2000 ports around the world, from single berth locations handling a few hundred tonnes a year to some of the world's largest ports who handle multiples of this. Different ports claim the title of world's busiest port (depending upon what is being measured which typically is: arriving shipping tonnage, number of containers handled, cargo volumes handled). Singapore, for example, is ranked the number-one port in the world in terms of cargo volume handled—515 million tonnes in 2008 (American Association of Port Authorities 2010). In terms of container volumes, most of the top-ten container ports are in Asia, with Singapore again in the number-one position handling almost 30 million TEUs each year (*ibid*). Rotterdam, Europe's largest container port, is ranked number nine, handling 10.8 million TEUs each year. In the United States, the leading container ports are Los Angeles, ranked number sixteen, followed by Long Beach at number seventeen (*ibid*).

It is important to first define exactly what is meant by the term "port." According to Stopford (2009) a port is "a geographical area where ships are brought alongside land to load and discharge cargo—usually a sheltered deep water area such as a bay or river mouth." Often ports comprise multiple terminals, a terminal being "a section of the port consisting of one or more berths devoted to a particular type of cargo handling" (Stopford 2009, p. 81). Ports handle various different categories of freight, some ports handle all categories of freight, while others focus on particular categories; different types of handling equipment at ports are usually required for these different categories of freight.

Ports can play varying roles in the supply chain, a role which can vary from that of simple transshipment hub to important logistics node, and which in turn is heavily dependent upon the supply chain strategies of those who use these ports. There is much discussion, at least in the academic literature anyway, concerning the role and function of the port. Robinson (2002, p. 252) notes that past paradigms (of ports) are "of interest, but not of problem-solving relevance. Indeed to persist with an inadequate paradigm is to find the wrong answers." Robinson suggests that "the role of ports and the way in which ports position themselves … must be defined within a paradigm of ports as elements in value-driven chain systems, not simply as places with particular, if complex, functions." Another contribution, aptly titled 'Rethinking the port,' by Olivier and Slack (2006), noted that the emergence of port operating TNCs (a trend discussed further below) requires a fundamental epistemological shift in reconceptualizing the port. Heaver et al. (2000), in their research into the European ports and shipping sectors, noted that "the role of the port and port authorities has to be redefined to guarantee that it remains a fully fledged player in this fast evolving market."

A number of trends impacting the ports sector will now be considered.

18.4.1 Port Development and Ownership

Traditionally most ports acted as simple transshipment hubs where freight passed between ships and landside transport. This was typically a very labor-intensive activity, but technological developments (such as increased use of containers and more sophisticated cranes) together with reform of dock labor schemes led to significantly decreased employment at many ports. This was further exacerbated by the fact that for reasons of geography some ports are located in peripheral locations (to allow short sea crossings) where alternative employment is scarce. With changes in ship type and the nature of freight being transported new facilities were developed either at alternative locations within ports, or in some instances on greenfield sites, with the result that many original port areas fell into disrepair. Some port areas however, leveraging their waterfront location, have benefited from significant developments in areas such as residential property and, with the growth of the leisure sector, marina development.

Recent years have seen significant reform concerning the ownership and management of the ports sector with, according to Brooks and Cullinane (2007), the relationship between ports and government changing profoundly over the past quarter of a century. They note that many governments have moved to extract themselves from the business of port operations and have concentrated

TABLE 18.1   Four Models of Port Administration

Models	Port functions		
	Land ownership	Regulation	Cargo handling
1. Pure public sector	Public	Public	Public
2. Public/private	Public	Public	Private
3. Private/public	Private	Public	Private
4. Pure private sector	Private	Private	Private

Source: Baird 1995.

on monitoring and oversight responsibilities. Baird (1995) put forward four models of port administration (Table 18.1) for the various aspects of a port's activities.

Most of the world's top container ports are public/private although examples of the other models also exist (Cullinane and Song 2002). Shanghai is an example of a pure public port, Hong Kong is an example of a private/public port, while a number of the U.K.'s largest ports are pure private ports. Privatization (of some or all of the activities outlined in Baird's model) of ports has been a popular strategy for a number of ports, although Cullinane and Song (2002) caution that privatization is only a partial cure for what ails the world's ports and that, if implemented in isolation, it simply cannot deliver the much-needed panacea for all of the industry's woes. Other drivers for improvement thus include areas such as reform of dock labor, a favorable regulatory climate and contestable markets for ports and their services.

18.4.2   Global Port Operators (GPOs)

Globalization of shipping and trade is resulting in increasing pressure on ports to reduce container terminal costs and improve operational efficiency. Mega shippers of freight are generally seeking single supplier contracts with carriers that can provide efficient and cost effective services (we noted already in Section 18.3 the trend among many consignors to source single logistics service providers). In turn, the carriers are seeking cost reductions and efficiency gains at the ports they utilise, with single sourcing across ports in terms of port terminal operations becoming more common. In response to this and to the need for integration in international supply chains, a number of global port operators (GPOs) have emerged who manage an increasing number of the world's ports. This has been helped by port deregulation and changes in ownership in many countries.

Leading GPOs include Hutchison (Hong Kong), PSA Corporation (Singapore), Dubai Ports World (Dubai), AP Moller–Maersk (Denmark, although the group's terminal operations are headquartered in The Netherlands) and COSCO (Beijing). These companies have expanded their operations extensively beyond their home ports. For example, following a bidding war with PSA, Dubai Ports World acquired in 2006 the port operations of P&O, a company headquartered in London and listed on the London Stock Exchange. In fact, the takeover caused initial controversy in the United States, as it also involved P&O's port assets in the United States moving into new ownership.

18.4.3   Interport Competition

Notteboom and Winkelmans (2001) noted that interport competition has intensified, even among more distant ports, and point out that for example the competition between European ports situated in different port ranges has increased considerably in recent years. Such interport competition challenges the traditional assumption where each country has to have its own port(s). Delays in new container port development in Britain for example have led some commentators to note that Britain could "find itself in danger of becoming little more than an appendage to the major North European continental ports" (Asteris and Collins 2007) (the implication being that Britain's international traffic would transit to and from deep-sea routes via ports, such as Rotterdam and Antwerp). Of course it should be added that ports can cooperate as well as compete! A good example is the merger of the European ports of Copenhagen and Malmo. Fleming and Baird (1999) noted that there have been many recent remarks and written comments that the real future competition will not be between ports and individual transport carriers per se, but between a handful of "total logistics chains." Indeed Goss (1990), drawing upon Verhoeff (1981), discussed five different forms of competition which ports are subject to, namely—competition between whole ranges of ports or coastlines; competition between ports in different countries; competition between individual ports in the same country; competition between the operators or providers of facilities within the same port; and competition between different modes of transport.

18.4.4 Port Clusters and Port-Centric Logistics

In some regions, port clusters (see, for example, De Langen 2002) have evolved, which Haezendonck (2001, p. 136) defines as "the set of interdependent firms engaged in port related activities, located within the same port region and possibly with similar strategies leading to competitive advantage and characterised by a joint competitive position vis-à-vis the environment external to the cluster." Attempts have been made to define the conceptual and other boundaries of a port cluster, for example the Busan port cluster in Korea (Roh, Lalwani, and Naim 2007). Some ports are actively encouraging companies to locate distribution centers and other value-adding logistics activities at ports rather than in their traditional locations, which tend to be in geographically central, inland locations. They argue that current patterns of (inland) distribution centre location ignore the fact that most of the freight that passes through these distribution centers first passes through a port. Therefore they argue that it is logical (and often times easier in terms of land cost, lack of congestion, etc.) to site such distribution centers at ports. They also recognize that higher profit margins can be made on some noncore port activities and this is driving them to engage in activities beyond simply providing berths for ships and other core port services. This approach has come to be labeled "port-centric logistics" (see for example Falkner 2006; Wall 2007; Analytiqa 2007); it can be defined as the provision of distribution and other value-adding logistics services at a port (Mangan, Lalwani, and Fynes 2008). Looking beyond the port's physical boundaries, Notteboom and Rodrigue (2005) argue for a new phase in port development, viz regionalization, where the reach of the port extends beyond the port perimeter and involves, inter alia, market strategies and policies linking it more closely to inland freight distribution centers.

18.4.5 Ports and Economic Growth

Increasingly, ports are recognized as key components in determining the overall competitiveness of national economies. Cullinane and Song (2002) point out that ports constitute a critical link in the supply chain and that their level of efficiency and performance influences, to a large extent, a country's competitiveness. Similarly, Sanchez et al. (2003) in the context of a number of Latin American countries, showed that port efficiency is a relevant determinant of a country's competitiveness and interestingly they add that, unlike most other relevant variables, port efficiency can be influenced by public policies. Bryan et al. (2006) provide a comprehensive review of the literature generally on ports and regional economic development and, taking the case of ABP's port activities in South Wales, they quantify the economic significance of that set of ports activities on the region. These issues then have generated the drive today to improve port efficiency, lower cargo handling costs and integrate port services with other components of the global distribution network.

Port-Centric Logistics in the United Kingdom

One advantage of establishing a distribution centre at a port is that it cuts down on the number of empty (return) containers on roads by 'stripping' (i.e. emptying) imported containers at the port. This also allows faster repositioning of containers to another port where they are required. In the port of Felixstowe in the United Kingdom for example, the BAP Group operates half a million square feet of on-port warehousing and is a major logistics provider for the retailer Sainsbury's. They cite a variety of examples where port-centric logistics has been effectively employed (*Ship2Shore* 2007):

- Sainsbury's previously took imported containers to an inland RDC, but now the containers are stripped at the port, eliminating a return leg of empty containers. They estimate that this saves 700,000 road miles for every 5000 TEUs handled.

- Many imported containers are not completely full because of weight restrictions on UK roads. However if the containers are to be emptied at the port, and not travel on the roads, then the containers can be filled to capacity, which they estimate can in some instances be up to 40 percent more.

Other ports are also developing port-centric logistics offerings. At the PD Ports owned port of Teesport in the North of England the retailer ASDA-Walmart has constructed a 500,000 square feet import centre. PD Ports have also established a company known as PD Logistics to provide various logistics and supply chain services for a diverse range of clients including companies such as Corus Steel and Lidl. PD Logistics has its own transport fleet which includes nearly 300 trailers, over 400 staff and 280,000 square meters of warehousing across the United Kingdom. The company has also diversified into a number of other areas such as waste and recycling.

18.5 MARITIME TRANSPORT PROVIDERS

One of the most common forms of measurement of the world fleet is in terms of vessel tonnage (all data here is from UNCTAD 2009, p. 37). In 2008 oil tankers and dry bulk carriers comprised the largest share of the world fleet at 71.2 percent of total tonnage. General cargo ships comprised 9.1 percent and container ships comprised 13.6 percent. Other types of vessels include offshore supply vessels, ferries and passenger ships (just 0.5 percent of total world tonnage), and liquefied gas carriers (one of the vessel categories with very strong growth rates).

A feature of the ship building sector (and related sectors such as shipping finance) is that it progresses through regular boom-bust cycles (see, for example, Stopford 2009). For example, as recently as 2007 demand for "new builds" (i.e., new ships) in the sector was far outstripping supply, driven in part by the increased demand for export capacity out of Asia. This in turn was leading to capacity constraints, especially in container traffic (*Port Strategy* 2007). In contrast, by the time of writing (early-2011), there was anecdotal evidence of new builds going straight to scrap on the basis that this provided the highest potential (but obviously not sufficient) financial return. A similar scenario is evident when one looks at the Baltic Dry Index (BDI), an index that tracks worldwide international shipping prices of various dry bulk cargoes (it is not restricted to the Baltic region). On May 20, 2008, the index reached its record high level since its introduction in 1985, reaching 11,793 points. Half a year later, on December 5, 2008, the index had dropped by 94 percent, to 663 points, the lowest since 1986.

In the container trades, vessel size has increased dramatically in recent years (a trend termed by some as "gigantism"). Some commentators (Levinson 2010) however caution that while vessel size may not have reached a potential design peak, it has reached an operational peak. Maersk Line's new flagship vessel the Emma Maersk, which won the title of ship of the year at the 2007 Lloyd's

List awards in London, is said to be one of the world's largest container vessel with an operating capacity of some 11,000 TEUs (20-ft equivalent units) (BBC 2006). The Emma Maersk is one of five such sister vessels and their actual maximum potential capacity is said to be up to 14,500 TEUs. Only certain ports can handle such ultra large vessels however, and consequently many container vessels in routine operation are much smaller than this. With fewer ports able to handle larger vessels, there is growing traffic concentration at certain ports. Increasingly, many midsized ports are playing a feeder role to the very large ports as hub and spoke networks have emerged. In these networks the larger vessels ply between the major transshipment hubs, with the result that the prosperity of the smaller ports is increasingly dependent on the route strategies of the major shipping lines. Indeed alterations in route network design by the major lines can also have significant impacts on the fortunes of the larger ports. Such lines tend to favor ports which: are in a good geographical position relative to other ports of call for best vessel transit/steaming time and port rotation; are close to marketplaces; have the necessary facilities, services and infrastructure; and are sufficiently flexible to allow service to be maintained if ships are out of schedule (Livey 2005). It is not uncommon for vessels plying long international routes to get delayed and a challenge for ports is to ensure that they manage the utilization of their available berthing capacity effectively, including to accommodate delayed vessels if necessary.

Global shipping companies have also increased significantly in scale. The largest container ship operator in the world is AP Moller Maersk which operates over 600 owned and chartered container vessels (the company, which earned revenues of just under £19 billion in 2005 and employs over 110,000 people, also operates bulk vessels and container terminals, and has various other interests also). Other global container companies of considerable scale (but not as big as Maersk) include Mediterranean Shipping Company, Evergreen, American President Lines, COSCO and NYK Line.

18.6 CONCLUSION

This chapter has hopefully highlighted both the role played by the various actors in the maritime transport chain, and the contribution which this mode of transport makes to global commerce. Ports, for example, have evolved from being simple transhipment points and many now provide a range of services and activities to support the wider supply chain. The maritime transport industry has many unique characteristics, involves disparate actors, is truly global in nature, and from an economic perspective is highly cyclical. In recent years, increased regulation is also a growing feature of the sector, particularly with regard to environmental and (post-9/11) security issues (see, for example, Psaraftis 2005). In July 2004, a new international security code known as ISPS (International Ship and Port Facility Security Code) came into effect and was designed to detect and deter threats to international security. The maritime mode is generally regarded as being more environmentally friendly than other modes of transport. In some areas there has for example been an apparent *renaissance* in short sea shipping, especially in Europe, in response to congestion and other issues affecting contiguous land transport networks. At present it is estimated (International Maritime Organisation 2009) that shipping accounts for 3.3 percent of global anthropogenic CO_2 emissions; however, it is predicted that shipping would account for between 12 to 18 percent of global CO_2 emissions by 2050 if no action is taken to reduce emissions from shipping (allowing for no greater than 2°C global temperature rise by 2100). Various efforts are underway to both reduce and mitigate the environmental impact of shipping (see, for example, www.lowcarbonshipping.co.uk). These are not confined only to CO_2 (other emissions such as sulphur and nitrous oxides also need to be considered), similarly as well as looking at vessel (hull and propulsion) design, the role of shipping within wider supply chains is also being considered. After all, if the CO_2 emissions from shipping are significantly reduced, this may not be of much benefit if the contiguous landside legs for the maritime freight have significant environmental footprints. This then takes us back to the starting point for this chapter, which was to see maritime transport within the context of the wider supply chain; shipping after all, like other modes of transport, is a derived demand.

REFERENCES

American Association of Port Authorities (2010), *World Port Rankings—2008*. Retrieved 6 April, 2010 from http://www.aapa-ports.org/Industry/content.cfm?ItemNumber=900&navItemNumber=551.

Analytiqa (2007) *Portcentric Logistics—The Supply Chain of the Future*, April 2007. Retrieved December 10, 2007 from www.analytiqa.com.

Asteris, M., and Collins, A. (2007). "Developing Britain's Port Infrastructure: Markets, Policy and Location." *Environment and Planning A*, 39:2271–2286.

Baird, A. (1995). "Privatisation of Trust Ports in the UK: Review and Analysis of the First Sales." *Transport Policy*, 2 (2):135–143.

BBC (2006). "Giant Christmas Goods Ship Docks." *BBC News online*, November 5, 2006.

Brooks, M., and Cullinane, K. (2007). *Devolution, Port Governance and Port Performance*, Oxford: Elsevier, p. 3.

Bryan, J., Munday, M., Pickernell, D., and Roberts, A. (2006). "Assessing the Economic Significance of Port Activity: Evidence from ABP Operations In Industrial South Wales." *Maritime Policy and Management*, 33(4):371–386.

Christopher, M. (1992). *Logistics and Supply Chain Management* (1st ed.). London: Financial Times/Pitman.

Cullinane, K., and Song, D. (2002). "Port Privatisation Policy and Practice." *Transport Reviews*, 22(1):55–75.

Davies, G. J., and Gunton, C. E. (1983). "The Buying of Freight Services: The Implications for Marketers." *The Quarterly Review of Marketing*, Spring:1–10.

De Langen, P. (2002). "Clustering and Performance: The Case of Maritime Clustering in the Netherlands." *Maritime Policy and Management*, 29:209–221.

Dicken, P. (2007). *Global Shift: Mapping the Changing Contours of the World Economy* (5th ed.). London: SAGE Publications.

Fabbe-Costes, N., and Colin, J. (2001). "Formulating Logistics Strategy." In D. Waters (Ed.), *Global Logistics: New Directions in Supply Chain Management*. London: Kogan pp. 33–54.

Falkner, J. (2006). "A Better Place to Do Logistics?" *Logistics Manager*, May.

Ferdows, K., Lewis, M., and Machuca, J. (2004). "Rapid-Fire Fulfillment." *Harvard Business Review*, 82(11): 104–110.

Fleming, D., and Baird, A. (1999). "Some Reflections on Port Competition in the United States and Western Europe." *Maritime Policy and Management*, 26(4):383–394.

Gattorna, J. (2006). *Living Supply Chains*. FT Prentice Hall, London, England.

Goss, R. (1990). "Economic Policies and Seaports: 4. Strategies for Port Authorities. *Maritime Policy and Management*, 17(4):273–287.

Haezendonck, E. (2001). *Essays on Strategy Analysis for Seaports*, Belgium: Garant, Leuven.

Heaver, T., Meersman, H., Moglia, F., and van de Voorde, E. (2000). "Do Mergers and Alliances Influence European Shipping and Port Competition." *Maritime Policy and Management*, 27(4):363–373.

International Maritime Organisation (2009). *Prevention of air pollution from ships* (Second IMO GHG Study 2009), Reference: MEPC 59/INF.10, 9 April 2009: Geneva.

Larson, P., Poist, R., and Halldorsson, A. (2007), "Perspectives on Logistics vs SCM: A Survey of SCM Professionals." *Journal of Business Logistics*, 28(1):1–25.

Lee, H. (2004). "The Triple-A Supply Chain." *Harvard Business Review*, 82(10):102–112.

Levinson, M. (2010). "Why Container Ships May Downsize." Accessed 6 April 2010, available at: http://news .bbc.co.uk/2/hi/business/7707901.stm.

Livey, P. (2005). "Inland Transport by Container: Why Think Water?" *Sea and Water Conference*, Hull.

Mangan, J., Lalwani, C., and Butcher, T. (2008). *Global Logistics and Supply Chain Management*. Chichester: Wiley.

Mangan, J., Lalwani, C., and Fynes, B. (2008), "Port-Centric Logistics." *International Journal of Logistics Management*, 19(1):29–41.

Marshall, A. (2005). "US-Asia Trade Boom Blighted by Logistics Bottlenecks." *CILT World*, Issue 11.

Notteboom, T., and Winkelmans, W. (2001). "Structural Changes in Logistics: How Will Port Authorities Face the Challenge?" *Maritime Policy and Management*, 28(1):71–89.

Notteboom, T., and Rodrigue, J. (2005). "Port Regionalisation: Towards a New Phase in Port Development." *Maritime Policy and Management*, 32(3):297–313.

Olivier, D., and Slack, B. (2006). "Rethinking the Port." *Environment and Planning A*, 38:1409–1427.

Panayides, P. (2006). "Maritime Logistics and Global Supply Chains: Towards a Research Agenda." *Maritime Economics and Logistics*, 8:3–18.

Port Strategy (2007). "Viewpoint: Capacity Crunch." July/August: 3.

Psaraftis, H. (2005), "EU Ports Policy: Where Do We Go from Here?" *Maritime Economics and Logistics,* 7(1): 73–82.

Robinson, R. (2002). "Ports as Elements in Value-Driven Chain Systems: The New Paradigm." *Maritime Policy and Management*, 29(3):252.

Roh, H., Lalwani, C. and Naim, M. (2007), "Modelling a Port Logistics Process Using the Structured Analysis and Design Technique." *International Journal of Logistics: Research and Applications*, 10(3):283–302.

Saldanha, J., Tyworth, J., Swan, P., and Russell, D. (2009). "Cutting Logistics Costs with Ocean Carrier Selection." *Journal of Business Logistics*, 30(2):175–194.

Sanchez, R., Hoffmann, J., Micco, A., Pizzolitto, G., Sgut, M., and Wilmsweier, G. (2003). "Port Efficiency and International Trade: Port Efficiency as a Determinant of Maritime Transport Costs." *Maritime Economics and Logistics*, 5:199–218.

Ship2Shore (2007) (customer magazine of Hutchinson Ports UK), *Port-Centric Logistics*, Issue 1, June 2007.

Stopford, M. (2009). *Maritime Economics*. London: Routledge.

UNCTAD (2009). *Review of Maritime Transport 2009,* Geneva.

Verhoeff, J. (1981) "Seaport Competition: Some Fundamental and Political Aspects," *Maritime Policy and Management*, 8(1):49–60.

Wall, G. (2007). "Heading for the Coast: Is Port-Centric Logistics the Way Forward, *Focus*, October, 42–44.

World Bank (2001). *The Evolution of Ports in a Competitive World*. Washington: October 2001.

CHAPTER 19
FREIGHT TRANSPORTATION PLANNING*

Kathleen Hancock
Civil and Environmental Engineering Department
University of Massachusetts at Amherst
Amherst, Massachusetts

19.1 INTRODUCTION

The transportation system consists of a vast network of multiple modes that carry both people and goods locally, regionally, nationally, and globally. The goals of transportation professionals to maintain mobility, improve safety, and ensure sustainability are increasingly challenged as demand increases.

As part of the transportation system, the effective movement of freight also plays a large economic role. When congestion on the transportation system increases, businesses and customers are affected in two ways. First, movement of freight becomes less productive, causing the price of moving goods to increase and, second, more freight transportation must be consumed to meet the needs of an expanding economy. To ensure that freight movement remains viable and competitive, transportation planners must include freight in planning activities, and ideally, they must incorporate it in an integrated manner into the overall planning process.

19.1.1 Legislative Requirements

A primary driving force behind today's freight planning activities occurred with the passage of the Intermodal Surface Transportation Efficiency Act (ISTEA) of 1992. Within ISTEA, both the metropolitan and statewide planning requirements were modified, and states were mandated to create a series of management systems, two of which address freight planning: the intermodal management system (IMS) and the congestion management system (CMS). This mandate was changed in 1995 by the National Highway System Designation Act to an operational recommendation, and many states that had begun the process under the mandate continued to develop systems and planning techniques to improve planning within their states. The Transportation Equity Act for the 21st Century (TEA-21), passed in 1998, continued an emphasis on implementing and integrating freight planning. In addition to these highway acts, the 1990 Clean Air Act Amendments (CAAA), followed by the Environmental Protection Agency's (EPA) 1993 General Conformance Regulations, have also influenced the need for transportation planning.

---

*Reprinted from the First Edition.

For metropolitan areas, section 1024 of ISTEA required the inclusion of the following factors into the transportation planning process (FHWA and FTA 1993):

- Methods to enhance the efficiency of freight
- International border crossings and access to ports, airports, intermodal transportation facilities, major freight distribution routes, national parks, recreations areas, monuments, historic sites, and military installations

For statewide planning, the law required that the following factors be considered (FHWA and FTA 1993):

- International border crossings and access to ports, airports, intermodal transportation facilities, and major distribution routes
- Methods to enhance the efficiency of commercial motor vehicles

Although the management systems are no longer mandated by the federal government, many states are continuing to follow their philosophy. Understanding their intent is therefore important to understanding the basis of current planning activities. The CMS is a "systematic process that provides information on transportation system performance and alternative strategies to alleviate congestion and enhance the mobility of people and goods" (FHWA 1994). Because of this definition, freight movement becomes a major factor, even though many argue that freight represents only a small portion of vehicle flow. As one measure of success or failure, planners are supposed to analyze their actions in terms of the enhancement of the mobility of goods.

The IMS is a systematic process that (FHWA and FTA 1994):

- Identifies key linkages between one or more modes of transportation where the performance or use of one mode will affect another
- Defines strategies for improving the effectiveness of those modal interactions
- Evaluates and implements these strategies

Unlike the CMS, which has a clearly defined goal of reducing congestion, the IMS requires the measurement of performance of certain critical parts of the total transportation system, most of which have not previously been measured. Defining these performance measures has taken two different paths: (1) measuring the efficiency of the entire freight system, and (2) measuring the efficiency of specific points of intermodal connection.

Guidelines, with associated funding from FHWA and FTA, have been in place for passenger transportation and, indirectly, for freight moving on our highways. However, major freight facilities, including seaports, rail lines, and airports, were not considered until the CAAA as clarified by the 1993 EPA General Conformity Regulations was established. The EPA requires that conformity for carbon monoxide and ozone include both direct on-site emissions and indirect emissions caused by vehicles coming and going from freight facilities (EPA 1993). The inclusion of indirect emissions requires planners to document impacts of freight facilities at a much greater level of detail.

19.1.2 Practical Considerations

Transportation is essential for development and growth and the primary need for transportation is economic. For people, this need is to have access to work and food. For freight, this need is to supply the goods by moving raw materials to processing to distribution to the point of final consumption to disposal. People and goods, for the most part, share transportation facilities and compete for services and resources.

Today's society is moving more freight than ever before, and the total is projected to grow by nearly 70 percent by 2020 (FHWA 2002). In 1997, the United States moved, on average,

41 million tons of freight per day valued at $23 billion. This represents 14.8 billion tons and $8.6 trillion dollars and translates to 3.9 billion ton-miles of freight movement per year (BTS 2001). Many components of the transportation infrastructure are already under stress due to heavy traffic congestion, safety and environmental concerns, and the recent emphasis on national security. With the projected growth and increasingly limited available resources, planning for freight will play an increasingly important role in the future of transportation and economic systems.

Congestion and Capacity.   Because freight typically moves in large vehicles with reduced operating characteristics, particularly on roads, increases in freight traffic have a greater impact on the capacity of the transportation system. With congestion increasing due to increases in both passenger and freight demand, conflict between these two groups is exacerbating the problem. Growth in international trade is creating greater pressures on ports, airports, and border crossings in addition to creating and expanding high-growth trade routes over facilities that are already congested by domestic traffic. When demand exceeds supply, resulting congestion impacts speed and reliability. The Texas Transportation Institute estimated that in 2000 3.6 billion person-hours of delay occurred on the highways of 75 urban areas (TTI 2002). This system delay has increased the value of the travel time of freight between $25 and $200 per hour, depending on the product being shipped. Unexpected delay for trucks can increase this amount 50 to 250 percent (FHWA 2001). Therefore, congestion increases the cost of goods, which has a direct effect on the U.S. economy.

Financing.   Traditionally, investments in freight transportation facilities, whether public or private, have met the needs of moving freight. With existing conditions and expected growth, this is no longer the case and neither sector has the necessary resources to meet these needs. Both sectors must look for creative ways to finance projects jointly. In addition, improvements that will maximize productivity must be identified and targeted for these limited resources. Planners must have adequate tools and information to identify which improvements to invest in.

Safety.   Safety is the top priority for public and private sector transportation professionals. For public sector professionals, safety relates to reduced crashes and associated societal costs on publically maintained facilities. For private sector professionals, safety relates directly to the cost of doing business. As freight movement increases across all modes, the interaction and competition of moving people and goods increases, and consequently the concern about safety increases. Although separation of freight and people has been considered, the practicality of doing so in the foreseeable future is nonexistent. This means that planners will need to continue to consider and plan for the very different characteristics of each.

National Security.   Recent events have changed concerns about security from controlling theft and reducing contraband to preventing attacks and enhancing security while keeping commerce moving. This involves protecting assets—facilities and vehicles as well as the supporting communications and power supply—both from direct attack and from use as instruments to deliver an attack. Although most of the focus by the profession is on screening goods and tracking their movement, consideration for facility enhancements is also important, particularly as space and specialized equipment are needed for this screening and other security activities.

Environment.   Since the passage of the National Environmental Policy Act of 1969, all major transportation projects have been required to include environmental impact assessments. Because of increased awareness of the impact that transportation, particularly freight transportation, has on air and water quality and land use and development patterns, transportation professionals must ensure that transportation projects are environmentally sound and economically sustainable. Air pollution, dredging, and noise are key issues facing the freight transportation professionals. An understanding of these issues when planning facility improvements or expansions will reduce their impacts and reduce the planning times required for conducting environmental reviews.

19.2   *FREIGHT TRANSPORTATION OPERATIONS TODAY*

Freight operations consist of everything required to move an item of freight from its origin or shipper to its destination or receiver. In the United States, most freight operations are handled by the private sector and are considered part of the supply chain management of business operations. The public sector owns and manages many of the facilities, including the highway system, that are required to move freight. It also regulates and taxes freight movement. This division in ownership and responsibility creates some unique challenges for freight planning.

19.2.1   Public versus Private Sector

Movement of freight is unique because it is predominantly managed by the private sector across facilities owned and maintained by the public sector. The most telling distinction between how these entities view transportation is in their respective definitions of transportation. The public sector defines transportation as the effective and efficient movement of people and goods from one place to another. The private sector defines transportation as the creation of place and time utility, where *place utility* means that goods or people are moved to places of higher value and *time utility* means that this service occurs when it is needed.

In the public sector, planning for transportation improvements occurs from 6 months in the short term to 20 years or more in the long term. In the private sector, short term planning is delivery of tomorrow's goods, while long-term is considered 6 months to a year in the future. This difference in time frames has made partnering between public and private stakeholders difficult, particularly when the players do not understand this difference.

Transportation goals also vary between the public and private sector. Goals of the public sector are to provide safe, reliable, and sustainable transportation to all users. Because transportation is considered a public utility in the sense that it is vital to the overall public interest, it is the responsibility of the public sector to ensure that it operates effectively and fairly. Goals of the private sector are to provide reliable, cost-effective service to specific customers in a competitive environment of providing a better service than the competitors.

19.2.2   Logistics

Movement of freight is a derived demand, meaning that goods are moved only in response to a need. A firm needs a commodity, either as input to a product or as an item to sell to a consumer. The provider of the commodity ensures delivery of the product to the customer. Several players can be involved in accomplishing this, including transportation firms, wholesalers, and third-party logistics firms that specialize in providing logistics management services.

Logistics management determines how and where freight moves, and because the goal of logistics management is to minimize costs, an understanding of these movements and the associated costs is important to the planning process. Movement of a product from point of production to point of consumption consists of several separate movements, each of which has costs associated with it. Factors to consider include the number and location of storage sites, storage time at each site, transport modes used between sites, and shipment sizes. Associated costs include building and operating storage facilities costs, inventory costs, shelf-life costs, transport costs, loading and unloading costs, loss and damage, order costs, and stock-out costs resulting from late deliveries (Cambridge Systematics et al. 1997).

Freight transport decisions are based on cost and also on customer satisfaction. Therefore, reliability and the ability to deliver goods undamaged, on time, and when needed is as important as minimizing the cost associated with that delivery.

19.2.3 Influences

Evolution of the movement of freight since World War II has occurred quickly due primarily to the influences of four exogenous factors:

* Globalization of business
* Deregulation of transportation and a changing governmental infrastructure
* Organizational changes in business
* Rapidly changing technology

Globalization has impacted freight movement in many ways, from foreign sourcing of procurement to selling goods to other countries to multifaceted international distribution, manufacturing, and marketing. With improved transportation services, use of land bridges—moving goods across countries without those countries being either the origin or destination for those goods—has increased.

Deregulation of air, motor, and rail carriers in the 1980s dramatically changed the way freight is transported. Overall, the cost and/or quality of transportation service has improved for shippers. Additional changes in governmental infrastructure that have impacted freight movement include deregulation of banking and communications, deregulation of motor carrier transportation in Canada, and changes in the European economic community. The opening of Eastern Europe and the dissolution of the U.S.S.R. have changed freight transportation patterns. The North American Free Trade Agreement (NAFTA) has redefined the north-south freight transportation corridors in the United States.

The restructuring of business, including mergers, acquisitions, and related activities, has changed how businesses ship goods. In some instances, logistics functions have been consolidated; in others, they have been outsourced to third-party logistics organizations. This has directly impacted the sizes and modes of shipments.

Improvements in technology, particularly related to computers and communication, have allowed businesses to improve inventory control and scheduling leading to increased use of just-in-time (JIT) delivery of goods and improved service quality. Motor carrier companies are now able to meet narrowly defined windows for pickup and delivery. Effects of the growth of the internet and e-commerce are just beginning to be seen and will change how business is conducted.

19.2.4 Commodities

Freight is made up of commodities that vary from raw materials to finished goods. Most raw materials are shipped as either dry bulk, such as coal or grain, or liquid bulk, such as oil or milk. Processed goods can be shipped in almost any form, either containerized or noncontainerized. The diversity of types of freight can be seen from examples like mail, automobiles, machine parts, scrap, garbage, hazardous materials, computers, pressurized liquid natural gas (LNG), clothing, and fresh flowers. How this incredibly diverse population of goods is grouped for transportation planning analysis is still an area of discussion and research.

For accounting, either to keep track of where goods are going or what goods are being manufactured and exported or imported, commodities have been grouped and classified by various agencies in the U.S. government. The U.S. Census Bureau used the Standard Industrial Classification (SIC) until 2002, when the SIC was replaced by the North American Industrial Classification System (NAICS) (http://www.census.gov/epcd/www/naics.html). The Federal Railroad Administration uses the Standard Transportation Commodity Classification System (STCC) and the Bureau of Transportation Statistics, in cooperation with Canada, uses the Standard Classification of Transported Goods (SCTG) (http://www.bts.gov/cfs/sctg/advants.htm). The U.S. Department of Commerce uses Export Control Classification Numbers (ECCN) (http://www.bxa.doc.gov/licensing/facts2.htm). When working with commodities, one or more of these commodity classifications have been and may be used to group them for analysis purposes or for determining national and international flows of goods.

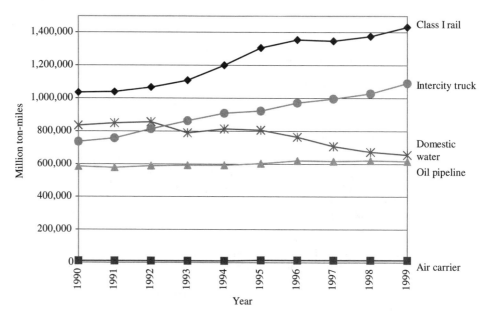

FIGURE 19.1   Freight movement by mode.

19.2.5 Modes

Freight, more so than people, travels by many modes including rail, water, air, and pipeline as well as truck. Figure 19.1 shows growth by mode of freight movement over the last 10 years of the 20th century. For almost all movements by modes other than truck, freight trips must also be multimodal. In other words, freight must change modes during a shipment from its origin to destination.

Each mode has unique operating characteristics that require that planners understand and account for these differences.

Highway.   Highway transportation is the dominant mode for passenger transportation and is a rapidly growing mode for moving freight. The highway network includes approximately 4 million miles of roads, ranging from freeways to unpaved rural roads. The majority of highways are paid for and maintained by public funds and are under the jurisdiction of public agencies. The highway system carries approximately 30 percent of intercity ton-miles of freight, which represents about 75 percent of intercity freight revenue.

The trucks using these highways vary in size, ownership, and use. About 9 percent of all trucks are used for over-the-road intercity freight and are classified by gross vehicle weight. These trucks are required to conform to standards for height, width, length, gross weight, weight per axle, and number of axles, with the current maximum weight at 80,000 lb. The trucking industry is a major force in moving freight but it is fragmented and diverse. Privately owned trucks transport their own company's products and are not regulated except for meeting federal and state safety requirements. For-hire trucks transport goods under contract and are generally regulated with some exceptions, such as those carrying exempt agricultural products or operating in a single state. For-hire trucking is also divided between truckload and less-than-truckload (LTL) services, with the latter including package carriers. Truckload carriers generally move directly from the shipper to the consignee. Less-than-truckload shipments are usually picked up by one truck, taken to a nearby terminal, reloaded on the line-haul truck, taken to a terminal near the consignee, reloaded onto a delivery truck, and delivered to the consignee.

Costs associated with trucking are primarily variable costs. These include labor; depreciation of vehicles; and repair and maintenance of vehicles, including tires, user fees, and fuel and lubrication, including taxes. Fixed costs, which account for about 25 percent of trucking costs, include license fees, property taxes, management, and terminals (Wood and Johnson 1996). Direct costs associated with highways, except for tolls and some user fees, are provided by the public and are not part of trucking costs.

The diversity in size and configuration of trucks provides the trucking industry with greater flexibility in moving goods, since the size and characteristics of the vehicle can be more easily tailored to the size and nature of a shipment. Motor carriers also have the reputation of providing a high level of customer service since they have more flexibility in scheduling and the ability to reach most origins or destinations. Also, as mentioned earlier, almost all freight shipped by other modes requires a truck movement at one or both ends of the shipment.

Highway operations, including traffic control and an understanding of congestion, are well known to transportation planners, and the incorporation of trucks into models used in the planning process is relatively straightforward. The biggest challenges come from predicting the number of trucks to be included and determining their impact on the capacity of the roadways, since capacity calculations include a fixed percentage of trucks in the vehicle mix and the purpose of freight planning is determining the number of trucks to include.

Rail. U.S. railroads are privately owned, although the original development of the rail system was supported by land grants that allowed for the quick expansion and dominance of rail as the freight transporter in the 19th and early 20th centuries. Track mileage has been reduced to approximately 113,000 miles, less than half of its peak. Individual rail companies own their track and have contractual agreements with other companies, including the federal government, to use this track.

The rail industry is grouped into three classes based on revenues: class I carriers with greater than $253.7 million in revenues, class II railroads with revenues between $20.3 and $253.7 million, and class III railroads with less than $20.3 million in revenues. Class I carriers, now less than seven companies, provide the majority of line-haul services and account for approximate three-quarters of total rail mileage and 90 percent of freight revenue (Muller 1999). Class II railroads are regional and usually connect to class I or other class II railroads. Class III include switching and terminal railroads.

Costs associated with use of railroads are approximately 50 percent fixed and do not vary with the volume of business. These include rights-of-way, classification yards, general management, and maintenance. Variable costs include purchase and maintenance or usage charges of equipment, usage charges of rights-of-way, and labor and fuel and lubrication costs. Because of the high fixed cost, rail can recognize significantly increased economies of scale: as volume increases, total cost of production decreases on a per-unit basis.

Railroads are especially suited to carrying large, bulky shipments that are not time-sensitive, such as milk, over long distances. Raw materials and other bulk commodities, motor vehicles, intermodal containers or containers on flat cars (COFC), and piggyback truck-trailers or trailer on flat car (TOFC) are the most common items shipped by rail. The biggest limitation to rail service is the fixed track, which limits the routes available, and the locations of the beginning and ending points of a rail movement.

Planners face a major challenge when incorporating rail movements into existing planning models. Most public sector planners do not have the same understanding of rail operations that they do of highways. Rail capacity is based on the number of trains that can safely traverse a section of rail, which is based on the type of signaling and control on that segment and whether more than one set of tracks is available or where and how often sidings are available to allow for passing in either direction. The speed of a train is dependent on the type of track control, the geometrics of the route, and the makeup and weight of the train. Another complicating factor when considering rail movements is that freight moves in cars that are combined to make up trains. From the beginning of a rail trip, a given car can change trains at classification yards any number of times before reaching the termination of that trip.

Pipelines.   For the most part, planners do not consider pipelines in their planning activities. Pipelines are privately owned and maintained, and freight that uses pipelines flows in bulk without containerization. Truck, rail, or tanker movements to or from pipeline terminals are generally modeled in those modes.

Domestic Water Carriers.   Moving freight by water falls into two categories, domestic and international. This section summarizes the characteristics of domestic movements.

Domestic waterways used for moving freight include navigable rivers, canals, the Great Lakes, and the Gulf and Atlantic Intracoastal Waterways. The U.S. Army Corps of Engineers maintains navigable rivers and harbors, including dams and locks. The U.S. Coast Guard also plays a role in the safety of waterways, including providing navigational aids such as buoys to guide vessels on their course. Terminals very greatly and can be as basic as a privately owned dock on a waterway up to a complex maritime port that serves both domestic and international shipments. Ports are usually public/private partnerships that are managed by a publicly based authority overseeing multiple public or private terminals consisting of several docks. Movement of domestic freight can also occur from port to port using open water, although this has substantially declined over time.

Carrier vessels consist of barges controlled by tugboats and towboats on inland rivers and canals, lakers or vessels that are not considered seaworthy for operation on ocean waters used on the Great Lakes, and both barges and seagoing vessels on the intracoastal waterways.

Maintenance and management of waterways and ports are generally publicly funded, while specific terminals may be privately or publicly funded. Costs for using these facilities are covered by shippers/carriers through user fees based on the services provided. Variable costs for shipping goods by water include maintenance or usage charges of vessels, usage charges, and labor and fuel costs.

Like railroads, domestic water carriers are suited to carrying large, bulky shipments that are not time-sensitive over long distances. Petroleum products, raw and construction materials and other bulk commodities, and intermodal containers are the most commonly shipped items. The biggest limitations to domestic water service are the geographically limited network and terminal locations, shipment time, and seasonal and weather effects.

The domestic waterway is primarily a long-distance mode that is of interest in regional and corridor planning. However, ports and terminals are of interest to local planners since most of the freight is offloaded into trucks, with some freight going to rail and pipeline depending on connectivity and commodity type.

Domestic Aviation.   The fastest-growing mode for moving freight is domestic aviation. Although expensive, the ability of aviation to deliver goods across long distances faster than any other mode has increased demand, particularly for parcel delivery and just-in-time delivery backup. Freight can travel in the belly of a passenger aircraft, on a dedicated cargo aircraft, and on a super cargo aircraft. Security concerns resulting from September 11, 2001, have affected this mode more than any other, particularly for the first option. Facilities associated with air freight include freight terminal(s) at the airport and access to the airplane. These are usually managed by a private entity, either the carrier or a freight forwarder. Individual airlines, including Federal Express and UPS, own and maintain aircraft used to move freight.

Like water, air is primarily a long-distance mode that is of interest to national planners while airports are of interest to local and regional planners. For this mode, landside freight transportation is almost exclusively truck.

Intermodal Transportation.   Over the past decade, the distinction between intermodal and multimodal transportation has become blurred. *Intermodal transportation* is commonly used to refer to movement of freight containers across one or more modes. Where planners need this distinction is at intermodal terminals that transfer containers from truck to rail or rail to truck. Although container transfer can occur at maritime ports or, less commonly, at airports, these activities occur within the port itself, are generally considered to be part of port operations and are usually represented in local planning as a generator of truck traffic. Truck/rail intermodal activity has a more direct impact on local planning. Characteristics of inter-modal terminals that are of interest include transfer capacity and rate, storage capacity, and hours of operation along with associated costs for normal operation and for delayed operation.

19.3 PLANNING

Freight transportation improvements are planned and implemented by both the public and private sectors. This has traditionally occurred independently. In general, publicly funded improvement planning follows a lengthy structured process with prescribed involvement by many stakeholders. In contrast, private sector planning has a short response time without external involvement and is driven by market trends. The freight planning process presented in this section focuses on planning performed by the public sector. However, effective planning cannot ignore the latter, since the business aspects of moving freight play an important role in how decisions are made.

The extent to which freight is considered in public planning both at the state and local metropolitan planning organization (MPO) level varies from being integrated into the prioritization and funding process to not being considered at all. This section presents the state of the practice in public sector planning for freight.

19.3.1 Purpose

The purpose of public sector planning is to determine systematically the use of available resources to meet the goals of the public being served. Some factors used to justify selection of projects include improvements to traffic flow and safety, savings in energy consumption, economic growth, increased accessibility, employment stimulation, competition with other cities or states, politics, or personal benefit. Although planning has focused in the past on moving people, the public sector must now include freight in the process and, ideally, allocate resources to the projects that most closely meet the goals of the overall system, irrespective of the user—people or freight—or mode. This implies having a planning process that considers all of the components of the transportation system related to those goals. However, this ideal has yet to be achieved in all but very unusual cases, and the state of the practice is to plan independently for freight and people.

19.3.2 Process

Because of the complexities outlined early in this chapter and the limited availability of data, freight planning procedures for the public sector remain extremely challenging. This section attempts to summarize some of the procedures currently available and accepted for use by planners.

The planning process consists of seven basic steps, which are interrelated and not necessarily sequential: defining the situation, defining the problem, exploring solutions, analyzing performance, evaluating alternatives, choosing the project, and constructing or implementing the selected project.

Situation and Problem Definition. Because of competing needs of the public and private sectors for freight, defining the situation and problem usually requires interaction between these sectors. This can occur through the establishment of a formal relationship, usually through a freight advisory group. Members of this group could include staff from the local planning agencies, port authorities, major carriers from rail and trucking companies, package delivery companies, and the region's major shippers. Private sector associations, such as chambers of commerce or economic development agencies, are also important players. FHWA's *Public-Private Freight Planning Guidelines* provide more extensive information about establishing and using these groups to assist with freight planning activities (FHWA 2002). If the group is structured appropriately, it can have an active role in all parts of the planning process.

After identifying the problem, planners must then quantify it, which consists of inventorying facilities and collecting related information. This information can include existing geometry or facility configuration, traffic control, traffic counts, land use, employment, economic activity, and origins and destinations. This is generally readily available for traditional planning but does not necessarily include details required for freight planning. The freight advisory group can be very beneficial in assisting with this activity.

Exploring Solutions, Analyzing Performance, and Evaluating Alternatives.   As with problem definition, freight advisory groups can provide valuable information for exploring solutions by identifying lists of cost-effective efforts that can be implemented quickly and easily to improve freight movement. This is often as simple as retiming a signal to account for left-turn movements by large trucks. For larger-scale studies, these groups can assist with providing the expertise to correctly model freight activities.

Once all possible solutions have been narrowed to those that will be considered in more detail, planners begin to analyze and evaluate these alternatives. The analysis of the performance of possible alternatives is often referred to as the planning process, but it is really the process that integrates system supply on a network with travel demand forecasts to show equilibrium travel flows. Data required for this process and models that are currently used are presented in the following sections.

The evaluation phase considers the effectiveness of each alternative under consideration to achieve the objectives of the project as defined by performance measures. The performance data from the analysis phase are used to determine costs and benefits for each alternative. Performance measures for freight facilities and for evaluating freight on the transportation system are presented in the Performance section.

19.3.3 Data

Probably the biggest issue for freight planners is the availability of data about freight movements that can be used for planning activities. Data collection is expensive and resource-intensive and most planning activities do not include the time or money to obtain these data. In addition, data on freight movements are usually considered proprietary and are held closely by shippers and carriers to maintain their competitive advantage. As a result, planners often select analysis techniques based on available data instead of determining the best techniques to meet their planning goals and acquiring the necessary data to support those techniques. This will remain an issue for planners, at least in the near future, and the models and methods discussed in the following section reflect this limitation.

Publicly available sources for freight data are outlined in the following paragraphs. Web pages for each data set are included after the name of the data set (Meyburg and Mbwana 2001).

*Commodity Flow Survey (CFS) (**http://www.bts.gov/ntda/cfs/**).*   Data were collected on shipments by domestic establishments in manufacturing, wholesale, mining, and selected other industries for 1993 and 1997 by the U.S. Census Bureau in partnership with the Bureau of Transportation Statistics. This is an origin shipper survey and does not have detailed origin-destination data for commodity by mode. Locational aggregation is by state or by National Transportation Analysis Region (NTAR). Data in the CFS are best used to determine what is sent out by states or NTAR to the rest of the country, the mode used, and the distance traveled. Commodities coming into a state or NTAR and market area dynamics are not effectively presented. Other limitations include lack of geographical detail, lack of industry coverage for industries that are moving very fast, absence of several kinds of movements, lack of vehicular flow information, focus only on domestic movements, and difficulty with matching with other data sources.

*Transborder Surface Freight Data (TSFD) (**http://www.bts.gov/ntda/tbscd/**).*   North American merchandise trade data by commodity type, by surface mode of transportation (rail, truck, pipeline, mail, and other), and with geographic detail for U.S. exports to and imports from Canada and Mexico have been collected and available monthly since 1993 by the U.S. Customs. This data set is a subset of official U.S. international merchandise trade data. The TSFD has been used to monitor freight flows and changes since NAFTA and for trade corridor studies. This is a customs data set, not a transportation survey data set, and is focused primarily on revenues.

*Vehicle Inventory and Use Survey (VIUS) (**http://www.bts.gov/ntda/tius/**).*   Information on trucks domiciled within a state and owned by businesses and individuals, ranging from multitrailer combination vehicles to pickups, vans, and minivans, is collected and available every 5 years by the U.S. Census Bureau. These data provide limited information about where trucks are used by zones or ranges of motion. No information is provided on what commodity trucks carried.

Motor Carrier Financial and Operating Statistics **(http://www.bts.gov/ntda/mcs/).** Annual and quarterly data are collected on industry, financial, employee, and operating for motor carriers of property and passengers.

Waterborne Transportation Data **(http://www.wrsc.usace.army.mil/ndc/).** A series of databases and statistics pertaining to waterborne commodity and vessel movements are collected by the U.S. Army Corps of Engineers and made available annually.

Additional information includes domestic commercial vessel characteristics, prot and waterway facilities, lock facilities, lock operations, and navigation dredging projects.

State Freight Transportation Profiles **(http://www.bts.gov/ntda/sftp).** State profiles present information of freight transportation for each of the 50 states, combining major federal databases into tables. Reports give a description of the database and contact points.

TRANSEARCH Data (Reebie Associates) **(http://www.reebie.com/).** TRANSEARCH is an integrated, multimodal freight flow database constructed from public and proprietary data sources by Reebie Associates. Market-to-market freight movements are provided for the United State, Canada, and Mexico for the year 2000. Tonnage moved by market pair, by commodity, and by seven modes of transportation are available at the county, business economic area (BEA), metropolitan area, state, or province level. Other information includes secondary traffic and freight rehandled by truck from warehouse and distribution center. Modal coverage includes for-hire truckload, for-hire less-than-truckload, private truck, rail carload, rail/truck, intermodal, air, and water. A cost is associated with using these data.

Rail Waybill **(http://www.ntis.gov/fcpc/cpn8441.htm).** An annual sample of freight movements terminating on railroads in the United States is collected by the Surface Transportation Board (STB). Sample size is approximately 2.5 percent of all rail traffic. Two files are created. The master waybill file contains confidential information on specific stations, railroads, and revenue and is not available to the public. The public-use file provides information on freight movements at the BEA level at five-digit Standard Transportation Commodity Code level. Origin, destination, intermediate railroads and junctions, commodity, type of car, number of cars, tons, and revenue are provided.

National Roadside Survey—Canada **(http://strategis.ic.gc.ca/SSG/ti01101e.html).** A 1991 survey of commercial vehicles was performed by the Canadian Conference of Motor Transport Administrators to provide information to assist with identifying the impact of changes within the Canadian trucking industry. Complementary surveys were performed in 1995 and 1999. Information includes carrier type, vehicle configuration, trailer configuration, capacity utilization, and drive category for inter-provincial movements.

Employer Database **(http://www.doleta.gov/alis/edbnew1.asp).** The State Employment Security Agencies collects information about employers, including who they are, where they are, what activities they do, and how much freight they generate or attract. The America's Labor Market Information System (ALMIS) Employer Database is available to the general public. Information about accessing this information is available at the website.

Many local planning agencies have collected information related to freight at different levels of aggregation for specific needs. Private companies are often willing to share information if confidentiality is ensured and they are convinced that the result will provide a benefit to their operations, i.e., improvements in their movement of freight. Also, cost-effective data collection on a smaller scale can be accomplished for a localized project. Contacting the local planning organization can often identify what data may be available for use on planning initiatives.

19.3.4   Planning Models and Methods

Analysis of the performance of planning options requires valid models for forecasting the demands for freight movement and how these demands change over time. Traditional efforts have focused on growth in types of commodities based on time series data to predict future commodity flows or in using the traditional four-step urban transportation planning model adjusted appropriately to consider the unique characteristics of freight. Although criticism exists of using the four-step process, current practice for public sector planning is to use some form of this approach (Maze 1994). The basic four-step approach, as it relates to freight, consists of (1) determining how much and where goods are generated and how much and where goods are used, (2) estimating volumes of freight moving from each origin to each destination, (3) selecting the mode or modes moving the freight, and (4) determining the route to be used. Either in step (2) or (3) the volume of goods must be converted to numbers of vehicles.

When determining freight demand, several factors make it unique and more complex than determining passenger demand. *Units of measure* can be easily determined for passengers. However, freight may be measured in units, by weight, by value, or by volume. The *value of time* is very different where the value of commodities being shipped can be directly impacted by time, such as fresh cut flowers. Passengers generally require no assistance for *loading and unloading,* while effective movement of freight is highly dependent on being loaded and unloaded both at the origin and destination and at transfer points along the trip. *Types of vehicles* also vary substantially for freight. Passengers, for the most part, only require seats in vehicles. Freight requires a wide variety of accommodations, from refrigerated containers to dry bulk hoppers to liquid bulk tank cars to flat beds to specialty vehicles. These differences should be included in considerations as planners use the models presented in the next sections.

Freight Generation and Attraction.   Base data of the amount of freight generated from or attracted to an area or location are often difficult to acquire or collect. When these data are available, the units are in average numbers of tons or average dollars of value and are usually based on nationally or regionally based samples. In addition to the data sets listed in the previous section, Fischer and Han (2001) have compiled other major sources and types of truck trip generation data.

If the planner has the option to perform data collection as part of the planning process, the following methods are in use today (Southworth 2003):

- Vehicle classification counts: traffic loop counters or videos or other traffic sensors
- Vehicle intercept and special traffic generator surveys: counting classifying or surveying vehicles as they enter and leave a specific location or area
- Truck trip travel diaries: daily travel surveys completed by drivers or dispatchers
- Carrier activity surveys: regulated surveys for safety or user fee legislation
- Commodity flow surveys: shipment inventories completed by shippers or establishments

These data have been used to develop commodity- or vehicle-based freight trip generation rates in different ways. The most straight-forward is to combine vehicle traffic counts or tons with employment or land use values to develop simple trip rates or tons moved per employee or unit of land (Cambridge Systematics 1995; Fisher and Han 2001). Care should be used in transferring these rates to other applications.

Another relatively simple procedure is to assume that demand is directly related to economic indicator variables such as those that measure goods output or demand in physical units. If these measures are not available, constant-dollar measures of output or demand or employment can be used. Indicator variables are used to derive either annual growth rates or growth factors of ratios of forecast-year to base-year values (Cambridge Systematics et al. 1997).

A more robust method is to fit least squares regression models to traffic counts or tons by commodity. This approach consists of identifying one or more independent variables, usually measures of economic activity, which can be used to predict the dependent variable, usually a measure of freight activity such as tons or ton-miles. For forecasting, regression techniques are applied to time-series data for both the independent and dependant variables. The resulting model of this relationship is then applied using forecasts of independent variables to future time periods. The *Quick Response Freight Manual* (Cambridge Systematics 1997) summarizes several models that have been used for different planning applications.

With all forecasts, several factors should be considered in their development, including the level of aggregation of geographic area, the level of aggregation of commodity, and whether the planning activity is for a new or existing facility.

Freight Trip Distribution. Once supply (generation) and demand (attraction) values have been developed, planners must determine the pattern of freight movements. One method for modeling commodity flows is to develop commodity-specific spatial interaction (SIA) models (Southworth 2003). The volume of freight for commodity c, V_{ci}, originating from area i is allocated to destination j using the general SIA model (Wilson 1970):

$$T_{cij} = V_{ci} * A_{ci} * W_{cj} * B_{cj} * f(c_{cij}) \tag{19.1}$$

where
T_{cij} = volume of freight of commodity c allocated moving from origin i to destination j
V_{ci} = volume of freight of commodity c in area i
W_{cj} = volume of freight of commodity c of interest in area j
$f(c_{cij})$ = inverse function of costs, c_{cij}, of transporting a unit of commodity c from i to j
A_{ci} and B_{cj} = balancing factors that ensure compliance with observed activities

Specifically,

$$A_{ci} = [\Sigma_{cj} * W_{cj} * B_{cj} * f(c_{cij})]^{-1} \qquad \forall_{ci} \tag{19.2}$$

and

$$B_{cj} = [\Sigma_{ci} * V_{ci} * A_{cj} * f(c_{cij})]^{-1} \qquad \forall_{cj} \tag{19.3}$$

These factors are solved using an iterative proportional fitting procedure that ensures

$$\Sigma_{cj} S_{cij} = V_{ci} \quad \text{for all } i \text{ and} \quad \Sigma_{cj} S_{cij} = W_{cj} \quad \text{for all } j \tag{19.4}$$

This is a doubly constrained SIA model (Wilson 1970). The origin-to-destination freight costs, c_{cij}, are derived directly from empirical data or by econometric modeling. These models are applied to zonally aggregated data for zones ranging from traffic analysis zones (TAZs) for urban areas to counties for regional or statewide planning.

More detailed analyses can be performed using advanced models such as a logit choice model (see Southworth 2002).

Several issues should be considered when applying trip distribution models to freight movements. First, a significant differences exists between modes and the associated cost values need to be carefully thought out to represent these differences effectively. In addition, backhauling of empty vehicles/containers and the associated costs should be included. If more than one mode is used for an individual shipment, the costs associated with changing modes, as well as mode-specific costs, should be included in the cost fuction used for distribution.

Freight Mode Choice. Moving freight short distances tends to be accomplished by a single mode, truck, without any real competition from other modes. Also, incorporating transportation improvements on the landside of maritime ports, airports, or intermodal terminals focuses on highway planning. For these types of planning activities, mode choice is not a consideration. However, for longer-distance freight movements, expanding terminal activities, and moving high-valued time-sensitive freight, planning across modes is important and mode choice is a part of the planning process.

This very quickly becomes an involved and difficult process because mode choice usually relates to the type of service required for moving freight which is not independent from a mode. Examples of questions to consider include whether the freight be moved on scheduled service or whether the commodity being shipped needs special packaging or handling.

The most common approach to analyzing mode split is use of a discrete choice model such as a logit model (Cambridge Systematics 1997). The selection is then based on the mode or combination of modes that provides the lowest cost. The general formulation for selecting mode K from k available modes between the origin i and destination j is:

$$P_{K/ij} = \exp(-\lambda c_{ijk})/\Sigma_k \exp(-\lambda c_{ijk}) \qquad (19.5)$$

With a given modal cost sensitivity parameter, λ, the averaged modal cost is:

$$c_{ij} = -1/\lambda \ \ln(\Sigma_k \exp(-\lambda c_{ijk})) \qquad (19.6)$$

The cost terms determine the effectiveness of the model, and it is therefore crucial that the modeler understand them and the context in which they are used. One cost that is difficult to include in the cost formulation is the value of reliability, which is a major factor from the business perspective in moving freight. In some cases, costs may consist of trade-off values instead of actual costs. Wigan et al. (2000) provide some quantitative insight into this. Additional methods have been proposed for determining mode split (for example Jiang et al. 1999), although these are currently more academic than applied.

Because of the complexity and interrelated nature of selecting between modes, which is based not only on cargo type, type of firms involved, and the nature of the geography, but also on values of service measures and perceptions of those involved, mode choice modeling is still very much in its infancy for freight planning.

Converting Volume to Vehicles.   For many planning activities, freight movements must be converted from volume or economic value to number of vehicles or vehicular equivalents. Although this sounds straightforward, several factors complicate this process, including variability in vehicle sizes within and across modes, variability in shipment sizes, variability in the relationship between volume and weight of different commodities, variability in packaging, and the amount of backhauling (returning an empty vehicle to its origin) involved. Estimated backhauling in the United States with current operating practices ranges from 15 to 50 percent of truck miles traveled (BTS 2001).

Freight Traffic Assignment.   Assigning vehicles to the transportation network provides the information required for determining most transportation-related impacts, such as congestion, air quality, and physical deterioration of facilities. For multiple origins and destinations several models are available, from the straightforward all-or-nothing assignment to comprehensive nonlinear programming optimized assignment.

Truck assignments on highways are the most common analysis currently performed. Several methods have been used by planners, depending on the purpose and the complexity of the problem. Many MPOs have included truck movements as a specific vehicle size class which is represented as a passenger car equivalent (PCE). Large single-unit trucks are considered as 1.5 PCEs while a tractor-trailer combination may be 3 PCEs for impacts on travel speeds and congestion.

The Wardrop equilibrium assignment model has been used for mixed passenger and freight highway traffic (Southworth et al. 1983). This approach assumes that all routes that are used have the same travel cost or impedance while routes that are not used cost more. Logit route choice models can be used for small networks with identifiable routes, while least-cost routing models can be used for long-distance movements. Because of the trip-chaining behavior of many truck trips, adjusting the "cost" using circuity factors may be required to adequately model truck flows. Validating assignment models can be problematic depending on the availability of track traffic counts for the analysis region.

19.3.5 Performance

Performance measures must be established to assess the current freight transportation system and to evaluate planning alternatives. This section presents several different performance measures that have evolved or been developed to address the needs of freight planning in the public sector.

Performance of a transportation network is evaluated based on the mobility of its users. In the past, traffic has been measured by the movement of people, rather than freight, which was consistent with the focus of older analytical procedures. As the awareness of the importance of goods movement increased, freight-based performance measures have been developed. Because both freight and people should be considered when evaluating a transportation network, freight-related measures should also relate to measures designed for the flow of people.

Traditional measures have been based on the level of service (LOS) concept, which was related to the ratio of volume to capacity (V/C). This was used as a measure of the success or deficiency of a transportation segment. One measure that came out of this approach is *delay,* which measures congestion conditions.

Accessibility has gained support as a measure that more effectively measures performance. This is the achievement of travel objectives within an acceptable time limit and can be applied at any level of evaluation. The advantage to this measure is that it focuses on the goal of the transportation system instead of its symptoms.

Additional measures have been proposed by different studies for specific applications. Hagler Bailly Services, Inc. developed a set of measures (interchangeably called indicators) for the FHWA to evaluate productivity and efficiency improvements in goods movement by motor vehicles on highways (Hagler Bailly 2000). Indicators compiled from previous efforts were ranked in terms of descriptive value and technical appropriateness. Descriptive value indicates the comprehensibility of the indicator for general audiences, and technical appropriateness relates to the usefulness of the indicator as a measure of freight movement productivity. Consideration was also given to data availability and costs. Thirteen indicators were identified as first-tier, meaning they are potentially valuable measures of the quality or cost of freight service or of the performance of facilities important for goods movement. The first four indicators relate to the quality or cost of freight service to shippers. The others address travel time and reliability of system performance significant to freight movements.

- Freight service:
  - Cost of highway freight per ton-mile
  - Cargo insurance rates
  - Fuel consumption of heavy trucks per ton-mile
  - On-time performance for highway-freight deliveries

- Highway system performance:
  - Crossing time at international border crossings
  - Point-to-point travel times for selected freight-significant highways
  - Hours of delay per 1,000 vehicle-miles on freight-significant highways
  - Ratio of peak period travel time to off-peak travel time at freight-significant nodes
  - Ratio of variance to average minutes per trip in peak periods at freight significant nodes
  - Hours of incident-based delay on freight-significant highways
  - Annual miles per truck
  - Customer satisfaction
  - Conditions on connectors between the National Highway System and intermodal terminals (pavement and traffic conditions)

Czerniak, Gaiser, and Gerard (1996) analyzed performance measures used by 15 state departments of transportation in their intermodal management systems. The measures are classified into six goals.

- Goal 1: Accessibility of intermodal facilities
  - Level of service
  - Conditions of a transportation route
  - Bridge restrictions (e.g., weight restrictions)
  - Queuing of vehicles
  - Turning radius into facility
  - Deficiencies of the facility (e.g., number of structures lacking 21-foot vertical clearance)
- Goal 2: Availability of intermodal facilities
  - Volume-to-capacity ratios
  - Railroad track capacity
  - Storage capacity
- Goal 3: Cost and economic efficiency
  - Cost per ton-mile by mode
  - Revenue ton-miles
  - Expenditures (e.g., for freight rail, to retire deficiencies)
- Goal 4: Safe intermodal choices
  - Number of accidents
  - Cost of accidents
  - Number of fatalities
- Goal 5: Connectivity between modes (ease of intermodal connection)
  - Number of intermodal facilities
  - Delay of trucks at facilities
  - Travel times
- Goal 6: Time
  - Total transfer time
  - Freight transfer time between modes
  - Average travel time

These are internal and external performance measures. Internal measures, such as those listed under goals 2, 3, and 4, focus on the operations of an intermodal facility. External measures address the overall system performance. Goals ranked below goal 6 include reliability of facility and environmental protection with measures of air quality and fuel usage.

The Volpe National Transportation Systems Center identified three levels of activity for defining performance measures for intermodal planning needs: system level standards relating to network connectivity, operational standards relating to service delivery, and facility level standards relating to terminal accessibility (Norris 1993) as shown in Table 19.1.

Depending on the application, planners should select the appropriate measures of effectiveness to evaluation options under consideration or to measure current performance of the system to be evaluated.

19.3.6 Case Studies

This section is meant not to provide detailed information about planning case studies, but to provide links to studies that have been performed with enough of a description for the reader to decide whether to pursue it further.

TABLE 19.1 Performance Standards by Analysis Level

System	Components	Measures
Network level	Access	Vertical clearance
		Capacity characteristics
		Bridge rehabilitation
	Safety	Grade crossings
		Waiting time at drawbridges
	Transferability	Cross-modal intersections
		Prioritization of track usage
		Legal/regulatory restrictions
Operations level	Service delivery efficiency	Line haul speed
		Door-to-door delivery time
		Customer service
		Real-time cargo information
		Percent on-time performance
	Economic efficiency	Cost per ton-mile
		Revenue per ton-mile
		Operating ratio
		Market share
	Environmental resource	Fuel use per ton-mile
	use efficiency	Emission per ton-mile
Facility level	Terminal accessibility	Access time to major transportation link
	Pickup delivery system	Gate queue length/time
		Container parking/storage capacity
	Interchange/transfer	Intermodal transfer time
		Average drayage length
		Average drayage cost
	Economic efficiency	Average annual revenue

Chicago **(http://www.fhwa.dot.gov/freightplanning/chicago.html).** Chicago has maintained a freight component in its regional transportation planning since the 1970s. In the years immediately before and after ISTEA, Chicago included freight sector input in its planning process through both formal data collection efforts and industry outreach.

Chicago is the hub of the nation's freight transportation system, with the largest inter-modal freight market in the nation. Class I railroads, as do a number of class IIs and class IIIs, operate within the region, which features 27 major intermodal yards, two waterborne freight facilities, three clusters of lesser-sized water terminals, and three auto transloaders. In 1986, trucks accounted for 12.5 percent of regional vehicle traffic (measured in vehicle equivalents).

Puget Sound **(http://www.fhwa.dot.gov/freightplanning/puget.html).** The focus of this case study is to (1) describe the environment for freight mobility in the Puget Sound region, (2) describe the structure and efforts of the Puget Sound Regional Council and the Regional Freight Mobility Roundtable, (3) chronicle the achievements of these organizations, and (4) capture key lessons that transportation planners might apply to other locales.

Puget Sound is a large U.S. port with growing containerized international trade coupled with increasing regional vehicle miles of travel and population growth. The metropolitan planning organization for the area, the Puget Sound Regional Council (PSRC), is addressing freight mobility issues and is using private freight sector input for decision making. In 1992, the PSRC and the Economic Development Council of Seattle and King County formed an advisory panel, the main purpose of which was to capture private freight sector input on freight issues. The Regional Freight Mobility Roundtable has developed into a model example.

Delaware Valley Regional Planning Commission **(http://www.dvrpc.org/transportation/ freight.htm).** Moving freight and stimulating economic development are appropriate and worthwhile goals for transportation planning. A region's vitality and businesses, jobs, and consumers all rely on a transportation system that can handle goods efficiently and safely. The Delaware Valley Regional Planning Commission (DVRPC) has committed significant resources and technical capabilities to examining freight issues in the Philadelphia-Camden region because of their strong support for economic development. This site provides abstracts to several of their freight planning studies.

The Delaware Valley region is a freight transportation gateway with a large freshwater port; freight service from three large class I railroads and 12 smaller short lines; an airport with expanding international cargo services; an excellent highway network, including inter-modal connectors; and numerous rail and port intermodal terminals that are equipped to handle all types of cargo. To capitalize on these resources, DVRPC has integrated these assets and placed freight directly in the transportation planning process.

Iowa Statewide Planning (**http://www.ctre.iastate.edu/Research/statmod/planning. htm**).   This study develops a matrix for the identification and development of tools and databases to support freight planning and modeling. Dimensions of the matrix include selected freight planning issues and scenarios and a prioritized list of commodity types for Iowa. Issues are identified by public and private sector practitioners and stakeholders and tools and databases are developed using GIS and Internet technologies.

This project investigates and establishes guidelines for the development of analytical methods to support the Iowa Department of Transportation's (Iowa DOT) intermodal statewide planning process. The Iowa DOT has been involved in freight transportation modeling efforts since the early 1970s, primarily on the development of grain forecasting models. The Iowa DOT is now developing multimodal techniques to mode movements of several commodities. The objective of the Department is to simulate the impacts of changes in transportation and nontransportation service variables on freight movements to investigate industry location decisions, the rationale behind commodity movements, and public policy impacts on freight movements.

19.4 THE FUTURE

With the rapidly changing environment for moving goods both in terms of business practices including integrated supply chain management and e-commerce, and transport practices including intelligent transportation systems, planning for freight will potentially undergo many changes in terms of both available data and modeling techniques.

19.4.1 Advanced Technologies

Advanced technologies will impact freight planning activities primarily in two ways. First will be the ability to obtain and make available data related to freight movement and traffic operations that impact its movement. Second will be the increasing ability of businesses to improve supply chain management using improved communications and inventorying technologies on the private sector side and the increasing ability of the public sector to manage traffic flow.

The potential for unobtrusive freight data-collection methods is high given the expansion of advanced technologies, including electronic data interchange (EDI), intelligent transportation systems for commercial vehicle operation (ITS-CVO), global positioning systems (GPS), Web-based data retrieval and assembly, and automated freight handling activities.

Private sector professionals are moving quickly to the use of real-time electronic information for managing and optimizing freight handling and movement to minimize costs, maximize service, and meet regulatory requirements. This same information could provide data required for public evaluation and planning purposes. Information from EDI and GPS includes what and how much is being moved, who is moving it, how it is being moved, why it is being moved, when it is moving, and where it is moving. The challenge is to harvest and disseminate this information in a way that does not compromise the confidentiality of individual firms and consequently their viability, competitiveness, and existence. Similarly, publicly gathered information, such as weigh-in-motion, ITS-CVO monitoring, and travel-time/volume data would also be collected and available to users (Hancock 2000).

Use of advanced technologies to link private and public sectors for information exchange and communications will increasingly affect where, when, and how freight is moved over transportation networks as vehicle and product tracking, in combination with traffic travel time, weigh-in-motion, and incident monitoring, provide on-the-fly intelligent decision-making and self-regulation. How and when the interaction develops will play an important role in freight planning.

19.4.2 Advanced Planning Techniques

Forecasting freight requires an understanding of freight demand and supply chain logistics. In addition to the traditional multistep model presented in this chapter, several additional approaches have been proposed, including Boarkamps, Binsbergen, and Bovy's GoodTrip model (2000), Nagurney et al.'s multilevel spatial price equilibrium modeling (2002), and Hancock, Nagurney, and Southworth's microsimulation modeling (2001). All three approaches emphasize multimodal network-based models and require an underpinning in decision-making of shippers, carriers, and receivers. The GoodTrip model expands the multistep process by explicitly including transactional stages and associated players involved in the supply chain before determining mode and route. The methodology proposed by Nagurney et al. models the components of the supply chain as a network of commodity flows, information flows, and prices for a given industry. Supply and demand equilibrium is obtained by iteratively solving across demands, production functions, and transactions costs for each movement in the supply chain. Transaction costs are dynamically updated to include costs of physically moving that freight as well as associated business costs.

High-speed desktop computing and more available data have allowed planners to consider detailed microsimulation of individual vehicle- or trip-based movements proposed by Hancock et al. (2001) as a feasible option. Individual movements are aggregated to obtain traffic volumes and modal information. This approach, as well as the multi-level network model, are both in the initial stages of prototyping. Although each shows promise, they are still a long way from practical applicability.

19.5 CONCLUSION

As presented in this chapter, planning for freight is a complex and difficult task. Because of the differences between public and private sector operations and planning and the variability of commodities and the means for shipping commodities, the ability of a public planning agency to grasp the breadth of this activity is problematic at best. However, if transportation planners are to make effective decisions in the current political and economic environment, a method for incorporating freight into the planning process is crucial. This chapter has attempted to provide a backbone on which to build a process for doing this while alerting the planner to the potential limitations and problems that will be encountered.

REFERENCES

Boarkamps, J. H. K., A. J. Binsbergen, and P. H. L. Bovy. 2000. "Modeling Behavioral Aspects of Urban Freight Movements in Supply Chains." *Transportation Research Record* 1725:17–25.

———. 2002. *National Transportation Statistics 2002*. U.S. Department of Transportation, BTS, Washington, DC. Available at http://www.bts.gov/publications/national_transportation_statistics/index. html.

Bureau of Transportation Statistics (BTS). 2001. *Transportation Statistics Annual Report 2000*. U.S. Department of Transportation, BTS, Washington, DC.

Cambridge Systematics, Inc. 1995. *Characteristics and Changes in Freight Transportation Demand: A Guidebook for Planners and Policy Analysis.* NCHRP Report 8-30. National Research Council, Transportation Research Board, Washington, DC.

———. 1997. *Quick Response Freight Manual.* Report DOT-T-97-10, U.S. Department of Transportation and U.S. Environmental Protection Agency, Washington, DC. Available at http://tmip.fhwa.dot.gov/clearinghouse/docs/quick/.

———. Leeper, Cambridge & Campbell, Inc., Sydec, Inc, T. M. Corsi, and C. M. Grimm. 1997. *A Guidebook for Forecasting Freight Transportation Demand.* Report 388, National Research Council, National Cooperative Highway Research Program, Transportation Research Board, Washington, DC.

Coogan, M. A. 1996. *Synthesis of Highway Practice 230: Freight Transportation Planning Practices in the Public Sector.* National Research Council, National Cooperative Highway Research Program, Transportation Research Board, Washington DC.

Czerniak, R., S. Gaiser, and D. Gerard. 1996. *The Use of Intermodal Freight Performance Measures by State Departments of Transportation.* Report DOT-T-96-18, U.S. Department of Transportation, Washington, DC, June.

Environmental Protection Agency (EPA). 1993. 40 CFR Parts 6, 51, and 93, "Determining the Conformity of General Federal Actions to State or Federal Implementation Plans."

Federal Highway Administration (FHWA). 2001. Creating a Freight Sector within HERS. White paper prepared for FHWA by HLB Decision Economic, Inc. U.S. Department of Transportation, FHWA, Washington, DC, November.

———. 2002. *The Freight Story: A National Perspective on Enhancing Freight Transportation.* U.S. Department of Transportation, FHWA, Washington, DC, November. Available at http://ops.fhwa.dot.gov/freight/publications/freight%20story/.

———. Public-Private Freight Planning Guidelines. U.S. Department of Transportation, FHWA, http://www.fhwa.dot.gov/freightplanning/guide12.html, accessed January 29, 2002.

Federal Highway Administration (FHWA) and Federal Transit Administration (FTA). 1993. 23 CFR Part 450 and 49 CFR Part 613, "Regulations—Statewide Planning; Metropolitan Planning."

———. 1994. 23 CFR Parts 500 and 626 and 49 CFR Part 614, "Regulations—Management and Monitoring Systems.

Jiang, F., P. Johnson, and C. Calzada. 1999. Freight Demand Characteristics and Mode Choice: An Analysis of the Results of Modeling with Disaggregate Revealed Preference Data, 158 *Journal of Transportation and Statistics,* December 1999.

Fischer, M. J., and M. Han. 2001. *Truck Trip Generation Data.* Synthesis 298, National Research Council, National Cooperative Highway Research Program, Transportation Research Board, Washington, DC.

Hagler Bailly Services, Inc. 2000. *Measuring Improvements in the Movement of Highway and Intermodal Freight.* Report DTFH61-97-C-00010, U.S. Department of Transportation, Federal Highway Administration, Washington, DC, March 20.

Hancock, K. L. 2000. *Freight Transportation Data in the New Millennium—a Look Forward.* National Research Council, Transportation Research Board, Washington, DC. Available at http://gulliver.trb.org/publications/millennium/00043.pdf.

Hancock, K. L., A. Nagurney, and F. Southworth. 2001. "Enterprise-Wide Simulation and Analytical Modeling of Comprehensive Freight Movements." National Science Foundation Workshop on Engineering the Transportation Industries, Washington, DC, August 13–14.

Maze, T. H. 1994. *Freight Transportation Planning Process (or Chaos).* Second Annual National Freight Planning Conference Report, February.

Meyburg, A. H., and J. R. Mbwana. 2001. "Data Needs in the Changing World of Logistics and Freight Transportation." Conference Synthesis, New York State Department of Transportation and the Transportation Research Board Freight Transportation Data Committee.

Muller, G. 1999. *Intermodal Freight Transportation,* 4th ed. Washington, DC: Eno Transportation Foundation, Inc., and Greenbelt, MD: Intermodal Association of North America.

Nagurney, A., K. Ke, J. Cruz, K. Hancock, and F. Southworth. 2002. "Dynamics of Supply Chains: A Multilevel (Logistical/Information/Financial) Network Perspective." *Environment and Planning B* 29: 795–818.

Norris, B. B. 1993. "Intermodal Performance Standards." Volpe National Transportation Systems Center, U.S. Department of Transportation, Integrating Transportation Management Systems into Transportation Planning and Operations National Conference, Nashville, TN, November..

Peyrebrune, H. L. 2000. *Synthesis of Highway Practice 286: Multimodal Aspects of Statewide Transportation Planning.* National Research Council, National Cooperative Highway Research Program, Transportation Research Board, Washington, DC.

Southworth, F. 2003. "Freight Transportation Planning: Models and Methods." In *Transportation Systems Planning: Methods and Applications,* ed. K. G. Goulias. Boca Raton: CRC Press, 4.1–4.29.

Southworth, F., et al. 1983. "Strategic Freight Planning for Chicago in the Year 2000," *Transportation Research Record* 920.

Texas Transportation Institute (TTI). 2002. *2002 Urban Mobility Report.* College Station, TX: TTI.

Transportation Research Board (TRB). 2002. *Freight Transportation Research Needs Statements.* Transportation Research Circular, Number E-C048, National Research Council, Transportation Research Board, Washington, DC, December. Available at http://gulliver.trb.org/publications/circulars/ ec048.pdf.

Wigan, M., N. Rockliffe, T. Thoresen, and D. Tsolakis. 2000. "Valuing Long-Haul and Metropolitan Freight Travel Time and Reliability." *Journal of Transportation and Statistics* 3(3):83–89.

Wilson, A. G. 1970, *Entropy in Urban and Regional Modelling.* London, Pion.

Wood, D. F., and J. C. Johnson. 1996. *Contemporary Transportation*, 5th ed. Upper Saddle River: Prentice Hall.

CHAPTER 20
MANAGEMENT OF TRANSPORTATION ORGANIZATIONS

George L. Whaley

San Jose State University, San Jose, California

ABSTRACT

This chapter starts with a discussion of the external and internal environments of organizations and the effect these environments have on managerial effectiveness. An overview of the four major management functions, how they relate to each other and typical transportation organizations follow. Since management involves the process of achieving organizational goals by engaging in functions that transcend roles, levels and often the industries of those involved in these ongoing activities, management contingencies are also discussed. The discussion focuses on key activities or functions that effective managers perform in transportation organizations. An opportunity to assess individual management and leadership style is provided and readers are encouraged to reflect on the impact style and role could have on their own managerial effectiveness. The chapter ends with a discussion of future trends for the transportation industry and major management challenges within the industry.

20.1 WHAT IS MANAGEMENT?

Management can be defined as getting work done through others, and it is both an art and a science (Cook and Hunsaker 2001; Gibson et al. 2009). The art or practice of management is thousands of years old; however, the formal study or science of management is less than 200 years old. George (1972) points out that managers who were involved in ancient military-related activities and transportation activities related to building large projects such as the great pyramids in Egypt or the great wall in China relied heavily on trial and error experiences to get the work done. Later in history, Henry Poor, editor of the American Railroad Journal from 1849–1862, established basic management theories and principles for large business enterprises (George 1972). In the latter part of the 1800s, F. W. Taylor, the "Father of Scientific Management," and a plethora of other practitioners helped to establish the science of management (George 1972). Many of the management theories, principles, and concepts from the ancient past are useful today in all organizations. Contemporary transportation managers have an advantage over historical managers because they can use the best practices of these predecessors and they have the opportunity to use appropriate management theories and principles that are supported by solid research evidence.

The effectiveness of transportation organizations depends on an intricate set of relationships between people, materials, machines and technology. Transportation engineers focus on these relationships from an equipment, materials, and systems perspective, and it is common to hear the term "management" applied to purely engineering activities. However, engineering, maintenance and operations departments in transportation organizations usually focus on the transportation system. Transportation systems involve inanimate components such as roads, bridges, materials, machines, vehicles, schedules, prices, and information systems. Some administrative areas such as transportation planning also focus on these components from an organizational goal, revenue and time perspective. Hence, the management of transportation systems overlaps but is not identical to management of the people to make the systems work. While the importance of transportation management systems should not be diminished, modern-day organizational effectiveness has increasingly hinged more on the people component than the other components mentioned. Thus, the focus of transportation management is the interrelationship between the people within the organization with less emphasis on the inanimate, economic or infrastructure components.

20.2 THE EXTERNAL ENVIRONMENT

The external environment contains forces and changes that impact all transportation organizations (Pouliot 2002; *Public Transportation Fact Book* 2008). These external forces often impact each other and serve as both constraints and opportunities for the organization (Gibson et al. 2009). The primary external items that foster change are: economic, geographic, environmental, social, political, and legal but not necessarily in this order. For instance, the decision by a transportation organization to use an alternative fuel such as liquid gas may be based on external funding from a federal grant. The grant may be initially viewed as an exciting opportunity to reduce carbon emissions (environmental), serve a new market (social/political) and become a new revenue source (economic) for the organization. However, management may soon find out there are unintended consequences to the grant decision related to both the external and internal environments. Use of an alternative fuel may require a major redesign of existing vehicle and monitoring equipment as well as development of new techniques for vehicle maintenance. These technical issues may involve operational and maintenance constraints and result in employee resistance to the change (social) because new job skills are required. In order to successfully implement changes related to any new technology, management needs to be aware of the interrelationships caused by changes that impact local interests (geographical) and other stakeholders. The federal grant (political) may require new regulations (legal) that impact external governmental reports, external and internal employee safety reports, and internal work procedures. External factors that increase management challenges in transportation more than other industries are: unionization, government regulation, safety and economic environment.

20.2.1 Unionization Issues

Moffat and Blackburn (1996) and Oestreich and Whaley (2001) suggest the degree of unionization impacts the transportation industry and many facets of management. The transportation industry is more highly unionized than many other industries and public sector employees are more highly unionized than private sector employees. Government statistics (*Public Transportation Fact Book* 2008) show that in 2007, total union membership percentages for U.S. *private* sector employees were 7.5 percent and 35.9 percent for *public* sector employees. Yet in the private sector of the transportation industry, union membership was 22.1 percent, which is almost three times higher than the total private sector union membership percentage and two-thirds of the level in the U.S. public sector. The high rate of unionization in the transportation field means that supervisors and managers need to have at least basic knowledge of labor relations to realize effective management. Although most large transportation organizations have specialists on-board in the area of labor relations or hire consultants, individual managers and supervisors are required to have general knowledge of work rules in areas covered by collective bargaining agreements. Oestreich and Whaley (2001) point out this level of labor relations knowledge is not only true for the legal aspects of traditional union-management relations, but it is also needed for the increasing trend of union-management cooperation in transportation areas. These managers and supervisors may also be called upon to serve on joint labor-management committees. The topics covered by joint labor-management committees range in complexity from how to handle a current budget crisis to how to implement a new federal law. Hence, the "knowledge" aspect of union-management relations is a basic management competency. Some transportation facilities require managers to have enhanced communication skills to explain labor-management issues to internal stakeholders and external entities such as public officials and the press.

20.2.2 Government Regulation and Promotion Issues

Throughout history the success of military campaigns depended on effective management, especially transportation management (George 1972; Gibson et al. 2009). The *Public Transportation Fact Book* (2008) points out the efficient functioning of the entire U.S. economy depend on an effective transportation sector. Wood and Johnson (1996) indicate that U.S. national defense also depends to some extent on management of the transportation system. During the Desert Storm and Desert Shield military operations, for example, 63 percent of the military cargo to the Gulf region was shipped by vessels of private maritime companies, and 64 percent of the military personnel were flown to the region using commercial airplanes (Wood and Johnson 1996). The new emphasis on rapid deployment of people and equipment overseas in the case of war and homeland security as well as survival during natural disasters inside the U.S. extends this view to contemporary transportation management. More recent military campaigns, suggest that U.S. national defense policy, segments of the economy and government regulations all underscore continued dependence on transportation management.

The *Public Transportation Fact Book* (2008) provides a summary of five decades of federal legislation ending with the current law, Safe, Accountable, Flexible, Efficient Transportation Equity Act: A Legacy for Users (SAFETEA-LU) enacted in 2005. Many of these laws regulate federal funding of transportation services, and different laws regulate various modes of transportation, such as railroads and domestic water carriers. Federal and state government agencies monitor and control the different modes of transportation. The key federal agencies are: the Federal Aviation Administration, the Federal Highway Administration, the Federal Railroad Administration, the Maritime Administration, and the Federal Transit Administration. The U.S. Secretary of Transportation is a cabinet level position under the control of a super-agency, the U.S. Department of Transportation (DOT), and DOT heads most of these agencies. One notable exception is the U.S. Coast Guard that transitioned from DOT to the newly formed Department of Homeland Security in March 2003. The federal SAFETEA-LU law extends through mid-2011 and discussions are underway in Congress to reauthorize or develop a new surface transportation law. Many states and local entities have their own regulations, transportation laws and administrative agencies that augment federal level coverage.

Transportation managers, particularly in the top echelons, are required to have a considerable amount of legal, communications and political skill to deal with the direct influences of federal, state and local agencies, as well as special districts, concerning the viability of their own organizations. In today's environment, competencies or the skills, knowledge, and abilities (SKAs) requirements trickle down from the top of the management hierarchy to first level managers or supervisors. Hence, the federal government not only has a compelling interest in promoting transportation, but also it has an interest in regulating transportation more than most other industries (Moffat and Blackburn 1996; Oestreich and Whaley 2001; Pouliot 2002).

20.2.3 Safety Issues

The Injury Facts (2007) published by the National Safety Council indicate commuter rail riders and transit bus riders are 25 times safer than car travel. Rod Diridon (2009), Senior Executive Director of the Norman Y. Mineta Transportation Institute states:

> *Mass transit has a much better safety record than the 43,000 people killed last year on the U.S. highways and we can improve our safety record in mass transit with smart use of existing regulations and use of technology such as Positive Train Control to reduce train collisions.*

Although public transportation continues to be one of the safest modes of travel in the U.S., there tends to be a higher concern, perhaps even a preoccupation, with safety in all modes of transportation than in most other industries. Legal and political pressures on carriers to maintain a solid safety record are enormous. A wide range of human resource management activities in transportation focus on company policies concerning safety related areas and government regulation. Oestreich and Whaley (2001) report one of the most severe external constraints upon transportation facilities is the legally mandated drug and alcohol testing of employees in safety-sensitive positions. The Omnibus Transportation Employee Testing Act of 1991 requires drug and alcohol testing in all areas of transportation. The U.S. Department of Transportation (DOT) has published detailed rules that require employers in the transportation industry to have programs in place that control drug and alcohol abuse. These rules include mandatory procedures for urine drug testing and breath alcohol testing. There are pitfalls for both union and management in transportation in this area of testing (Wytkin 1997). For union officers, supervisors and managers in the transportation industry who deal with employees in safety-sensitive positions, it is important to be familiar with the government's requirements of drug and alcohol testing. Even minor violations of these regulations can easily cause formal grievances for organizations with unions. Oestreich and Whaley (2001) indicate that while most unions in the transportation field strongly support safety measures, some are opposed to random drug and alcohol testing. Indeed, it is random testing that is the most controversial aspect of this issue, as opposed to not testing or testing for "just cause" reasons.

New and evolving technologies complicate the fragile balance between technology and regulation. Various forms of intelligent highway technology such as vehicle intelligence integration (VII) and positive train control (PTC) systems have been in existence for at least a decade to monitor vehicle use and prevent train crashes. Recent advances in smart cell phones and social networking technology such as Instant Messaging, Texting, Tweeting, and Facebook have created additional implications for safety and other transportation areas. This is particularly true when passenger safety is in question and where the carrier's vehicles have greater than normal potential for injuring the general public. One recent example is passage of the Rail Safety Improvement Act (2008) that requires widespread installation of PTC systems by December 2015 to prevent casualties due to unauthorized incursion by trains. Among new company policies, pressure has increased to restrict personal use of telecommunication devices such as cell phones and non-work-related use of computers and social networking by employees in safety-sensitive positions. For example, a train engineer who tweets for personal use while driving a train places the passengers, general public, and equipment at risk; however, this same social networking technology can be used by the engineer to update track and traffic conditions faster than most traditional monitoring equipment. VII systems may provide new methods to monitor the use of vehicles for safety and other public policy uses but, it raises additional questions concerning whether public records cover all communications at work as well as privacy implications of monitoring incidental personal use. As technology evolves, the potential for safety advances, new abuses develop and the laws and transportation organization policies need to keep pace. Most transportation organizations attempt to use existing human resource (HR) policies to cover abuses in these new areas. If abuses become more widespread, new policies could be needed to curtail non-job-related use of these new tools while operating equipment at work. These trends affect all transportation subindustries and sectors and complicate the role managers play, especially supervisors, because supervisors usually play a front-line monitoring and disciplinary role.

20.2.4 Economic Issues

One constant in transportation is the changing external economic environment. It was previously mentioned the country's economic development and quality of life for most individuals depend on the development of effective transportation systems (*Public Transportation Fact Book* 2008). Transportation metrics such as availability, sustainability, headway, average trip length and cost are part of the decision to locate, relocate, expand, and reduce organizations as well as determine the desirability of a location as a "good place to work." The U.S. general economic environment alternates between the three business cycles of growth, contraction and stability. The impact of these three business cycles are further complicated by the lag between planned economic activities and when the activities are implemented.

Stability is the easiest cycle to handle but seldom exists for long periods of time. In periods of rapid growth, transportation organizations experience more than normal external pressure to expand service levels. The expansion of service levels usually require capital investment, new legislation or regulations and result in other long term impacts on the transportation infrastructure. Often, the economic impact is not achieved because the environment changes before the required expansion of service levels can be implemented. When the previously mentioned growth cycle contracts or slows down, use of public transportation modes do not always slow at the same rate but begin to fluctuate and present different challenges for infrastructure, revenue and operating expenses. For instance, high unemployment that usually accompany recessions may decrease the use of highways, bridges and ferries by automobiles while increasing the use of other modes such as bus and rail because they may be less expensive options. These changes in transportation modes and lag in revenue flows challenges the budgets of local entities for infrastructure improvements. President Barack Obama signed the 2009 American Recovery and Reinvestment Act that was designed to help stimulate the U.S. economy and create new jobs while assisting local governments to update their transportation infrastructure and ease budget deficits. A large amount ($ 8.4 billion) was allocated for public transit projects, $8 billion for high-speed rail, $1.3 billion for Amtrak upgrades, $1.1 billion for new airport grants and $27.5 billion for highway infrastructure (www.plunkettresearch.com). The 2009 law indicates these funds can be used for capital projects such as bridge, track, signal and bicycle improvements, and purchase buses to improve the fleet and for preventive maintenance. However, this funding was not designed to help transportation organizations with expenses related to operational budget "deficits." If economic slowdowns extend into long recessions, transportation organizations often react by reducing service levels, increasing prices, and reducing investment in infrastructure that usually result in reduced transportation demand and revenue. The circular relationship between growth, contraction, and stability in business cycles is usually further complicated by the geographical location of the transportation organization and the reality that local economies and U.S. economy often move in opposite cycles of growth, contraction, and stability. Labor costs, revenue potential, and political considerations vary widely across the United States and help to support the traditional idea of local funding rather than federal government help with operating transportation deficits. If the current operating environment of transportation organizations is one of fiscal austerity and contraction, then prudent management suggest the need for contingency planning based on three different plans ready to implement at all times: growth, contraction, and stability.

Successful transportation organizations gravitate toward this contingency planning approach over time and throughout their management hierarchy while focusing on strategic pricing, cost-reduction, and productivity improvements. This contingency or situational management approach to the circular nature of external economic pressures places a premium on proactive transportation managers. Moreover, these managers are called upon to demonstrate a proclivity to scan the external environment for changes and development of management acumen in all four management functions: planning, organizing, directing and controlling.

20.3   *THE INTERNAL ENVIRONMENT*

The internal environment is equally critical to the successful management of transportation organizations as the aforementioned external forces. Usually, it is easier for organizations to control their internal environment than external environment. This involves the control and utilization of three key internal factors: people, financial, and technical resources. The transportation management focus here will be on the people resource, namely, the organizational culture, demographics of the workforce and workforce planning.

20.3.1   Organizational Culture

One important aspect of the internal people resource is the impact of their commonly held values on the organization. Gibson et al. (2009), Lund (2003), Osland (2007), and Robbins, DeCenzo, and Moon (2008) indicate organizational climate or culture (OC) is the set of organizational values based on what people in organizations actually "stand for and do." Numerous studies point out OC is an important determinant of decision-making and what actually happens in transportation organizations (Boxx, Odom, and Dunn 1991; Lund 2003; Whaley 2001). The "do it by the book," "detailed," and "mechanistic" OC of many transportation facilities have been influenced by the history of these organizations, external pressures from the need for public safety, and regulations associated with the funding relationship to the government. The thinking and learning styles of the engineers, technicians and maintenance employees that form a large part of these organizations contribute greatly to OC. As a result, traditional transportation firms have organizational cultures that emphasize control, stability and order. Perhaps the things-oriented "mind set" that is helpful in managing the inanimate components of the transportation systems influences the management of people. Paul, Niewoehner, and Elder (2007) indicate the clarity, precision, accuracy and logical "thinking and reasoning mindset" of engineers and related technical employees tend to lead them to focus on "things" rather than "people." This mindset supports mechanistic organizational cultures found in many transportation organizations as well as serve to attract, reward, and retain employees with traditional, things-oriented management styles. However, not all employees in transportation organizations support this alignment with mechanistic organizational cultures, and subcultures with different mindsets do exist in transportation facility departments. For example, Lund (2003) reports in his study the job satisfaction of administrative and professional employees in subcultures such as marketing was negatively related to mechanistic cultures. Gradually, transportation organizations have moved away from reliance on mechanistic cultures and have moved toward more flexible and spontaneous organic cultures or mixed cultures. Several reasons for these changes in OC to gain traction in transportation are the increasing complexity of tasks, changing demographics of the workforce, and the increased hiring of managers outside the transportation sector. As evidence grows that both mechanistic and organic organizational cultures can help to achieve organizational goals, transportation facilities have trained managers to take a situational approach, matching the work, styles and goals of organizational subcultures (Lund 2003; Whaley 2001).

20.3.2   Demographics of Workforce

The demographics of an organization are controllable to some extent through internal decisions related to hiring, promotion and retention. Toole and Toole (2007) indicate the transportation industry is facing the retirement of a large number of "baby boomers" that have been the foundation of many transportation organizations. Haas (2007) reports that industry sources estimate 50 percent of managers in transportation will be eligible to retire by 2009. They report a smaller number of students are pursuing disciplines that are critical to transportation. Economic recessions usually slow down these trends but do not fundamentally change them and this education and demographic gap in the transportation workforce comes at a time when the skills needed to work in the field are changing and expanding (Toole and Toole 2007). Hence, transportation organizations need to give more attention to the internal environment than in the past.

20.3.3   Workforce Planning

Workforce planning can help to mitigate some negative impacts of these aforementioned demographic trends and assist with alignment with the internal and external environments. Wayne Tanda (2009), Resource Management Director of Monterey County Transportation indicated:

> *Some transportation organizations have initiated workforce planning programs such as phased retirements, rehire of retired employees on a temporary basis as well as hiring of retired employees from other transportation organizations to counter demographic trends.*

Additionally, Haas (2007), indicated that while large employee retirements in the transportation sector may reduce management experience levels and repositories of organizational history, these trends could be used as an opportunity to promote the changed performance and diversity goals of transportation organizations, change the organizational culture and train employees for new skills. Since minorities and females in transportation organizations generally exist in larger numbers in the lower levels than senior management, promotion from within HR policies may help to meet one component of diversity management level goals. These new managers will need additional education and training to meet the new challenges such as technological advances and fast changing economic cycles as much as any other new managers. Transportation organizations should consider more emphasis on succession planning, community outreach, and use other workforce planning tools to match their internal environments with the realities of the external environment and the organization's diversity goals. Recent interviews with transportation leaders indicate increased emphasis on workforce planning, training, transportation certificate programs and formal college degree programs in transportation management are needed to improve the internal environment (Diridon 2009; Haas 2009).

20.4   FUNCTIONS OF MANAGEMENT

The effective management of organization takes into consideration both the external and internal environments as it seeks to accomplish organizational goals. Management functions include activities that often transcend the previously mentioned environmental forces and sometimes the industries involved to achieve organizational goals. Functions are the critical activities that effective managers actually perform. Over the last one hundred years, management scholars have identified from four to fourteen overlapping principles, processes or functions that effective managers perform. Most agree that planning, organizing, directing, and controlling are the four basic management functions. The "directing" function is often called "leading" or "influencing" while the other three function names are constant (Cook and Hunsaker 2001; Gibson et al. 2009; Robbins, DeCenzo, and Moon 2008).

20.4.1   Planning

Planning is the management function that focuses on goal setting and how best to achieve goals (Cook and Hunsaker 2001; Gibson et al. 2009; Robbins, DeCenzo, and Moon 2008). There are various levels of goals ranging from the vision and mission usually set at the top of the organization down to rules and procedures at the bottom. For example, the vision of a transportation agency might be, "To be the transit agency with the highest ridership and fare-box recovery in our region by year 2012." An example of a procedure for an individual bus operator might be, "Bus operators are required to call out all stops on each trip in a manner they can be understood by all riders." A procedure for an engineering manager at the transportation agency might be, "All engineering hours charged to a project must have a valid charge number from an approved budget and approved by an engineering supervisor." Unless the goals mentioned in these examples are integrated across all levels and areas of the organization and understood by stakeholders, there is little chance they will guide organizations toward success. Effective goal statements help with understanding and implementation, and one way to measure the effectiveness of the previously mentioned goals is to make them specific, clear, and challenging (Cook and Hunsaker 2001). The SMART method: specific and simple (S), measurable (M), achievable (A), realistic (R), and timed (T) is one popular technique. Therefore, a transportation organization with a goal, "To provide mobility to the public" is most likely an ineffective goal statement because it does not adequately address the five components of the SMART method.

Most organizations consider planning to be the responsibility of line management and not the planning department that advises management. Strategic planning is goal oriented, usually transformative, and most likely conducted by top-level managers. Tactical planning is usually less pervasive, shorter in duration, focused on one task or area and conducted by lower-level managers. If a bus company decides to change its mission to include other modes of transportation, this would most likely be considered a strategic change in their goals, developed by top-management, affecting the entire organization, and taking several budget cycles to implement. While important to the bus company, a small change to the fare charged on one bus route is usually a tactical goal change, less pervasive and transformative than a change in mission. Goals at all levels in an organization serve to guide the actions of employees and managers, motivate them to achieve the plans of the organization, help in resource allocation, act as constraints and help to measure results.

Management researchers and practitioners, such as Cook and Hunsaker (2001), Gibson et al. (2009), and Robbins, DeCenzo, and Moon (2008), indicate that planning is the most important management function and all other organizational activities should support the organizational goals. The planning department or unit is a more recent addition to transportation than many other industries. Although many larger transportation organizations have a planning department, the employees in these units are considered to be support staff and do not have organizational accountability for strategic organizational goals they help to establish. In transportation organizations, short run and operational activities often become urgent and take priority over both long run and more strategic activities. Moffat and Blackburn (1996) indicates a common shortcoming of bus field supervisors in transportation is their focus on what is urgent and short range, and there is a propensity to ignore items that are goal oriented and future oriented. This suboptimization of goals sometimes extends beyond first-level managers to higher management levels and across each level in transportation organizations. Consider the following traffic control scenario: A budgeted city project to replace an outdated traffic signal over the next ninety days that could improve traffic flow receives less priority than a request to install a new traffic signal where pedestrian jaywalking has increased and one death at the intersection has occurred. If the goal of the city traffic department is "to reduce traffic congestion and increase the flow of traffic by twenty percent within one year without increased cost and reduced benefit," why does suboptimization occur? Urgency, outdated procedures, internal and external politics, lack of training, poor direction, and lack of clear performance metrics may lead the traffic department supervisor to ignore the strategic goals and metrics of the organization before making a tactical decision. One example of the assistance appropriate metrics and transportation infrastructure approach can provide is the City of San Jose, California. Jim Helmer, Director of Transportation for the City of San Jose reports the "injury crash rate per 1000 residents" is used as a traffic safety performance metric that helps to align traffic safety with the overall safety goals for the city (Helmer, 2008). Wood and Johnson (1996) suggest that lack of a profit motive, the number of stakeholders, structure and a plethora of laws involved in goal setting in the public sector, makes planning a longer, less integrated and more involved process than in the private sector. Moreover, if a transportation organization has a mechanistic organizational culture, is hierarchical, makes all decisions "by the book" and communicates strictly through the chain of command, supervisors and other managers may feel a disconnection from goal setting at any level. This feeling of disconnection often leads to alienation or "going it alone behavior" and "cutting corners" rather than advocacy for improving the planning process. The need to prioritize internal activities as well as the previously mentioned external forces such as the SAFETEA-LU law serve to align local, regional and national transportation goals. Additionally, they help to increase the importance, integration, and effectiveness of planning at all levels of transportation organizations. These improvements are examples of the contribution planning could make to the transportation management function and confirm it as the most important management function. Plans need to be implemented effectively, and integrating the other three management functions (organizing, directing, and controlling) with planning is crucial.

20.4.2 Organizing

Management effectiveness also involves organizing activities related to the structure of the organization in terms of the work content and process at the macro- and microlevels (Gibson et al. 2009). Traditional, formal organizations focus on organizational structure more than the other functions of management (Cook and Hunsaker 2001; Gibson et al. 2009). However, effective organizations align organizational structure with the external forces and the vision/mission/strategy as well as other internally controlled factors. The larger, higher, or macro-level organizing activities focus on arranging jobs, people, and other resources into units that are efficient and that are most likely to help achieve the goals of the organization. When these structural arrangements are displayed in a picture or chart format, it is called an organizational chart. Moffat and Blackburn (1996) and Wood and Johnson (1996) suggest the typical structural units at the top of most transportation organizations are: operations, maintenance, and administration, and each unit has subunits in larger organizations. Arrangement of transportation organizations into these three basic units at the top is reinforced by the method some federal funds are earmarked in legislation for specific subunits such as "operations." Fundamentally, operations is usually a separate unit in transportation because of the day-to-day focus on service levels and support. Maintenance is a separate unit within most transportation organizations because it focuses on keeping equipment in service, meeting important safety goals, and it requires different employee skills than the operations and administration units. The administrative unit is devoted to activities that support the other units and the customer. Administration is often divided into smaller subunits such as Human Resources and Finance. Planning, Information Technology and Marketing are more recent subunits found inside administrative units and they frequently report directly to the top manager in larger transportation organizations.

Many organizational structure formats exist and each one has advantages and disadvantages (Cook and Hunsaker 2001; Gibson et al. 2009; Osland et al. 2007; Robbins, DeCenzo, and Moon 2008). Organizational structures are called "functional" structures when similar jobs inside these units are based on similar disciplines, activities, or outputs such as operations, planning, maintenance, marketing, engineering, human resources, finance, and information technology. Other distinctive structural formats are: project, product, process, territory, and customer structures. A project structure is said to exist when a key activity or department starts and ends at a specific time. If a transit organization's structure at the top is based solely on bus service and light rail service, this is a form of product/service organization structure. A trip or route-based transit organization is often referred to as territory or customer organization structures. Newer organizational structures called virtual, modular, network or hollow structures are distinct because they farm out part of traditional activities to brokers or other organizations. If a traditional functional transportation organization consisting of maintenance, operations, administration, and planning departments subcontracts all or part of the planning activities to another organization, this is an example of a virtual structure. Combinations of these distinct organizational structures are called hybrid formats or structures.

The matrix structure is a popular hybrid structure that usually exists when temporary projects are combined with any other organizational structure. Transportation agencies with marketing and planning activities that include several time-sensitive projects and the balance of the organization is structured along functional or product lines could be considered a pure project or perhaps a hybrid matrix organizational structure. Pure project structures exist when employees are 100 percent allocated to organizationally sanctioned projects. Matrix structures require selected department employees to work on temporary projects that are managed by employees such as project managers outside their own department and department employees temporarily assigned to these projects part-time continue to report to their own "home" department mangers. When used properly, there is less inclination to over staff the organization in matrix structures and responsibility for activities such as performance appraisals and career planning with the home departments is maintained. Although matrix structures can create a staffing resource savings to transportation organizations, employees who work in matrix structures often feel they actually report to more than one manager and do not feel committed to any single project. Since each organizational structure has advantages and disadvantages, the management challenge is to find the structural format that best suits and assists the organization to meet its goals.

Many factors could be involved in the decision regarding how to structure a specific transportation organization. For example, if a city transportation department wants to start a "campaign" to make its streets safer, how should it structure activities to meet organizational goals? Should the campaign activities be assigned to an existing administrative department such as community relations, assigned to a project office along with other traffic engineering projects, organized as a temporary project as part of a matrix structure or serve as a vehicle to start a separate marketing department as a new function? If the size, urgency and criticality of the campaign are small, the activities would be most likely placed in an existing department. Which existing department selected would depend on factors such as tradition, existing employee skills and political clout of the subunit managers. If the campaign is viewed as temporary, a project or matrix structure would be appropriate. If an on-going need for campaigns exists as well as the need for more administrative than technical skills exist in the organization with a view toward marketing as a key activity, this would favor starting a new marketing subunit. Typically, engineering is a separate unit reporting to the top manager of the transportation organization when it is focused on strategic technologies; however, it can be disbursed throughout operations or maintenance units at the top. When the engineering activities are more tactical, routine or focused on the goals of the operations and maintenance units, engineering is sometimes disbursed and assigned to these two units. Planning departments that provide different types and levels of support to other units could be easily placed in the units they support or other administrative units. However, planning activities are often placed in a separate unit and report directly to top management when planning is a more strategic priority for the organization.

At the smaller, job or microlevel of the organization, the organizing function involves the grouping of tasks and placing the right people into the right jobs. This could involve conducting task, work or job analyses to write appropriate task, work or job descriptions and associated specification statements. Organizations may need to change the structure of tasks, jobs, departments and the entire enterprise in order to be effective. Often the structure of jobs, departments and the entire enterprise are altered or redesigned to incorporate these changes. The four common forms of job redesign based on work content found in organizations are: job enrichment, job enlargement, job rotation, and job simplification. Although all four methods can provide motivation through changes to the job content, job enrichment usually provides more employee motivation than the other three methods (Cook and Hunsaker 2001; Osland et al. 2007). Job enrichment accommodates increased "task complexity" without changes in the amount of work required and fulfills the higher level needs of the employee. The level of motivation and job satisfaction is based on the way each employee experiences three psychological states: meaningful work, responsibility for outcomes, and feed back on the results of their work (Bernardin 2010; Cook and Hunsaker 2001; Osland et al. 2007). Therefore, the degree of additional work motivation achieved is based on each jobholder's perception of the changes involved with the job redesign. A difference in perception might exist as a result of the external and internal forces faced by the organization, the role played by the individual and the individual's past experience with the task. Changes in job content based on emergency conditions, compliance with federal regulations such as the previously mentioned 1991 DOT regulations, the Americans with Disabilities Act (ADA) law or recent advances in information technology such as the Internet could impact most jobs in transportation. Would a design engineer who works for a state transportation agency on a highway overpass improvement project perceive an increase in task complexity for earthquake retrofit design work as job enrichment or simply more work to do (job enlargement)? Either a design engineer with a different skill set or a private sector consulting engineer may view the complexity of the same retrofit design work differently. The perception of the task, together with buy-in of the employees and organizational support for changes are the keys to successful job redesign.

Reengineering is a different form of redesign that focuses on changes to core business processes. Champy (1995) defines reengineering as a radical redesign of business processes to achieve dramatic improvements in critical, contemporary measures of performance such as cost, quality, service, and speed. The key to adding value to the organization based on reengineering is connected to how tasks relate to each other in order to meet the organizational goal. In the previously mentioned job enrichment for an engineer, a state highway overpass design department may involve job reengineering if the redesign effort involves the interrelated tasks and roles of many departments and jobs within the organization. Alternatively, the impact of reengineering on a different state highway design department, the purchasing department and even one document, a manual parts catalog could be examined. If the manual parts catalog is identified as a problem area, the engineers may desire an automated parts catalog to increase the speed and accuracy of buying parts for the organization while lowering cost. In order to implement this change, the engineers may discover the effects on activities such as purchasing, information technology, finance, accounting and facilities. In this example, the workflow in the form of changed work rules, roles and responsibilities may be more important for achieving the goals than the change in job tasks. If the cost of implementing this change is less than the benefits, a cost-benefit analysis suggests the change should be implemented. On the other hand, if the cost exceeds the benefits, the change should not be implemented in the manner suggested by the reengineering workflow analysis. Reengineering is effective when it improves core processes, competencies, and competitive advantage, and it falters when it becomes simply a label to substitute for restructuring, downsizing, and cost-reduction (Bernardin 2010). Therefore, the overall organizational structure and job structure needs to be aligned and any job redesign format should be carefully analyzed to make sure it benefits outweighs associated costs and fits the plan or goals of the organization.

20.4.2 Directing

This management function is sometimes referred to as leading or influencing (Cook and Hunsaker 2001; Gibson et al. 2009; Hersey, Blanchard, and Johnson 2001; Robbins, DeCenzo, and Moon 2008). Directing or leading involves influencing others through change, motivation and communication and its activities include more than simply giving instructions to employees. Directing often includes interpretation of the organization's goals such as visions, missions, and strategic plans and policies. This functional area's activities may range from changing new ways of thinking about mobility management in general to explaining how access to fewer resources will impact an employee's job and perhaps the impact of implementing a new provision in the union-management labor agreement. Often, directing entails a change in environment, resources, job or skill level and employees may resist the required change. Employees are usually motivated to change quickly and temporarily based on management tactics such as intimidation and fear but longer lasting changes are usually based on items that are more positively perceived such as trust, self-interest and education (Cook and Hunsaker 2001; Osland et al. 2007). Some may suggest that money is a motivator, however, modern motivation research (Hersey, Blanchard, and Johnson 2001; Osland et al. 2007) suggests that rewards such as money are seldom motivators for professional and managerial employees. In unionized settings where seniority rather than merit is used for rewards and mobility, individual monetary rewards are not a motivating factor for most employees. Modern motivation research suggests that generally speaking, money is really a hygienic or maintenance factor. Maintenance factors can easily demotivate employees especially when they are not properly allocated, administered, and communicated but do not motivate the employees. Experienced managers note that employees may use money as a status symbol or substitute money in the conversation when another workplace factor is the real issue bothering them. Employees do not usually work harder or smarter after a salary increase or promotion but they usually complain and produce less output and quality when these salary increases and incentives do not occur equitably (Cook and Hunsaker 2001; Osland et al. 2007). When job goals are unclear, money is often substituted as the goal and appears to motivate through reinforcement. If money is tightly linked to job goals, such as incentive-based compensation plans, it could become a motivator. As previously mentioned, in highly unionized settings such as transportation, pay for many employees is set by the collective bargaining process and most raises and promotions are determined by seniority, thus, the salary-performance link to motivation is less clear. However, one should not conclude money is unimportant to employees, but it is seldom the primary reason to perform job tasks.

Most successful change strategies require effective organizational, group, and individual communications. Communication skills such as persuasion and conflict resolution are required in the directing function. If your city's engineering department is accountable for a new "traffic calming" project, as the department supervisor, what communication techniques would be most effective to explain the goals of the project to engineers with more traditional traffic control experiences? One-on-one communication to all field located employees by their managers who are located elsewhere would be costly. Placement of key employee information on the organization's Intranet may save money and improve access. Use of social networking tools based on computer or smart telephone technology could expand this access. One downside is employees may not have access to computers in the field and fail to notice placement of the information only on the organization's Web site and use of mobile communications devices such as tablets and cell phones may be problematic in safety-sensitive jobs. In this situation, slower and more traditional methods such as voicemail, e-mail, or regular telephone and mail may result in the most cost-effective solution. In addition to the tangible costs and benefits of each communications method, the decision-maker needs to consider the intangible aspects of employee communication style and aforementioned organization culture before making a choice based on the total cost of communications. Good ideas often fail because of poor verbal and nonverbal communications.

Communications choice and style are key components of interpersonal skills, but effective listening; diversity management, impression management, and emotional intelligence are also at the top of the list. Cook and Hunsaker (2001) indicate effective listening is one of the most important communication skills for managers. Diversity based on legally protected status such as age, gender, race, national origin, religion, and disability will most likely continue to be covered under federal law in the future. States, cities, and other local entities may have additional legally protected areas such as sexual orientation. Employers that value differences should also consider the role of other types of diversity such as learning styles and leadership styles discussed in this chapter that have an influence on how work is performed. Not only is consideration for other forms of diversity a good thing to do, it could also provide the organizations with a competitive advantage when customers value diversity as a goal and they are also diverse. If diversity becomes the future organizational goal or norm then effective listening will be a key element for managers and organizations to effectively address diversity.

IQ is a common individual or personal intelligence measure and a new measurement of interpersonal skill is emotional intelligence. Emotional intelligence (EQ or EI) is a different concept from IQ, and it seeks to measure several types of interpersonal skills (Osland et al. 2007). Goleman (1998) suggests EQ can be used to enhance the performance of the manager and the organization. Chen, Jacobs, and Spencer (1998) go further and indicate that close to 90 percent of success in leadership is attributable to emotional intelligence. Five components of EQ that transportation managers need to keep in mind are: (1) self-awareness, (2) managing emotions, (3) motivating oneself, (4) empathy, and (5) social skill. Self-regulating emotional control is helpful so that constant change does not "get the manager off-track." If doing more with fewer resources is the wave of the future, then expecting employees to perform at a rate higher than stated in the job description is the next logical step. This expectation takes many different forms in the transportation industry. Some organizations include these items in their performance management system as "stretch" goals for employees. Coleman and Borman (2000) and Osland et al. (2007) identify this performance expectation as "Organizational Citizenship Behavior (OCB)." While both formats include EQ, the level of job performance expectation involved in OCB requires organizational support and makes implementation of OCB vulnerable to inappropriate self-promotion by employees. Some degree of "office politics" exists in all organizations but self-promotion and impression management are not automatically synonymous with dysfunctional organizational politics. If employee performance is high and job complexity is low, persuasion skills and impression management are usually perceived by others as a positive influence on goal attainment as opposed to organizational politics (Cook and Hunsaker 2001).

Traditional labor management relationships are based on short-run interpersonal items such as power and fear. Bernardin (2010) and Oestreich and Whaley (2001) indicate Interest Based Bargaining or Negotiations (IBN) is based on quite different interpersonal expectations and norms and IBN is gaining acceptance in transportation settings. They conclude IBN is most likely to succeed when the survival of both union and management is at stake and when the aforementioned values of trust, self-interest and education exist between the parties. Hence, directing assists in the planning, organizing, and controlling functions and requires effective leadership skills.

20.4.3 Controlling

Effective management entails comparing the goals and standards of organizations to actual results (Gibson et al. 2009). The key activities in the controlling function are monitoring and regulating organizational activities and taking corrective action where necessary. Although controlling activities and tools are distributed throughout organizations, the most common monitoring tool in transportation is the budget. The budget is the organizational plan for people, equipment, operations and services expressed in monetary terms. The actual expenses are usually compared to the budget in the form of a variance report. Managers are expected to monitor results by explaining variances to each budget item and making the necessary changes to meet or alter the budget. Effective controlling could require making changes to the existing budget or development of a different type of budget. The organizational structure can influence the type of budget used for management decision-making. If a transportation facility has either a project or product organization structure, a project budget or product line budget should be considered. Time emphasis, future expectations and other nonfinancial goals could dictate the type of budget and monitoring required. What should be the appropriate budget format in a situation where a transportation facility has "time period" flexibility? Should management advocate "buying ahead" by increasing construction expenses in the current year and exceeding the current year's budget in order to reduce the total construction costs over several years? If a transportation organization considers downsizing the organization based on an unfavorable economic environment, what is the appropriate budget format impact? Should the organization reduce the budget of each department on a fixed percentage, reduce the budget based on reduced headcount, remove departments in total or in part based on outsourcing or simply "start from scratch" with a zero-based budget approach? In each situation or question mentioned, the constraints and opportunities posed by the external and internal environments on the controlling function suggest a different form of budget and monitoring. Buying ahead is an accepted purchasing practice to reduce total cost and downsizing may require a "before and after" budget for future comparisons. Where nonfinancial goals are on par with traditional financial metrics, an organizationally focused balanced scorecard approach may be useful (Bernardin 2010; Osland et al. 2007). This scorecard approached could be extended to include transportation specific items. Many public transportation agencies today are faced with reduced financial resources stemming from unfavorable economic and governmental conditions. These dwindling resources result from many factors such as reduced fare-box recovery, lower ridership, less income due to less taxing authority, less income based on lower public priority of services provided or funded projects that have been completed. Whatever economic climate exists today, current trends suggest that transportation managers will be increasingly asked to do more with fewer resources. Moreover, the previously mentioned rapid business cycle changes and austere economic environments will require higher levels of budgeting and financial analysis skills for all levels of transportation managers than previously required.

Examples of significant controlling activities in transportation are maintenance schedules, accident prevention, safety training and numerous human resource areas such as sick leave usage, workers compensation claims, overtime and vacation. Reengineering projects that were mentioned earlier as part of the organization function discussion, overlap sometimes with planning, directing, and controlling. For instance, a shipping organization may use several modes of transportation such as air, trucks, rail, and ocean to move cargo to customers. The tasks and roles involving planning, budgeting, billing, and claims could overlap and cause major inefficiencies. Reengineering the business processes of the shipping unit or department to be more efficient could change the tasks, roles, and reporting metrics. In this example, enterprise resource planning (ERP) software could be used as a control "tool" that provides a single, integrated application across all modes of transportation. This ERP tool could assist the change process and integrate all four management functions. Most likely, increased financial acumen and increased information technology knowledge are required across organizational subunits in transportation and from the top down to the supervisory level. Thus, the increased need to control costs causes all four management functions to overlap.

20.5 CONTINGENCY APPROACHES TO TRANSPORTATION MANAGEMENT ACTIVITIES

The writings of Classical Era management scholars and practitioners espouse that all managers should use management functions and principles in precisely the same manner (George 1972; Osland et al. 2007). Most modern research and contemporary managers advocate that effective organizations depend on situational or contingency approaches to management. Cook and Hunsaker (2001), Gibson et al. (2009), Osland et al. (2007) and Robbins, DeCenzo, and Moon (2008) suggest key situational factors that make a difference to organizational effectiveness are: goals of the organization, role of the manager, level of the manager, structure of the organization, technology of the organization, and the environment of the organization. Additionally, the unique characteristics of the industry where the managers work require consideration. The goals, roles and management levels are examined here as key determinants of managerial effectiveness in transportation organizations. One assessment instrument related to management or leadership style and role is provided so that individuals can assess their own style and reflect on the potential impact it could have on organizations where they work.

20.5.1 Organizational Goals

Goals and goal setting were previously discussed as key components of the planning function. Additional discussion of goals is warranted because the type, priority and measurement of goals can determine whether one transportation organization is more successful than another similarly situated organization and goals become critical contingent success factors. Efficiency and effectiveness are different yet related types of contingency goals (Cook and Hunsaker 2001; Robbins, DeCenzo, and Moon 2008). Efficiency is the ability to make the best use of available resources and in most instances it is measured as the ratio of outputs to inputs of the organization. There are a number of efficiency scenarios demonstrated in Figure 20.1 concerning outputs and inputs (Spray 1976). For example, efficiency is doing more with less but doing less with even less resources is also a form of efficiency.

Drucker (1967) has described efficiency as simply "doing things right" and noted efficiency is important but deceptive at the same time. Efficiency is easier to measure in private sector transportation firms because quantifiable marketplace results such as profit and revenue are used as metrics. However, efficiency in the public sector can be measured in terms of quantity and quality metrics such as service levels, stakeholder complaints, and commendations and bond ratings. According to

EFFICIENCY = OUTPUT / INPUT

1. OUTPUT INCREASES FASTER THAN INPUT (+/+)

2. INPUT DECREASES MORE THAN OUTPUT (–/–)

3. PRODUCE SAME OUTPUT WITH LESS INPUT (0/–)

4. PRODUCE MORE OUTPUTS WITH SAME INPUTS (+/0)

5. OUTPUT INCREASE AS INPUT DECREASES (+/–)

FIGURE 20.1 Efficiency metrics.

Drucker (1967), "doing things right" is important but "doing the right thing" is more important. An organization can be efficient and yet unsuccessful, that is, the organization can efficiently pursue the wrong goal. Hence, effectiveness is a balance between efficiency and goal priorities (Cook and Hunsaker 2001; Gibson et al. 2009). Norman Augustine, former CEO of Martin Marietta tells the following story concerning a bus company where it is efficient and is not effective (Dess and Lumpkin 2003):

> *I am reminded of an article. I once read in a British newspaper which described a problem with the local bus service between the towns of Bagnall and Greenfields. It seemed that, to the great annoyance of customers, drivers had been passing long queues of would-be passengers with a smile and a wave of the hand. This practice was however, clarified by a bus company official who explained, it is impossible for drivers to keep their timetables if they must stop for passengers.*

Goals expressed as metrics, such as "Percentage Sick Outs," "Overtime Used," "Ridership," "Fare-box Recovery," and "Time Schedule," require a critical balance for a bus company to survive and thrive. Prioritizing goals is not an easy task, and in the transportation setting, competing goals usually exist. For instance, a bus company with the goal to reduce equipment maintenance overtime expenses can achieve that goal by hiring more maintenance employees that might result in over-running the department budget because of increased total compensation. The right balance of competing goals across situations in transportation organizations is as important as efficient and effective goal implementation. Measurable goals were one component of the previously mentioned SMART method for measuring goal effectiveness. All other factors remaining equal, goals that are measurable tend to be remembered and used in decision-making.

20.5.2 Managerial Roles

It was pointed out earlier that management functions need to be altered based on different situations such as external and internal constraints and opportunities. One key internal constraint or opportunity is the type of jobs or roles in organizations. Cook and Hunsaker (2001), Gibson et al. (2009) and Hersey, Blanchard, and Johnson, (2001) define role as an organized set of expected behaviors of a particular position. Kotter (1990) describes the differing roles of nonmanagement, managers and leaders in organizations and the skill, knowledge and abilities that make them successful in these roles. Figure 20.2 shows examples of the impact role changes could have on individuals and the organization.

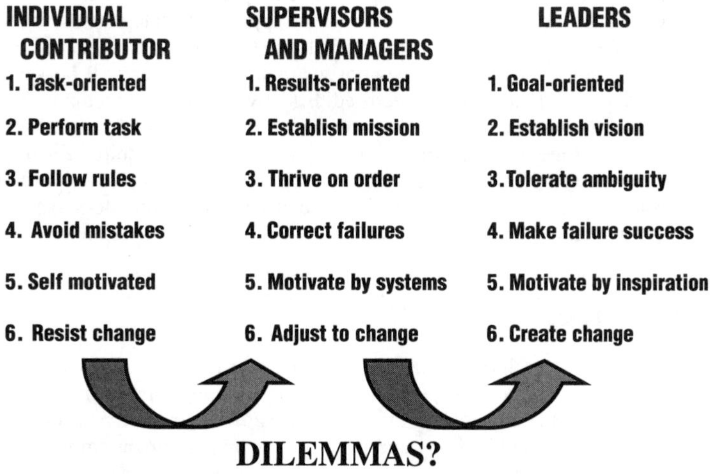

INDIVIDUAL CONTRIBUTOR	SUPERVISORS AND MANAGERS	LEADERS
1. Task-oriented	1. Results-oriented	1. Goal-oriented
2. Perform task	2. Establish mission	2. Establish vision
3. Follow rules	3. Thrive on order	3. Tolerate ambiguity
4. Avoid mistakes	4. Correct failures	4. Make failure success
5. Self motivated	5. Motivate by systems	5. Motivate by inspiration
6. Resist change	6. Adjust to change	6. Create change

DILEMMAS?

FIGURE 20.2 Transition in roles.

❏ **INTERPERSONAL**
1. **FIGUREHEAD**
2. **LEADER**
3. **LIAISON**

❏ **INFORMATIONAL**
4. **MONITOR**
5. **DISSEMINATOR**
6. **SPOKESPERSON**

❏ **DECISIONAL**
7. **ENTREPRENEUR**
8. **DISTURBANCE HANDLER**
9. **RESOURCE ALLOCATOR**
10. **NEGOTIATOR**

FIGURE 20.3   Mintzberg's management roles.

Success in one role does not automatically prepare employees for success at another role. Thus, individuals and organizations need to recognize additional preparation is needed as employees transition from one organizational role to another role. One transportation example of role as a contingency factor is based on the previously mentioned research study by Moffat and Blackburn (1996). This study reports most bus field supervisors are promoted internally from bus operator jobs and very little is usually done by the bus companies to prepare and support these supervisors in the new role. The study indicates new bus supervisors especially lack training for the transition to leadership roles and their previous role as bus operators provided little opportunity to develop leadership and management skills. Mintzberg (1990) and Robbins, DeCenzo, and Moon (2008) identify three general roles in Mintzberg's research that describe what managers typically do and what makes them effective. The three managerial role categories are: interpersonal, informational, and decisional. These three role categories are further divided into ten managerial roles: figurehead, leader, liaison, monitor, disseminator, spokesperson, entrepreneur, disturbance handler, resource allocator, and negotiator. Each role is placed under the appropriate category in Figure 20.3.

Effectiveness in each role contributes to the overall effectiveness of the organization, but success in a "spokesperson" role does not guarantee a manager will be successful in a "negotiator" role because different skills are required. Effective transportation managers recognize different roles are required for different situations and each role needs to be fulfilled by them within the resource constraints of the organization. For example, the skill set required for one's role as a negotiator during collective bargaining is much different from representing a transportation agency in a spokesperson role with the TV media to explain the effects of a work stoppage. Modern research confirms the contingency approach to the roles of managers in organizations.

20.5.3 Managerial Levels

Managers at each organizational level need critical skills, knowledge, and abilities (SKAs) to carry out the four basic management functions: planning, organizing, directing, and controlling and the right SKA mix depends on the situation. Supervisors or first-level managers are the lowest level in the management hierarchy and have only nonmanagement employees reporting to them. Examples of first-line managers in the transportation industry are: bus supervisor, dispatch supervisor, project engineering manager, planning department manager, and financial department manager. Middle-level managers usually have other managers or supervisors reporting to them. Since most medium and large size transportation organizations are composed of at least four major subunits: transportation planning, operations, maintenance, and administration that report to top management, middle-level managers are typically the heads of these functions. In smaller transportation organizations fewer subunits usually exist, the middle-level managers usually report to the top manager and may take on some activities of first-line managers. In larger transportation organizations, the top management level is usually composed of vice presidents or C-level managers of the major functions and the top manager of the organization with middle-level managers reporting to the top level. When the top manager has the title chief executive officer (CEO), the typical titles of managers reporting to the top are: VP of Human Resources, VP of Planning, chief operating officer (COO), chief information officer (CIO), comptroller or chief financial officer (CFO), and so on. Different levels of management emphasize a different set of skills, knowledge, and abilities that are needed to be effective. Hersey, Blanchard, and Johnson (2001) and Katz (1974) identified three key skills: technical, human, and conceptual in Figure 20.4 that is needed for successful managerial performance.

Although managers at all levels use all three skills, Katz's (1974) research indicates that effective top managers emphasize conceptual skills, middle-level managers emphasize human or interpersonal skills and lower level managers emphasize technical skills. Hence, management level in the organizational hierarchy is an important contingency factor for managerial effectiveness.

If transportation managers are promoted from engineer to supervisor and higher levels of management, they need to prepare for these changes in skill level through education and/or planned job experiences. Technically trained employees such as engineers may survive as supervisors of technical departments where they can rely on their technical expertise but may encounter difficulty with the interpersonal relations and conceptual aspects of planning, organizing, directing, and controlling demanded by higher levels of management. Small transportation organizations and transportation consulting firms do not usually have these three distinct levels of management and SKA flexibility and managerial training become even more important because goals, roles, and level considerations are intertwined. When budgets for transportation operations expand, management training opportunities usually increase; however, when economic times are difficult, the trend toward less supervisory training, emphasis on cost-effective computer-based training and the delay of management training is evident in most transportation areas. Moffat and Blackburn (1996) indicate bus companies have reduced their support for training over time, and the budget shortfalls across transportation organizations today continues this trend. This trend toward lower support for internal training requires

Management Levels
Top Level Managers

Skill Emphasis
Conceptual: Ability to grasp the big picture and analyze abstract issues

Mid Level

Human: Ability to demonstrate positive interpersonal skills

Low Level Managers

Technical: Ability to demonstrate competence and expertise in a particular field

FIGURE 20.4   Managerial roles, skills, and levels.

individuals, bus companies and other transportation organizations to seek cost-effective training from outside sources such as local universities, private consultants, the National Transit Institute (NTI) and internal training through on-the-job training (OJT). Hence, education and training are intertwined with management level as a contingency factor (Diridon 2009; Haas 2009; Tanda 2009).

20.6 LEADERSHIP IN TRANSPORTATION MANAGEMENT

Leadership was previously mentioned as a key part of the Directing function of management and many management scholars actually call this function, leadership (Gibson et al. 2009; Robbins, DeCenzo, and Moon 2008). It is widely recognized that leadership is helpful for nonmanagement employees but it is a critical skill for managers. Blanchard and Johnson (2001), Covey (1990), and Hersey et al. (2009), Kotter (1990), and Robbins, DeCenzo, and Moon (2008) identify leadership as an influence process and key success criterion for managers in all industries and all levels of management. Mintzberg (1990) identifies leadership as one of the ten basic management roles, and Kotter (1990) identifies leadership as a distinctly different role and set of activities from management. Although management and leadership roles and activities overlap to some degree, leadership is unique in many ways. Leadership is regarded as such an important ingredient of individual and organizational success that it is examined here in more detail. Moffat and Blackburn (1996) identify leadership as key success criterion for the bus supervisor. Do these research findings apply to other transportation management jobs? Let us suppose a brilliant traffic engineer working for the same organization as the bus supervisor had poor peer group rapport before promotion internally to supervisor of the traffic department. The new traffic department supervisor may have appropriate content knowledge for the traffic supervisor job based on past experience, but lack the skill, knowledge, and ability to influence subordinates because of poor interpersonal relationships. For instance, this new supervisor may take classes in how to develop and administer performance appraisals yet fail to develop "poor performing employees" or influence them to change behavior in line with job and organizational goals. Interpersonal skills such as effective communications emphasize problem solving, trust, and follow-up instead of personalities are the key. Although it helps to have good relationships with subordinates, it is not necessary that they "like you" to influence them in a positive manner. Most likely the above traffic supervisor will not work out as an effective leader in this department, both management and leadership are important success criteria.

20.6.1 Leadership Style

A consistent pattern of behavior is commonly referred to as a "style." It is difficult to be a successful manager in transportation settings with only one leadership style. For example, a government-owned and -operated carrier may require different leadership behaviors in its top management than a private carrier would require. A small transportation organization may call for a different type of leadership than a large organization. Influence on an experienced group of engineers working on a complex task may require a different leadership style than influence on inexperienced engineers working on a simple task. Perhaps in the earlier traffic supervisor scenario, if the supervisor had acceptable interpersonal skills, but had an inflexible leadership style that did not match the task situation and development level of subordinates within the department, positive influence and effectiveness would remain elusive. In order to be effective, leadership style and match of style to the situation is needed to be flexible and effective (Hersey, Blanchard, and Johnson 2001). Of course, appropriate interpersonal skills and organizational resources are needed to successfully implement the activities involved.

The match between leadership style and organizational culture is another important ingredient for individual and organizational effectiveness. Moffat and Blackburn (1996) suggest a traditional autocratic leadership style is prevalent in repair and maintenance operations, and in bus and trucking operations. In this example, the organization expects supervisors to be control-oriented, to enforce rules and punish noncompliance. The research findings indicate bus operators frequently complain the only time they hear from their supervisor is when they are in trouble. Engineers in the same environment frequently complain that supervisors micromanage their work and get in the way of productivity (Moffat and Blackburn 1996). Moreover, first-line supervisors lack the leadership training needed to relate well to the people they supervise. This type of internal organizational environment or culture and lack of management training makes innovative practices and changes in leadership styles very difficult. Increasingly, unanticipated changes in organizational goals or resources and planned change programs challenge managers in transportation to change their leadership style to match the organizational culture or change the situation in some other manner.

A popular view of leadership that is not supported by modern research is that leaders are born and not made. Bennis and Thomas (2002) contradict this view of leadership and indicate leaders are made and not born. They report that every leader regardless of age, had at least one intense, transformational experience. Traits or characteristics-based leadership theories were also popular for a time but with a few exceptions are not generally supported by modern research. There are numerous theories of leadership based on items such as style, power, change and the situation (Gibson et al. 2009). Traditional style theories of leadership suggest one uses the same approach to influencing others in all situations. Modern research supports a plethora of multiple style categories and situational approaches to leadership. Two basic styles, "autocratic" and "democratic" appear in some form in both multiple style and situational approaches to leadership. The autocratic style is similar to Theory X and the democratic style is similar to Theory Y identified by McGregor (1960). Figure 20.5 shows assumptions that people in general make about human behavior are related to Theory X and Y styles.

THEORY X ASSUMPTIONS

1. The average human being has an inherent dislike of work and will avoid it if he can.

2. Because of this human characteristic of dislike for work, most people must be coerced, controlled, directed, and threatened with punishment to get them to put forth adequate effort toward the achievement of organizational objectives.

3. The average human being perfers to be directed, wishes to avoid responsibility, has relatively little ambition, and wants security above all.

THEORY Y ASSUMPTIONS

1. The expenditure of physical and mental effort in work is as natural as play or rest.

2. External control and the threat of punishment are not the only means of bringing about effort towrd organization objectives. Man will exercise self-direction and self-control in the service of objectives to which he is committed.

3. Commitment to objectives is a function of the rewards associated with their achievement.

4. The average human being learns under proper conditions not only to accept but also to seek responsibility.

5. The capacity to exercise a high degree of imagination, ingenuity and creativity in the solution of organizational problems is widely, not narrowly, distributed in the population.

FIGURE 20.5 Assumptions about human behavior.

20.6.2   Leadership Style Assessment

Pfeiffer and Jones (1972) provide a short exercise shown in Figures 20.6 and 20.7 that allows individuals to assess their leadership style based on assumptions supervisors generally make about human behavior in work settings. You can participate in this exercise by placing a checkmark (X) somewhere between 10 and 40 points on the line provided in Figure 20.6 indicating whether you believe your style is Theory X or Theory Y.

Figure 20.7 provides 10 workplace scenarios and the exercise requires respondents to address each situation as they would actually respond on the job in the role as a supervisor. Each response is given a "weight" or score between 1 and 4 and this results in a total score between 10 and 40 based on "summing" your responses to each scenario.

A total score between 10 and 20 points suggests respondents would use a Theory X style across all situations and a score between 30 and 40 points suggests respondents would use a Theory Y style

Indicate on the scale below ✔ where you would classify your own basic attitudes toward your subordinates in terms of McGregor's Theory X and Theory Y assumptions.

Theory X _____ Theory Y

    10   20   30   40

FIGURE 20.6   Management style.

Directions: The following are various types of behavior which a supervisor (manager, leader) may engage in relation to subordinates. Read each item carefully and then put a checkmark in one of the columns to indicate what you would do.

IF I WERE A SUPERVISOR :	Make a Great Effort to Do This	Tend to Do This	Tend to Avoid Doing This	Make a Great Effort to Avoid This
Points:	(1)	(2)	(3)	(4)
1. Closely supervise my subordinates in order to get better work from them.				
2. Set the goals and objectives for my subordinates and sell them on the merits of my plans.				
3. Set up controls to assure that my subordinates are getting the job done.				
4. Encourage my subordinates to set their own goals and objectives.				
5. Make sure that my subordinates' work is planned out for them.				
6. Check with my subordinates daily to see if they need any help.				
7. Step in as soon as reports indicate that the job is slipping.				
8. Push my people to meet schedules if necessary.				
9. Have frequent meetings to keep in touch with what is going on.				
10. Allow subordinates to make important decisions.				

**Award the number of points (1–4) for each response as indicated above, depending on your response to each scenario. Reverse the point scale for scenario #4, #10. Add the total points for all ten scenarios.

FIGURE 20.7   Supervisory style: the *x-y* scale.

consistently. The middle area or score between 20 and 30 points would suggest the use of a situational style. There may be a gap between the "guess" of your overall style and your summarized score for the 10 workplace scenarios. Whaley (2003) indicates supervisors and higher level managers who have responded to the exercise in Figures 20.6 and 20.7 find their style based on the 10 questions in Figure 20.7 is much closer to their actual style at work than the initial "guess" provided in Figure 20.6. These successful transportation supervisors reported a wide range of scores (10 to 40) in the exercise (Figure 20.7) that supports the situational leadership approach rather than the traditional, "one style fits all" approach (Whaley 2003).

20.6.3 Situational Leadership

Modern research suggests it is impractical and ineffective to use the same leadership styles in all situations if the goal is to become a successful supervisor or higher-level manager. Researchers tend to agree the situation is critical to the selection of an appropriate leadership style but they disagree concerning which aspects of the situation (i.e., task, power, relationships, goals, groups, values, and change) to take into consideration. Hersey, Blanchard, and Johnson (2001) indicate key situational factors that require consideration are the development or maturity level of subordinates and the nature of the task. If you are the supervisor of bus maintenance and mechanics reporting to you perform simple brake repairs, a more "directive," "task-oriented," "authority-oriented," "Theory X" style is usually more effective. On the other hand, if the fuel used for the coaches is a new fuel such as liquid gas or bio diesel, this usually suggests more task complexity and a more "supportive," "people-oriented," "participative," "Theory Y" style is usually more effective. If the mechanics performing the simple brake job are inexperienced and result in them being "unwilling/unable" to perform the task, a Theory X style is more effective and if the mechanics are highly experienced (willing/able), a Theory Y style is more effective. A popular situational theory model of leadership (SLT) is based on a task-by-task match of the follower's four situations to one of four appropriate leadership styles of the leader (Gibson et al. 2009; Hersey, Blanchard, and Johnson 2001). There are several versions of the SLT model depending on whether a follower's four situations are described by "developmental" level, "readiness" level, or "maturity" level. The version shown in Figure 20.8 suggests the four leadership styles are associated with four different development levels based on competence and commitment for each employee on each specific task performed.

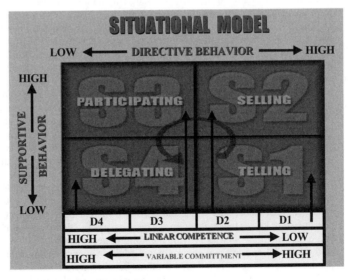

FIGURE 20.8   Situational model.

The Situational Leadership Model is attractive for transportation supervisors because it appears to place more emphasis on the development level of the subordinate and the specific tasks that may be easier to change rather than the style of the manager that is supported by the organizational culture. If a mismatch exists between the development level of the subordinate and the style of the leaders, several approaches are shown in Exhibit 20.1 that could work in your organization depending on ease of change:

Exhibit 20.1

1. The leader can change styles.
2. The leader can focus on developing the employee around a specific task.
3. The leader can temporarily use others in organization to influence the employee.
4. The leader can focus on changing the content of the task to match the employee.
5. The leader can select a different employee or different situation.

Many other credible situational leadership models exist. Some situational models emphasize the role, power, or style of the leader while others emphasize leadership in terms of change, values and interaction of the team (Bales and Cohen 1994; Fielder 1967; Gibson et al. 2009; House 1971; Kouzes and Posner 1996; Robbins, DeCenzo, and Moon 2008).

Since change has become an important factor in transportation today, leadership models that involve change are discussed here. Several modern leadership theories are built around change and values of the leader as situational factors and transformational leadership is the most popular of these models. Gibson et al. (2009) and Osland et al. (2007) indicate transformational leaders are value-driven change agents who make followers more conscious of the importance and value of task outcomes; they provide followers with a vision and motivate them to go beyond self-interest for the good of the organization. As mentioned earlier in the discussion of the external environment, transportation organizations along with the U.S. economy are undergoing rapid change. Hence, if employee self-interest is the major reason for resistance to change concerning major budget revisions, a transformational approach to leadership should be strongly considered to help managers to lead change and balance organizational budgets.

The SYMLOG model (Bales and Cohen 1994) provides a very flexible values and behavior based tool for measuring the effectiveness of a leader using: Most Effective Person (MEP) score, each team member's score and the score of the entire team across different internal and external settings. SYMLOG is more metrics oriented than most other situational models and it can be used to determine an overall team effectiveness score and compare individual scores to other critical organizational concepts and settings. For instance, consider comparing your individual or team score to: the perception of current organizational culture, the perception of future organizational culture, the perception of values employees are rewarded for in the organizational culture and the perception of values for which your customers believe you demonstrate in the culture. SYMLOG allows the measurement of differences between the leader's value profile and other concepts that may be important to the organization such as organizational culture and how to move the leader's profile toward the desired effectiveness metric (MEP). If a transportation organization is structured around different team formats, SYMLOG can be used to compare and match the leader's value profile to the value profile of team format such as cross-functional, local, or virtual teams.

Since most leaders are made and not born; increasingly, organizations need to identify leaders, convince employees that leadership is a necessary skill, identify job situations to develop and demonstrate leadership, as well as provide the training necessary to develop leaders. Additionally, changes in the external and internal environments could influence the usefulness of leadership approaches. As the work in transportation organizations become more complex, change-driven, metrics-oriented and team-oriented, the team, transformational, situational, and SYMLOG-based leadership models and associated assessment instruments become more useful.

20.7 TYPICAL CHALLENGES FACED BY NEW MANAGERS IN TRANSPORTATION

New managers within the transportation industry today are beset with more challenges than their counterparts decades ago (Toole and Toole 2007). The sources of these management challenges come from outside the organization, the organization itself and the employees. The results of interviews and discussions with new managers are shown in Exhibit 20.2 and there are eight typical challenges faced by new managers in transportation (Whaley 2003):

Exhibit 20.2

1. Subordinates do not respect your authority because they remember you broke the work rules as a nonmanagement employee.

2. Managers are not familiar with the transportation or organizational culture.

3. Several of employees have more task knowledge and experience than the manager and they are vocal about it.

4. Managers are asked to enforce organization rules and regulations and they do not personally agree with the organization's policies and procedures.

5. Managers do not have sufficient breadth of management training or experience.

6. New managers do not identify with management because they were union members in the past.

7. What the manager does and says is usually interpreted as organizational policy.

8. Constraints prevent managers from treating everything and one individually.

20.7.1 Unchanging Challenges for New Transportation Managers

New managers need to understand the challenges listed in Exhibit 20.2 and become comfortable with their new role before they can expect their employees to be comfortable with them in this new role. When new managers are uncomfortable with organizational policies and procedures, it is suggested they do not undermine company goals by telling employees about disagreements with the policies or procedures. First, managers should seek to understand the employee's reasons or behavior as well as the benefits of the policy from management's viewpoint and then explain these points to the employees in terms of their self-interests and in a neutral manner. When employees remind the new manager they made mistakes in the past as a nonmanagement employee, the new manager can take this negative communications as a "teachable moment" to discuss the new role. This approach could start the relationship toward transparency and trust by admitting past behavior was not appropriate. New transportation managers need to be especially aware of these challenges and to become proactive. The new manager can become proactive by asking these same employees to change their own inappropriate behaviors and help align their behavior with the goals of the organization. New supervisors are often shocked to discover employees interpret most things supervisors say or do at work as company policy. These situations in which the supervisor's statement or actions are not company policy can be used as an opportunity to explain to employees the multiple roles you play at work and provide concrete examples regarding when you are and are not speaking for the organization. This clarification builds trust and is especially important in employee discipline situations. Oestreich and Whaley (2001) point out it is important in discipline situations where the employee is covered by a collective bargaining agreement for the employee to know when you are merely giving a personal opinion versus playing a management role and giving a direct order. The specific contexts in which each of the eight situations in Exhibit 20.2 arises are usually unique. However, these situations happen often enough to new managers that new managers should anticipate them and use each situation to help build proactive strategies to deal with them. The most important thing for any manager to remember concerning building employee trust and respect is "authenticity," "consistency," and "dependability," and the manager can demonstrate these critical items by "walking their own talk."

20.7.2   Changes in Transportation Environment

Nearly two decades ago, Kanter (1990) suggested the complexity and rate of change organizations face began to accelerate. Kanter stated this observation in the form of a management challenge:

The major challenge management faces today is living in a world of turbulence and uncertainty where new competitors arrive on the scene daily and competitive conditions change. We can no longer count on a stable world that is unchanging and unvarying and manage accordingly.

Hersey, Blanchard, and Johnson (2001), Moffat and Blackburn (1996), Whaley (2003), and Wood and Johnson (1996) agree that a quicker pace of change is expected for organizations now and in the future. Change is a persistent theme in transportation today and it is likely to accelerate and become more ubiquitous in the future.

External and internal environmental forces on organizations were covered earlier; however, the sources of these changes and a summary of how they relate to future challenges in transportation are emphasized here. The sources of environmental change have become multifaceted, as economic, legal and political forces have become intertwined in transportation. The amount of federal government stimulus increased in 2009 along with the number of sources of funding for transportation projects. Moreover, the emphasis on alternative fuels and sustainability in economic decisions has increased. The trend in public policy is toward mixed transportation modes, infrastructure development and integration. The SAFETEA-LU law ended in 2009 and was extended through the middle of 2011 with a series of extension bills in Congress. Permanent legislation is likely to replace SAFETEA-LU in the second half of 2011 and this new transportation law will help to shape the economic, legal, and political landscape for at least the next 5 to 6 years. The previously mentioned Moffat and Blackburn study was a comprehensive survey of stakeholders such as transit agencies, directors, managers, and supervisors of operations concerning the role and responsibilities of the bus field supervisor and the findings remain pertinent today(Moffat and Blackburn 1996). Findings from the bus field supervisor study are summarized in Exhibit 20.3 and point to the role traditional, mechanistic organizational cultures play in the management of transit organizations and the pressure to change.

Exhibit 20.3

1. Almost all supervisors are hired from the ranks of bus operators.
2. Knowledge of transportation systems and experience with them is critical but little is done to prepare supervisors for this new role.
3. Little or no formal training is provided to supervisors in people management.
4. Increased interaction is required of supervisors with bus operators and public.

Many transportation organizations are changing their business models and management approaches based on a shift in the values or culture of the organization. The formal, top-down, "do it by the book" cultures in transportation are being slowly replaced with more open, bottoms-up, "do it by the planned goals" cultures, leading to a blending of the old and new approaches to organizational culture. Another noteworthy change is the previously mentioned growth of a teams approach in transportation organizations. If an organization embraces the values related to teams, it should be easier for the new supervisor to make use of the critical SKAs of each employee and not feel the need to be the expert on each task the subordinate employee performs. This in turn changes the leadership approach of each manager and encourages the use of the previously discussed situational leadership models. Whaley (2001) reported these findings related to the organizational culture in transit could be extrapolated to many parts of the transportation sector and support Exhibit 20.3 as well as the eight typical challenges for new transportation managers mentioned in Exhibit 20.2.

Individual employees provide much of the impetus for internal changes today because they expect different things from managers and organization than in previous generations (Hersey, Blanchard, and Johnson 2001; Toole and Toole 2007). Generation X (born between 1980 and 1990) employees and transportation managers alike expect to participate in organizational decision-making (Gibson et al. 2009). This could make employee empowerment and change easier to implement. If participation is an organizational cultural value, in addition to an individual expectation, employee buy-in to the changes they help to create makes the changes easier to implement. Therefore, many management challenges remain as change accelerates and the future become more uncertain.

20.8   THE FUTURE OF TRANSPORTATION MANAGEMENT

The *Public Transportation Fact Book* (2008) lists key changes that have occurred over the last ten years in public transportation as well as provides stakeholder viewpoints toward future changes. Exhibit 20.4 lists the major changes anticipated in transportation:

Exhibit 20.4

1. Less resources available and expanded service requirements.
2. Expanded privatization and emphasis on technology, safety and security.
3. Change in the organization culture from authoritarian to participative.
4. Increase use of lower cost and nonunion labor.
5. Less reliance on gasoline and diesel fuel and increased exploration of alternative fuels.
6. Increased use of technology across the board from route planning and communications to intelligent highways and computerized services.
7. Increased workforce diversity in U.S. and abroad.
8. More interest in labor-management cooperation processes such as IBN.
9. Shift of power from transportation companies to stakeholders.
10. Knowledge will be the primary source of power within the organization.
11. More participation of stakeholders and quicker expected responses due to expanded use of technology such as the Internet, Intranet, and social networking.
11. Greater link between transportation, the environment, and local economic vitality and planning.

Many ideas included in the items listed in Exhibit 20.4 appeared in the 1999 *Public Transportation Fact Book*; however, recent interviews with transportation leaders suggest some items have changed in terms of priority and language (Diridon et al. 2009; *Public Transportation Fact Book* 1999). The issues surrounding resources and service levels have remained the top priority for the transportation sector. The urgency to improve on public policy integration and implementation in this area has been increased due to the recent economic down turn and the scientific and societal connection of global warming, infrastructure improvements, and alternative energy technology to public policy.

Over 14 years ago, Wood and Johnson (1996) suggested a comprehensive and integrated national transportation policy was needed. The transportation specific, Intermodal Surface Transportation Efficiency Act (ISTEA) legislation in 1991 increased federal support to mass transit; however, it was not integrated very well with environmental aspects of laws such as the Occupational Safety and Health Act (OSHA) and North American Free Trade Agreement (NAFTA). The SAFETEA-LU legislation in 2005 is the replacement for ISTEA, and it provides management training support but it is not well coordinated with training under existing transportation legislation, and many organizations have cut back training because of budget constraints. Today's economic environment has resulted in many publicly supported transportation organizations across the country to suffer large budget cuts at the time when new investments are needed in transportation training, education and infrastructure. The President Obama administration is looking at new methods to fund transportation and regulate across modes of transportation and industries.

NAFTA is not a transportation policy per se but it impacts occupational safety and health and the carbon footprint by emissions from the increased movement of carriers across country borders in North America. Vehicle emission is the second-highest contributor to poor air quality in the United States, and in California's urban areas it is the number-one contributor to the greenhouse effect (Helmer 2008). This study reports the federal highway trust fund is slowly being depleted and its purchasing power eroded as congestion continues to grow, air quality worsens and aging infrastructure becomes more fragile (Helmer 2008). There is an urgent need to integrate sound economic and environmental policy at the international, national, state and local levels with effective regulatory policies in U.S. transportation. Closer alignment of city, county, and state planning with federal goals would help to achieve the goals of each in an environmentally sensitive manner. For example, it will not be uncommon in the future for a city agency to perform a transportation study to convince a private firm in a different industry to locate in their city or not to leave their city. This trend could also lead to greater marketing of services provided by local transportation agencies.

The coordination of stakeholders is important in formulating governmental policy and no stakeholder should be overlooked. One positive example of coordination among multiple stakeholders at the local level involved rewriting the City of San Jose's level of service policy to accommodate walking, cycling and transit. The Director of Transportation and his staff worked with key stakeholders including the development community, and the city's Housing and Economic Development departments to rewrite the city's level of service policy on major transit corridors, in a city where the auto is the predominant mode choice for commuters (Helmer 2008). Not only was such a shift in policy needed to break from past practice; it was also a necessity if the City wanted to meet its goals to reduce dependency on the auto and to improve air quality. Community improvement plans were designed by the surrounding neighborhoods, developers would then be required to construct sidewalk, bus-stop, bike lane, accessibility ramp, street tree, or lighting improvements as desired by the community instead of widening intersections and adding travel lanes that was the past practice based upon assumed trips for new development. With the benefit of an override to conduct a traffic assessment, the development community now sees the benefit of such a break from past practice, saving time and money to plan their projects and existing infrastructure gets replaced with modern, more sustainable products. Moreover, the surrounding residents get a more walkable environment, and residents moving into these new higher density housing projects, are attracted to walk or ride to local retail and transit stops.

Osterman et al. (2001) indicate unions tend to be an overlooked stakeholder in the planning process. Unions can play an even larger and more positive role in public policy-making at the national level and this role is also important for the survival and future growth of unions. Some transportation organizations have enlisted their labor unions to help survive current financial crunches. For example, union wage, benefit and work rule concessions together with a new business model for U.S. automakers in crisis, environmentalists pressure for alternative energy use and favorable public policy could lead to the rebirth of the U.S. automobile industry. The new Obama administration indicates parts of a national transportation policy exist today but the parts need to be integrated with legal, environmental, safety, and current global realities for overall effectiveness.

The priority of many items in Exhibit 20.4 depends on the location of stakeholders, local resources available, political clout and pressure for change from the public. For instance, item numbers 5 and 12 concerning alternative uses of energy have different environmental implications for states where automobiles are manufactured as compared to states where auto emissions are high or states where alternative energy is likely to be available and inexpensive. Item numbers 2, 6, and 11 in Exhibit 20.4 concerning the use of advanced technologies could have the greatest potential impact over the next ten years; however, the cost and general acceptance of new technologies together with organizational policies, legal constraints, and political clout of the stakeholders will determine their priority and impact. What is clear, is employees and organizations expect more from transportation managers today because they see behavior in this area as increasing safety; lowering cost and increasing revenue or other metrics that represent increased societal, organizational, and customer value. These increased expectations require improved change management and associated leadership skills from future transportation managers because change will be expected, faster, pervasive, intertwined, and more valued. Therefore, an increase in cost-effective education and training based on traditional as well as on-line delivery, self-learning, continual learning, speed of learning, and e-learning with an emphasis on interpersonal skills is needed. These future trends in transportation will alter the required management skill set for future transportation managers and will most likely transform what was in the past a "luxury skill set" to a "minimally expected skill set."

REFERENCES

American Public Transportation Association. 1999. *Public Transportation Fact Book.* Washington D.C.: APTA.

American Public Transportation Association. 2008. *Public Transportation Fact Book.* 59th Edition, Washington D.C.: APTA.

Bales, R. F. and Cohen, S. P. 1994. *SYMLOG: A System for the Multiple Level Observation of Groups.* New York: Free Press.

Bennis, W. G. and Thomas, R. J. 2002. *Geeks and Geezers: How Era, Values, and Defining Moments Shape Leaders.* Boston: Harvard Business Review Press.

Bernardin, H. 2010. *Human Resource Management: An Experimental Approach,* 5th edition. New York: McGraw-Hill Companies, Inc.

Boxx, R. W., Odom, R. Y., and Dunn, M. G. 1991. "Organizational Values and Value Congruency and Their Impact on Satisfaction, Commitment, and Cohesion: An Empirical Examination Within the Public Sector." *Public Personnel Management,* 20, Summer, 195–205.

Champy, J. 1995. *Reengineering Management: The Mandate for New Leadership.* New York: Harper-Collins Publishers.

Chen, W., Jacobs, R., and Spencer, L. M. 1998. *Calculating the Competencies of Stars. Working with Emotional Intelligence,* New York: Bantam Books.

Coleman, D. and Borman, W. C. 2000. "Investigating the Underlying Structure of the Citizenship Performance Domain." *Human Resource Management Review,* 10, 25–44.

Cook, C. W. and Hunsaker. 2001. *Management and Organizational Behavior.* New York: McGraw-Hill Higher Education.

Covey, S. R. 1990. *Principle-Centered Leadership.* New York: Schuster and Simon.

Dess, G. G. and Lumpkin, G. T. 2003. *Strategic Management: Creating Competitive Advantages.* New York: McGraw-Hill Higher Education.

Diridon, R. Personal Interview. July 21, 2009.

Drucker, P. F. 1967. *The Effective Executive.* New York: Harper and Row.

Fiedler, F. E. 1967. *A Theory of Leadership Effectiveness.* New York: McGraw-Hill.

George, C. S. Jr. 1972. *The History of Management Thought.* Englewood Cliffs, NJ: Prentice-Hall, Inc.

Gibson, J. L., Ivancevich, J. M., Donnelly, J. H., and Konopaske, R. 2009. *Organizations: Behavior, Structure, Processes.* New York: McGraw-Hill/Irwin.

Goleman, D. 1998. *Working with Emotional Intelligence.* New York: Bantam Books.

Haas, P. Personal Interview. July 30, 2009.

Haas, P. 2007. "Diversity in American Subnational Transportation Agencies: Challenges and Opportunities." *The International Journal of Diversity in Organizations, Communities and Nations,* Vol. 7.

Helmer, J. R. 2008. "Building a Sustainable TMC in San Jose: Leading the Way to Create a Safer and More Sustainable Transportation Infrastructure System." 12th Annual Transportation and Infrastructure Summit. Irving, Texas.

Helmer, J. R. Telephone Interview. August 14, 2009.

Hersey, R. E., Blanchard, K. H., and Johnson, D. E. 2001. *Management of Organizational Behavior: Leading Resources*. Upper Saddle River, NJ: Prentice-Hall.

House, R. J. 1971. "A Path-Goal Theory of Leadership." *Administrative Science Quarterly,* 16, 321–338.

Injury Facts. 2007. Itasca, IL: National Safety Council.

Johnson, S. 1998. *Who Moved My Cheese?* New York: G. P. Putnam's Sons.

Kanter, R. M. 1990. *The Planning Forum Network*, 2 (1), 1.

Katz, R. L. 1974. "Skills of an Effective Administrator." *Harvard Business Review*, September-October, 52, 90–102.

Kotter, J. 1990. "What Leaders Really Do." *Harvard Business Review*, 68, 103–111.

Kouzes, J. M. and Posner, B. Z. 1996. *The Leadership Challenge*, San Francisco: Jossey-Bass.

Lund, D. B. 2003. "Organizational Culture and Job Satisfaction." *Journal of Business and Industrial Marketing*, 18, 219–236.

McGregor, D. 1960. *The Human Side of Enterprise*. New York: McGraw-Hill

Mintzberg, H. 1990. "The Manager's Job: Folklore and Fact." *Harvard Business Review*, 90, 168.

Moffat, G. K. and Blackburn, D. R. 1996. "Changing Roles and Practices of Bus Field Supervisors." *Transportation Research Board: Synthesis of Transit Practices 16*. Washington, D.C.: National Academy Press.

Oestreich, H. H. and Whaley, G. L. 2001.*Transit Labor Relations Guide*. Mineta Transportation Institute Report 01-02. San Jose: Mineta Transportation Institute, San Jose State University.

Osland, J. S., Kolb, D. A., Rubin, I. M., and Turner, M. E. 2007. *Organizational Behavior: An Experimental Approach*. Upper Saddle River, NJ: Pearson Education Inc.

Osterman, P., Kochan, T. A., Locke, R. M., and Piore, M. J. 2001. *Working in America: A Blueprint for the New Labor Market*. Cambridge: MIT Press.

Paul, R., Niewoehner, R., and Elder, L. 2007. *The Thinker's Guide to Engineering Reasoning*. Dillon Beach, CA: Foundation for Critical Thinking.

Pfeiffer, J. W., and Jones, J. E. 1972. *The 1972 Handbook for Group Facilitators*. Iowa City: University Associates.

Pouliot, M. 2002. "Transport Geography on the Web, Hofstra University." Department of Economics and Geography.

Robbins, S. P., DeCenzo, D. A., and Moon, H. 2008. *Fundamentals of Management: Essential Concepts and Applications*. Upper Saddle River, NJ: Pearson Education Inc.

Spray, S. L. (Ed.). 1976. *Organizational Effectiveness: Theory, Research, Utilization*. Kent, OH: Kent State University Press.

Tanda, W. Telephone interview. August 6, 2009.

Toole, C. L., and Toole, J. S. 2007. "Preparing Tomorrow's Transportation Workforce Professional." Washington, D.C.: Transportation Research Forum. Proceedings of March 15–17, 2007 conference, Boston University.

U.S. Department of Commerce, Bureau of the Census. 1998. *Statistical Abstracts of the United States 1998*. 118th Edition, pp. 444.

www.bls.gov, retrieved August 1, 2009.

www.fra.dot.gov, retrieved July 10, 2009.

www.plunkettresearch.com, retrieved July 3, 2009.

Whaley, G. L. 2001. Unpublished Study of Organizational Culture in Transportation Agencies, 1997–2001.

Whaley, G. L. 2003. Interviews of New Transportation Managers, 1994–2003.

Whaley, G. L. 2009. Interviews of Transportation Managers, 2009.

Wood, D. F., and Johnson, J. C. 1996. *Contemporary Transportation*. Upper Saddle River, NJ: Prentice Hall, Inc.

Wytkin, E. 1997. *Toward a Cooperative Future*. Mineta Transportation Institute Report 97-2. San Jose: Mineta Transportation Institute.

INDEX